ENCYCLOPAEDIA OF
BIOCHEMISTRY

ENCYCLOPAEDIA OF
BIOCHEMISTRY

Dr. Bethany J. Mathews

ANMOL PUBLICATIONS PVT. LTD.
NEW DELHI - 110 002 (INDIA)

ANMOL PUBLICATIONS PVT. LTD.
H.O.: 4374/4B, Ansari Road, Daryaganj
New Delhi - 110 002
Ph.: 23278000, 23261597
B.O.: No. 1015, Ist Main Road, BSK III Stage
III Phase, III Block
Bangalore - 560 085 (India)
Visit us at: www.anmolpublications.com

Encyclopaedia of Biochemistry

First Published, 2006

ISBN 81-261-2864-X

PRINTED IN INDIA

Printed at Mehra Offset Press, Delhi.

Preface

For some time there has existed a widespread consciousness of the need for the interpretation of the biochemistry terms to the layman. The average reader is deluged continually with a quantity of facts in the field that we call biochemistry. A possible aid to the solution of this problem is the compilation of a volume, which endeavors to explain to the ordinary reader terms which are new to him and which have not yet become part of his general vocabulary. Existing contributions of the kind are limited in scope to some such field as biochemistry. This volume attempts in brief compass to cover broadly the whole realm of biochemistry.

The aim of this volume is to fill the need for definitions of many new terms that have come into usage during recent years and also to include many of the old terms that are used in current literature. Many words from fields closely related to biochemistry are included because of their wide usage in biochemistry literature.

Clarification of the meaning of terms, precision in their use, and uniformity in usage among workers in biochemistry and related fields are essential in the growth of a science. When a concept or process can be expressed precisely in ordinary language it appears unnecessary, and indeed detrimental to the growth of a science, to coin a new term. It is worth while to make biochemistry literature intelligible to as wide a field of readers as possible.

The aim of this volume is to provide a concise repository of number of major concepts in this field. In preparation of this book, the author has freely consulted large number of books and journals so no authenticity is claimed. Author is specially thankful to M/S Anmol Publications for shaping this book in its final form.

Suggestions for further improvement of this book are not only welcome but also greatly appreciated.

Author

A

Abderhalden test for cystine A reaction depending on the formation of naphthalenesulfocystine by treating cystine with betanaphthalenesulfochloride in NaOH.

Abderhalden's anhydride theory In the 1920's, Abderhalden pointed out that much of the behavior of proteins is better explained on the basis of a structure of diketopiperazine rings held together by latent valences, than on straight chain polypeptides. He was able to obtain a number of these cyclic anhydrides of amino acids from protein hydrolysis products.

Abderhalden-schmidt test for proteins and their decomposition products One to two drops of reagent, made by dissolving 0.1 gm. of ninhydrin in 300 cc of water, is added to the test solution and heated. The presence of a protein, polypeptide, amino acid or peptone is indicated by a blue color. Reagent may also be used in pregnancy diagnosis.

Abelin test for neoarsphenamine Neoarsphenamine in urine is detected by splitting off formaldehyde. Urine, not older than 3-4 hours and preferably collected within ½-1 hour after injection is used. 10-15 cc. are treated with a freshly prepared and filtered 1% solution of phenylhydrazine hydrochloride. After cooling 1 cc. of freshly prepared 5% potassium ferrocyanide solution is added. If formaldehyde is present, a fuchsin-red color is developed upon the addition of 5 cc. of concentrated hydrochloric acid. Dilution with an equal volume of water and extraction with ether renders the test more sensitive, a yellow color being obtained in the extract which turns red when hydrochloric acid is added.

Abietic Acid A dterpene carboxylic acid, and the isomeric neoabietic acid, can easily be interconverted. These two resin acids are the main components of rosin (up to 90%), from which they could be obtained by treatment with heat or acids, possibly as product of the rearrangement of other diterpene carboxylic acids. Amber contains derivatives of A.a.

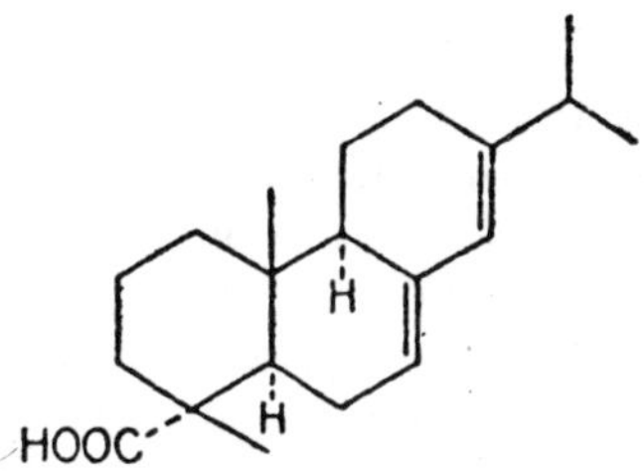

Fig. Abictic Acid

Ablation experiment An experiment designed to produce an animal deficient in one or a few cell types, in order to study cell lineage or cell function. The idea is to make a transgenic mouse with a toxin gene (often diphtheria toxin) under control of a specialized promoter which activates only in the target cell type. When embryo development progresses to the point where it starts to form the target tissue, the toxin gene is activated, and that specific tissue dies. Other tissues are unaffected.

Absorption, intestinal e.g. of carbohydrates, it is a function of the species, selectivity of mucosa, presence of other substances, enzyme activity involving phosphorylation, circulation of the blood supply as well as the nature of the substance absorbed.

Accretion Process through which the gravitational force of an object in space collects matter.

Acetaldehyde reductase An enzyme which catalyzes the reduction of acetaldehyde to alcohol, similar in structure to the yellow respiratory enzyme.

Acetoacetic acid CH_3COCH_2COOH; diacetic acid; a keto-fatty acid found in fermentation and as an intermediate of metabolism; b.p. 180°.

Acid Water-based solution containing more free hydrogen ions (H^+) than free hydroxide ions (OH^-); when acids are added to a neutral solution (such as water), they increase the solution's ratio of hydrogen ions to hydroxide ions

Acid metaprotein A metaprotien soluble in acid and formed when hydrolysis takes place in acid solution (also by pepsin).

Acid number The number of milligrams of potassium hydroxide needed to neutralize the free fatty acids of 1 gram of a fat, oil, or wax.

Acidosis A condition of lowered alkali reserve with unaltered pH of the blood; a symptom due to excessive acid formation as in surgical anesthesia, gastro-intestinal disturbances, febrile diseases or insulin deficiency; excessive loss of alkali as in cholera or burns; defective acid elimination as in uremia or asphyxia; or excessive ingestion of acid as in drugs.

Ackerman reaction for guanidine When 3 parts of guanidine are heated with 4 of benzenesulfonchloride, 6 of 33% sodium hydroxide solution and 30 of water, the cooled reaction mixture yields white needles melting at 212°C.

Acrylamide gels A polymer gel used for electrophoresis of DNA or protein to measure their sizes (in daltons for proteins, or in base pairs for DNA). Acrylamide gels are especially useful for high-resolution separations of DNA in the range of tens to hundreds of nucleotides in length.

Activation centre One of the two organisation centres in the insect egg.

Activation energy The initial investment of energy in a reaction. Activation energy (Ea) is required to energize the bonds of the reactants to an unstable transition state that precedes the formation of products.

Active site That specific portion of an enzyme that attaches to the substrate by means of weak chemical bonds.

Acylase A peptidase which hydrolyzes polypeptides with a terminal tyrosine group, postulated for the pancreas by Abderhalden.

Adamkiewicz reaction for proteins A reddish-violet color and a slight green fluorescence is produced upon the solution of protein substances in glacial acetic acid followed by the addition of concentrated sulfuric acid.

Delicacy of the test is increased by addition of an equal volume of glacial acetic acid to the test solution and floating the mixture on an equal volume of concentrated sulfuric acid. Either free or combined tryptophane is indicated by the reaction.

Adams' method for milk-fat A quantitative method for determining milk-fat by absorption of sample on fat-free paper coil, drying, followed by extraction with ether. The fat in ether extract is weighed after drying.

Adapter A short chemically synthesized DNA double strand which can be used to link the ends of two DNA molecules.

Adaptive radiation The emergence of numerous species from a common ancestor introduced to an environment presenting a diversity of new opportunities and problems.

Adaptive Traits Any aspect of form, function, or behavior that helps an organism survive and reproduce under a given set of environmental conditions.

Addison's disease Chronic suprarenal insufficiency; a disease characterized by pigmentation of the skin, muscular and nervous asthenia and anemia, due to insufficiency of adrenal (suprarenal) glands, especially of the cortex; asthenia is shown by easy tiring, soft pulse.

Adenase An enzyme which deaminates adenine (6-aminopurine) to hypoxanthine (6-oxypurine).

Adenine (A) One of the purine bases found in DNA and RNA (6-aminopurine), one member of the base pair A- T (adenine- thymine).

Adenosine A nucleoside consisting of one molecule of adenine and one molecule of d-ribose. May be considered as derived from inosinic acid through the loss of phosphoric acid. The linkage is through C_1 of the sugar to an N of the adenine.

Adhesion Sticking togethter due to the contact of dissimilar substances, as glue to wood.

Adipic acid $C_6 H_{10} O_4$; m.p. 153°; a dicarboxylic acid present in beet juice.

Adipocere (Corpse wax) Waxy material present in corpses removed from watery graves probably consisting largely of insoluble calcium, magnesium and ammonium soaps of saturated fatty acids. Formerly, it was believed that adipocere originated from protein.

Adler benzidine test for blood A blue color is produced when blood in water is acidified with acetic acid, treated with benzidine solution and hydrogen peroxide.

Adme An anacronism used in drug development, meaning absorption, distribution, metabolism and excretion.

Admiralty test A test for the efficiency of excremental disinfectants in which gelatin and starch are employed as organic test substances.

ADP (adenosine diphosphate) A ribonucleoside 5'-diphosphate serving as phosphate group acceptor in the cell energy cycle.

Adrenal cortex hormone extract (Units) No official standards. Rat unit is based on the percentage of rats that can be kept alive by the preparation after adrenalectomy. The dog unit depends on the maintenance of a normal nitrogen concentration in the blood.

Adrenal gland Part of the endocrine (glandular) system responsible for secreting *hormones* instrumental in stress-related and vital metabolic functions; it sits near the kidneys

Adrenaline Epinephrine; suprarenin; $C_9 H_{13} O_3 N$; a hormone secreted by the adrenal glands serving to increase blood pressure, to strengthen the heart beat and to raise the blood sugar.

Adrenergic Pertaining to nerves that liberate an adrenalin-like substance on stimulation.

Adrenotropic (adrenotropic hormone) A pituitary hormone stimulating the secretion of adrenalin by the medulla of the adrenal gland.

Adrenoxine An oxidizing substance thought to be formed in the lungs by the combination of oxygen and the internal secretion of the adrenals.

Adrephine A. mixture of adrenalin and ephedrine; an astringent for the nose, throat and larynx.

Adsorption compounds Compounds formed at interfaces in heterogeneous systems, not involving primary valence bonds.

Adsorption The phenomenon of concentration at interfaces manifested by substances which decrease interfacial tension, usually, but not necessarily between liquid-liquid or liquid-solid interfaces; manifested in drying solutions over a solid drying agent, or decolorizing and deodorizing with charcoal. Positive adsorption: the concn. of the solute in the interfacial film. Negative adsorption: the concn. of the solvent in the interfacial film.

Adult The mature stage of an organism, usually recognized by the organism's attaining the ability to reproduce.

Affinity The extent to which tissues from early embryos tend to cling together when in isolation from their normal surroundings.

Agaricin Agaric acid; laricic acid; a substituted

lauryl citric acid, found in a fungus of larch trees; used to control night sweats.

Agarose gel electrophoresis A method to analyze the size of DNA (or RNA) fragments. In the presence of an electric field, larger fragments of DNA move through a gel slower than smaller ones, producing different migrating "bands". Usually, these are visualized by soaking the gel in a dye (ethidium bromide) which makes the DNA fluoresce under UV light. This is the gel of choice for DNA or RNA in the range of thousands of bases in length, or even up to 1 megabase if you are using pulsed field gel electrophoresis.

Agglutinate To clump together, as in the case of bacteria, blood cells, etc.

Agglutination process of *antigens* bunching up together to make observable clumps

Aggressin Toxic substances formed by organisms which paralyze the host's defenses. (obsolescent) rom the pecullar viewpoint ot agriculture which may be described as the economic utilization of the farm for the production of food. There is also a growing tendency to include the biochemical aspects of chemurgy, which is the utilization of farm products for industrial purposes, such as the production of power alcohol, solvents, drugs, etc.

Where soils and crop improvement are concerned biochemistry is involved in the so-called "classification of soils," their preparation, the working over of fertilizers and their production by the aid of living organisms, and even genetic studies, in which chemicals, e.g. Colchicine, are used to produce large or new plants. The effects of heat and cold, chemical reagents like ethylene, on storage and ripening have also biochemical aspects. The ethylene is supposed to react with starch in the ripening process. This field also should include aquaculture, or the growth of plants on synthetic nutrients in solution.

A large division of this subject deals with animal nutrition with the practical objective of obtaining more and better eggs, butter, milk and the like. An agricultural biochemist's interests might lead him to administer female sex hormones in order to induce virgin animals to secrete large amounts of milk. Fertility is an important consideration, and involves the usual attention to vitamin E and the like. Where animals are produced for food the conditions for rapid growth are a subject of serious study. Diets are investigated for special effects, e.g. The kind of fat to develop on a hog.

The pathological aspects of the lives of animals and plants occupy a great deal of the attention of the biochemist in this field. Plant diseases are of a very great variety and attack selectively seeds, roots, stems or leaves. The toxic agents have been extracted in many cases. The first viruses to be studied were plant disease viruses. An entire subject, "forest pathology," had grown up, which can be listed here as a close relative of agricultural biochemistry.

Aglucone Aglycone; the name given to the group of the non-sugar residues of the glycosides.

Agmatine ä-guanidino-butylamine. A poisonous amine formed by bacterial decomposition of arginine; also derived from herring spawn or ergot; lowers blood sugar.

Agnosterol $C_{30}H_{48}O$; m.p. 162°C. A pentacyclic sterol with 3 double bonds found in lanolin.

Agostini test for glucose Five drops of the test solution are treated with 5 drops of 0.5% gold chloride solution and 2 drops of 1:20 potassium hydroxide solution, heated to boiling and cooled. A violet color indicates glucose; the test is sensitive to 1:10000.

Agricultural biochemistry Agricultural biochemistry is a branch of applied biochemistry which concerns itself with any and all possible situations involving life processes in the pursuit of agriculture. These may include numerous issues such as the cycles of soil organisms, the nurture of plants and their entire metabolic history, animal husbandry from birth to death, pathological conditions to both plants and animals, the biochemical aspects of genetics in the service of breeding, the preparation of animal and plant foods, the control of insects, the processing of farm products, quick ripening, storage, etc

Agrostology Division of botany dealing with grasses.

A-homochelidonine $C_{12}H_{23}O_5N$; stout colorless needles, m.p. 170-171°; alkaloid found with β-homochelidonine in Adlumia, Bocconia, Corydalis, Dicentra, Sanguinaria and other plant genera.

A-hormone A substance which influences genetically a number of characters, colors of eyes, testes, skin and brain, of wild-type moths; found in the testes, but not species specific.

Air-bladder The organ of pressure regulation of a fish.

Air-cells Air-filled spaces in bones or lungs.

Akropeptide theory A theory of protein structure advanced by Fodor. Enolization takes place at the peptide linkage of a dipeptide leaving a free valence bond on a carbon and nitrogen atom, which combine with similar bonds on another molecule of the dipeptide yielding a tetrapeptide. This process is repeated, yielding a di-tetrapeptide.

Alanine metabolism Alanine is converted to a-ketopropionic acid and NH_3, the latter going to urea and the former probably losing CO_2 to form the next lower fatty acid. Alanine may be converted into glucose. It is thus anti-ketogenic. Under certain conditions it is synthesized in the body.

Albén reagent for albumin in urine Five gms. of tannic acid are dissolved in a mixture of 240 cc. of 50% acetic acid. A turbidity or precipitate when urine is treated with the reagent indicates albumin.

Albinism Absence,of pigment where it normally occurs.

Albuminoids In the American Classification of proteins—simple proteins having essentially the same chemical structure as the other proteins but highly insoluble in all neutral solvents, nd in general, in dilute acids and alkalies; scleroproteins. Examples: keratin, spongin, fibroin and collagen.

Albumins In the American Classification of proteins—a subclass of the simple proteins, crystallizable, soluble in water and salt solutions, and coagulated by heat. E.g. egg, contains carbohydrate groups and should actually be classified with the conjugated proteins.

Albuminuria The older, less accurate but common name for proteinuria.

Alcock's theory of protein anabolism Amino acids as such do not build up proteins, but are first broken down to simpler structures. The essential amino acids are necessary, not for proteins, but because of some special function they have.

Alcohol (1) Ethyl alcohol made by fermentation or synthetically;

(2) a class of organic substances characterized by an OH group of very slight acid properties due to union with certain carbon hydrogen radicles.

Aldehyde mutase An enzyme of mammalian liver and kidney which promotes the simultaneous conversion of 2 molecules of acetaldehyde to one of ethyl alcohol and one of acetic acid; also acts on other aldehydes and promotes the oxidation of aldehydes by a number of alphaketonic acids.

Aldehyde An organic molecule with a carbonyl group located at the end of the carbon skeleton.R-HC=O.

Aldose sugar A sugar containing a group which is an aldehyde or is potentially an al-dehyde.

Alecithic Having little yolk; said of e.g. echinoderm eggs.

Aleuritic acid $C_{13} H_{26} O_4$ an acid extracted from shellac.

Alginate Component of the cell walls of many rhodophytes and kelps. Alginates have an affinity for water, and so help to slow dessication when the algae are exposed to the air; they are commercially important in the production of paper, toothpaste, beer, and frozen foods.

Alimentary glycosuria The presence of glucose in the urine following the ingestion of very large amounts of sugar.

Alimentary secretions Saliva, gastric juice, bile, pancreatic juice, intestinal juices.

Aliphatic Those carbon compounds characterized by straight, branched, or cyclic carbon chains, and lacking the benzenoid ring structure.

Alizarin $C_{14}H_8O_4$; Anthraquinone dye, m.p. 290°, found as a glucoside in madder.

Alkali metaprotein A metaprotein soluble in alkali; formed when protetin is hydrolyzed by alkali or trypsin.

Alkaline phosphatase An enzyme which catalyzes the hydrolysis of phosphomonoesters of the 5' nucleotides. Used to dephosphorylate (remove phosphate groups from) the 5' ends of DNA or RNA molecules, to facilitate 5' end-labeling with 32P added back by T4 polynucleotide kinase; or to dephosphorylate the 5' ends of DNA molecules to prevent unwanted ligation reactions during cloning.

Alkaloid An organic base in which at least one nitrogen atom forms part of a cyclic system, usually having marked physiologic action. The term is frequently restricted to substances of plant and animal origin.

Alkalosis A condition of increased alkali reserve with unaltered pH of the blood; due to excessive loss of acid from the system as carbon dioxide in excessive pulmonary ventilation or HCI of

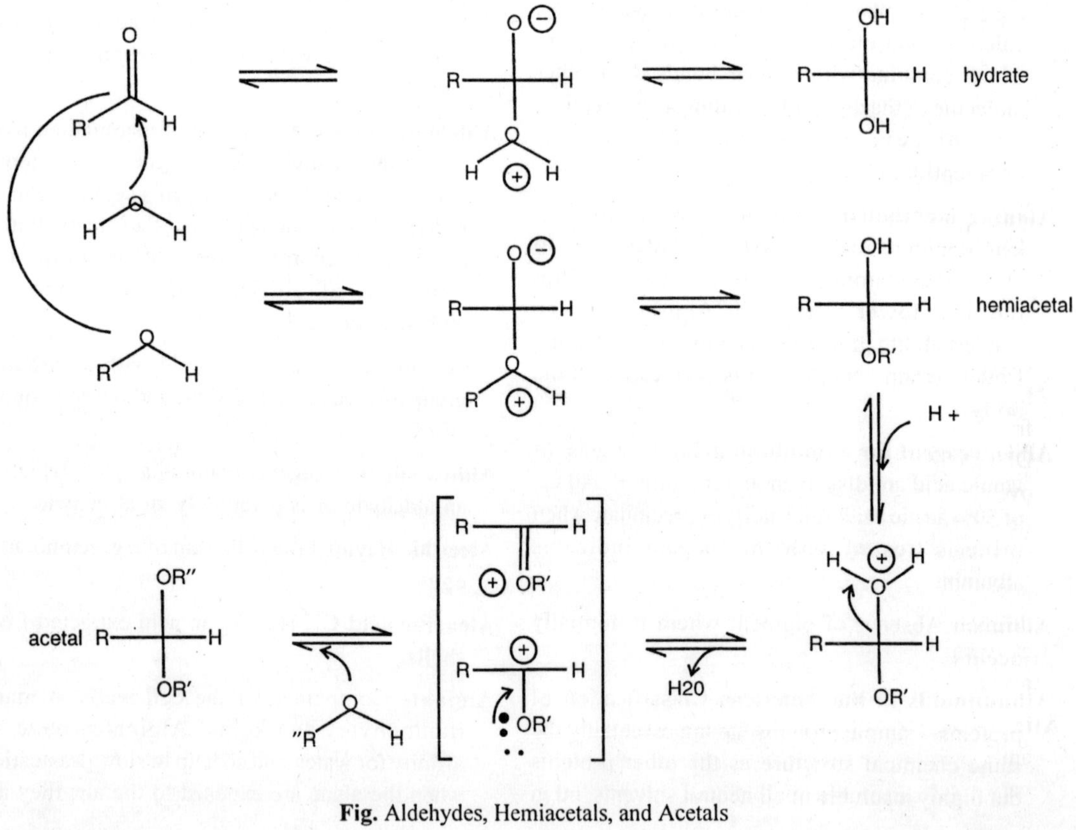

Fig. Aldehydes, Hemiacetals, and Acetals

stomach, or excessive administration of alkali in food or drugs.

Allantoicase The enzyme which catalyzes the conversion of allantoic acid into urea and glyoxylic acid; found in fishes associated with uricase and allantoinase.

Allantoin Glyoxyldiureide, $C_4H_6O_3N_4$, m.p. 235°. Oxidation product of uric acid found in urine of lower animals, as end product of purine catabolism, in allantoic fluid, fetal urine and in plants. Used in the treatment of wounds and ulcers as it clears dead tissue and promotes the growth of new cells.

Allantoinase An enzyme which catalyzes the formation of allantoic acid from allantoin; found in fishes with uricase and allantoicase, also in amphibia and echinodermata.

Allantois A vascularized extraembryonic membrane; in reptiles and birds, it functions in excretion and respiration in placental mammals, it functions in oxygen transport by way of the umbilical cord.

Allele An allele is one of the alternative (two or more) forms of a particular gene inherited separately from each parent; usually found at the same locus on homologous chromosomes. Equivalent genes in the two sets might be different, for example because of single nucleotide polymorphisms.

Allele-specific ligation Technique permitting discrimination of two allele at locus by providing two short synthetic oligonucleotides that would bind ajjacent to teach other on amplified DNA fragment, and would be ligated in presence of DNA ligase; if one of alleles containing mutation overlapped by 3'-end of one oligonucleotide, its ligation to oligonucleotide bound 3' to it would be prevented; to deduce identity of unknown allele, differentially labelled oligonucleotide pairs may be designed for each allele and their ligation efficiency compared in presence of unknown allele.

Allele-specific PCR (AS-PCR) Refers to amplification of specific alleles, or DNA sequence varients at same locus; specificity is achieved by designing one or both PCR primers so that they partially overlap site of sequence difference between amplified alleles.

Allelomorph A Mendelian character as considered in the light of a contrasting one; one of the two or more genes which "compete" for a particular locus on a chromosome and which influence in different ways the same body characteristics.

Allergen A substance inducing allergy.

Allergy A changed, hostile sensitivity to the ingestion or inoculation of a substance to which there had been a previous exposure.

Allometry Allomorphosis and Heterauxesis.

Allomorphosis The relation of parts of organisms at some definite age to wholes or parts also at some definite age, but of different groups (races, varieties, species, genera); e.g. egg size or hatching weight to adult size or weight.

Allopelagic Insensitive to cold or heat in the ocean.

Allosteric site A specific receptor site on an enzyme molecule remote from the active site. Molecules bind to the allosteric site and change the shape of the active site, making either more or less receptive to the substrate.

Allotetraploid An individual or species resulting from the doubling of the chromosomes of a species hybrid.

Allotropy Existence of a substance in different forms possessing a different set of predominantly physical properties.

Alloxantin $C_8H_6O_8N_4 \cdot 2H_2O$; uroxin; formed in the hydrolysis of the glycoside, convicin, from soya bean.

Almén test for blood in urine Equal volumes of tincture of guaiac and oil of turpentine are shaken until emulsified, the urine is added and presence of blood indicated by precipitation of an intensely blue colored resin. Test is an application of the Van Deen reaction.

Aloy-valdiguie reaction for codeine & morphine The addition of 3-4 drops of reagent to 2 cc. of concentrated sulfuric acid, followed by the addition of the alkaloid or its salt, produces a color within a few minutes. Reagent — One cc. of 0.1% formaldehyde solution added to 100 cc. of 1% uranium acetate solution. Sensitivity — 0.03 mg.

Alpha (a) helix A spiral shape constituting one form of the secondary structure of proteins, arising from a specific hydrogen-bondng structure.

Alternation of generations Occurences of multicellular diploid form, the sporophyte, with a multicellular haploid form, the gametophyte.

Alternative splicing Refers to fact that certain genes retain or omit particular exons in the final spliced transcript.

Alu sequence A family of sequence-related elements about 300 bp in length approximately 500 000 copies of which are scattered along the human genome.

Alvarez reactions for phy ruvic acid and β-Naphthols Ten drops of a solution of 0.05 gm. of naphthol in 1 cc. concentrated H Cl with 1 drop of pyruvic acid. Alpha napthol produces a yellow color in the cold, an orange color on warming and no change on dilution with water. Beta naphthol yields a red to blue color, changing to yellow on dilution with water or alcohol. The napht˙ols may be used to test for the pyruvic acid.

Alveolus A small cavity; tooth socket.

Amandin A vegetable globulin protein of kernels of plums, peaches, and almonds;

Amaurosis Loss of vision not explained by ocular lesions, caused in the same manner as Amblyopia.

Amber Fossilized plant resin.

Ambergris A cholesterol-like waxy secretion of the sick sperm-whale.

Ambisexual Ertaining to both sexes; ambosexual.

Amblyopia Dimness of vision not explained by ocular lesions or refractive error. It may be due to traumatism, diseases like syphilis, malaria, diabetes, toxic materials such as tobacco, alcohol, quinine, opium, thyroid extract, dinitrophenol, strabismus, intestinal parasites, shell shock (psychic blindness), nyctalopia (night blindness). It may be congenital or acquired, transient or permanent.

Amboceptor A specific immune body necessary for the action of complement on a toxin.

Ambrine A cholesterol-like product found in whale gallstones.

Amenorrhea Abnormal absence or cessation of menstruation. Primary amenorrhea refers to nonappearance of menstruation at puberty, secondary amenorrhea to the same phenomenon after normal puberty. Anatomical causes sometimes occur as in surgical or X-ray removal or destruction. Otherwise the principal causes are anemia and tuberculosis. Occurrences have been reported in avitaminosis, insanity, shock, exposure, phantom pregnancy, and many other conditions of an endocrine character, as thyroid hyperfunction or subovarianism. The endocrinological type has been treated with some success by the use of the anterior pituitary sex hormone extracts (e.g. antuitrinS).

Amicrons In the Siedentopf and Zsigmondy classification of particle size, those particles not visible in the ultramicroscope.

Amidases Enzymes that break the linkage between carbon, phosphorus, nitrogen, etc., e.g. as in acid amides.

Amino acid oxidase An enzyme isolated from liver, kidney or intestinal mucosa, which oxidizes natural amino acids and the non-physiologically occurring alpha-amino acids with the exception of beta-hydroxyglutamic acid; resolved into a yellow prosthetic group flavinadenin dinucleotide and a specific protein; maintains a dissociation equilibrium in the oxidized form.

Amino acid Any of a class of 20 molecules that are combined to form proteins in living things. Consisting of the basic formula NH_2-CHR-COOH, where "R" is the side chain which defines the amino acid:The sequence of amino acids in a protein and hence protein function are determined by the genetic code.

Amino acids, physiology of Amino Acid Structure of the Proteins. The tissues of our bodies, skin, muscle, tendon, are chiefly protein substances. The number of proteins in the animal and vegetable world appears to be infinite. Yet they are all constructed of about twenty-one units, called the amino acids. These have an extraordinary ability to link together in chains in numbers up to thousands. One definition of infinity might be the possible number of different protein molecules that could be built by permutations and combinations of the amino acids. The extraordinary thing, in fact, is that nature ever succeeds in duplicating a protein molecule. Perhaps she never does exactly. But she comes so close to it that so far as we can tell the casein of cow's milk is always the same, the proteins of muscle seem to be constant in their properties, and so on through the list of proteins that make up the familiar animal and vegetable structures of which we are constructed and on which we live.

All the amino acids except proline and hydroxyproline have Structure I, while these two amino acids have II. Each amino acid has an amino group, NH_2 or $NH{\cdot}CH_2$, which has an alkalinity about equal to that of ammonia; and each has a carboxyl group, COOH, which has the acidity of an unusually strong organic acid. The R represents a chemical group which is different in each amino acid, and gives it its character as an individual.

In proteins Emil Fischer demonstrated that the amino acids are joined, by what he termed peptide linkings, each NH_2 group condensing with the COOH of another amino acid, with elimination of the elements of water. Simple chains of a few amino acids, Fischer termed peptides.

The proteins are peptides of tremendously long chains. These protein chains seem usually to be rolled or folded into balls or otherwise made to take a globular or ellipsoid or sausage shape. Their rates of diffusion were found by Northrop and Anson to approximate the rates that would be calculated for spheres, and the asymmetries calculated from ultra centrifugation are not great. Exceptions are the fibrous proteins, such as silk and wool, in which the molecules appear to be extended into straight wavy chains, long bundles of which make the visible fibres.

Path of Amino Acids through the Body. Except for the transient supply of proteins with which we are born, all those in our bodies are obtained from the proteins of other animals and vegetables, which we eat and digest into their constituent amino acids or simple peptides, and then build into our own tissues. However, only a fraction of the amino acids that we invite into our bodies really find acceptance there as naturalized citizens, integral units of our own structures. Many other fates beset the immigrant amino acid; it may be disintegrated to make some entirely different product; or it may simply be burned for fuel. We shall try to follow some of the paths in the body that are taken by amino acids after digestion and absorption in the alimentary tract.

Methods for Determination of Amino Acids. In studies of protein digestion and of the nature and fate of the digestion products, methods for measuring the amounts of amino acids present in blood and other parts of the body are indispensable tools. Two such methods developed in our laboratory have contributed part of the physiological information that will be discussed.

The first was the "nitrous acid method" which depends on the reaction:

$$RCH(NH_2)\,COOH + HNO_2 \rightarrow$$
$$RCH(OH)\,COOH + H_2O + N_2$$

The N_2 gas is a measure of the amount of amino acid present. The reaction is not entirely specific for amino acids, because other amines with NH_2 groups also react; but such amines are usually not present in important amounts or can be removed.

A later and more specific method depends on reaction with a mild oxidizing agent called ninhydrin. Its effect is indicated by the equation:

CH(NH$_2$) COOH+O® RCHO+NH$_3$ +CO$_2$

The analysis consists merely of heating the mixture for a few minutes and measuring the CO_2 evolved. This reaction is so specific for free amino acids that it serves to pick out and measure them in the most diverse mixtures of other amines, organic acids, peptides and other biological products.

Digestion and Absorption. In the human stomach the chief visible change, as noted a century ago by Beaumont through the bullet hole in the stomach of Alexis St. Martin, is that food proteins which enter as insoluble matter, such as meat or coagulated egg white, are dissolved. Chemical studies show that the long protein chains are unrolled and broken into relatively short peptide chains, which still, however, are fairly long. No absorption of the products occurs in the stomach; absorption begins only after the chyme enters the intestine.

In the intestine the chyme meets the enzymes secreted by the pancreas and the intestinal wall. These enzymes hydrolyze the long peptides of the chyme to short peptides containing only 2 or 3 amino acids in the molecule, and to free amino acids. Also, any unchanged protein particles that have escaped the gastric juice are digested. The ability of the intestine to digest, not only gastric peptides, but also intact proteins, makes possible the nutrition of people with achylia gastrica and even of persons who have had the stomach completely removed.

Van Slyke, Cullen and McLean found that in dogs during digestion the amino acid content of the blood rose about 20 per cent, as the blood perfused the intestines, and that the greater part of the absorbed amino acids was removed by the liver. In return, the liver poured into the blood of the hepatic vein an amount of urea nitrogen that had been taken up. One could watch the work of the liver in taking up the amino acids and destroying them, turning their nitrogenous parts into urea for excretion by the kidneys. Unreasonable and wasteful though it seems, a large part of the amino acids absorbed from the intestine appears to be captured and destroyed by the liver, and never to have a chance to reach and nourish the other tissues.

Other experiments, performed with Dr. Gustav Meyer, showed that the liver did not get all the absorbed amino acids, but that some escaped, and could be absorbed by other tissues. It was found that even in the fasting animal the amino acid concentration in the tissues was about 10 times as great as in the blood plasma, viz., about 40 to 60 mg. of amino acid nitrogen per 100 grams of tissue, compared with 5 mg. per 100 grams of plasma. When amino acids were injected into the circulation they were quickly taken from the blood by the tissues, where the amino acid contents might be increased to 2 or 3-fold their former values. In one experiment the amino nitrogen of the liver rose to 150 mg. per 100 grams; in the muscles the increase was never so great. During the next three hours the amino acids in the muscles and kidneys remained practically unchanged, but the amino acids in the liver fell almost back to their original level, and an equivalent of urea nitrogen appeared in the circulation.

Fate of Amino Acids in the Liver. The evidence in these experiments, that the liver is the organ where urea formation takes place, supported an old but much contested hypothesis that the liver is the only organ that forms urea. Its unique distinction in this power was confirmed by Bollman, Mann and Magath of the Mayo Clinic, who showed that removal of the livers from dogs led to an accumulation of amino acids in the blood, and entirely stopped the formation of urea.

Graham Lusk in the early part of the century showed that when protein was catabolized by dogs made totally diabetic by phloridizin poisoning, about 60 grams of glucose were formed and exreted from each 100 grams of protein catabolized. Lusk and his collaborators also showed that when certain amino acids were fed their carbon was partly or entirely turned into glucose when their nitrogen was turned into urea.

Mann[8] and his colleagues showed that no glucose formation from proteins or amino acids occurred when the liver was excluded.

Furthermore, the acceleration of the body's heat production that occurs during assimilation of protein digestion products was shown by Mann and his colleagues not to occur when the liver was excluded. This accelerated heat production, called by Rubner the "specific dynamic action," apparently either represents energy produced by the reactions which the amino acids undergo in the liver, or is caused by other reactions in the cells which are stimulated by the presence of products formed in the liver. Such substances must be other than the urea and glucose, for neither of these causes the observed amount of heat acceleration.

Not all the treatment met by the amino acids in the liver is destructive. During periods of heavy protein feeding the body stores considerable amounts of protein in the liver and, in less amounts per gram of tissue, in the other tissues. The reserve protein seems to be different from the structural proteins of the tissues. In the liver Berg had shown that it can in fact be differentiated with the microscope by its droplet structure in the cells. Functionally it is characterized by the readiness with which it is metabolized at the onset of starvation, and with which it is used to replace blood proteins depleted by hemorrhage, as found by Whipple and his colleagues.

The liver also appears to be the place where plasma fibrin and albumin are formed. It was demonstrated thirty years ago by Whipple that injury of the liver retarded or prevented formation of fibrin. Work by the same author and others had accumulated evidence that the liver is essential also for the formation of the albumin of the plasma. These proteins are presumably formed from free or combined amino acids taken out of the blood by the liver.

Transamination in the Tissues. Some of the amino acids can be synthesized in the body. One of the reactions by which the synthesis occurs was discovered by two Russian biochemists, Braunstein and Kritzman. It is called "transamination," and it enables the cells to change keto-acids, RCOCOOH to amino acids, R $\cdot$ $CH(NH_2)COOH$ by replacing the oxygen atom of the ketone CO group with the elements of ammonia. This ammonia, however, must be transferred to the ketone acid from the $CH(NH_2)$ group of one of the dicarboxylic amino acids, aspartic or glutamic. Since at least some of the keto acids can be produced by partial oxidation of carbohydrates, transamination appears to be one of the processes by which the body can construct a portion of its own amino acids. The reaction of transamination is caused by an enzyme that occurs in all the tissues, and is rich in the muscles.

Tranamethylation. The discovery of du Vigneaud that methyl groups can be transferred from methionine, $CH_2 \cdot S \cdot (CH_2)_2 \cdot COOH$, to other substances in the body had opened a field rivaling in interest that of transamination. Du Vigneaud proved that choline could be formed with the aid of methyl groups from methionine. He administered methionine in which the methyl of the $CH_3 \cdot S$ group was marked by hydrogen in the form of deuterium. When this labelled methionine was fed to rats on a choline-free diet choline could be isolated from their tissues with part of its methyl groups containing deuterium. This observation showed that methyl groups had been taken from the methionine to make choline, presumably by methylating ethanolamine. It was found that this marked methyl group could be further transferred from the choline to creatine. The methyl group of the amino acid, methionine, could therefore be used by the body in synthesizing two of its essential non-amino acid constituents, choline and creatine.

The Continual Replacement of Amino Acids in Living Tissue Proteins. The extent to which nitrogen fed in the form of amino acids is synthesized into tissue proteins, both in the form of the fed amino acids and of other amino acids formed in the body from the fed material by transamination or other reactions, had been studied brilliantly by Schoenheimer and his colleagues. They have synthesized amino acids with heavy nitrogen, N_{15}, have fed the amino acids to rats and mice, and have hydrolyzed the proteins in their bodies and isolated various amino acids from the hydrolysates. Finally, they have analyzed the isolated amino acids for N^{15} to determine the amounts of the ingested nitrogen that were built into the body proteins, both in the form of the administered amino acid and in the form of other amino acids derived from the administered one by transamination or other processes. They have found that incorporation into tissue proteins began almost immediately, and that, while more of the marked N^{15} was incorporated in the form of amino acid with which it was administered, than in any other amino acid, nevertheless a considerable proportion of the marked nitrogen was found distributed among other amino acids in the proteins.

The results of Schoenheimer and his colleagues have dispelled the view that the tissue proteins when once laid down remained as unchanging structural blocks until eventually destroyed by the wear and tear of metabolism. It appears that every protein molecule in the living body is itself alive in the sense that it is continually changing and renewing its structure.

Essential and Non-essential Amino Acids for Animal Nutrition. From the fact that by the transaminase action the body can synthesize some of its own amino acids, it would follow that not all the amino acids in the body proteins must be provided readymade in the food proteins, but that some can be made in the body. The task of finding which of the 21 amino acids are indispensable parts of the diet, and which ones the animal body can build for itself, has occupied some of the leading biochemists since the time of Magnus-Levy's demonstration that the rabbit could make glycine. The names of F. Gowland Hopkins in England and of Lafayette Mendel, T. B. Osborne and W. C. Rose in this country have been especially brilliant in the list of those who have unravelled this problem step by step. As the result of work by all the investigators in the field Rose finally was able to divide the amino acids into the two groups, which must be supplied in the food of growing rats, while the other group can be made by the rat and need not be supplied in the food. The list does not apply to all animals; for example, glycine is indispensable for the chicken. However, it appears probable, although not yet proven, that the necessities of man and the rat are the same or nearly so.

Detoxifying Effects of Amino Acids. It has long been known that various herbivora and man use glycine to combine with and detoxify benzoic acid by forming hippuric acid: $CH_2(NH_2)\cdot COOH + C_6H_5COOH = CH_2(NHCOC_6H_5)COOH + H_2O$

The synthesis appears to occur in the liver, and the ability to form hippuric acid after glycine feeding is used as a test of liver function.

Form. This peculiar effect of the two sulfur-containing amino acids was discovered by Miller, Ross and Whipple. They had observed that a heavy feeding of meat would protect a dog from the effects of a dose of chloroform that would have led to fatal liver degeneration if administered in the fasting state. Investigation of the different types of amino acids yielded by protein digestion proved that only two containing sulfur had the protective effect.

Nutrition by Intravenously Injected Amino Acids. Henriques and Anderson in Copenhagen demonstrated in 1913 that nitrogen equilibrium could be maintained with intravenously injected amino acids as the sole nitrogen intake. They placed a cannula in the neck vein of a goat kept in a stall, and maintained the animal in nitrogen equilibrium for several weeks by giving the necessary nitrogen in the form of a protein digest hydrolyzed completely to amino acids and injected in a slow stream through the cannula. The therapeutic application did not follow till 25 years later, when Elman began the regular use of intravenous injections of predigested protein. Farr and MacFadyen showed that the injected amino acids were assimilated fully as well as the nitrogen from proteins digested in the alimentary tract. A large part of the nitrogen required can be given in this form for a period of weeks. Whipple and his colleagues and Elman found that when the reserve tissue proteins of dogs had been depleted by fasting and the plasma proteins had also been depleted by bleeding, the proteins could be rapidly restored in tissues and plasma by administering a mixture of amino acids containing all those essential for nutrition, but not if essential amino acids were omitted. It appears that intravenous amino acid administration can be used to nourish the tissues at times when feeding by mouth is impossible or inadvisable.

Formation of Non-Protein Nitrogenous Constituents of the Body from the Amino Acids. Some of the amino acids in the tissues are condensed into peptides. Of these glutathione, which is glutamyl-cysteinyl-glycine, is active in oxidation-reduction reactions. Carnosine, or beta-alanyl-histidine, occurs in mammalian muscle, and presumably has a physiological function. Another compound of beta-alanine is the vitamin, pantothenic acid. Serine combines with phosphoglycerides to form a constituent of nerve tissue. Furthermore, animals have the power to transform the elements of amino acids in more profound ways than by condensation with their amino or carboxyl groups. Animals can be reared with no other nitrogenous foods than proteins and slight amounts of certain nitrogenous vitamins. Hence it appears that all the nitrogenous constituents that occur in large amounts in the animal organism can be formed from the amino acids yielded by the digestion of proteins. Among such constituents are the purines which form part of the nucleic acids of cell nuclei and the creatine of muscle tissue. The hormones thyroxine and adrenalin, from their organic structures, are presumably derived from tyrosine or phenyl alanine.

Amino group A functional group that consists of a nitrogen atom bonded to two hydrogen atoms. It can act as a base in solution, accepting a hydrogen ion and acquiring a charge of +1.

Aminoethylphosphoric acid Occurs in relatively large amounts in malignant tumors.

Aminophorases A family of enzymes involved in transaminations e.g. glutamic acid aminophorase, aspartic acid aminophorase; dicarboxylic acids transfer ammonia to mono- and dicarboxylic alpha-ketonic acids, mono-carboxylic acids need the intermediation of some dicarboxylic alpha-ketonic acid; amines and peptides cannot function as ammonia donators and aldehydes, ketones and hydroxy ketones cannot function as acceptors; rates vary for active substances and with source of enzyme, as pigeon muscle.

Aminopolypeptidase A group of polypeptide-hydrolyzing enzymes that hydrolyze the polypeptide chain at the amino end.

Ammonia tolerance test The determination of the amount of urea formed after ingestion of ammonium citrate (diagnostic for liver cirrhosis).

Amnion In land vertebrate, an extraembryonic membrane in the form of a fluid filled sac around the embryo; it absorbs shocks and keeps the embryo from drying out.

Amniotic fluid fluid that surrounds, cushions, and feeds the fetus in mammalian reproduction

Amphigony Bisexual reproduction.

Amphimixis The process of blending of heredity carrying chromatin of the nuclei during cell conjugation in certain protozoa.

Amphioxus Lowest vertebrate; lancelet; the probable ancestral marine precursor of all vertebrates.

Amphipneust An animal using both gills and lungs for breathing.

Amphipyrenin The substance of the nucleus membrane.

Ampholyte A substance capable of giving off or taking on protons; acts as acid toward bases and as base toward acids.

Amphoteric A name applied to compounds having both basic and acidic groups, or capable of ionizing both as a base and as an acid.

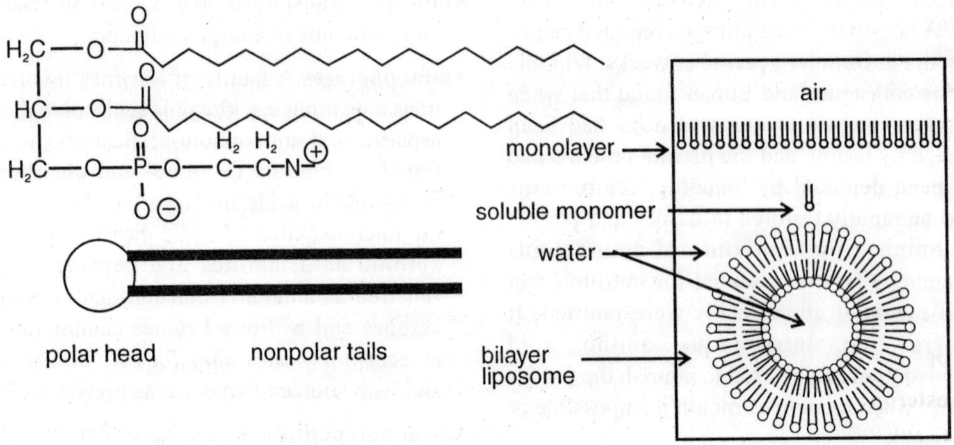

Fig. Double chain Amphiphiles

Amphiphilic A molecule having water soluble polar group attached to a water-insoluble hydrocarbon chain.

Amplification An increase in the number of copies of a specific DNA fragment; can either be *in vivo* or *in vitro*.

Amygdalin $C_6H_5CHCNOC_{12}H_{21}O_{10}$ A glycoside found in the kernels of Rosaceae fruits. It is mandelonitrile amygdalose; crystallizes as rhombs from water, m.p. 200°.

Amygdalose $C_{12}H_{22}O_6$; the disaccharide composed of 2 molecules of glucose, obtained by hydrolysis of amygdalin.

Anabasine The chief alkaloid of Nicotiana glauca; Used as an agricultural insecticide.

Anabolic Pathways Metabolic pathways that build up complex molecules from simpler compounds, requiring energy input usually provided by catabolism.

Anabolism Collectively, the chemical processes by which organisms build complex molecules from smaller ones (compare *catabolism*); the constructive part of metabolism

Analogy The similarity of structure between two species that are not closely related; attributable to convergent evolution.

Anaphase The stage of mitosis when sister chromatids of each chromosome separate and move to opposite poles of the spindle.

Anaphylaxis Allergy; the name given to the protein shock phenomenon shown when a minute amount of foreign protein is injected and a week to a month later another small dose of the same protein is injected. Death frequently occurs a few minutes after the second injection.

Anastomosis Natural or operative joining of structures, as of nerves or blood vessels.

Anatomy The study of the structure of organisms as made out by dissection.

Anchor Sequence A hydrophobic amino acid sequence which fixes a segment of a newly synthesized, translocating protein within the lipid bilayer membrane of the endoplasmic reticulum.

Anchylosis Bony fusion causing stiffness as of tooth to jaw bones, etc.

André reagent for alkaloids Crystalline precipitates are produced upon the addition of potassium dichromate to many alkaloids — brucine, cocaine, codeine, quinine, etc.

Andreasch reaction for cysteine and iron The addition of ferric chloride and ammonia to a solution of cysteine hydrochloride produces a reddishviolet color. The reverse test may be used for iron.

Androgen A chemical substance with activity similar to that of the male sex hormones.

Androgenesis The fertilisation of an egg, the nucleus of which has been so treated that nuclear fusion cannot ensue, by normal sperm.

Androgenic hormones The two male sex hormones, androsterone and testosterone, produced in the interstitial cells of the testis. They control secondary male sex characteristics and accessory sex organs, motility of sperm, composition of semen, and comb growth in the capon.

Androgynal Hermaphrodite, androgynous.

Androsterone $C_{19}H_{30}O_2$; m.p. 182-183°C.; a male sex hormone subsidiary to testosterone produced in the interstitial cells.

Anemias Deficiencies in the number of red blood cells and in the hemoglobin content of the blood, which constitute a complex of symptoms, such as pallor, cardiovascular disturbances characteristic of oxygen deprivation, exertion dyspnea and palpitation, tachycardia, arrhythmia, low blood pressure, nervous symptoms, digestive symptoms, menstrual disturbances, loss of libido and albuminuria. The principal types are

(1) normocytic (orthochromic) anemias

(2) microcytic (hypochronic) anemias and

(3) macrocytic (hyperchromic) anemias.

In the normocytic type the number of erythrocytes and the hemoglobin concentration are reduced in strict proportion. It is characteristic of hemorrhage, hemolyses due to poisoning and most infections or it may be "aplastic anemia" due to exhaustion of bone marrow. Blood transfusion is used as emergency treatment. Microcytic anemias are characterized by a reduction of corpuscular hemoglobin below 30% and a decrease of the mean volume of erythrocytes with a color index below one. This is usually caused by iron deficiency, inadequate digestion, exhaustion. The treatment is dietary.

Macrocytic anemias, including "pernicious anemia," have a high color index, with a mean corpuscular hemoglobin content above 30% and a high volume. They are caused by a lack of antianemic principle of the liver or the "intrinsic factor" secreted by the stomach. There is often failure of the stomach to secrete free hydrochloric acid. The missing factors have to be supplied as well as hydrochloric acid.

Aneuploid A cell containing a number of chromosomes that is not an even multiple of the haploid number (n).

Aneurism Arterial abnormal dilatation.

Angelic acid $C_5 H_8 O_2$, a monocarboxylic unsaturated acid which occurs in the roots of Angelica Archangelica; m.p. 45-46°.

Anidian A blastoderm on which no trace of embryonic axis appears.

Anionogens Elements in an unionized state which can become anions in the body.

Anisotropic Having unequal axes.

Ankylostomiasis Hookworm disease; uncinariasis; Egyptian chlorosis; a disease due to infection by nematode worms, ankylostoma duodenale or Nector americanus, attacking the young and undernourished by invasion through the skin. There is a severe dermatitis accompanied by anemia, perverted appetite for dirt, lethargy. Carbon tetrachloride, tetrachlorethylene and thymol are used to drive out the worms.

Anlage (Rudiment, Ébauche) A group of cells, not necessarily distinguishable from their surroundings, which will later give rise to a given organ.

Anneal The act of two nucleic acid sequences hydrogen bonding through complementarity of the bases determining their sequences. Generally synonymous with "hybridize".

Anoestrus A period of sexual inactivity in animals other than monkeys and men.

Anomers Two stereoisomers (α and β) of a given sugar that differ only in the configuration about the carbonyl (anomeric) carbon atom (carbon 1 for pyranoses and carbon 2 for furanoses).

Anserine Beta-alanylmethyl histidine, a dipeptide isolated in 1929 from goose muscle; m. p. 238°-239°, present in muscles of birds and fishes.

Anterior pituitary-like principle Gonadotropic, glycoprotein hormones found in human pregnancy urine (where it is called P.U.), and in pregnant mare serum; forms basis of Aschheim-Zondek test; similar to luteinizing hormone, but formed in the chorionic tissue of the placenta; can cause ovulation, prolongation of pregnancy, diminution of estrogen secretion, but no follicular growth; in the male causes premature descent of the immature testes of primates.

Antheroidiogen A phytohormone which is chemically a diterpenoid. M, 330. It is derived structurally and biogenetically from the gibberellins.

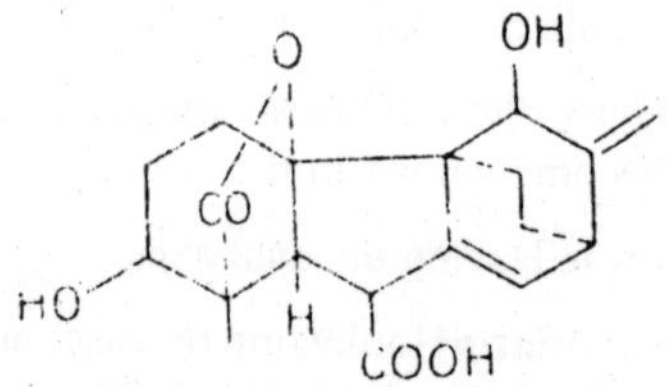

Fig. Antheridiogen

Anthocyanins (anthocyans) A class of violet, red and blue pigments of fruits, blossoms and leaves of many plants, having the pelargonidin nucleus, which is a substituted benzopyranol nucleus. Some representatives are: delphinidin, pelargonidin, cyanidin, peonidin. Through them inheritance of color in plants may be studied. The pigments attack iron and tin and have caused trouble in canning.

Anthrax Splenic fever; charbon; malignant pustule; woolsorter's disease; a contagious disease caused by Bacillus anthracis which may enter through the skin or lungs or sometimes through the intestinal canal. It is found in both animal and human form. Treatment is by antianthrax serum subcutaneously and intravenously if septicemia sets in.

Anthropodesoxycholic acid Chenodesoxycholic acid.

Anthropoid Man or ape.

Antiberiberi vitamin Thiamin Chloride; vitamin B_1.

Antibiotic resistance Resistance conferred to the host the ability to survive a given antibiotic by plasmids containing resistance genes. If such a plasmid is present in a host, that host will not be killed by (moderate levels of) ampicillin or tetracycline. By transforming plasmids containing antibiotic resistance genes, one can get rid of all the contaminating bacteria which does not have such plasmid, thus ensuring that the plasmid will be propagated as the surviving cells divide.

Anticoagulin A substance which prevents coagulation of shed blood (as hirudin, snake venom).

Anticodon A specific sequence of three nucleotides in a tRNA, complementary to a codon for an amino acid in a mRNA.

Antidromic Against the normal direction—as a nerve impulse sent peripherally in an afferent nerve.

Anti-enzyme (1) old term to designate supposed substance which prevents the splitting of tissues of stomach by its own proteolytic enzymes.
(2) anti-body produced by injection of an enzyme, e.g. antiureaae on injection of urease.

Antigen A foreign macromolecule that does not belong to the host organism and that elicits an immune response.

Antigenic variation Mechanism to ensure rapid sequence variation of the gene(s) encoding homologues of an individual protein antigen; usually involving multiple, related gene copies.

Antihormones Substances which counteract the action of hormones.

Antiketogenesis The capacity of certain compounds, notably glucose and glycerol, to prevent the accumulation of acetone bodies by oxidizing them as they are formed. The oxidation of the molecule of glucose facilitates the oxidation of two molecules of fatty acid, the liberated glycerol taking care of the third fatty acid.

Antineuritic Preventing symptoms of lack of vitamin B, such as polyneuritis.

Antisense construct Plasmid that contains coding

region of gene placed adjacent to promoter in such orientation as to cause gene to be transcribed from DNA strand complementary to that from which RNA is normally transcribed.

Antisepsis The prevention of the breeding of microorganisms.

Antiseptic Substance that kills microorganisms or prevents them from growing; typically used in medicine to clean surgical equipment and wounds

Antithrombin An active principle of the liver preventing clotting of blood in blood vessels.

Antitoxin A substance opposed in action to a toxin.

Antuitrin growth factor A commercial preparation of anterior pituitary extract.

Antuitrin-S A proprietary preparation of chorionic gonadotropin.

Anus Lower outlet of alimentary canal.

Aorta Principal large artery leading from heart.

AP-1 site The binding site on DNA at which the transcription "factor" AP-1 binds, thereby altering the rate of transcription for the adjacent gene. AP-1 is actually a complex between c-fos protein and c-jun protein, or sometimes is just c-jun dimers. The AP-1 site consensus sequence is (C/G) TGACT(C/A) A. Also known as the TPA-response element (TRE). (TPA is a phorbol ester, tetradecanoyl phorbol acetate, which is a chemical tumor promoter).

Apiin $C_{27}H_{32}O_{16}$ a glycoside of the leaves of parsley and celery, consisting of apiose and apigenin; exists as glossy needles of yellow crystalline powder, m.p. 228°C.

Apituitarism Absence or dificiency of pituitary; hypopituitarism; hypohypophysism.

Apocodeine An alkaloid formed by the dehydration of codein; an expectorant orally and a purgative on injection.

Apocrine Pertaining to glands; pertaining to part of cell contents (mucus).

Apposition Growth by deposit in succesive layers.

Araban A hemicellulose; a pentosan of gum arabic; an anhydride of arabinose.

Arabinose An aldopentose, with the hydroxyl group on the opposite side of the carbon chain from the β and ±-hydroxyls; produced as prisms, by boiling gum arabic or beet-root chips with dilute H_2SO_4.

Arachidic acid n-Eicosanic acid; a saturated fatty acid, $CH_3(CH_2)_{13}COOH$ found in peanut oil; m.p. 77°C.

Arachidonic acid A 20-carbon fatty acid with four double bonds, found in lecithin and cephalin.

Arachnids Arthropods with four pairs of legs, as spiders, scorpions, mites.

Araroba The source of chrysarobin, a parasiticide, of Andira araroba, a Brazilian tree.

Arbutin $C_{12}H_{16}O_7 \cdot H_2O$; a glycoside of Arbutis Uva Ursi, consisting of glucose and hydroquinone; a diuretic and an antiseptic. Crystallizes as needles, m.p. 165°C.

Archaean era Oldest geologic era of life; includes laurentain period, confined to simple marine invertebrates.

Archaebacteria The more ancient of the two major lineages of prokaryotes, represented today by a few groups of bacteria inhabiting extreme environments.

Archaeopteryx A transtional form on (or very near) the evolutionary road leading from reptiles to birds.

Archetti test for caffeine or uric acid The substance is added to a nitric acid solution of potassium ferricyanide and heated to boiling; a precipitate of Prussian Blue is obtained if caffeine or uric acid is present.

Archipallium The primitive mantle or cortex of the cerebrum, primarily concerned with smell.

Area opaca The opaque yolky extra-embryonic blastoderm of the sauropsids, which later forms the yolk-sac.

Area pellucida The transparent central part of the sauropsidan blastoderm, from which the axiated embryo arises.

Area A morphogenetic area is one of the constituent regions on a Fate-Map.

Arecoline $C_8 H_{13} O_2 N$ or methyl ester of lmethyl tetrahydropyridine carboxylic acid; a parasympathetic stimulating alkaloid of the areca nut; colorless oil, b.p. 209°; large doses cause central paralysis, small doses used as a diaphoretic and anthelmintic.

Areolar Spaced as in fibro-elastic tissue.

Argenteum Strong light reflecting area of skin of fishes on sides and belly.

Arginase An enzyme found in mammalian liver and in traces in other organs and tissues, that splits d-arginine into ornithine and urea; is activated by sulfhydryl—heavy metal combinations, particularly cysteine with iron or copper.

Arginine metabolism In mammals and other urea-excreting animals, arginine is converted to omithine and urea by the enzyme arginase, found chiefly in the liver; synthesized in limited amounts by the body; also a potential source of guanidine acetic acid which can be methylated to form creatine; convertible to glucose in vivo.

Arnica Extract of flowers of Arnica montana, used as a remedy for bruises.

Arnold test for acetoacetic acid in urine Urine is mixed with an equal volume of reagent; the addition of ammonia produces a brownishred color or precipitate, changing to purple-violet with excess concentrated hydrochloric acid. Reagent—

(1) 1 gm. of p-aminoacetophenone in 2 cc. concentrated hydrochloric acid mixed with 100 cc. of water;

(2) 1 gm. of sodium nitrate on 100 cc. of water. In use, mix 2 volumes (1) with one of (2).

Arnold-mentzel test for hydrogen peroxide When an excess of reagent (1 gm. of vanadic acid in 100 cc. concentrated sulfuric acid) is added to the test solution, a permanent red color is produced. Sensitivity-0.0006%.

Arny-dimler test for lactic acid in pharmaceuticals The substance is extracted with 20 cc. of ethyl acetate; the extract is washed with 5 cc. of water and 5 cc. of dilute sulfuric acid. After filtration through a dry filter paper and evaporation on a steam bath until the solvent is completely evaporated, the residue is dissolved in 5 cc. of 1% resorcinol solution. Production of a red color when the solution is floated on 5 cc. of concentrated sulfuric acid indicates lactic acid.

Aromadendrene The principal sesquiterpene of eucalyptus, b.p. 121° at 10 mm.

Arrayed library Individual primary recombinant clones (hosted in phage, cosmid, YAC, or other vector) that are placed in two- dimensional arrays in microtiter dishes. Each primary clone can be identified by the identity of the plate and the clone location (row and column) on that plate. Arrayed libraries of clones can be used for many applications, including screening for a specific gene or genomic region of interest as well as for physical mapping. Information gathered on individual clones from various genetic linkage and physical map analyses is entered into a relational database and used to construct physical and genetic linkage maps simultaneously; clone identifiers serve to interrelate the multilevel maps. Compare library, genomic library.

Arrectores pilorum Tiny hair erecting muscles in the skin.

Arsenoxide A compound formed by the oxidation of trivalent As atoms without change of configuration, e.g. as the oxide of arsphenamine.

Arsphenamine Salvarsan; arsenobenzene; arsenobenzene dihydrochloride; Ehrlich's benzene dihydrochloride; Ehrlich's 606; used in syphilis and other spirillum affections (relapsing fever, frambesia, malaria).

Arteries Vessels that carry blood away from the heart to organs througout the body.

Arterioles Branches of the arteries present within organs.

Arteriosclerosis "Hardening of the arteries"; a chronic disease affecting arterial walls with inflammatory and degenerative changes as thickening, hardening and loss of elasticity. It is generally expected in old age, essential hypertension, bacterial infection, diabetes, gout, lead poisoning and any abnormal cholesterol metabolism. Symptoms are present in the brain (dullness, headache, vertigo) the eye, the heart (angina pectoris, coronary disease, arrhythmias), the kidney, and the extremities (obliteration of the pulse) depending on the case. The treatments are directed toward easing symptoms. Iodides and thiocyanates have a rational basis in cases where cholesterol deposition is marked.

Artery A blood vessel carrying oxygenated blood from the heart.

Arthaud-butte reagent for uric acid The reagent is a solution of 40 gm. Rochelle salt, 20 gm. of sodium thiosulfate and 1.484 gm. of copper sulfate in a liter of water; one cc. precipitates 0.001 gm. of uric acid.

Arthritis A chronic disease of the joints involving changes in cartilages, synovial membranes and bone destruction and new bone formation. It may be brought about by general or focal infection combined with predisposing factors of age and lowered resistance. Two principal forms are

(1) the atrophic (rheumatoid) form and the hypertrophic (osteoarthritis). The former attacks the young usually with much pain, the latter is accompanied by bony outgrowths. Many types are described as acute rheumatic arthritis, gonorrheal arthritis, septic arthritis, tubercular arthritis (white swelling), gout, syphilitic arthritis, Charcot's joint, fibrositis (muscular rheumatism). Treatments are palliative and stress physiotherapy combined with rest and exercise.

Arthropods Invertebrates with jointed legs and chitinous or calcareous exoskeleton.

Artificial selection Deliberate human breeding of living forms.

Aschheim-zondek test A test for pregnancy depending on swelling and hemorrhage of the ovaries of immature female mice following the subcutaneous injection of urine of pregnant women.

Ascites Disease of peritoneum marked by accumulation of large amounts of fluid.

Ascorbic oxidase A plant enzyme which specifically catalyzes the oxidation of ascorbic acid by molecular oxygen to dehydroascorbic acid; optimum pH 6; may be a Cu protein; attacks ascorbic acid homologues.

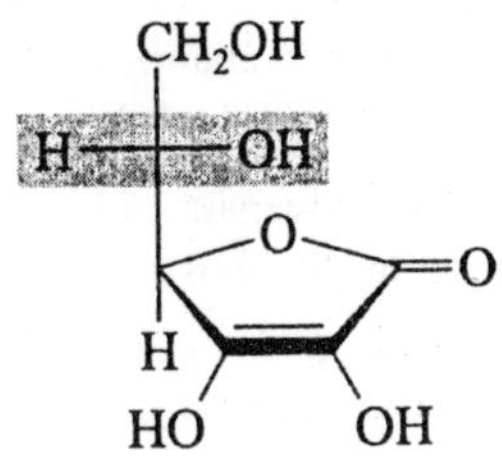

Fig. Ascorbic acid

Aseptic Free of living organisms in any form.

Asexual reproduction Mode of reproduction in which a single parent that produces genetically identical offspring by budding or division of a single cell or the entire organism into two or more parts.

A-Site The binding site on a ribosome that holds the tRNA carrying the next amino acid to be added to a growing polypeptide chain.

Asmacher reaction for histamine Treatment of a solution of a histamine salt with sodium hydroxide, extraction of the free base with hot chloroform and the addition of 0.2 gm. anhydrous copper sulfate gives rise to a dark blue color.

Asparaginase A specific enzyme, found in barley, yeast, bacteria and liver, that hydrolyzes amide N from asparagine, giving aspartic acid; optimum pH 8.

Asparagine Aminosuccinamic acid; the acid amide of aspartic acid found in many proteins; $CONH_2CH_2CH(NH_2)COOH$.

Asparagosin A fructosan from the tubers of Asparagus officinalis.

Aspartase An enzyme found in Bacterium Coli that

splits out ammonia from aspartic acid yielding fumaric acid.

Aspartic acid metabolism Probably forms ketosuccinic acid. The body can synthesize it, so its ingestion is not essential to health. Can be 75% converted to glucose in the body.

Aspartic aminophorase A transamination enzyme of animal and plant tissue which promotes the conversion of aspartic acid and pyruvic acid to oxaloacetic acid and alanine and vice versa; also can do the same with hydroxyaspartic acid.

Assimilation The conversion of foreign material into living matter in the constructive phase.

Assymetric carbon A carbon atom covalently bonded to four different atoms or groups of atoms.

Astacene Astacin, a carotenoid found mainly in Crustacea; its structure is 3,'-4,'-tetraketo-ß-ß'-carotene; $C_{38} H_{44} O_4$; occurs in combination with protein in the shell and eggs of the lobster, salmon, shrimp, etc.; is also present in the chicken retina and concerned there with color vision.

Astaxanthin An animal xanthophyll found in crustacea and among other invertebrates; is a diketodihydroxybeta-carotene and oxidizes to a tetraketo-beta-carotene, astacin or astacene; as astaxanthin dipalmitate the prosthetic group of a conjugated protein which is the blue color of lobster shell, and is set free on cooking, producing its own red color.

Astigmatism Blurred vision due to lack of sharpness of focus of light rays on retina.

Atabrin Chinacrin; $C_{23} H_{31} ON_3$, a powerful synthetic antimalarial.

ATG or AUG The codon for methionine; the translation initiation codon. Usually, protein translation can only start at a methionine codon (although this codon may be found elsewhere within the protein sequence as well). In eukaryotic DNA, the sequence is ATG; in RNA it is AUG. Usually, the first AUG in the mRNA is the point at which translation starts, and an open reading frame follows - i.e. the nucleotides taken three at a time will code for the amino acids of the protein, and a stop codon will be found only when the protein coding region is complete.

Atopen An allergen; the substance to which an allergic person is sensitive.

Atopy The condition of being sensitive to an atopen (allergen).

ATP Adenosine triphosphate". A relatively stable, high energy molecule used to fuel chemical reactions within cells.

ATP synthase An enzyme complex that forms ATP from ADP and phosphate during oxidative phosphorylation oin the inner mitochondrial membrane or the bacterial plasma membrane, and during photophosphorylation in chloroplasts. Uses a proton gradient to chemiosmotically drive the synthesis.

Atria The chambers of the heart that receive blood returning to the heart.

Atrophy Decrease in size and function; emaciation; arrested development.

Atropine $C_{17} H_{23} O_3 N$; an alkaloid found only in traces in solanaceous plants, but obtained by racemization of hyoscyamine; elongated prisms from alcohol, m.p. 118°; causes dilation of the pupil of the eye; hydrolizes to tropine and tropic acid; checks flow of saliva, mucus, and of HC_1 of the stomach, and relaxes intestinal spasms.

Attenuation A form of gene regulation wherein termination of transcription is controlled to regulate overall levels of gene expression.

Atwater-rosa-benedict respiration calorimeter An apparatus for measuring animal heat production and O_2 -CO_2 exchange simultaneously. It consists of a well-insulated constant temperature chamber surrounded by running water at a definite temperature. The rise in temp. of the water is a measure of the heat produced. The heat loss from evaporation of water is determined by the gain in weight of a concentrated $H_2 SO_4$ absorber. The CO_2 produced is absorbed by soda lime.

Aucubin $C_{15} H_{24} O_9$; glycoside of the common spotted laurel.

Auditory ossicles The 3 small bones-malleus

(hammer), incus (anvil) and stapes (stirrup), which convey vibration from the external ear drum.

Auflagerung The laying of competent ectoderm upon a dead inductor in order to test inductive power.

Auricle (1) upper chamber of heart; (2) external ear.

Ausculatation A system of listening to sounds of the organs.

Autocatalysis A reaction in which a product of the reaction accelerates it.

Autochtonous Indigenous; native.

Autocoids Substances which act as chemical messengers in the blood; hormones (stimulating) and chalones (soothing).

Autointoxication Poisoning by toxins absorbed from alimentary canal.

Autolysis Autolysis may be defined as the self digestion of tissues, post mortem. Unless otherwise specified the digestion is assumed to be concerned with the proteins of the tissue. In a broader sense it may also refer to changes which occur to any of the components of a tissue, post mortem.

While the softening of meat on standing, without putrefactive changes, must have been common knowledge antedating recorded history, this phenomenon was first studied scientifically in 1889 by Salkowski He showed that the process was one of protein digestion carried on by enzymes already present in the tissues, and in every way analogous to the digestion of proteins in the gastro intestinal tract. The broader implications of the process were recognized by the pathologist Jacoby who conceived of the mechanism as operating in the living tissues to cause such changes as he observed in the liver in experimental phosphorus poisoning. It was Jacoby who in 1900 gave the phenomenon its name, and who anticipated in his studies much of the subsequent development of the subject. We believe today that the autolytic mechanism is a physiological one, in constant activity, mediating the changes in cell and tissue masses both in the direction of atrophic shrinkage or protein mobilization and in the reverse direction of growth, and hypertrophy.Most of the studies of autolysis have been made on mammalian tissues post mortem. The factors which initiate the proteolysis, and those which accelerate or inhibit it, have been investigated in great detail. The much more difficult field of demonstrating the reverse or synthetic process and the conditions which favor or control it is still far from being completely solved or understood.In the following paragraphs we shall attempt to summarize the significant advances which have been made in this field and the methods used to secure the data. There are extensive reviews of the literature available to the student who wishes to pursue the subject in greater detail and no attempt will be made here to present a complete bibliography. Illustrations. Perhaps the field under discussion can best be understood if a few typical examples of autolysis are first described.

1. When a muscle system is rendered inactive by severing the nerve, or by artificial immobilization (as in placing a broken limb in a cast), the muscle mass shrinks rapidly at first, more and more slowly as the reaction proceeds, until finally a static situation is attained in which the muscle mass remains constant. The curve of weight loss resembles in every respect a digestion curve which approaches an equilibrium. There is good reason to believe that such an equilibrium is attained between the metabolic requirements of the remaining muscle tissue and its supply of blood and lymph. The muscle system may lose as much as 50% of its mass without numerical decrease of cell population and with no significant changes in water or salt content. While muscular strength is diminished, the essential contractile machinery appears to remain intact. This is a typical disuse atrophy and upon removal of the cast and resumption of contractile function the muscle hypertrophies and eventually regains its original mass and strength. In disuse atrophy we see the gross summation of individual cell shrinkages throughout the tissue, and this is essentially a reduction of the

cell proteins and protoplasm. With reduction in active cell mass goes reduction in metabolism to the point at which supply and demand and the removal of waste products are again in equilibrium. It is probable that all autolytic phenomena are based fundamentally upon blood supply and that shifts in cell mass are the result primarily of shifts in blood supply, or its equivalent.

2. In its migration to the spawning grounds the salmon takes no food. It may travel hundreds of miles upstream to reach its destination. Muscle fat is consumed in making this journey, and in addition as much as thirty per cent of the muscle tissue. This mobilization is accomplished slowly and without loss of function—though obviously with reduction of contractile strength. During this period the gonads are developing and increasing in mass. The amino acids required for gonadal maturation and for fuel can only be derived from the active muscle tissue, and the only known mechanism for such mobilization is the autolytic process which is known to be present and active in the muscles of this fish.

3. During metamorphosis of the tadpole into the adult frog the muscular tail is slowly and completely absorbed. In this case autolysis alone is first involved in the process and is due to decreased blood supply to the tail. Eventually cells die, and the process of removal may very well be intensified by phagocytic invasion. The fact that even the skeletal connective tissue structures are removed, makes it clear that additional agencies are required to explain the process, since these proteins are not digested by the autolytic enzymes themselves. Since the process is slow the number of phagocytes present in a given area need not exceed the normal count.

4. A very striking example of mobilization of structural and stored protein is seen in insects during pupation. In the larval honey bee investigated by Bishop. the so called "fat body" develops into a large tissue mass. During pupation this undergoes an orderly disintegration. At the same time larval structures are being replaced by those of the fully developed insect—with all the highly specialized equipment required for its new type of life. The whole process takes place in a closed system, since no food is taken and the only exchange with the outside is by respiration. Bishop was able to demonstrate autolyais of the "fat body," and found that its rate increased during pupation.

5. In the pregnant mammal the mammary gland undergoes hypertrophy in preparation for active secretion of milk. At the close of lactation the gland atrophies again to its resting mass. The first phase is under hormonal control which very likely leads to increased blood supply as a primary step toward hypertrophy. The second or atrophic phase is correlated with a decreasing blood supply and may be hastened by pressure due to partial emptying of the gland. The cycle hypertrophy, secretion and atrophy to the resting condition may be repeated many times by the same gland.

6. In starvation the whole organism loses weight as tissue proteins are mobilized. Some tissues like the liver lose protein rapidly; skeletal muscles lose protein more slowly; heart, brain and the respiratory muscles are maintained with only very slight loss. From the metabolic balance sheet it is clear that after the first easily mobilized reserves of carbohydrate and fat have been burned, the tissue proteins are being converted into amino acids and used an fuel, and that the more essential organs are maintained at the expense of less essential. It is not clear in just what ways this protein mobilization is initiated, but the mechanism by which it is accomplished is autolytic. With return to normal conditions the process reverses and the animal regains its lost tissue proteins from the food ingested.

7. In advanced age there is a gradual shrinkage of the whole organism which seems correlated with the slowly decreasing vigor of the circulatory apparatus. no acids and used an fuel, and that the more essential organs are maintained at the expense of less essential. It is not clear in just what ways this protein mobilization is initiated,

but the mechanism by which it is accomplished is autolytic. With return to normal conditions the process reverses and the animal regains its lost tissue proteins from the food ingested. 8.In advanced age there is a gradual shrinkage of the whole organism which seems correlated with the slowly decreasing vigor of the circulatory apparatus.

Materials and Scope Most of the data upon autolysis comes from studies of mammalian tissues-chiefly liver, kidney and spleen. Practically all mammalian structures have however been studied sufficiently to indicate that the mechanism is the same wherever there are active cells. Certain tissues, like mature white fibrous, elastic, cartilaginous and osseous show practically no autolysis. Such material once laid down does not appear to ever digest itself and is permanent unless invaded by phagocytes. Certain types of leucocytes function as digestive cells, and are provided with special digestive enzymes in addition to the autolytic, just as are the stationary digestive cells of the intestinal tract.

In addition to mammalian tissues, quite extensive studies of the other vertebrate phyla indicates an autolytic mechanism identical with that of the mammal. Among the invertebrates the studies of Bishop already cited indicate an autolytic mechanism in insect tissue, not sufficiently studied as yet to be accurately characterized. Chen and Bradley reported the very slow autolysis of certain molluscan muscles which resembled vertebrate autolysis qualitatively. Further work in this field (about to be published) shows that molluscs, crustaceae and arachnids whose tissues can be isolated and studied separately, have a digestive pattern nearly identical with that of the vertebrates. The only difference thus far disclosed in this preliminary survey lies in a slight shift of the optimum pH for autolysis toward greater acidity. It seems now almost certain that the autolytic mechanism is universally distributed in animal tissues and is at least superficially the same. It is quite as constant a finding as is respiration itself.

In the plant world less is known of the autolytic mechanism and its distribution, though analogous processes are also found in such forms as the yeasts.

Methods of Study In general the methods that have been used are much alike. Fresh tissue is finely divided and suspended in water. Bacterial growth is prevented by some bacteriostatic agent like toluene or merthiolate. The pH is adjusted with buffers or followed as digestion proceeds, since it affects the results very considerably. Activators or inhibitors may be added. Foreign protein may be added as extra substrate. Finally the tissue may be extracted with various solvents and the enzymes separated by fractional precipitation, or by adsorption techniques, which are designed to isolate them in pure form. Such enzymes, more or less purified and separated from the other cell proteins, may then be studied with naturally occurring pure proteins as substrate material, or with synthetic substrates of known structure designed to discover the specific linkages attacked.

Most of the earlier work was done with either the tissue proteins alone or with such added proteins as casein, fibrin, gelatin and hemoglobin. Gelatin has been much used among the European investigators. Since it is not a native protein, and has been shown by Anson not to be hydrolyzed by his purified cathepsin, the interpretation of the results must be open to serious question. Undenatured hemoglobin has been shown by Anson to be readily cleaved by cathepsin, to be strictly reproducible, and to be insusceptible to cleavage by secondary proteases of the autolytic group. It may be anticipated that the use of purified hemoglobin will make it possible to determine the amount of activity of the autolytic mechanism in various tissues and species. Such a comparative study cannot yet be compiled because the tissue proteins themselves vary greatly in fragility. Thus the keratins are highly resistant to the common proteoclastic enzyme; collagen is digested by pepsin but not by trypsin or cathepsin; elastin is slowly digested by both pepsin and trypsin while its behavior to cathepsin is not known; liver proteins and hemoglobin are readily fragmented. Thus in comparing the over-all autolytic activity of two tissues directly there are several factors which influence the result—the enzyme concentration and activity, and the protein concentration and fragility. Where extractions of the enzyme are resorted to in attempting such an assay, with the use of a known substrate like hemoglobin, the difficulty of assuring a complete extraction is serious. Willstiitter showed that a part of the cathepsin was easily obtained in solution—the "lyo" form—while the rest was firmly bound by insoluble proteins as the "desmo" form. The same difficulty exists in making cathepsin assays with the use of small molecular substrates.

Autolysis is measured by the rate and extent of cleavage. The methods employed are various and not sufficiently standardized so that the results from one laboratory can always be compared with those from another. It is generally agreed that trichloracetic acid is a convenient precipitant of the native proteins, and at proper final concentration will remove the unchanged protein while allowing even the larger cleavage products to pass through into the filtrate. A final concentration of about 5% has been found satisfactory. As digestion proceeds more soluble nitrogen appears in the filtrates, and the tyrosine color reaction increases in intensity.

The soluble nitrogen is determined by macro- or micro-Kjeldahl techniques, and the data so obtained probably provide the most direct method of measuring proteolysis we have, and are thoroughly reliable.

The color given by phenolic compounds with the Folin-Ciocalteu reagent has also proven satisfactory as a measure of primary cleavage. Although the color developed is not specific for tyrosine, it nevertheless follows very closely the soluble nitrogen curve, and has the advantage of speed. The color may be estimated by colorimeter or by the various types of photoelectric cells available.

The formol titration of free groups in the trichloracetic acid filtrate—or the equivalent titration in alcohol—has been much used in the past, while the gasometric methods of Van Slyke and changes in conductivity have had a limited use. Inasmuch as these methods measure total protein cleavage, they are not strictly comparable to the previously mentioned methods, and do not represent primary cleavage alone. The determinations are rapid and easily made and are useful in following total cleavage. They have been widely used in the past, particularly in Europe.

Post mortem changes affecting Autolysis. When a tissue hash is prepared as rapidly as possible after death of the animal, and sampled and precipitated at once, the trichloracetic acid filtrate is characterized by: a small but nearly constant amount of nonprecipitable nitrogen; a slight color test for tyrosine; a small amount of amino nitrogen by formol titration or the Van Slyke reaction with nitrous acid; and a nitro-prusside reaction for -SH compounds that varies from a trace to a fairly brilliant color. The figures are quite characteristic of the tissue and are so constant as to indicate an equilibrium between these components and the tissue proteins on the one hand and the blood and lymph on the other.

The reaction of the interior of the living cell lies between 6.8 and 7.2. Immediately after death and mechanical disintegration this reaction changes rapidly, and becomes more acid. Within an hour it may reach 6.2 in the case of liver, and thereafter slowly return toward neutrality. The acids produced in such a digest are H_3PO_4, carbonic, fatty acids and lactic in the order of their amounts. The production of these acids is evidently not dependent upon autolysis since in breis made alkaline to begin with, these acids may increase as much as 300% over the initial concentration, without any demonstrable cleavage of the proteins. They do however influence very largely the subsequent rate and extent of the digestion of the proteins.

Proteolysis appears to begin at once post

mortem, as measured by the appearance of initial cleavage products. If a latent period exists it is a matter of minutes only, during which the acidity is rapidly increasing.

Typical results After a tissue has been prepared in the cold, to minimize early changes, it is sampled and warmed quickly to the temperature selected for digestion. Samples are removed from time to time and analyzed for soluble nitrogen or tyrosine in the trichloracetate filtrate. Cleavage proceeds rapidly at first, and more slowly later until a stage of digestion is attained at which the reaction ceases, or increases very slowly. This equilibrium point may be reached in three to five days, or may require a much longer period depending on the nature of the tissue. In the case of liver, diluted to a 20% brei, and under the most favorable conditions, digestion of all available tissue proteins is complete in five to ten days. Only the nondigestible stroma proteins remain. In the case of mammalian muscle, about thirty percent of the total protein is digested. When game fish muscle like mackerel is used, a considerably larger fraction of the total protein is digested. The extent of the digestion is characteristic of the tissue. If hemoglobin is added in the late stage, digestion is resumed and continues till the added hemoglobin is also completely hydrolyzed. Cessation of digestion therefore appears to be determined by the amount of available substrate present in the brei, and under optimum conditions this is digested completely, or very nearly so. Furthermore under the usual experimental conditions for rapid autolysis, cessation is not due to destruction of the enzymes, for these have been shown to survive for many weeks. Cessation of cleavage represents the exhaustion of available substrate.

Effect of H ion concentration The earlier literature on autolysis described many substances which increase autolysis of a tissue like liver. It eventually became clear that the common factor in all of these accelerators was the production of an acid reaction. (In the case of H_2S and the colloidal sulphides which slowly hydrolyze to produce H_2S, there is a second factor present—the -SH effect).

The average pH of a freshly prepared liver brei is usually between 6.5 and 7.0. If the pH is maintained at 7 by means of buffers there is little or no digestion of the proteins. If the H ion concentration is allowed to rise, or is increased by the addition of small increments of acid, digestion is progressively increased and there is less and less undigested material remaining when equilibrium is reached. At pH 4 digestion is maximal, and the only material remaining is the stroma shreds which resist digestion indefinitely. If the H ion concentration is raised beyond pH 4 digestion decreases again and reaches zero in the neighborhood of pH 2.5.

At this level the enzymes are slowly destroyed, and at pH 2 they are destroyed at once.

At pH 7 there is little enzyme activity and no substrate to be attacked. As the H ion concentration rises the enzyme grows more active and the amount of available substrate is increased, as shown by the greater percentage of the total protein hydrolyzed when the reaction has become static. The effect of acid is therefore to increase the actual or susceptible substrate till all that can be made available is thus converted at about pH 4. It is concluded that the acid-protein salt, or the protein cation, is the actual substrate. The base salt, or the protein anion, is not digestible. Certain proteins are rather resistant to cleavage. Egg albumin and the albumin and globulin of serum belong in this group. At a pH greater than 5 these proteins inhibit the digestion of liver itself. At pH 5 they neither increase nor decrease digestion, while at greater acidity they are themselves digested and increase the total cleavage products obtained. The isoelectric points of these proteins are close to pH 5. In more alkaline solutions they are believed to adsorb the enzyme strongly and thus inhibit autolysis, while in reactions more acid than 5 they become fragile and are themselves digested. These results are believed to confirm the hypothesis that it is the acid-protein salt that is the actual substrate material.

When hemoglobin is added to a liver brei it is digested to some extent between pH 6 and 7. The amount digested increases rapidly between 6 and 4. Since its isoelectric point is about 7 there should be available hemoglobin at all reactions more acid than 7. The rapid increase between 6 and 4, the potimum, lends weight to the idea that the enzyme itself grows more active and reaches its optimum activity in the neighborhood of pH 4.

Since studies with small molecular substrates also show optimum cleavage in this same pH region, we may tentatively conclude that the enzyme itself is most active in this reaction.

Effect of the oxido-reductive level and -SH This field is under such active investigation at the present time that any generalization must be considered tentative. Certain facts however seem thoroughly established. When a tissue is killed and dispersed, the particles show a nitroprusside reaction for -SH of varying intensity. This reaction normally disappears from the digest within two days. It persists best in the more acid breis. Oxidation to the -SS- form is believed responsible for this change. It is hastened by bubbling air or oxygen through the digest, especially if a trace of copper salt is present. It may be quickly abolished by adding such oxidants as, KIO_3,KIO_4 or I_2. When this is done autolysis proceeds at a greatly reduced rate but eventually attains the same level of digestion as the control. On the other hand, if cysteine or certain other -SH compounds are added to a brei, the digestion rate is usually increased over the control. Where this is not the case it is probable that the -SH reaction is already optimal.

From such data the conclusion has been reached that the oxidative level as indicated by the -SH reaction is an important determinant in the speed with which hydrolysis takes place. When the oxidative level is low and the -SH reaction strong and persistent, autolysis is optimal. There is no clear evidence that the substrate mass has been increased, but the enzyme activity is greater. This is the "activation" effect by -SH first described by Grassmann and Dyckerhoff for yeast, and soon after by Waldschmidt-Leitz et al for spleen. The phenomenon has been studied by a large number of investigators but is still imperfectly understood. The activating effect may be of the enzyme alone, and this is the usual assumption. It is also possible that the substrate is rendered more easily fragmented.

The work of Bergmann, Fruton and their associates indicates that cathepsin is a complex of proteinasee,some of which are activated by -SH and some of which require no such activation. These have been characterized thus far only with synthetic substrates having small molecules. Anson separated a single cathepsin from spleen which digests hemoglobin without cysteine activation. Bradley et al have presented evidence of two catheptic proteinases in the behavior of whole liver brei. In this study KIO_3 was added to abolish the -SH reaction in a large primary digest. Secondary digests to which cysteine was added were prepared and all maintained at the pH of 4. The enzymes were active at the end of thirty days as was shown by adding hemoglobin without cysteine to a secondary digest set up from the primary digest at that time.

Sense of more hydrolysis; the individual cells decrease in mass and as a result the tissue as a whole shows atrophy. The process continues until the metabolism of the diminishing tissue mass is again in equilibrium with the blood and oxygen supply. With return to the original blood supply the new equilibrium is again disturbed, in favor of synthesis; the cells and the tissue hypertrophy to the original mass and metabolic rate.

The above statement is evidently an attempt to correlate what is known of the autolytic mechanism with what is observed in atrophy and hypertrophy. New facts will undoubtedly require its revision. However, it is generally accepted that in some such way this widely distributed enzyme system functions to relate cell mass and tissue mass and their metabolic requirements, to the systems of supply. It's significance lies in the evidence that it functions at all times in maintaining the dynamic equilibria of the living organism, while its post mortem function is only incidental in initiating the removal of dead cells.

When shifts occur in the fundamental factors of supply and demand, the resulting shifts of tissue mass will be permanent or temporary depending upon these fundamental factors. Thus the shrunken muscles of poliomyelitis, of old age or of the tadpole's tail are evidence of permanent changes in the factor of supply. Disuse atrophies of immobilized muscles, the atrophy of the mammary gland as lactation is concluded, or the tissue losses incident to undernutrition are examples of temporary shifts in the supplydemand equilibrium. The growth of tumors or of the body as a whole are examples of more permanent shifts in the direction of synthesis.

The determining factor in all these changes appears to be the blood supply. Data to support this statement are far from adequate, and there may be other factors equally important. It is obvious however that a cell population situated along a given vascular system and in equilibrium with it, cannot maintain itself unaltered if that blood supply is decreased. Furthermore metabolism is known to be correlated more or less directly with tissue mass, which may imply that cells cannot reduce metabolism except by a decrease of active protoplasm. Thus the shrinkage of the individual units of the population will be the first effect automatically produced by the autolytic mechanism. If the decrease of the blood supply is too great or too prolonged some of the less favorably situated cell units will die, and the population will be numerically decreased. These cells are then removed by their own digestive enzymes and probably also by the intervention of phagocytic cells with their additional specialized enzymes. Both types of atrophy are well known.

The Autolytic Enzymes The tissue proteinases were classified by Waldschmidt-Leitz as consisting of cathepsin, carboxypolypeptidase, aminopolypeptidase and dipeptidase. The last three named make up together the "erepsin" of Cohnheim and were found to be widely distribtued in tissues first by Vernon. Cathepsin was believed to be a single proteinase, acting best at pH 4, activated by -SH and similar reducing agents and initiating cleavage of the native proteins of the tissue. The remaining peptidases were believed to act specifically on the polypeptide fragments of primary cleavage, and on dipeptides, with the production of amino acids. The reaction curves of the members of this group overlap sufficiently so that they could conceivably function together in the complete conversion of tissue proteins to their final cleavage units.

The classification above has recently been shown to be inadequate, as has been indicated in previous paragraphs. Based upon separations and studies of specific behavior with synthetic substrates, Bergmann, Fruton and their associates· Johnson and Berger, and others have shown that there are probably many more enzymes in tissues than those described by WaldschmidtLeitz, and that they will eventually be susceptible to an entirely different classification based upon more fundamental characteristics than were disclosed by the earlier work. At the present time it would perhaps be premature to do more than indicate that such detailed studies are in process. The reader is referred to articles which have recently appeared in Advances in Enzymology for summaries of this work.

Protein Synthesis Substrate hydrolysis has been the usual criterion of autolytic activity. On theoretical grounds we assume the reaction to be reversible and under proper conditions to lead to the synthesis of protein molecules from cleavage products. The experimental evidence for such reversals is obviously extraordinarily difficult to obtain in a tissue brei. Without the organization of the cell to protect and insure stability of a peptide chain once formed in order that it may go on to the next step in synthetic growth, the possibility of building up a native protein molecule seems remote. Voegtlin, Maver and Johnson have reported success in such an experiment however, where the environment of the tissue extract, glutathione and cleavage products was changed from anaerobic to aerobic conditions. There have been a few other reports of similar successful syntheses. There have also been failures reported. Perhaps the most convincing demonstration of synthesis was reported by Behrens and Bergmann who showed that liver or spleen cathepsin can catalyze the synthesis of a single specific peptide bond by the use of a well defined simple substrate. This experiment we believe finally completes the evidence which establishes the validity of the major concepts of the autolytic mechanism as functional in both atrophic and hypertrophic tissue changes.

Autonomic nervous system The part of the nervous system dealing with involuntary actions; A semi independent group of nerve cells and fibers distributed throughout the body and controlling the viscera.

Autonomous induction An induction in which the inductor takes no part itself in what is formed (opp. of *Complementary induction*). All chemical inductions are autonomous inductions.

Autophagy (1) Nutrition of the organism on tissue reserves as in fasting;

(2) consumption of one's own flesh in insanity.

Autophyte A plant deriving its food from inorganic sources.

Autoplastic transplantation Transplantation between different sites on the same individual.

Autoradiography A process to detect radioactively labeled molecules (which usually have been separated in an SDS-PAGE or agarose gel) based on their ability to create an image on photographic or X-ray film. This process does not result in a linear relationship between the intensity of the signal and the amount of radioactivity unless special steps are taken. There is now increasing use of phosphorimagers and other modern devices to detect and quantitate radioactive molecules which have been separated in gels.

Autosome A chromosome not involved in sex determination. The diploid human genome consists of 46 chromosomes, 22 pairs of autosomes, and 1 pair of sex chromosomes (the X and Y chromosomes).

Autotomy The spontaneous shedding of an appendage (arthropods).

Autotroph An organism able to build all the organic molecules it requires using carbon dioxide (present in air and in water) and energy from the physical environment. Photosynthetic autotrophs use sunlight; chemosynthetic autotrophs extract energy from chemical reactions involving inorganic substances. compare heterotroph.

Auxesis Growth in size by cell expansion without cell-division.

Auximone Accessory growth-promoting factor in plant food.

Auxochrome A group which when introduced into an organic compound containing a chromophor, converts the compound into a dye.

Avena einheit A measure of the potency of the of the plant hormones defined as "that amount of substance which when present in a block of 3% agar-agar, 2mm. × 2mm. × 0.5mm., and applied to one side of a decapitated Avena Coleoptile, will cause a curvature in the coleoptile of 10° in two hours, when kept at 22-23C° and 92% relative humidity."

Avertin A proprietary name for tribromethyl alcohol, a basal anesthetic; given rectally to produce deep sleep, not permitting operation without the addition of an inhalation anesthetic.

Avidin A glycoprotein which binds to biotin with very high affinity (Kd = 10-15).

Avitaminosis Diseased condition due to vitamindeficiency.

Axon Main elongated outgrowth of a nerve cell, which carries impulses from the cell body.

Azafrin $C_{27}H_{38}O_4$; the pigment of the Azafranillo root.

Azotobacter A group of aerobic bacteria found in soils and in water basins and capable of fixing atmospheric nitrogen, under aerobic conditions and in non-symbiotic state; mechanism of fixation believed to be through the ammonia stage, although still disputed; an enzyme, designated as azohydrase is claimed to bring about the reaction of fixation.

Azzolini test for epinephrine (adrenaline) 4-5 cc. of test solution are shaken with 1-2 drops of hydrochloric acid and 2-3 cc. of a mixture of 4 parts ammonia and 96 parts alcohol are superimposed. Presence of epinephrine indicated by rose-red ring at alcohol-acid interface. Metallo-organic salts do not interfere. Sensitivity—1:500000.

B

B. M. R. Basal metabolic rate, an average value of consumption of oxygen and production of carbon dioxide under standardized conditions of normal metabolism.

B.I. Buffer index; the ratio of a small increment of alkali and the small increment in pH brought about by its addition. If acids are used negative values can be used in both cases, thus always giving a positive index.

Babo reaction for uric acid Red precipitate of cuprous oxide is obtained when an alkali urate is boiled with dilute Fehling's solution. A white precipitate of cuprous urate is obtained with free uric acid, being converted to cuprous oxide on boiling with alkali.

Bachstez-cavallini reaction to differentiate ascorbic and isoascorbic acids Addition of 2 drops of 10% uranium acetate solution to a faintly alkaline 1% solution of ascorbic acid produces a slight brown color which is discharged by the addition of a few drops of concentrated sodium hydroxide, followed by precipitation of amorphous sodium uranate. A deep brownish-red color and a clear solution is obtained with isoascorbic acid.

Bacillus Rod-shaped bacterium which multiplies by division.

Back Mutation Reverse the effect of a point or frame-shift mutation that had altered a gene; thus it restores the wild-type phenotype.

Bacterial artificial chromosome (BAC) A chromosome-like structure, constructed by genetic engineering. BAC is a cloning vector capable of carrying between 100 and 300 kilobases of target sequence. They are propagated as a mini-chromosome in a bacterial host. The size of the typical BAC is ideal for use as an intermediate in large-scale genome sequencing projects. Entire genomes can be cloned into BAC libraries, and entire BAC clones can be shotgun-sequenced fairly rapidly.

Bactericidin A substance which kills bacteria without lysis.

Bacteriolysin Substance that destroys bacteria by dissolving them; bacteriophage.

Bacteriophage By definition "bacteriophage" (or phage) connotes a large group of sub-microscopic agents which induce transmissible lysis of bacteria and are capable of passing bacteriaretaining filters. They are widely distributed in nature, abound particularly in the intestinal tracts of man and animals and possess particulate diameters within the limits of 8 μ to 100 ii. Like the mosaic viruses, purified phage preparations have been found to be nucleoproteins. Molecular weight determinations by means of centrifugation analysis and diffusion have given figures of 400,000 to 300,000,000.

The first observations on bacteriophage were made in 1915 by Twort. He noted in colonies of

Staphylococci isolated from calf-lymph curious, glassy-looking areas within which there were masses of granular debris. He found that it was possible to form vitreous zones in normal cultures by the simple expedient of placing on the surface of a young agar culture a drop of filtrate from a suspension of bacterial growth in which this degenerative change already had taken place.

In 1917 d'Herelle called attention to the same phenomenon observed while he was studying cultures of dysentery bacilli recently isolated from the stools of patients suffering from bacillary dysentery. For no apparent reason ordinary turbid broth cultures of dysentery bacilli would suddenly become crystal clear and this spectacular lysis could be induced in a fresh culture of dysentery bacilli by adding a drop of the filtered, cleared culture. d'Herelle coined the term "bacteriophagy" to denote all phases of the reaction between bacteriophage and the susceptible bacterial host and for several years he studied the phenomenon intensively. He developed the view that phage is a single living organism, "Protobios bacteriophagum" which may adapt itself to live upon a wide variety of bacterial substrates. Others, notably Bordet, believed that phage is a bacterial enzyme of unspecified origin but possessing such properties that its introduction into the bacterial cell disturbs the normal metabolic processes with the result that the cell initiates production of the same agent. The essential difference between the two hypotheses is simply that in d'Herelle's view phages are small living units stemming from a single parent "race" and acquiring special characteristics by adaptation while Bordet and his followers believe they are modified bacterial proteins. The consensus of opinion among modern workers is that phages are bacterial viruses, protein in nature and altogether analagous to the animal and plant viruses.

METHODOLOGY

Isolation of phage Because phages are so ubiquitous their isolation is not especially difficult and the fact that they are commonly found in the intestinal contents makes crude sewage a good source material. The sewage sample mixed with broth is incubated overnight and then is filtered successively through a small amount of infusorial earth to remove solid debris and a Chamberland L-3 candle or some similar device to retain bacteria. One ml. aliquots of the filtrates are added to 10 ml. amounts of broth in test tubes and each tube is inoculated with sufficient amount of a young culture of the test organism to produce faint turbidity. The test tubes are held for 24 hours and are examined at intervals for clearing. Clearing indicates the presence of an active phage and the cleared culture, after filtering to remove any viable bacteria, may be set aside as a source of phage for the organism in question.

If instead of complete lysis, the test suspensions show only a demonstrable reduction in turbidity they are filtered through L-3 candles and the process repeated with the filtrate. At the same time 1 ml. of the filtrate should be added to 10 ml. of a faintly turbid suspension of the bacteria and 0.05 ml. of the mixture spread over the surface of an agar plate. After incubation the bacterial growth on the agar plate may reveal punched out areas, the 'taches vierges', or plaques, which represent loci of phage action on the bacteria. Microscopic examination will reveal that these plaques contain mostly granular debris resulting from cellular disintegration. Such a plaque may be used as the source material for inoculation of a young broth culture seeded into broth. If this first passage on a broth culture does not produce complete lysis, filtration and serial passage upon the susceptible bacteria should be resorted to.Since in nature phages commonly occur as mixtures it is desirable to derive 'pure line strains' before undertaking any systematic work. This is accomplished conveniently by spreading the mixed phage, suitably diluted, over an agar surface inoculated with the test organism. After 24 to 48 hours incubation an isolated plaque is selected and the surface is touched with a sterile needle. The minute amount of adherent phage obtained in this fashion is used to induce lysis in a young broth culture. When lysis has occurred the culture is filtered and the filtrate again is plated out for single plaque selection. This repetition is not always necessary although it helps to ensure the isolation of a pure line strain.

2. **Quantitative methods**Two general procedures for the quantitative determination of phage have been developed. These are: (a) the plaque count and (b) the activity-titration.

a *The plaque count*. The theory of this method is identical with that of the colony count for estimating numbers of viable bacteria. Serial dilutions of a phage-containing solution are plated out on a substrate of susceptible bacteria and the plaques are counted after a suitable period of incubation. Since each plaque is assumed to be the locus of action of a single phage molecule, the plate count multiplied by the dilution factor gives the number of phage particles in the original solution. Gratia has evolved a superior method of performing the plaque count. In this procedure the essential steps are:

a. Add 0.5 ml. aliquots of phage dilutions to 3.5 ml. amounts of a broth suspension of bacteria containing 1.6×10^8 organisms/ml.

b. Mix thoroughly and quickly with 1 ml. portions of 2.0% agar heated to 100^0 C.

c. Remove 1 ml. to the surface of a previously poured and solidified 2% extract agar plate and smear over the surface with a sterile 15 mm. loop.

d. Incubate and count the plaques.

The activity titration: The Krueger activity assay depends upon the fact that the time at which lysis occurs in a standard suspension of susceptible bacteria maintained under certain controlled conditions is a function of the amount of phage present at the start of the reaction. The quantitative activity unit employed has the dimensions of a velocity constant and in the case of the Staphylococcus and antistaphylococcus phage studied by Krueger and his co-workers there are about 1×10^{10} such units in each ml. of ordinary lysate. This corresponds to about 1/3 plaque per activity unit.To conduct a phage activity assay with the Staphylococcus and antistaphylococcus phage mentioned above, 4 ml. aliquots of appropriate broth dilutions of phage-containing solution are pipetted into test tubes of 15 mm. inside diameter. For each run there are included 4 ml. amounts of standard phage (1×10^{10} units/ml.) diluted to 10^{-3}, 10^{-4}, 10^{-5}, and 10^{-6}. To all tubes are added 1 ml. portions of a broth suspension of fresh staphylococcal culture containing 12.5×10^7 organisms per ml. After 1.5 hours turbidity readings are made at 0.2 hour intervals until lysis begins. The tubes are read against formalinized broth suspensions of staphylococci ranging from 5×10^7 bacteria/ml. to 20×10^7 bacteria/ml. Once lysis begins, turbidity comparisons are made at 0.1 hour intervals. The time at which each suspension is reduced to a turbidity end point of 8×10^7 cells/ml. is noted. The data obtained from the dilutions of standard phage are plotted as logarithms of initial phage concentrations against the time of lysis (time to reach the arbitrary turbidity end point). This gives a standard curve from which the time of lysis of any unknown solution can be directly converted into original concentration of phage.Northrop has devised a modification of the activity assay for use with concentrated non-sterile lysates. His unit is the quantity of phage necessary to cause lysis of 5 ml. of standard bacterial suspension (0.3 $mm.^3$ of centrifugally packed cells per ml.) in I hour, it is eqiuvalent to 10^{10} Krueger units.The activity determination has been used extensively by Kruger and his co-workers, Northrop, Clifton and Morrow, Lin, Fischer and others. With it one can obtain reproducible accurate results in three to four hours instead of twenty-four to forty-eight as in the case with the plaque method. However, it can not be adapted to all organisms inasmuch as it requires fairly rapid growth and complete lysis.The significance of quantitative data secured by application of phage analytical procedures depends upon the method employed. With the plaque count one cannot distinguish between a single phage molecule and a thousand such molecules grouped together. On the other hand the activity assay determines the total phage present no matter what the state of aggregation.

General Feature of the Phage-Bacterium 1. The Mechanism of Phage Action on a Susceptible Bacterial Substrate

The following general relationships have been found to obtain in studies on certain strains of Staphylococci, colon bacilli, dysentery bacilli and B. megatherium conducted by the investigators enumerated in the preceding section. **A.** Bacterial reproduction follows the normal growth curve just as though no phage were present up to the time lysis (massive dissolution) of bacteria begins. **B.** Phage production proceeds at a rate considerably faster than bacterial growth; therefore the ratio of phage to bacteria in a given mixture is constantly rising. **C.** When approximately 100 phage units/bacterium have accumulated, lysis of bacteria occurs very rapidly. This lyric threshold varies with the organisms studied and with the conditions under which the experiments are conducted. Terminal lysis is a spectacular phenomenon resulting in the complete destruction of millions of bacteria in each cubic centimeter of broth within a period of half an hour. It involves actual cellular disruption accompanied by hydrolytic cleavage of bacterial proteins in some instances. Granular cellular debris may or may not remain after lysis has gone to completion depending on the nature of the organism lysed. **D.** The finished lysates may contain as much as 10,000,000 times the initial concentration of phage. **E.** Substances in solution during the reaction may alter the lytic threshold in interesting fashion. For example, small amounts of manganous chloride reduce the phage/bacteria ratio requisite for lysis from 100 to 12; sodium sulfate and sodium chloride have just the opposite effect raising the ratio to as much as 1,000. F. By adjusting environmental conditions it is possible to produce phage in the absence of bacterial growth or to cause organisms to grow in the presence of phage without forming more phage. G. Delbrück and Ellis have made a careful study of the reaction between a strain of B. Coli and its homologous phage. They observed that the process of phage production is discontinuous and occurs in welldefined steps. The increases (in phage) develop very rapidly and seem to depend upon liberation of phage by cellular lysis.

Ancillary Phenomena Phages exhibit decided limitations in the range of their activity; that is, no single phage is capable of lysing all organisms. Usually it is found that a particular phage will act on one goup of bacteria but not on others, e.g., a coli phage will not lyse staphylococci. Even within the group for which it displays activity there occur resistant strains able to grow in concentrated phage solution. Apparently phage specificity is linked with the antigenic structure of the bacterial host; certain definite chemical components must be present on the cell surface for phage fixation to occur and if these materials are lacking this highly essential initial step in the phage-bacterium reaction cannot be consummated. Phages have been isolated for species of the following genera: Actinomyces, Corynebacterium, Vibrio, Micrococcus, Bacillus, Bacterium.

Likewise there exist tremendous differences between phages which act on the same host species. These differences extend to particle size, dimensions of plaque produced, antigenic composition and resistance to physical and chemical agents.

When phage-lysed broth cultures are maintained in the incubator for long periods of time secondary growths of bacteria may sometimes develop. These consist of phageresistant bacteria which usually exhibit a wide variety of altered characters. The dissociated strains may possess entirely different structural, biochemical, pathogenic and immunologic characteristics. Such alterations are reflected in changed colonial structure, fermentative capabilities and host invasiveness. The transfer of a secondary culture to bouillion results in a growth in which resistant bacteria and phage coexist—the so-called "mixed" culture. Secondary cultures do not represent a permanent resistant state, for the capacity to resist phage action may readily be lost if the bacteria are repeatedly transferred to new media or if phage is eliminated from the mixture in some way.

Some strains of bacteria are phage carriers (lysogenic). That is, they live in close association with phage and when the cultures are filtered, phage can be detected in the filtrates. Despite the fact that lysogenic organisms constantly produce phage in the course of their metabolic activities, they display no evidence of damage. Systematic studies have shown that the lysogenic state is not at all uncommon and that the host cannot be divested of its phage-forming function. Every cell of a lysogenic strain produces phage and in spore-formers this tendency is carried through the spore stage.

General Properties of Phage Phages are readily inactivated by a wide variety of chemical agents. In some instances chemical inactivation has been proven to be reversible. For example staphylococcal phage completely inactivated in 2.8% $HgCl^2$, may be restored to its original titre by precipitation of the Hg^+ ion.

When phage solutions are heated the loss of activity proceeds logarithmically with time and in some instances heat inactivation is reversible. The critical thermal increment for anti-staphylococcal phage is 101,000 calories. Exposure of phage to large doses of X-rays results in loss of activity by two mechanisms, an indirect inactivation due to some decomposition product of water and a direct effect ascribable to ionization within the phage molecule. Since the indirect effect can be suppressed by the addition of a protective foreign protein, it is possible to calculate the sensitive volume of the phage molecule with minimal inaccuracy. Values of the sensitive volume determined in this manner agree well with data derived from diffusion, centrifugation and ultrafiltration experiments. Phage inactivation induced by sonic waves follows the equation for a monomolecular reaction. The presence of homologous organisms during treatment of phage with sonic waves tends to inhibit phage destruction.

The specific inhibition of phages by bacterial extracts and cellular autolysates has been the subject of considerable study. Treatment of the inhibiting extracts with acid enhances the inactivating capacity. The polysaccharide nature of the inhibitor has been proven in some instances.

There is no evidence that phages possess an independent metabolism. Within the range from pH 3.4 to pH 12.0 they carry a negative electrical charge and below pH 3.4 the charge tends to become positive although measurements are made difficult in this zone by the inactivating effect of H+ ions.

Of particular significance in connection with the therapeutic use of phage (vide infra) is the inactivating effect of those products which occur in tissues as a constant part of the phenomenon of inflammation. It is known that blood serum, white cells, whole dead bacteria, tissue debris and in fact colloidal suspensions of many sorts exert either a direct destructive action on phage or an indirect inhibitory effect on the interaction of phage and bacteria.

Phage lysates serve as antigens, i.e., when injected into animals they evoke the formation of antibodies. These antibodies are of two kinds, those which act on phage and those which are specific for the cellular components. The antibacterial antibodies may be absorbed by bacteria but the phage antibodies cannot be so absorbed. Of great theoretical importance is the fact that the antigenic structure of a phage, i.e., its serological character is independent of the substrate on which it is grown.

Experimental evidence has been advanced for the view that bacteria synthesize an inert protein precursor which is converted into active phage protein by an autocatalytic reaction. While such a precursor has not been isolated as yet, its presence can be demonstrated in "activated" bacteria, i.e., cells which have completed a phase of rapid growth and subsequently have been brought to a resting state by storage in salt solution at low temperature. The addition of these cells to phage results in a very prompt increase in phage titre to levels as high as ten times the original value. The cellular component responsible for this increase is very labile and can be inactivated by several chemicals such as iodoacetic acid and methylene blue light in concentrations that are not lethal for the cells themselves. The same thing holds for the heat inactivation of precursor; it proceeds at temperatures too low to injure the bacteria and the rate of inactivation rises abruptly with small temperature increments. The critical thermal increment for heat inactivation of phage precursor is 90,000, indicating that the reaction involves protein denaturation.

Pictures of phages taken with the electron microscope show that they consist of a round head and a much thinner tail which gives them a curious sperm-like appearance. There appear to be patterns of granules within the head. The phages adhere to susceptible bacteria either by the head or by the tail.

THE USE OF PHAGE IN CLINICAL MEDICINE

As a Diagnostic Aid Craigie and Yen have developed a method for typing typhoid bacilli which is based on certain peculiarities of a strain of phage specific for the Vi form of the typhoid bacillus. It has a selective affinity for the particular strain of E. typhosa on which it is propagated and for epidemiologically related strains. This phage, known as type II Vi phage does not act on the W form of E. typhosa nor on other pathogenic or normal inhabitants of the intestinal tract of man. By using this phage it has been possible to divide freshly isolated cultures into a limited number of readily distinguished types.

The typing of typhoid strains is of real epidemiological importance. All clinical cases arising from a single carrier will show the same type of organism as the carrier and if all carriers in a given area are typed during an outbreak it is possible to immediately eliminate some of them from consideration and to direct attention to the activities of other carriers. Typing will permit the detection of two or more concurrent outbreaks and in addition will effect the separation of endemic and epidemic cases. In a given epidemic if all cases show the same type of organism, a carrier is probably responsible whereas the detection of a variety of types would imply sewage contamination.

The natural occurrence of antidysentery phage in stools at times during the course of bacillary dysentery has been recommended as a diagnostic aid. In one outbreak of institutional dysentery due to the Hiss-Y strain the homologous phage was found in 33% of cases in the first week, 80% in the second, and 45% in the third. Experience in this field has not given uniformly good results and further work is required before the procedure can find widespread application.

As a Therapeutic Agent Recently a survey of the applications of phage to treatment of disease was made by Krueger and Scribner at the request of the Council of Pharmacy and Chemistry of the American Medical Association. They concluded that, with certain special exceptions, the lyric destruction of bacteria by phage in vivo is subject to serious limitations. In their estimation there are four significant mechanisms attending the use of crude lysates; namely non-specific protein shock, local immunization, specific immunization by means of antigenic substances in the lysate, and promotion of phagocytosis. The reported data on the use of phage in various diseases caused by bacteria are, for the most part, insufficient to establish phage therapy as a method of choice. Only rarely have these studies included an adequate bacteriologic background, control groups or careful comparison with accepted therapeutic procedures. There is evidence, however, that a properly prepared lysate can serve satisfactorily as: **A.** vaccine for the treatment of certain diseases, e.g., some types of staphylococcic lesions. **B.** An aid for the induction of non-specific protein shock in syndromes in which at times such shock may be used to good purpose, e.g., typhoid fever. **C.** A measure for enhancing the general resistance of an infected area when appiied topically. This depends on the non-specific action of the lysate in mobilizing macrophages and microphages.

Phage solutions possess no measurable degree of superiority over well-known and accepted preparations employed for the same purposes; for example, bacterial vaccines and toxoids in carbuncles and furunculosis or typhoid vaccine in non-specific shock therapy. Modern chemotherapeutic approaches to the treatment of a variety of conditions for which phage has been recommended (bacteremias, gonorrhea, cystitis) offer more chances of success than does phage.

The literature cites numerous accounts of reactions ranging from mild to severe following the injection, local application or ingestion of phage and animal experimental work has shown that lysates may contain enough soluble toxin or Reynal's spreading factor to be actually dangerous. Thus the use of phage in any individual case should be undertaken with caution and with the above mentioned limitations in mind.In bacillary dysentery and cholera there is some evidence that phage may be of value. Here again much additional work is required before phage can be established as a therapeutic agent of choice.

Bacteriophage A virus that infects a bacterium and which is often used in molecular genetics experiments as a vector, or cloning vehicle. Recombinant phages can be made in which certain non-essential l DNA is removed and replaced with the DNA of interest. The phage can accommodate a DNA "insert" of about 15-20 kb. Replication of that virus will thus replicate the investigator's DNA. One would use phage l rather than a plasmid if the desired piece of DNA is rather large. Bacteriophage l and M13 are ones commonly used in cloning and/or subcloning of small genes or DNA fragments in *E. coli.* Bacteriophage P1 is one that is used for fragments up to 95 Kb in size.

Bacteriotropin Substance in blood serum which makes bacteria more susceptible to engulfment by phagocytes; opsonin.

Baeyer-drewsen test for acetone Acetone produces a blue color when a solution of it is treated with o-nitrobenzaldehyde and heated to boiling, followed by the addition of a few drops of sodium or potassium hydroxide solution.

Baicalin $C_{21}H_{16}O_{11}$, a glycoside of glucuronic acid and 5, 6, 7-trihydoxyflavone, present in the roots of Scutellaria baicalensis.

Bakhar An enzyme used in the manufacture of rice beers.

Balata Rubber-like material from the latex of S. American Mimusops balata.

Baljet reaction for digitalis glucosides and strophanthins Digitoxin, gitalin and g- or k-strophanthin yield a red to orange color, at room temperature, with the following reagent — a freshly prepared mixture of a 1% alcoholic solution of picric acid with an equal volume of carbonate-free 10% sodium hydroxide solution.

Balsam of peru Extract of Myroxylon Pereirae containing benzyl benzoate and cinnamate, used as antiseptic.

Balsam of tolu A semi-solid extract of Myroxylon toluifera, used as an antiseptic expectorant.

Band A pattern of light and dark regions by Giemsa staining that can serve as landmarks on chromosomes.

Baptisin Glycoside of the roots of Baptisia tinctoria, which upon hydrolysis yields rhamnose and a complex flavone. $C_{26}H_{32}O_{14}$.

Barac reaction for ascorbic acid Ascorbic acid, in alkaline solution, produces an orange color with diazotized sulfanilic acid. The acid functions as a reducing agent, evolving nitrogen from the reagent.

Barbaloin The active principle of aloes, a glycoside of d-arabinose and alphahydroxy-anthraquinone.

Barbital Veronal; barbitone; an odorless, slightly

bitter, white powder, $C_8 H_{12} O_3 N_2$, diethylmalonyl urea or diethylbarbituric acid, m. p. 188-190°, used as a hypnotic.

Barbituric acid Malonyl urea; a soporific and parent substance of many soporifics, m. p. 253°.

Bardach test for blood in urine 5 cc. of urine are treated with 3 drops of 30% acetic acid, allowed to stand 2 minutes and 2 drops of saturated alcoholic solution of guaiac resin added. After shaking, 0.5 gm. sodium perborate are added, 3 cc. 80% acetic acid quickly added, the solution shaken and overlaid with 2-3 cc. alcohol and 1-2 drops pyridine. Blue color, forming near pyridine, indicates blood.

Barfoed reagent and test for glucose Glucose forms a red precipitate of cuprous oxide when a solution of it is boiled with a few drops of reagent for a short time. Reagent — 13.3 gm. of neutral crystalline cupric acetate are dissolved in 200cc. of 1% acetic acid. 0.1% glucose in dextrin may be detected.

Barger-berger-todd thiocrome reaction for vitamin B_1 A dilute solution of vitamin B_1, treated with a very dilute solution of potassium ferricyanide followed by sodium hydroxide, gives rise to blue fluorescence, due to the formation of thiocrome.

Barnacles Crustaceans which attach themselves to surfaces in adult stage.

Barral test for albumin and bile pigments in urine Filtered urine is floated on a 20% solution of o-phenolsulfonic acid. Formation of a green ring indicates bile pigments, albumin forms a white ring. Sensitivity 5 mg. albumin per liter.

Barret test for acetone in urine Immediate formation of a precipitate when urine or blood filtrates are slowly distilled into dilute (7:1) Nessler's solution indicates the **Bartley test for bile in urine** Bile pigments in urine impart a green color to the clear filtered solution when treated with hydrochloric acid and ferric chloride. Shaking with chloroform removes any blue color due to indican.

Basal ganglia Masses of gray matter at base of cerebral hemispheres.

Basal metabolism The minimum respiratory metabolism which can be expressed most satisfactorily in terms of calories per square meter of body surface per hour. The measurement is made on completely relaxed subject lying down following preliminary rest period, 15 to 18 hours after eating in a warm room. The subject should not be asleep.

Base Pair (bp) One pair of complementary nucleotides within a duplex strand of a nucleic acid. Under Watson-Crick rules, these pairs consist of one pyrimidine and one purine: i.e., C-G, A-T (DNA) or A-U (RNA). However, "noncanonical" base pairs (e.g., G-U) are common in RNA secondary structure.

Base sequence analysis A method, sometimes automated, for determining the base sequence.

Base sequence The order of nucleotide bases in a DNA molecule.

Base In molecular biology, this term refers to the purine bases adenine and guanine, and the pyrimidine bases uracil, thymine, and cytosine, or modification of these bases.

Batyl alcohol An active principle of bone marrow, found in fish oils; octadecyl monoether of glycerine.

Bauchstück The ventral half of an amphibian gastrula, from which all the organiser region has been removed; hence it never forms any neural axis.

Bauer liver function test In abnormal liver function galactose, orally administered, is excreted in the urine.

Bayer 205 Germanin; Foumeau 309; a modified trypan blue dye with peptide linkages, used in treating trypanosome diseases.

Bayer reaction for adrenaline A solution of 2 cc. adrenaline, 1 cc. sulfanilic acid solution, 2 cc. sodium biiodate solution and 1 cc. 10% phosphoric acid solution is colored reddish-yellow if adrenaline concentration is 1:50000; yellow at dilution of 1:625000; yellowish at 1:830000. Yellow color is visible up to 1:5000000 dilution.

B-DNA Conformation of the Watson-Crick double helix in which the two strands form a right-handed double helix.

Becher test for kidney insufficiency 2 cc. of filtered blood serum, made protein-free by precipitation with trichloracetic acid, is boiled with 0.5 cc. concentrated nitric acid for 0.5 minute, cooled, treated with 1.5 cc. 33% sodium hydroxide solution and diluted to 4 cc. with sodium hydroxide. Positive test is indicated by yellow color which may be compared colorimetrically with 0.03874% potassium dichromate solution.

Bedeutangsfremde Selbstdifferenzierung Self-differentiation in accordance neither with the tissue's original presumptive fate nor with the presumptive fate implied by the new surroundings, i.e. neither selfwise nor neighbourwise.

Behenic acid N-docosanic acid; a saturated fatty acid, $CH_2 (CH_2)_{20}$ COOH, found in oil of ben from seeds of Moringa pterygosperma, m.p. 84°.

Beilstein test for halogens in organic compounds A halide-free copper spiral is dipped into the organic substance, ignited, and the substance allowed to burn outside the flame. A green color, when the spiral is heated in the non-luminous portion of the flame, is a positive test.

Bell test for alum in flour or bread 5 gm. of flour are mixed with 5 cc. water and 1 cc. saturated ammonium carbonate solution and 1 cc. of reagent are added. Lavender to blue color, depending upon amount of alum, is developed. Pure flour yields light red color, rapidly becoming dirty brown. Reagent— Freshly prepared logwood tincture from 5 gm. logwood and 100 cc. alcohol. Belladonna Deadly nightshade extract containing the active principle hyoscyamine, an antispasmodic.

Bence-jones protein A protein of bone marrow and leukocytes and found in the urine in leukemia, eczema and multiple myelomata, coagulates under 55° C. and redissolves on boiling; molecular weight about 35,000.

Benedict reagents for reducing sugars The reagent consists of two solutions.

1. (a) 69.3 gm. copper sul fate dissolved in 1000 cc. of water. (b) 346 gm. Rochelle salt and 200 gm. anhydrous sodium carbonate in 1000 cc. of water. (c) 200 gm. potassium thiocyanate in 1000 cc. of water.

(2) Equal volumes of the solutions are mixed in the order indicated for use. 2.5 to 5 gm. of anhydrous sodium carbonate are added to each 30 cc. of the resulting solution. II. 173 gm. of sodium citrate. and 100 gm. of anhydrous sodium carbonate are dissolved, by heat, in 600 cc. of water, the solution filtered and diluted to 850 cc. 17.3 gm. of copper sulfate are dissolved in 150 cc. of water and added slowly, with stirring, to the carbonatecitrate solution in a large beaker. The solution is ready for use.

(3) To test for glucose, 8 drops of urine are treated with about 5 cc. of reagent, the solution boiled for 1-2 minutes and cooled. A voluminous red, yellow or green precipitate indicates glucose.

Benedict-denis method A method for the determination of total sulfur, frequently applied to urine.

Benedict-hopkins-cole reagent 250 cc. of saturated solution of oxalic acid are added slowly, while cooling, to 10 gm. of magnesium powder. The solution is filtered, made acid with acetic acid and diluted to 1 liter.

Benzedrine ß-phenyl isopropylamine; a compound related to ephedrine and epinephrine; orally acts as a stimulant, increasing rapidity and intensity of heart action; on inhalation causes shrinkage of mucous membrane.

Benzoyl peroxide test for 1, 2,-BENZPYRENE A carcinogenic hydrocarbon found in coal tar; shows characteristic fluorescence spectrum bands at 4180, and 4400,A.

Berberine An alkaloid from various plants, relatively inactive, $C_{20} H_{19} O_5 N$.

Beriberi Avitaminosis B_1 ; hypovitaminosis B_1, a disease due to lack of vitamin B_1 in the diet or to failure to absorb it. The symptoms are loss of strength, neuritis, edema, myocardial degeneration and muscular atrophy. There is a dry or paralytic type (atrophic) and a dropsical (wet) type, also cardiac forms and "infantile beriberi," which is acute. The treatment is dietary or by direct administration of the vitamin.

Bertrand's reagent for alkaloids Alkaloids form insoluble precipitates with silicotungstic acid, the sensitivity varying with different alkaloids from 1:8000 to 1:500000.

Beta particle High velocity electrons ejected in nuclear decomposition.

Beta Pleated Sheets A zig-zag shape, constituting one ofrom of the secondary structure of proteins, formed of hydrogen bonds between polypeptide.

Betaine The simplest member of the group of betaines; it is derived from glycine and has the formula fate dissolved in 1000 cc. of water. (b) 346 gm. Rochelle salt and 200 gm. anhydrous sodium carbonate in 1000 cc. of water. (c) 200 gm. potassium thiocyanate in 1000 cc. of water.

Equal volumes of the solutions are mixed in the order indicated for use. 2.5 to 5 gm. of anhydrous sodium carbonate are added to each 30 cc. of the resulting solution. II. 173 gm. of sodium citrate. and 100 gm. of anhydrous sodium carbonate are dissolved, by heat, in 600 cc. of water, the solution filtered and diluted to 850 cc. 17.3 gm. of copper sulfate are dissolved in 150 cc. of water and added slowly, with stirring, to the carbonatecitrate solution in a large beaker. The solution is ready for use.

To test for glucose, 8 drops of urine are treated with about 5 cc. of reagent, the solution boiled for 1-2 minutes and cooled. A voluminous red, yellow or green precipitate indicates glucose.

Betaines A group of completely methylated heterocyclic ±, β, and -amino acids found distributed in the plant kingdom.

Betitol Tetritol obtained from beet molasses; $C_6 H_8 (OH)_4$; M.P. 224°.

Bezssonov reaction for antiscorbutic extracts and polyphenols Reagent—2.7 gm. phosphomolybdic acid and 44 gm. sodium tungstate heated with 5 cc. 85,% phosphoric acid, 60 cc. 5 normal sulfuric acid and 400 cc. water for 2 hours at 50-60°.

A blue color is obtained with all vegetable antiscorbutic extracts. Of the monophenols, only guaiacol reacts, forming a yellow-brown color. Hydroquinone gives an intense blue; catechol red-violet turning blue; orcinol and resorcinol no color; protocatechuic acid, pyrogallol and tannin yellowbrown; phloroglucinol gives no reaction.

Bial reagent and test for pentoses in urine Not more than 1 cc. of urine is added to 5 cc. of reagent, after the latter has been boiled; a green pigment, extractable with amyl alcohol, indicates pentoses. Reagent —1-1.3 gm. orcin are added to 300 gin. 30% hydrochloric acid, to which has been added 25-30 drops of 10% ferric chloride solution.

Biennial A plant fruiting every two years, storing food in the interim.

Bile A bitter yellow fluid, secreted by liver cells of practically all vertebrates and stored in most species in the gall bladder, where it is considerably concentrated by absorption of water, with a change of its reaction from slightly alkaline to weakly acidic, it is eventually excreted into the duodenum, whence its essential constituents, the bile acids, return to the liver. The most important components of this aqueous solution, besides the alkali salts of the bile acids, are alkali carbonate, bile pigments, lipids, including cholesterol, and mucin. Bile is an emuslifying agent for lipids, thus essential for their digestion.

Bile acids Those constituents of bile that keep water insoluble fats in solution or dispersion; they exist in the bile as the sodium salts of the acids conjugated through a peptide linkage with glycine and taurine; they are all derived from the same nucleus, cholanic acid, but have hydroxyl groups at different positions.

Bile pigments A series of tetmpyrrole pigments, found in the bile, and resulting from the destruction of hemoglobin by oxidative splitting of the porphyrin ring system at one methine (—CH=) group and removal of the iron and globin component.

Bile salts The alkali salts of the bile acids; they form soluble addition products with fatty acids which are diffusible; in addition, they lower surface tension, allowing for more facile diffusion and attack of lipids by lipases.

Bile tract disease Cholecystitis; cholangitis; cholelithiasis; dyskinesia of the bile passages; a group of disorders of the bile passages which may be characterized by the inflammation of the gall bladder (cholecystitis) or of the bile passages (cholangitis) or the formation of "stones" or calculi in the gall bladder or bile ducts (cholelithiasis) in both acute and chronic forms. The origin of these diseases may be due to infection by various bacilli, e.g. streptococcus, typhoid, bile stasis, dietary upsets leading to separation of cholesterol, especially in "gallstones" (cholelithiasis). An entire science of cholecystography has been developed to visualize the stones by means of dyes opaque to Xrays, which usually contain high percentages of iodine. Jaundice is frequently an accompaniment of biliary disease because of liver involvement. The symptoms include "biliousness" which is characterized by nausea, headaches, drowsiness, belching. Gall stones are largely cholesterol with protein inclusions. In treatment the flow of bile can be increased only cautiously. The agents used to promote the flow of bile, cholagogues or cholekinetics, are usually magnesium salts. Substances which tend to increase the amount of bile secreted, or choleretics, may be salicylates or cinchophen combined with the bicarbonates and sulfates.

Numerous tests have been developed in connection with bile tract and liver diseases besides cholecystography. Bile pigments are determined in the blood and in the urine, e.g. the van den Bergh test. For urobilinogen one uses the Ehrlich aldehyde reagent. Various types of jaundice (hemolytic, obstructive, parenchymatous) are thus distinguished. Liver function is also tested by ability of the liver to store glycogen from galactose, blood sedimentation tests, amino acid tests to determine breakdown of tissue, and special tests on serum proteins, e.g. Takata-ara reaction.

Bilirubin (bilifulvine, biliphaein) $C_{33}H_{36}N_4O_6$; a porphyrin derivative with the 4 pyrrole nuclei in a straight line arrangement; it is formed by the destruction of hemoglobin; on oxidation the red bilirubin yields green biliverdin.

Biliverdin $C_{33}H_{36}N_4O_8$; a green bile pigment found in the uterus and in feces; it is a porphyrin derivative with the 4 pyrrole nuclei in a straight line arrangement, resulting from the oxidation of bilirubin; on oxidation biliverdin yields blue bilicyanin.

Binding site A place on cellular DNA to which a protein (such as a transcription factor) can bind. Typically, binding sites might be found in the vicinity of genes, and would be involved in activating transcription of that gene (promoter elements), in enhancing the transcription of that gene (enhancer elements), or in reducing the transcription of that gene (silencers). Note that whether the protein in fact performs these functions may depend on some condition, such as the presence of a hormone, or the tissue in which the gene is being examined. Binding sites could also be involved in the regulation of chromosome structure or of DNA replication.

Biochemistry Definitions and Classifications of Branches. Biochemistry is a hybrid science. It arose from the overlapping of more or less purely chemical issues and more or less biological issues. Consequently it is subject to a great variety of definitions reflecting the scope of the pure sciences and the manner in which they are combined. The manner of combination results either in an emphasis on biology or on chemistry. "Biochemistry" is a term which stresses chemistry and to most biochemists, at least of two or three decades ago, it meant ""the chemistry of biologically important substances." The emphasis has been shifting to "the chemical aspects of living processes" and, we venture to predict, will lead to the employment of the expression "chemical biology." This is a trend in all scientific development and has led, for example, in the borderland between physics and chemistry, to the emergence of two subjects, "physical chemistry" and "chemical physics." Thus, we can expect another pair, "biological chemistry" (or biochemistry) and "chemical biology," to divide the field now customarily described by the former term. The term "physiological chemistry," designed possibly to call attention to the dynamics of the biological processes has been falling into disuse, because physiology is after all only one branch of biology. If it is retained it may become a branch of "biological chemistry."

It is possible also that "chemical biology" might receive a name like "cheobiology." The trend toward renaming in the direction in which "chemical" serves as the adjective is illustrated by the relatively recent rise of "chemical embryology" (Needham). If we are right, this in itself may become a branch of "chemical evolution" which in turn would be a branch of "chemical biology."

Other definitions of biochemistry emphasize, consciously or unconsciously, the philosophy of the definer. The author of the definition, may be an outright materialist, in which case he would stress the physical basis of life, or an idealist of one kind or another who believes in entelechies, life principles and the like. B. Moore illustrated the latter in his denfiition of biochemistry as "the study of the chemical substances produced by biotic energy." One hates to think of what would become of the subject according to such a definition if there were no special distinctive "biotic energy."

The modern trend is toward what the author has called "dynamic biochemistry" before favoring wholeheartedly "chemical biology." What is left behind may well go under the name of "descriptive biochemistry" which is a large part of the Dictionary, which catalogues and describes many of the substances found in living sources.

A convenient, but theoretically poorly founded, distinction is that of a "pure" and an "applied" science. The varieties of "applied biochemistry" include what the medical student gets as "pathological chemistry" or what the botanist might take up as "phytochemistry," the zoologist as "zoochemistry."

There are even more complicated hybrids in such subjects as Vernadsky's "biogeochemistry" and the loosely defined field of "comparative biochemistry"' which ranges over all living types in an evolutionary manner.

We suggest that order in this creation of sectors of research might be set up by listing the divisions of biology as

(1) morphology

(2) distribution

(3) physiology

(4) aetiology and the customary divisions of chemistry as physical, inorganic, organic, analytical, etc., with subdivisions such as "colloidal" and the like, and then blending them into hybrid sciences. This will account for the appearance of such terms—used in England—as "aetiological chemistry" meaning the division of biochemistry dealing with "cause and effect in the determination of development" and thus naturally including as a part of itself "chemical embryology."

Biochemistry, divisions of No rigid, conventional divisions of biochemistry exist, as in the case of chemistry and physics proper, where a distinction between organic' and "inorganic," for example, is rather sharp. Still dominant is the treatment where the class of substance involved is the name of the subject, as "Carbohydrates," "Fats," "Proteins. Inorganic Constituents," "Accessory Substances," and now '"Hormones." Almost all texts describe these groups of substances and then take up what is known of their ingestion or synthesis in the body, their role, their metabolism and the further course, final consumption and elimination of the products of their metabolism. That at least the principal classes of materials serve as equivalents of one another and are mutually convertible tends to upset any such classification. The latter part of many a text, therefore, resorts to a series of chapters where the physiological process, e.g. digestion, or the function of an important organ, e.g., the kidney or the liver, serves as the division of the subject. Occasionally a text starts with a phsico-chemical introduction in place of the organic chemistry of the principal materials. In such cases a stress on colloid chemistry is imperative. A time is coming when introductions will have to stress a certain amount of new theoretical material from the field of physics, e.g. spectroscopic theory, electron diffraction and its applications in the electron microscope. From this point of view the fields of biochemistry shape up as applications of sectors of the other sciences to living systems.

Biogeography The science of the geographical distribution of living things.

Bioluminescence The production of light by a chemical reaction within an organism. The process occurs in many bacteria and protists, as well as certain animals and fungi.

Bioluminescence Bioluminescence is a word applied to the light emitted by various living organisms. At least 40 different orders of animals contain luminous species and two groups of plants, the bacteria, responsible for the luminescence of flesh and dead fish, and the fungi, which live in phosphorescent wood. Luminous bacteria are so small that individuals cannot be seen by their own light but colonies are visible. They are easily cultured and are nonpathogenic to man but may infect living animals, giving rise to a luminescent disease of sand-fleas, shrimps and midges, which is eventually fatal. Luminous bacteria may also live symbiotically in special organs of certain fish, notably Photoblepharon and Anomalops, of the Banda islands.

Bacteria and fungi emit light continuously day and night, while all other forms luminescence only when stimulated. The phosphorescence of the sea appears when many different kinds of small organisms are disturbed by the breaking of waves or motion of a boat. Among the groups of animals containing luminous species are flagellates, radiolaria, sponges, jelly-fish hydroids, sea pens, ctenophores, nemerteans, earthworms and many marine worms, shrimp, ostracods and copepods, myriapods, several groups of insects, monuses, squid, brittle stars, balanoglossids, tunicates and fish.

Bioluminescence is never dependent on a previous illumination of the cells or a previous radiation of any kind, nor is it connected with crystallization, friction, or rubbing, but is the result of oxidation by molecular oxygen of a definite substance produced in the luminous cell. It is a chemiluminescence. The luminous material or photogen is almost universally manufactured by living cells as granules, which may normally undergo oxidation within the cell, as in the fire-fly, (intracellular luminescence) or be extruded as a luminous slime or secretion (extracellular luminescence) as in a small ostracod crustacean, Cypridina. Most is known concerning the chemistry of extracellular luminescence, especially that of Cypridina.

In Cypridina the granules in the secretion dissolve on contact with sea water and the homogeneous luminescence is emitted by the resultant colloidal solution. Two kinds of granules are distinguishable in the cells of Cypridina, one large and yellow, the other small and colorless. In fact, in five of twenty-five different groups of luminous animals tested, it can be demonstrated that luminescence is due to two chemical substances, luciferin (yellow) and luciferase (colorless), which can be easily separated because of a difference in resistance to heating and other properties.

Crude luciferin solution is prepared by making a hot water extract of a luminous organ. Heating destroys the luciferase but does not harm the luciferin. Crude luciferase solution is prepared by making a cold water extract of a luminous organ, when both luciferin and luciferase dissolve and luminescence occurs. The extract is then allowed to stand in the air until the light disappears, evidence that the luciferin has been completely oxidized, leaving the luciferase, an enzyme, in solution. A luciferase solution, by virtue of this mode of preparation, must contain the oxidation product of luciferin as well as luciferase.

Luciferin and luciferase are quite specific. Luciferin from one animal will not luminesce if mixed with the luciferase of another luminous form unless the animals are closely related, such as two species of the same genera or two genera within the same order. Even if the color of the luminescence is different in the two different species light will appear, provided the species are closely related. In this case it is interesting to note that the color of luminescence of the resultant "cross" is determined by the animal supplying the luciferase. Luciferase must be the source of the light. It is convenient to designate the luciferins and luciferases by prefixing the name of the animal from which the substances are obtained.

Most luminous animals, if dried rapidly will again luminesce on moistening. Dried Cypridinae have been kept for 26 years without deterioration and can be used for preparing luciferin and luciferase.

Cypridina luciferin is purified by extraction of the dry Cypridinae with methyl alcohol. Ten per cent of butyl alcohol is then added and the methyl alcohol removed in vacuo. The supernatant butyl alcohol extract is chilled and benzoylated with benzoyl chloride. After fifteen minutes this solution is diluted with ten volumes of water and the new inactive benzoyl luciferin derivatives extracted with pure ether. After removing the ether in vacuo the residual liquor is hydrolyzed with hydrochloric acid in absence of oxygen. The free active luciferin is left in the acid solution and can be extracted with butyl alcohol. By repeating the benzoylation and hydrolysis, a product purified 2,000-fold, as compared with dry Cypridinae, can be obtained.

To purify luciferase it is usually sufficient to dialyze a cold, wellstirred water extract of rapidly dried, powdered Cypridinae against cold running water for twenty hours. Dialysis removes pigment, and a precipitate forms which can be filtered off. A few drops of toluene added to this solution will preserve it for months with little loss in activity if kept in a refrigerator.

Cypridina luciferin is slowly dialyzable; not destroyed by trypsin; soluble in water, absolute methyl, ethyl, and propyl alcohol but insoluble in acetone, benzene and ether; readily adsorbed on fine particles. It does not act as an antigen.

Cypridina luciferase is non-dialyzable; destroyed by trypsin; soluble in water, but insoluble in alcohols and all fat solvents; readily adsorbed on surfaces. It is capable of forming an antiluciferase when injected into the blood of a rabbit. Other luciferins and luciferases have different properties.

The oxidation of luciferin with production of light in the presence of luciferase gives an oxidation product which cannot be reduced, whereas the oxidation without luminescence by oxidants like potassium ferricyanide is reversible.

This Product is called oxidized lucifenn. The change with ferricyanide occurs in two steps, one of which is the reversible oxidation previously referred to; the second is irreversible and probably also an oxidation, although this has not been definitely demonstrated. The spontaneous oxidation of luciferin without emission of light in crude solutions (without luciferase) is probably catalyzed by traces of heavy metals in the solution and proceeds much more slowly when the luciferin has been purified. Both the non-luminescent oxidation and the luminescent oxidation undoubtedly take place simultaneously when luciferin is mixed with luciferase.

Cyanide does not affect luciferase but forms an irreversible combination with purified Cypridina luciferin. Azide combines reversibly with luciferin whereas urethane, sulfanilamide, sulphathiazol, sulphapyridine and p-aminobenzoic acid probably act reversibly to inhibit luciferase activity.

Biome One of many naturally occurring ecological communities characterized by distinct climatic, geological, and biological features.

Biophysics, scope, principles, and methods The domain of discussion may be indicated by defining biophysics as the application to biological research of the methods and content of the physico-mathematical sciences. "Method" is advisedly placed first. It has become clear that the experimental and theoretical procedures usually associated with physics are not the exclusive property of that science, but represent rather adequately the developed and mature form of the scientific method itself. Biophysics is the adaptation of this methodology to the problems of biology.

It appears from this that the word "biophysics" is in part a misnomer. Biophysical research is biological research with a particular emphasis—concern for the logical ordering and quantification of biological theory, for the more intimate unification of theory and experiment, for the introduction into experimental biology of techniques and measuring devices which have been developed in the so-called physical sciences, and for the reduction-when it is possible and useful, and only then-of biological problems and concepts to already existent laws and concepts of the physical sciences. Jacques Loeb, a founder of biophysics, certainly devoted most of his efforts to formulating biological problems in terms of presumably underlying physical laws. But when he came to discuss theory, he stressed quantification almost exclusively—he did not assert that biology cannot become a science until it is physical, but that it cannot become a science until it is mathematical.

That physics—the science of matter—has in fact preoccupied itself with those objects which used to be termed "non-living" is logically accidental, not intrinsic. Taken in this sense, the "physicalizing" of biology merely means the attempt to analyze certain complex phenomenal patterns in terms of somewhat simpler patterns which have already been the object of intensive study. In this sense it is neither novelty nor heresy, but an inevitable extension of the classical biological procedure of analyzing organisms into organs, organs into tissues, tissues into cells, and cells finally into their discernible parts.

It is only fair to add that "biophysics" is usually used in a sense which contrasts it with biochemistry—that is, as a study of the physical as distinct from the chemical aspects of biological systems. This viewpoint would osmotic pressure, surface tension, the electrical and mechanical properties of biological systems as physical, while the composition and metabolic activities of such systems are the concern of chemistry.

This criterion has been serviceable in the past, of course; but it is so narrow as to be both cramping to research and rather difficult rigidly to maintain. Thus, the study of bioluminescence must in this view separate sharply into two parts: the analysis of the luciferinluciferase system is biochemistry, while the emission of a photon by the reduced enzyme is biophysics. It is valid to say, "For the moment, let us consider only the physical aspects of this phenomenon, and ignore the chemical and purely biological." But the moment passes quickly, and the essential inseparability of the different aspects demands the broader interpretation.

It remains to consider the best way of subdividing biophysical activity into its branches according to some specified criterion of classification. From the preceding discussion it is clear that one apparently natural division-into theoretical and experimental—is not a desirable one. The nature of biophysics, as here defined, demands the same close interweaving, mutual support, and mutual stimulus of theorizing and experimenting as is now practised so successfully in physics.

To adopt the existing fundamentum divisionis of biology is no more desirable; for the fences around bacteriology, embryology, histology, physiology, botany, and the like, in view of their haphazard historical origin, exhibit small regard either for practical utility or for logical consistency. Some of these fields are organized around a particular function, some around a particular set of structures, some around a particular class of organisms, and some are mere catch-alls for topics which do not fit handily into the remaining cells.

If we accept the cell as fundamental concept in biology, a useful though by no means perfect mode of classification suggests itself—the hierarchy: cell-parts, cells, cell-aggregates. These further subdivide in a fairly natural way, principally according to functions and activities.

Thus, under Cell-parts, we think of: Protoplasmic Structure; Enzyme Systems of the Cell; Nucleus and Chromosomes; Cell Membranes and Permeability; Golgi Bodies and Mitochondria.

Under Cells would come: Growth; Form; Motion; Division; Differentiation; Metabolism; Senescence; Stimulus-Response.

Under Cell-aggregates there are similar categories: Organic Form and Differentiation; Growth; Metabolism; Senescence; StimulusResponse. These would intersect with the triple classification: Tissues; Organs; Organisms. Thus, for example, we would find Secretion under Metabolism, Organs (perhaps also under Cells, Metabolism); special senses under Stimulus-response, Organs; and the nervous and endocrine systems under Stimulus-response, Organisms. Another category under Cell-aggregates would be Populations, which would cover the material of biometrics, ecology, and much of bacteriology.

Such a topic as The Virus is not an insuperable obstacle to this arrangement; there are many good reasons for placing it under CellParts. But it is obvious from the topics named that a better method than either equal-ranking classes or a hierarchy would be a multidimensional network, in which topics may be different distances apart, but in which each topic is linked to many others directly, and to all others indirectly. Protoplasmic Structure would be close to if not precisely at the centre of such network. With the further development of biophysics, the appropriate ordering will doubtless become more evident.

Biophysics in the broadest sense, especially on the theoretical side, is not represented with complete adequacy by any one text. One excellent study in phenomenological terms is "Physical Biology," by Alfred Lotka. At the other extreme are the studies, too numerous even to begin citation, which involve simply the borrowing of apparatus or instruments from physics for use in biological research. Closely related are the use of radioactive isotopes as tracer substances in metabolic studies, of X-ray and other optical techniques for studying the struc- ture of biologically interesting substances.

Less easy to classify, but definitely biophysical, and usually combining theoretical and experimental work, are such examples as the classical studies of A. V. Hill on muscular contraction, of S. Hecht on vision, and (most recently) of Hoagland in "Pacemakers and Certain Aspects of Behavior."

Applied biophysics is exceedingly scanty. There are a few scattered medical applications, such as the gold number or its modern equivalents in the analysis of cerebrospinal fluid, the use of X-rays and radium in the treatment of cancer. But biophysics is yet too young to show applications on any large scale.

There are many fields of biology in which the rather young biophysical method has not been active, so that a sufficient amount of representative material is not available for presentation. Some topics of biophysical interest, e.g., metabolism, are still referred to biochemistry for the same reason.

Bios Vitamin necessary for yeast growth; found to be a mixture of i-inositol, vitamin (B_1) and "biotin" and/or pantothenic acid.

Biosterin A highly active vitamin A concentrate, first prepared in 1925 by Takahashi.

Biotechnology A set of biological techniques developed through basic research and now applied to research and product development. In particular, the use by industry of recombinant DNA, cell fusion, and new bioprocessing techniques.

Biotin The term "biotin" should now replace several old names, to wit: coenzyme R (of Allison, Hoover and Burk), protective factor X (of Boas), the protective factor against egg white injury (of Parsons), the vitamin H (of György) and a part of the Bios II b or the adsorbable factor of the yeast growth promoting substance (of Kögl and Tönnis). It was isolated in 1936 by Kögl and Tönnis from the Bios II b factor of yeast as a methyl ester of the empirical composition $C_{11} H_{18} O_3 N_2 S$ and having a melting point of 166°-167° when highly purified. Saponification with cold alkali gives free biotin of the empirical composition $C_{10} H_{16} O_3 N_2 S$ and melting at 230°-232°. Biotin is a simple monocarboxylic acid, a derivative of valeric acid. The nitrogen atoms form a urea structure which can be opened up with a loss of one carbon to form a diaminocarboxylic acid, from which in turn one can reform biotin by reaction with phosgene. The sulfur is functioning in a thioether structure, since a sulfone can be formed by oxidation. The most probable structure is that of 2'-keto-3,4-imidazolido-2-tetrahydrothiophene-n-valeric acid.

Looked upon as an active principle of yeast growth biotin is active even at a concentration of one part in 500,000,000,000. As the active substance stimulating several species of the root-nodule bacterium Rhizobium and obtained from concentrated cultures of Azotobacter, and previously known as coenzyme R (R=respiration) it shows effects in concentration of I part in 100 billion. As Vitamin H (H for "Haut" or skin) it counteracts the effects of avidin of both raw and dried egg white to the extent of 10,000 units per milligram by a rat assay method (i.e. 0.1 gamma of biotin per rat per day for 30 days protects against egg-white injury). The biotin may be extracted from liver, dried eggs, potato starch, fresh eggs, dried yeast, autolyzed yeast.Inactivation of biotin has been accomplished both under acid and alkaline conditions. Aeration is not effective but stronger oxidizing agents destroy activity quickly and completely. Nitrous acid inactivates without a loss of nitrogen.Biological effects of biotin other than those used in assay of the foregoing effects concern the fatty infiltration of the livers of rats on a low fat diet with high biotin. The fermentation of yeast is increased as well as its respiration and growth both aerobically and anaerobically, in the presence of plentiful nitrogen in the form of ammonia. Butter-yellow tumor formation brought on by avidin-containing diets should be counteracted by biotin.

Bitter Acids Refer to a mixture of substance in the resin fraction of hops. They are monoacylphloro-glucides. The compounds are chemically very labic. During the brewing of beer, the undergo a reduction of rign size to produce the isoacids, which also taste bitter.

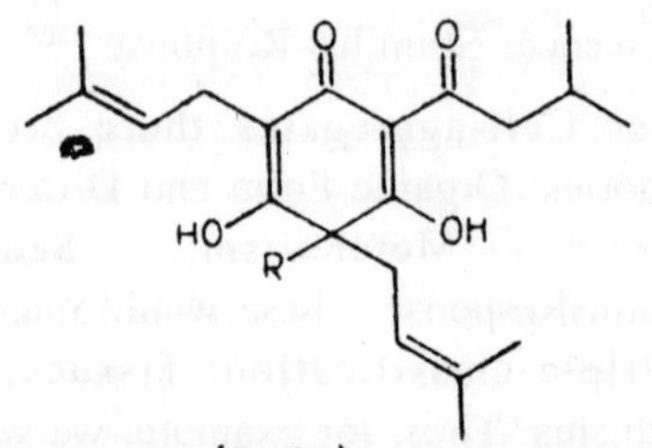

Humulon (R = OH)
Lupulon (R = —CH_2—$CH(CH_3)_2$)

Fig. Bitter Acids

Biuret reaction A general test for a protein or polypeptides containing three or more amino acids; to a solution of the substance tested add a dilute solution of $CUSO_4$, and then a dilute solution of sodium hydroxide. A blue-violet color is formed if a long polypeptide is present, and a violet purple color if a short one is present.

Black's test A qualitative test for betahydroxybutyric acid.

Blastema. The small bud composed of competent cells from which begins the regeneration of an organ or appendage.

Blastocele The hollow internal region of the blastula, filled with a colloidal fluid probably exuded from the cells.

Blastocolla A gummy substance coating certain buds.

Blastocyst The mammalian embryo at the time of implantation into the uterine wall.

Blastocyst In mammalian development, a modified blastula stage consisting of a hollow ball of surface cells and an inner cell mass.

Blastoderm The flat sheet formed by the embryo's cell-layers when cleavage is discoidal.

Blastomere Term applied to a cell forming part of a blastula, the first formed by zygote division.

Blastomere One of the cells into which the egg divides during cleavage; the term is not usually applied after the blastula stage. When blastomeres or sets of blastomeres differ in size, the terms *macromere* and *micromere* are used.

Blastula A single-layered, hollow ball of cells, an early stage of embryonic development, formed from the fertilized egg by cleavage and cell migration.

Blastula The stage in development at which a central cavity is formed in the mass of dividing cells (the *Blastocoele* cavity).

Blind spot Area of retina where optic nerve originates.

Bloch dopa reaction Frozen skin, treated with a 1% dopa solution (3, 4-dihydroxyphenyl alanine), gives rise to a dark brown or black pigment. Reaction due to a specific oxidizing enzyme, dopaoxidase. Used to indicate leukoderma or vitiligo.

Blood A red fluid, sp. gr. in man 1.055, composed of red and white blood cells, the erythrocytes and leukocytes, suspended in a straw colored liquid, the serum or plasma, dissolved in which are fats, amino acids, carbohydrates, vitamins, hormones, etc. The blood is essential to the transport of O_2 and CO_2 to and from the tissues, the food to the cells, waste products to the excretory organs, etc. The amount of blood is about 9% of the body weight, and contains bewteen 40—45,% of the vol. of the red blood cells.

Blood crystals Haemoglobin, haemin, etc. crystals formed on shaking blood with ether or chloroform.

Blood platelets Colorless bodies about 1/3 size of red blood corpuscles.

Blood pressure pressure exerted by the blood on the walls of blood vessels; caused by the heart's muscular pumping

Blood serum Fluid left after removal of corpuscles and fibrin.

Blood sugar, regulation of Blood sugar is an index of a metabolic process which does not proceed evenly. It is supplied from more than one source and it is utilized at many places. The dynamic balance, therefore, is subject to possible violent fluctuations which are difficult to trace back because of a multiplicity of causes.

From Claude Bernard to F. C. Mann the demonstration has become cumulative that the liver is the prime factor for maintaining the normal blood sugar level. It does so directly from its own glycogen stores and indirectly from muscle glycogen by converting muscle lactic acid to its own glycogen.

Techniques of study have been developed around the study of diabetic animals, so rendered by surgery or by phlorhizination, and even more direct demonstrations, demonstrating the formation of liver glycogen not only from glucose but from protein and fat and the hydrolytic products of protein and fat. The concept of the D: N ratio or the ratio of the amount of glucose excreted to the amount of nitrogen excreted, e.g. in a depancreatized or phlorhizinized animal, has been very helpful in studying the changes. The D in this ratio really represents a difference between the hepatic contribution and extra-hepatic utilization. Proper interpretation leads to the conclusion that diabetics utilize carbohydrates quite well and that the entire dietary and hormonal picture must be balanced quantitatively, that ketosis is due to heightened formation of acetone bodies in the liver and not through failures in the extra-hepatic tissues. Analysis of data on the respiratory quotent (R.Q.) leads to more or less similar conclusions. The breakdown of fatty acids in ways other than at the beta carbon has eased up the arguments for the conversion of fat to glucose. This was further bolstered by direct experiments with animal tissues, such as perfusion of the isolated liver.

Modern research tends to emphasize the role of the liver in maintaining blood sugar levels in balancing out the utilization of the tissues. The contribution of insulin by the pancreas is important, but not all important, because conditions can be set up where one can do fairly well without the pancreas, e.g. the "Houssay" animal, lacking both the anterior pituitary and the pancreas and comparatively free of diabetic symptoms for many weeks. This endocrine balance is not the only one, because the thyroid and the suprarenal cortical hormones are also involved in conjunction with their dependence on the pituitary for stimulation. All these effects, however, are cries-crossed and complicated by hormonal effects on the tissues, if only to a smaller degree under normal conditions, and to a large degree under emergency conditions (e.g. secretion of epinephrine in sudden calls for defense). Many disturbances may be due to overproduction of sugar rather than breakdown in utilization. Similarly "underproduction" is relative to "over-demand." This can only result in finer differentiations of types of disturbances of the homeostatic "mechanisms" for maintaining blood sugar at optimum levels.

Blood transfusion human, immunological aspects of The presence of isoagglutinins, i.e., antibodies that agglutinate cells of the same species is an important factor in human blood transfusion. Human serum contains two main agglutinins for erythrocytes designated α and β, while the two agglutinogens occuring in the red cells are A and B. This gives rise to 4 groups of human blood. One group (AB) contains A and B only; another (O) contains only α and β. A third group (A) contains A and ß only, and a fourth group (B) contains B and a only. Individuals of group AB containing no agglutinins in the serum are sometimes called universal recipients, while individuals of the second group, having no isoagglutinins in the cells are described as universal donors.

Blood, acid-base balance of The normal pH of the body is 7.4 and the maximum tolerated variations are 0.4 pH units. Blood serves as a buffer medium. It helps the tissues to maintain an acidbase balance by virtue of ventilation and kidney action. Forced respiration may cause enough lose of carbon dioxide to step up the pH 0.2 to 0.3 units. The tissue fluids are considerably stronger buffers than even the blood. About 2/3 of the buffering action of blood against strong acids is carried by bicarbonate. A decrease in the pH and in the alkali reserve stimulates increased ventilation which tends to undo the decrease; an increase in the pH brings about decreased ventilation which tends to undo the increase. The buffer system of the blood is not mathematically the most efficient but has the advantage of working well with easily accessible carbon dioxide and difficultly accessible alkali. Buffering capacity, however, has to be rebuilt by the kidneys. This may occur through the excretion of acid phosphates or the formation of ammonia from urea. A small change in pH of the blood is accompanied by large changes of pH in the urine. Acid urines bespeak the building up of an "alkali reserve" in the blood. Ammonia formation is a secondary defense and leads to the sparing of bicarbonate, e.g. in diabetes, diarrhea with loss of alkali, etc. Excess of bicarbonate is simply taken care of by excretion.

The buffering action of the blood keeps the difference of pH between arterial and venous blood down to 0.02 units. The carbon dioxide is all accounted for as HCO_3 and $H_2 CO_3$. The buffering action leads to a conception of acidosis not as a condition marked necessarily by a low pH but by an accumulation of a high amount of buffered acid tending to destroy the alkali reserve. Similarly alkalosis need not show directly in the pH. Thus one might ingest a certain amount of strong HCI without a material change of pH but at a loss of alkali reserve. An acidosis, e.g. in diabetes, starts a chain of events as the reduction of the ratio of HCO_3 to $H_2 CO_3$. The CO_2 has to be reduced and that means increased ventilation. Thus a condition of hyperpnea is set up. This is followed by dyspnea. The urine will be acid because of excretion of ammonium salts. Acidosis may also be due to retention of carbon dioxide for some reason.

Blotting A technique for detecting one RNA within a mixture of RNAs (a Northern blot) or one type of DNA within of a mixture of DNAs (a Southern blot). A blot can prove whether that one species of RNA or DNA is present, how much is there, and its approximate size. Basically, blotting involves gel electrophoresis, transfer to a blotting membrane (typically nitrocellulose or activated nylon), and incubating with a radioactive probe. Exposing the membrane to X-ray film produces darkening at a spot correlating with the position of the DNA or RNA of interest. The darker the spot, the more nucleic acid was present there.

Bloxham's test An old qualitative test for urea based on conversion to barium cyanurate.

Blunt End A terminus of a duplex DNA molecule which ends precisely at a base pair, with no overhang (unpaired nucleotide) in either strand. Some but not all restriction endonucleases leave blunt ends after cleaving DNA. Blunt-ended DNA can be ligated nonspecifically to other blunt-ended DNA molecules (compare with Sticky End).

5'→3'

NNNCCC GGGNNN Smai cut, no overhang

NNNGGG CCCNNN

3'←5'

Bogomoloff reaction for urobilin Treatment of an alkaline solution of urobilin with copper sulfate solution gives rise to a red or violet color which may be extracted with chloroform.

Bolliger reactions for creatinine I. Creatinine, in sodium hydroxide solution, gives a purple with an approximately 10% solution of sodium 3,5-dinitrobenzoate.

II. A pink to rust-brown color, depending upon the creatinine concentration, is obtained when the substance is treated with an aqueous alcoholic solution of 1, 3, 5-trinitrobenzene and dilute sodium hydroxide.

Bondi-schwarz test for acetoacetic acid in urine Acetoacetic acid decolorizes Lugol's iodine solution, the reaction product having a sharp characteristic odor. This is a modified reaction.

Bone Part of skeleton tissue hardened by calcium salts, such as calcium phosphate.

Bone, influence of diet on the composition of On reviewing the literature, dealing with the composition of bone, it may be stated that there has been only scant effort to interrelate diet to the composition of bone. It is generally recognized that the inorganic composition of bone may be expressed as $[Ca_3 (PO_4)_2]n.CaCO_3$. Where n is between 2 and 3 except that approximately 6 per cent of the Ca is replaced by Mg, Na. and K. It has been known, from a few papers, that n decreases with age, in high calcium—low phosphorus rickets and in a pathological condition called "Marble bones".

In experiments recently reported' an attempt was made to interrelate the above findings to the ratios of dietary calcium to phosphorus (which influence the ratios of serum calcium to inorganic phosphoruss.) It was found that the bones of rats fed high calcium low phosphorus had significantly higher carbonate to phosphate ratios than the group of rats on the high phosphorus low calcium diet.

The rats which received vitamin D supplements had denser bones, but the ratio of carbonate to phosphate in the bones was similar to those animals which did not receive the anti-rachitic vitamin. An examination of the blood revealed that the calcium to inorganic phosphorus ratio was higher of the high calcium-low phosphorus fed rats than the high phosphorus and low calcium fed animals. In the presence of vitamin D these differences still prevailed but were less marked. The serum Ca x P products, however, was higher explaining the increased density of the bones.Thus the most important factor in determining whether there was a "high carbonate" bone (low n) or "low carbonate') bone (high n) depended upon the ratio of calcium to phosphorus in the diet. These experiments, therefore, indicate that the dietary calcium to phosphorus ratio has an important influence on the composition of bone Bone salt • Name sometimes applied to the inorganic matter of bone; essentially $2Ca_3(PO_4)_2.CaCO_3$, i. e. dihalite.

Borinski test for peroxidase in milk A bright blue color is obtained when 5 cc. raw milk are mixed with 10 drops of a solution made by dissolving 0.85 gm. of finely powdered guaiac resin in 85 gm. 75% alcohol, to which is then added 10 cc. liquid phenol and 5 cc. 3% hydrogen peroxide.

Börnstein reaction for saccharin A yellow color changing to red and then to dark green is obtained when saccharin is heated with excess resorcinol and a few drops of concentrated sulfuric acid. Dilution and addition of excess alkali yields a solution reddish by transmitted light. Sensitive to 1 mg. saccharin in 5-6 liters of water.

Bose reaction for reducing sugars The test solution is added to the reagent which has been heated for less than a minute; a deep violet color indicates reducing sugars. Reagent - 2 cc. 25% sodium carbonafe solution mixed with 1 drop of 1% o-dinitrobenzene'solution.

Botulin botulism Allantiasis; poisoning due to a toxin of B. botulinus. The toxin, botulin, may be found in meats and vegetables insufficiently cooked. There is a variable period of incubation followed by eye symptoms and motor depression. Antitoxins have been developed but must be used very early. Botulin is so powerful that mere tasting of spoiled food may be fatal, yet it is destroyed by boiling for several minutes.

Bradshaw test for myelopathic albumose in urine Presence of Bence-Jones proteinmyelopathic albumose - is indicated by the formation of a precipitate when concentrated nitric acid is added to cold urine. Precipitate dissolves on heating and reprecipitates on cooling, in contrast to ordinary albumin.

Brazilin Brasilin; $C_{16}H_{14}O_5$; an ambercolored flavone from Brazil wood; used as an indicator.

Brieger reaction for differentiating choline and neurine With phosphotungstic acid, choline but not neurine will precipitate; with tannic acid, neurine but not choline will precipitate.

Bright's disease Nephritis; nephrosis; a term for a set of kidney disorders with inflammatory and degenerative lesions. Nephritis stresses inflammation of the kidneys, while nephrosis stresses the degenerative changes. Vascular changes may be called scleroses. The main groupings are

(1) acute glomerulonephritis

(2) the nephroses (lipoid nephrosis, chronic parenchymatous nephritis and subacute nephritis with edema) and

(3) the nephroscleroses (chronic nephritis and chronic interstitial nephritis). In acute glomerulonephritis infection, such as by hemolytic streptococci, is. primarily responsible. Albumin in the urine, fever, edema may lead to hypertension, uremia or cerebral hemorrhage. Various forms of albuminuria have been differentiated to aid diagnosis of nephritis. Persistence of high-blood urea nitrogen indicates great kidney damage. Diets emphasize carbohydrate and fat. Diuretics of increasing power are used cautiously. The nephroses involve degeneration to necrosis of the parenchyma of the kidney with respect to albumen, fat, lipoid or carbohydrate. There is marked albuminuria and hypoproteinemia. Infections and poisoning, e.g. by mercury, are possible causes. There are symptoms of edema all over the body, digestive, circulatory, urinary and uremic symptoms. The blood shows hypercholesterolemia, a reversal of albumin to globulin ratio, an increase in sodium chloride, a diminution of total protein. A decrease in basal metabolism also takes place. The edema is difficult to remove by diuresis or diaphoresis, but this can be accomplished by raising the colloidosmotic pressure by intravenous injection of acacia or by increased protein intake. The

nephroscleroses are due to pathological changes in the vascular and interstitial tissues with symptoms of albuminuria, hypertension, edema, nitrogen retention, acidosis and uremia. The disease is preventable but hardly curable. Heart therapy is stressed.

Many tests have been developed in connection with Bright's disease, as tests for albumin and blood in the urine kidney function tests, blood chemistry tests such as N.P.N. (non-protein nitrogen or urea nitrogen) bilirubin, calcium, icteric index, serum albumin and globulin, total protein, van den Bergh, uric acid, carbon dioxide combining power, chlorides, cholesterol and cholesterol esters, xanthoproteic test for phenol.

Bromelin An intracellular proteolytic enzyme found in the pineapple. It is similar in action to papain.

Bromidrosis The unpleasant odor of perspiration due to rancidity of fatty con stituents.

Bromural Alpha-bromoisovaleryl urea, a sedative and hypnotic; "dormigene."

Brownian movement Rapid, random, oscillatory movement around a mean position, executed by any particle of small enough size to remain in more or less permanent suspension in a liquid; the result of the random impacts of molecules of the liquid upon the particle.

Brown-lum reaction for differentiating dextrose and levulose Ferric chloride solution is added dropwise to a solution of 0.5 gm. of the sugar in 10 cc. water until a permanent yellow is obtained. On warming, a slightly deeper color is formed with levulose; the solution is colorless with dextrose. On addition of 1 drop of 2% diphenylamine in sulfuric acid, an intense violet indicates levulose.

Brucellosis Undulant fever; Malta fever; an infection caused by one of the Brucella, e.g. Brucella melitensis (goats), Brucella abortus (cows), Brucella suis (hogs) via milk or through the skin after a period of incubation. Fever symptoms may be supplemented by many others. A brucellergen skin test, based on a nucleoprotein of Brucella, is diagnostic. Agglutination and opsonic tests, as well as cultivation of the organism from the blood, are used. Vaccines are available for treatment.

Brucine An alkaloid found with strychnine; dimethoxystrychnine; less poisonous than strychnine;$C_{23}H_{28}O_4N_2$.

rueckner test for ergosterol A solution of the substance in 2 cc. of benzene or chloroform is treated with 1 cc. acetic anhydride, 0.5 cc. acetone, a small crystal of copper acetate and 0.5-1 gm. anhydrous zinc chloride. A blueviolet color with reddish fluorescence is obtained. Sensitivity -0.001 mg. per cc.

Brugsch test for porphyrin in urine Acidify 10 cc. of urine with 2 cc. glacial acetic acid, extract with 2 cc. ether; treat ether extract with 5 cc. 5% hydrochloric acid and shake thoroughly. Viewed under ultra-violet light, a red fluorescence is obtained with porphyrin.

Brunner's glands Glands in submucous coat of small intestine; duodenal glands.

Budding: The developing shoot of mostly meristematic tissue; often protected by a covering of modified leaves.

Buelow test for acetone in urine Acetone reacts very rapidly at room temperature with a hydrochloric acid solution of 2,4-dinitrophenylhydrazine to form a precipitate. The test is quantitative and will detect as little as 0.1%.

Buffer Substances, which by their presence in solution increase the amount of acid or alkali that must be added to cause unit change in pH; the most efficient buffers are mixtures of weak acids or weak bases with their corresponding salts.

Bufodesoxycholic acid A compound of bufotalin and suberyl arginine; $C_{40}H_{62}O_{11}N_4$.

Bufotalin A sterol-like compound found in toad venom; $C_{26}H_{36}O_6$.

Bufotoxin Active principle of toad venom; a compound of bufotalin and suberyl arginine; $C_{40}H_{62}O_{11}N_4$.

Bulbocapnine $C_{19}H_{19}O_4N$; rhombic needlea, m. p. 1990; alkaloid of several genera of Corydalis and

Dicentra; it is a benzanthracene derivative; stimulates flow of saliva and tears, produces emesis, slows respiration, and produces a cataleptic condition.

Burchard reaction for cholesterol The addition of acetic anhydride and a few drops of sulfuric acid to a solution of cholesterol in chloroform, chlorobenzene, chlorotoluene or ethylene chloride produces a red to violet color, the acid showing a green fluorescence.

Burian reaction for xanthine bases Alkaline solutions of adenine, guanine, hypoxanthine and xanthine produce a red color with diazobenzene sulfonic acid. Except for caffeine, beteroxanthine, paraxanthine and theobromine, all methylxanthines react similarly.

Bursitis An inflammation of a bursa due to trauma or strain or infection, accompanied by pain when a joint is moved. Special forms have the special names: miner's elbow, tennis elbow, housemaid's knee, weaver's bottom, bargeman's bottom, Fleischmann's bursitis (sublingual), plantar bursitis (over instep), bunion, Albert's disease (foot). Besides casual therapy, injections of sodium morrhuate have been used. Physiotherapy, e.g. heat and massage, is helpful.

Butyn Dibutyl-aminopropyl-p-aminobenzoate sulfate, m.p. 98-100°, local anesthetic used in infiltration anesthesia, rhinology, urogenital surgery, tooth extraction, etc.

Byssus viscous threads whereby mussels attach themselves to objects.

C

Cachexia Emaciation due to wasting or malnutrition, cancer, or the like.

Cadaverine $NH_2 (CH_2)_5 NH_2$; an amine produced by the putrefactive action of bacteria on lysine or proteins containing it.

Cadinine A sesquiterpene, the main component of oil of cubeb and of cade; $C_{15} H_{24}$.

Caffeine Theine, guaranine, methyltheobromine, trimethylxanthine; m.p., 235-237°; sublimes, 178°; is a weak base; from tea, coffee, guarana, kola nuts; used as diuretic; cardiac, respiratory, vasomotor, reflex, and psychic stimulant; in small doses stimulates muscular contraction and lessens fatigues, large doses having opposite effect.

Caisson disease The bends; compressed-air disease; divers' paralysis; a condition due to too rapid return from pressures about 32 lbs. per square inch to atmospheric pressure which causes too rapid an evolution of gas from tissues. The gas is principally nitrogen. The pain is referred to as "bends," the vertigo as "staggers" and the dyspnea as "the chokes." Treatment is by recompression.

Calciferol A vitamin D, namely Vitamin D2; it is a crystalline material, m.p. 115-117°, made by irradiation of ergosterol. It is found in yeast, several fish oils, but not cod-liver oil, and is used in several commercial antirachitic preparations. It contains 3 rings and 4 double bonds.

Calcite A common crystalline form of natural calcium carbonate, CaCO3, that is the basic constituent of limestone, marble, and chalk. Also called calcspar.

Calcium carbonate A "salt" used by many marine invertebrates, such as corals and echinoderms, and by protists, such as coccolithophorids, to construct their exoskeletons.

Caldera Basin-shaped crater formed by the collapse of a volcanic peak with a diameter many times that of the original volcanic vent.

Callan-henderson reagent for copper Copper salts produce a brownishyellow color when treated with an amyl or isoamyl alcohol solution of sodium diethyldithiocarbamate.

Callose Extraneous material in cell wall, not cellulose.

Calomel electrode, normal A half-cell, composed of a layer of mercury-calomel paste moistened with N KCI solution, over which is a normal solution of KCI and saturated with salomel. An electrode of Pt in a glass rod dips into the mercury and leads out of the cell. It is a standard reference cell used as a standard against which other half cells are compared.

Calorie, great Kilogram calorie; food calorie; 1000 small calories.

Calorigenic effect Increase of animal's heat production resulting from intake of food or drugs.

Calorimeter An instrument for measuring total heat liberated or absorbed. Unit of standard metabolic rate in kilocalories per day.

Calorimetry The measurement of heat exchange.

Calorimetry, direct The measurement of an organism's heat loss by radiation, convection, conduction, and evaporation.

Calorimetry, indirect Calculation of an organism's heat production based an chemical measurements, mainly of oxygen consumption, or carbon dioxide production or both with or without consideration of urinary nitrogen.

Calory Calorie; amount of heat which increases the temperature of one gram of water one degree Centigrade (averaged between the freezing and boiling points of water).

Cambium Soft formative middle layer of a tree.

Campbell test for dihydroxyacetone Dihydroxyacetone reduces phosphomolybdic acid with the formation of a blue color which may be compared colorimetrically or titrated with 0.01 N potassium permaganate to the disappearance of the blue color.

Camphene $C_{10}H_{16}$; present in l-form in citronella, and in d-form in ginger.

Camphor $C_{10}H_{16}O$; constituent of camphor tree and laurel.

Canada balsam Oleoresin of the pine Abies balsamea; a solution in xylene is used for slide mountings.

Canaline Amino acid formed from canavanine by a liver enzyme.

Cancer: A disease process initiated by transformation of a cell to behavior lacking in normal proliferative control, and often involving invasive behavior and localized vascularization.

Candlefoot Intensity of illumination of distance of I foot by a standard light.

Cantharidin Active principle of Spanish (and Chinese) fly; used to stimulate hair growth and to blister skin; alleged to act as aphrodisiac.

Cap site Two usages: 1) In eukaryotes, the cap site is the position in the gene at which transcription starts, and really should be called the "transcription initiation site". The first nucleotide is transcribed from this site to start the nascent RNA chain. That nucleotide becomes the 5' end of the chain, and thus the nucleotide to which the cap structure is attached.

(2) In bacteria, the CAP site (note the capital letters) is a site on the DNA to which a protein factor (the Catabolite Activated Protein) binds.

Cap All eukaryotes have at the 5' end of their messages a structure called a "cap", consisting of a 7-methylguanosine in 5'-5' triphosphate linkage with the first nucleotide of the mRNA. It is added post-transcriptionally, and is not encoded in the DNA.

Capillary Minute blood vessel connecting an artery to a vein.

Capon unit The dose of androgenic material which on subcutaneous injection daily for 6 days into at least 5 capons, produces an average increase of 20% in the area of the combs.

Capsicum Dried ripe fruits of Capsicum minimum, used as digestive stimulant or for drawing blood to skin; the active principle is the pungent "capsaicin".

Carbamino reaction The reaction between amino acids, carbon dioxide, barium or calcium hydroxide. Salts of carbamino acids RCH COO > Ca NH COO

Carbhemoolobin A compound said to be formed of about the 5-10% of the total carbon dioxide of the blood which is not bicarbonate but combined with hemoglobin.

Carbohydrases A large group of enzymes that hydrolyzes di-, tri-, and polysaccharides to monosaccharides. R. D. C.

Carbohydrate tolerance The capacity of the body to assimilate carbohydrates, as determined by the blood sugar concentration; reduced in liver poisoning and diabetes.

Carbohydrate, formation from fat The glycerol moiety is converted by the liver to glucose; the conversion of the fatty acids is argued from a low R.Q., e.g. as less than 0.7 in hibernation, or 0.8 to 0.6 in feeding butter to rats, or in plants during germination, or in silkworm metabolism at certain starvation stages.

Carbohydrates A group of polyhydroxy alcohols, usually in multiples of straight 5 and 6 carbon chains, which are used as a source of energy and for structure in plants and animals. Formed of C, H and O in the proportions of one molecule of H_2O to every carbon.

Carbohydrates. Classification of Various schemes of classification have been proposed. One of the most convenient is:

I. Simple Sugars (Monosaccharides) Monoses, dioses, trioses, tetroses, pentoses, hexoses, heptoses, octoses, nonoses and decoses (formaldethyde is usually ruled out.)

II. Compound Sugars Disaccharides (dipentose, pentose-hexose, methyl pentosehexose, dihexose); trisaccharides, tetrasaccharides, polysaccharides (pentosans, methyl pentosans, hexosans, condensed amino saccharides, mixed pentosans, mixed hexosans).

III. The Cycloses or Cyclitols Inositols, quercitols, tetritols.

Carbonyl A functional group found in aldehydes and ketones that consists of a carbon atom joined to an oxygen atom by a double bond.

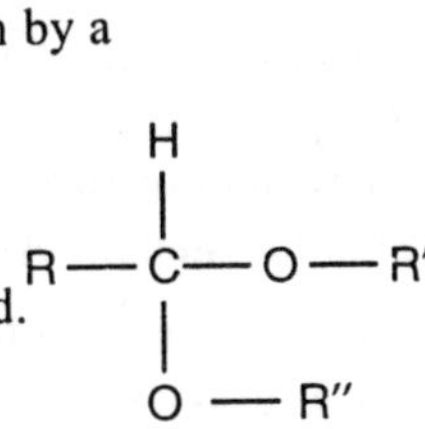

Fig. Carbonyl-protecting groups

Carboxyhemoglobin Compound of hemoglobin with carbon monoxide, incapable of reacting with oxygen and dissociating with greater difficulty than the analogous oxygen compound.

Carboxyl Terminus Refers to the COOH end of a peptide chain (by custom drawn at the right of a protein sequence).

Carboxyl A functional group present in organic acids, consisting of a single carbon atom double-bonded to an oxygen atom and also bonded to a hydroxyl group.

Carboxylase A thiaminoprotein enzyme of yeast which in the presence of Mg or Mn promotes the decarboxylation of pyruvic acid and other alphaketomonocarboxylic acids.

Carboxymethyl-scysteine $HOOC{\cdot}CH_2{\cdot}S{\cdot}CH_2{\cdot}CH(NH_2){\cdot}COOH$; is excreted as a mixed disulfide.

Carboxypeptidase A group of polypeptide hydrolyzing enzymes that split off the end of a polypeptide chain, the amino acid that has a free carboxyl group.

Carbromal Alpha-ethylbutyryl-carbamide, a sedative and hypnotic; uradal; adalin.

Carcinogen A chemical substance which has the property of giving rise to cancer when administered to animals by external or internal application.

Carnosine β-alanylhistidine. A dipeptide found in the skeletal muscles of reptiles and vertebrates.

Carotene Carotin; commonly described as existing in 3 forms, α, β, and γ. Their formulas have been worked out relative to vitamin A so that the ß form yields two molecules of the vitamin, while the others can furnish only one. For this reason the β form is often described as β, β'-carotene. All the carotenes have the formula $C_{40}H_{55}$. Many other isomers are being isolated by chromatographic analysis. Both plants and animal products contain them, e.g., carrots, alfalfa, liver, butter. An alcohol, cryptoxanthine, $C_{40}H_{55}OH$, is closely related to y-carotene, and also yields one molecule of vitamin A, $C_{20}H_{29}OH$, on oxidative fission. In the carotenoids various ring residues have been identified so that a-carotene has been renamed structurally as β-α carotene, and γ-carotene as β-lyco-β'-carotene. There is also a ä-carotene which turns out to be β-lyco-α -carotene with no provitamin value, which is due only to the beta ionone residue.

Carotene oxidase An enzyme of various beans

which promotes the oxidation of carotene, xanthophyll, various vegetable oils, probably by the addition of oxygen at a double bond.

Carotenoids Carotenoids are hydroaromatic or aliphatic fat soluble C,H or C,H,O compounds, which can be formulated as condensation products of eight isoprene groups. They contain seven to eleven conjugated double bonds, causing characteristic absorption bands, accounting for the yellow, orange or red color of all carotenoide. The chain, with a methyl group on every fourth carbon, may terminate at one or both ends in a tri-methyl cyclohexene ring. The carotenes comprise hydrocarbons, secondary alcohols, ketones and acids. They occur in the free state, as esters, as glycosides, and as "symplexes" with proteins (e.g. rhodopsin, visual purple; astacene, in crustacean pigments, etc.). Carotene itself (C_{40} H_{56}) and xanthophyll (C_{40} H_{56} O_2) accompany chlorophyll in all green plants. Other specific carotenoid pigments occur in algae and in special organs of higher plants (root of Daucus carrota, fruit of maize, red pepper, tomato, etc.). Some lower animals contain special carotenoids while the carotenoid in the normal and pathological (xanthomatous) deposits, corpus luteum and egg yolk in vertebrates are of alimentary origin. Axerophthol, a colorless split product of certain carotenoids, is Vitamin A, and crocetin and safranin act as "external hormones" in the sex cycle of certain algae; otherwise the biological significance of the carotenoids is still obscure.

Carotenoids, specific Light yellow to deep-red plant and animal pigments, extractable by fat solvents and having a structure resembling carotene, namely a carbocyclic or semi-carbocyclic configuration attached to an unsaturated polyene hydrocarbon chain. A relationship to phytol is indicated. The polyene hydrocarbon chain is C_{22} H_{26} and is common to the group. Besides the carotenes the following are of importance: Lycopene, found in tomato, watermellon, etc., is ß-ß-lycopen; xanthophyll or lutein, found in green leaves, flowers, corpus luteum, etc., is 3-3'-dihydroxy-βα-carotene; rhodoxanthin, from the of the yellow Zea mays is 3-3'fruit of the yew, is 3-3'-diketo-4-4'dehydro β–β carotene; zeaxanthin dihydroxy-β–β-carotene; fucoxanthin of the brown algae, Fucus sp., is a-a'-fucoxanthin; cryptoxanthin from papaya and maize is 3-hydroxy-ß-carotene-ß'-carotene; capsanthin from the ripe fruit of Capsicum annum is 3-hydroxy-β–carotene-5'-desoxy-a'-fucoxanthin; helenien from the flowers of the Helenium sp., is β-α-carotene-3-3'dioldipalmitate; physalien from asparagus is β–β-carotene-3-3'-dioldipalmitate; astacene from various crustacea is 3-3'-4-4'-tetraketo β–β carotene; capsorubin from Capsicum annum is 5,5'-desoxy-a-a'fucoxanthin; euglenarhodon of the Euglena sp., is 2-2'-4-4'-tetraketoß-ß'-carotene; rubixanthin from rose hips is 3-hydroxy-ß-carotene-ß'lycopene; and rhodoviolascin from the purple bacteria, Rhodovibrio and Thiocystis, is 3-3'-dimethoxyÄ $_{3',4}$ -dehydro-1'-2'-dihydro β–β lycopene.

The identification and purification of these compounds is a large field of investigation by itself, and represents one of the triumphs of the chromatographic technique.

The association of carotenoids with chlorophyll in plants suggests a role in photosynthesis, e.g. in controlling the equilibrium between different forms of chlorophyll. They are probably participants of an oxidation-reduction system.

Carotenols Hydroxy carotenoids, found esterified with fatty acids, combined with sugars or proteins; include rhodopsin, porphyropsin, iodopsin, the vitamins A.

Carr-price reactions for vitamin A I. Chloroform-washed and dried antimony trichloride is dissolved in chloroform to make a 30% solution. After standing, the clear solution is decanted and used from a burette. 2 cc. of reagent are added to 0.2 cc. of a 30% solution of the oil in chloroform; a billiant blue color, unchanged for 3 minutes, is obtained.

II. With stannic chloride in chloroform a deep blue color rapidly changing to purple is obtained.

III. The addition of anhydrous ferric chloride to

chloroform solution of the Vitamin A—containing oil gives rise to a fluorescent reddish-violet color.

IV. A red-violet color fading to brown is obtained when finely powdered anhydrous aluminum chloride is added to the oil. Addition of a minute trace of aluminum chloride to a solution of oil in chloroform containing dry hydrogen chloride gas or phosgene produces a fine purple color.

Cartilage Gristle; elastic tissue lining joint surfaces or forming special structures, like the ear.

Carvone $C_{10}H_{14}O$; d-form chief constituent of dill and caraway; 1-form chief conatitutent of spearmint; a ketone of dipentene.

Casanova reaction for lecithin An ether solution of lecithin is shaken with a 10% aqueous solution of ammonium molybdate and floated on sulfuric acid; a reddish ring is formed which changes to green and then to intense blue.

Cascara sagrada Dried bark of Rhamnus purshiana, very bitter and used as laxative; active principle is "emodin".

Casein A phosphoprotein found in the milk of every mammal. It has a particle weight ranging from 75000 to 375000, indicating that it is a reversible dissociable component system. It is a fairly strong acid, probably due to an unesterified H of the phosphoric acid, the esterified group being connected to a hydroxyl group of serine. In the presence of rennin it forms soluble paracasein, which in the presence of Ca++, forms the insoluble curd Ca paracaseinate.

Castor oil Oil of the seeds of Ricinus communis; the purgative effect is due to glyceryl esters of ricinoleic acid.

CAT assay An enzyme assay. CAT stands for chloramphenicol acetyl transferase, a bacterial enzyme which inactivates chloramphenicol by acetylating it. CAT assays are often performed to test the function of a promoter. The gene coding for CAT is linked onto a promoter (transcription control region) from another gene, and the construct is "transfected" into cultured cells. The amount of CAT enzyme produced is taken to indicate the transcriptional activity of the promoter (relative to other promoters which must be tested in parallel). It is easier to perform a CAT assay than it is to do a Northern blot, so CAT assays were a common method for testing the effects of sequence changes on promoter function. Largely supplanted by the reporter gene luciferase.

Catabolic Pathways break complex molecules into simpler compounds, releasing energy in the process.

Catabolism The sum of all processes in an organism by which the chemical energy of food or body substance is transformed to other forms of energy, mainly heat. (Break down processes.).

Catabolizable energy Energy which is available for animal heat production. Metabolizable energy minus energy in milk or eggs.

Catalase There are probably several catalases, although "catalase" of different sources has been crystallized. Nearly all living tissues contain it. It is high in liver and in sprouted seeds. Its biological function is to destroy any accumulation of hydrogen peroxide. Trypsin inactivates it. An anticatalase may be produced by immunological procedures. The molecular weight of crystalline catalase has been reported as 225,000 to 300,000. Biliverdin may be produced from it by reaction with acetone and HC1. It probably contains 3 molecules of iron porphyrin and 1 molecule of bile pigment Fe porphyrin per molecule. Its absorption bands are at 629, 544 and 506.5 mu. and its i.p. is 5.5.

Catalysis The acceleration positive or negative and possibly the initiation of a reaction, by comparatively small amounts of foreign substances called catalysts, these substances being recoverable unchanged at the end of the reaction. The catalyst acts as a source of surface energy for the reaction.

Catalyst A substance which changes the rate of a reaction actually in process without being consumed by the reaction.

Cataphoresis The migration of the colloidal micelle, through the dispersion medium, due to an imposed e.m.f. on the micellar particle.

Catarrh The inflammation of the mucous membrane in a cold.

Catechol Ortho-hydroxyphenol; obtained from catechu dye; m.p. 105°.

Catechu Dried extract of Uncaria gambier; used as general astringent and for diarrhea.

Catgut Filaments made from sheep intestines.

Cathemoglobin The compound of hematin with denatured globin, e.g. produced by denaturation of methemoglobin with chloroform.

Cathepsin A system of proteolytic enzymes present in animal and bacterial cells generally. It is active in slightly acid solution and contains components which are inactivated by copper salts and iodoacetate and are activated by SH groups and cyanide.

Cation A positively charged ion.

Cationogens Elements in an unionized state which can become cations in the body.

Caul Amniotic membrane; amnion; omentum.

Cna clone "complementary DNA"; a piece of DNA copied from an mRNA. The term "clone" indicates that this cDNA has been spliced into a plasmid or other vector in order to propagate it. A cDNA clone may contain DNA copies of such typical mRNA regions as coding sequence, 5'-untranslated region, 3' untranslated region or poly(A) tail. No introns will be present, nor any promoter sequences (or other 5' or 3' flanking regions). A "full-length" cDNA clone is one which contains all of the mRNA sequence from nucleotide #1 through to the poly(A) tail.

Cecidogen A chemical substance, usually secreted by an insect, which has the property of giving rise to a gall on a plant.

Cellobiose $C_{12}H_{22}O_{11}$; a reducing disaccharide, 4 - (Beta-D-glucosyl)-D-glucose, prepared by acetolysis of cellulose followed by deacetylation of the resulting octaacetate.

Cellulase A polyase which splits cellulose, found in Helix pomatia.

Cellulose Carbohydrate polymer of the simple sugar glucose. It is found in the cell walls of plants and green algae, as well as dinoflagellates. Cellulose is the most abundant compound on earth that is manufactured by living things.

Cellulose $(C_6H_{10}O_5)x$; a polysaccharide composed of β-D-glueopyranoside units. It forms the cell walls and fibrous structure of many plants, and is the chief constituent of paper and wood. It is insoluble in water and organic solvents but dissolves in ammoniacal copper solutions (Schweizer reagent) forming complex ions from which it may be regenerated by means of acid; it dissolves also in a mixture of alkali and carbon disulfide forming xanthogenates from which it is regenerated by the addition of acid to give various materials such as rayon and Cellophane. Cellulose eaters and ethers are widely used in plastics, lacquers and explosives, and as artificial silk, viscose, nitrocellulose, guncotton, celluloid.

Cellulose decomposition by micro-organisms The utilization of cellulose by micro-organisms takes place under a wide range of circumstances and is in general an uncontrolled, though not necessarily undesirable process. Its economic implications are great and many problems in this field remain to be solved. Decomposition almost always takes place under conditions that permit of its attack by a mixed microbial population and most usually the substrate is not exclusively cellulosic. As a result it is not easy to assess the role played by any individual organism, and pure cultural studies may be misleading in that such a condition is in fact abnormal.

Some distinction can be made, however, between those processes in which substantially pure cellulose is attacked, and those in which cellulose is the major constituent, not alone but accompanied by other, usually more available, plant constituents.

Microbial attack on pure cellulose is in practice frequently undesirable, but rarely proceeds to completion in the sense that much of the cellulose is fully decomposed. More usually the attack is manifest by spoilage of appearance or loss of

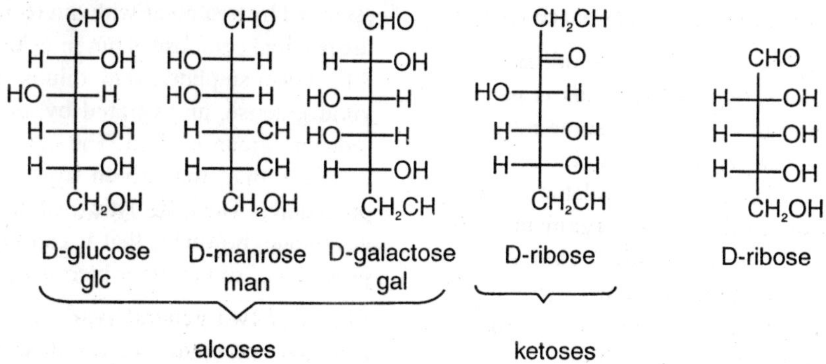

Fig. Monosaccharides[Carbohydrates]

strength. Paper is subject to discoloration as the result of fungal activity, and many common fungi can develop slowly on paper if the moisture conditions are suitable. The attack may be accelerated by the use of sizes or finishes of an available nature. Fibres such as cotton, flax, etc., are subject to considerable damage by "mildewing" if improperly stored, and again fungi are mainly responsible since they are capable of growing at low moisture levels. Textile fabrics similarly may be damaged by microbial attack and loss of strength may be more serious than pigmentation. Rot-proofing of fabrics or cordage exposed to moist conditions is designed primarily to prevent the initiation of such attack. In the last decade innumerable uses have been found for cellulose derivatives, particularly the esters and ethers. One of the valuable characteristics of these compounds is their greater resistance to microbial spoilage under ordinary circumstances. While it may appear that the utilization of pure cellulose by micro-organisms is generally an undesirable process, many attempts have been made to develop controlled fermentation processes that would yield products of commercial value. Wood pulp has usually been regarded as the most promising form of cellulose for this purpose, and certain anaerobic bacteria the most suitable agents in view of the vigor of their attack and the incompleteness of the oxidation accomplished. The direct products are fatty acids, ethyl alcohol, carbon dioxide and hydrogen, with methane frequently formed by a secondary process. The yields of acids, however, have not been too encouraging and the economic feasibility of the process is questionable. In some sewage disposal plants the combustible gases which arise in part from cellulose are employed as a source of power. Fermentation products have been obtained indirectly from cellulose after preliminary hydrolysis with acid, as in the Bergius process but the organisms then employed do not have to have the property of attacking cellulose.

Microbial attack on the cellulose of plant materials occurs in innumerable natural transformations, which are usually more complex than may appear at first sight owing to the heterogeneity of the substrate and the variety of environmental conditions encountered. Cellulose is utilized microbially in the digestive tract of herbivorous animals and in insects; cellulose is decomposed in soil as natural vegetation and crop residues decay and are incorporated; cellulose is decomposed in compost piles and manure heaps; cellulose is decomposed in marshes and swamps, and in rivers and lakes; cellulosic walls in living plants are penetrated by the action of pathogenic organisms, and in woods disintegrated by wooddecomposing fungi. In almost all these natural processes the cellulose is the constituent which suffers the greatest loss, and the removal of the cellulose is the change which produces the most significant alterations in the properties of the plant material concerned.

In the study of the microbiological utilization of cellulose, information of two kinds is needed. Firstly, there must be available some means of following the changes produced in the cellulose, and secondly, there must be available methods of recognition of the active organisms. The former, insofar as the overall changes are concerned, is the easier; the latter, involving as it does the isolation, separation, and culture of the organisms is the more difficult.

Cellulose is wholly insoluble in water, and accordingly in those cases in which substantially pure cellulose is appreciably attacked, the loss in weight after washing may be used quantitatively. If gummy or mucilaginous synthetic products are produced by the organism, dilute carbonate or ammonia may be used to aid removal. The determination of the extent of attack on fibers, fabrics, and paper is more difficult since the product may be spoiled without extensive and easily visible decomposition. In such circumstances microscopic methods are ordinarily employed. Tendered fabrics and fibers treated with dyes such as congo red and methylene blue color more intensely than unattacked materials. The behavior on swelling with strong alkali or with alkali and carbon disulphide is abnormal, and microbial damage can be distinguished from that produced chemically by these means.

The determination of the extent of attack undergone by the cellulose of plant materials is made usually by isolation of the cellulose by some modification of the Cross and Bevan procedure. Alternate chlorination and extraction with hot neutral sulphite results in the removal of associated cell wall constituents, such as the hemicelluloses and lignin. A cellulosic residue similar but not identical in composition may be obtained by direct treatment of the plant material with hot monoethanolamine, and the simplicity of this process has much to recommend it for comparative purposes. An approximate figure may be obtained by a hydrolytic method in which easily hydrolyzable polysaccharides are first removed by treatment with dilute acid, and then the residual cellulose is put into solution in cold 80 per cent sulphuric acid, diluted and boiled to yield glucose, and titrated by any convenient reducing group method. The figure so obtained is lower than that given by direct isolation procedures. The hydrolytic method has the great advantage, however, that it can be used in the presence of soil or other inert material.

Media of two general types are used for the isolation and culture of cellulose decomposing organisms, namely those consisting of some form of purified cellulose, such as filter paper, and those containing regenerated cellulose, or partially hydrolyzed cellulose. Many workers have employed filter paper or cloth laid on the surface of agar or silica gel containing the necessary salts. The paper or cloth is ordinarily moistened with nutrient solution and has to be kept quite moist if good growth is to be obtained. Such media are satisfactory for the original isolation of cellulose-decomposing bacteria in the mesophilic range, but are less satisfactory for the isolation of thermophilic organisms or fungi. Filter paper as a medium does not lend itself well to the normal dilution plating techniques by which authentically pure cultures are ordinarily obtained. Regenerated celluloses, such as those obtained after solution in cuprammonium, or hydrocelluloses obtained by brief treatment with warm strong sulphuric acid, may be incorporated in agar since the effects of the treatments are to destroy the macrofiber structure. The growth of many cellulose organisms on these media is very restricted but adequate for separation. Cellulose dextrins prepared by the controlled cold acid hydrolysis of cellulose may similarly be employed in agar. Opalescent media are obtained on which active cellulose decomposers make limited colony growth, but inasmuch as the colonies may be surrounded by a clear halo their recognition is easy. Cellulose dextrin media are well adapted to the enumeration of cellulose-decomposing bacteria in soil or sewage, and permit of far more satisfactory comparisons than the dilution method in which presence or absence of growth on filter paper strips is observed.

The ability to attack cellulose is not a property limited to a few highly specialized organisms. Many bacteria, actinomycetes and fungi utilize cellulose to a greater or less degree. The widespread use of pure cellulose, such as filter paper, in isolating and characterizing cellulose organisms has led to over emphasis of the specialized forms and neglect of those organisms which do not develop vigorously on such a highly purified insoluble substrate but which nevertheless do utilize cellulose when it is present with other cell-wall constituents of greater availability.

Among the bacteria, those falling within the anaerobic thermophilic group have perhaps been most studied because of their possible industrial use. The products usually given from cellulose by anaerobic organisms are acetic, butyric, and lactic acids, ethyl alcohol, carbon dioxide and hydrogen, though traces of other alcohols and acids may also be found. Methane, which has long been regarded as a product of the anaerobic dissimilation of cellulose, is probably not directly so, but is rather produced secondarily by methane organism which accomplish reduction of carbon dioxide by hydrogen transfer from one of the primary products. The most vigorous thermophilic cultures are more or less stable composites of several organisms, and are beat obtained from the dung of a herbivorous animal. Such composite cultures frequently do not give reproducible yields of fermentation products. Purification of such cultures usually results in a loss of fermentative power, and a change in the relative proportions of the several products. The crude culture may consist of several distinct organisms only one of which is capable of accomplishing the initial attack on cellulose. The associated forms are usually facultative and their presence enables the cellulose organism, a strict anaerobe, to be active under aerobic or only partially anaerobic conditions. Several anaerobes, the activity of which is optimum within the unusual range of 38° to 50°, have been isolated from the digestive tracts of man and animals. Best known perhaps are the Khouvine organism, Bacillus cellulosae dissolvens and Plectridium celluloiyticum.

The rapidity of action of the mesophilic cellulose bacteria is on the whole far less than that of the thermophilic forms. Few are limited to cellulose as a source of carbon, and the majority are capable of utilizing slowly and incompletely a wide range of carbohydrates. Many short rods with rounded ends have been described and placed in the genus Cellulomonas. Other curved rods with similar properties have been classified as belonging to the genus Cellvibrio. Most attention has been directed to the cytophagas, that are widely distributed in soil.

This is due partly to their specialized character, and partly to their remarkable morphology. The type species was incorrectly named Spirochaeta cytophaga because of its spirochaete-like appearance. Young cultures on filter paper consist of long, thin, flexuous rods, tapering or pointed at the ends. The cell wall is not rigid and the cell may move by a peculiar flexing movement. This is sufficient to exclude the group from the true bacteria. Older cultures may contain in addition a large coccus, which in fact is a sporoid or microcyst, and not a coccoid contaminant as has been maintained by as great an authority as Winogradsky. Species exist both with and without the microcysts. It is probable that the cytophagas should be included under the order Myxobacteriales. Cytophagas that utilize cellulose only feebly but attack a wide range of other compounds have been isolated from marine sources. It is clear that all species of this organisms are not obligate cellulose decomposers, and even the soil forms, that have been said to be so, can utilize the predominantly pentosecontaining cellulosans that are associated with the cellulose of most angiosperms more readily than the long-chain cellulose fraction. Most mesophilic cellulose bacteria are highly aerobic and few facultative or anaerobic forms are known though they must undoubtedly exist. Aerobic spore-formers have been described under the genus Cellulobacillus, but the justification for the creation of new genera on the basis of cellulose-decomposing ability alone is questionable.

The list of fungi capable either of growing on pure cellulose or attacking the cellulose of plant materials is a very long one. Many are more vigorous in the latter circumstances than in the former. Ample available nitrogen must be present because the synthetic abilities of fungi are far greater than of bacteria. As much as one-third the carbon utilized from cellulose may be present in the microbial tissue formed. A deficiency of nitrogen retards the rate of decomposition. There is some evidence that in soil, fungi may be more vigorous than bacteria in the decomposition of the cellulose of added plant material and that a fungal attack precedes a bacterial attack. Sequential changes in the population are an inevitable result of substrate heterogeneity. Single fungi do not seem able to accomplish either so rapid or so extensive removal of cellulose from plant materials as mixed populations of fungi and bacteria. Some fungi can develop at relatively low moisture levels and are responsible for "mildewing" of fabrics and paper. Growth in such circumstances is usually very restricted and the damage is often limited to a deterioration in strength, and staining due to the color of the fungal growth. The cellulose of dead limbs on trees, and intact lumber is utilized slowly by the higher fungi or Basidiomycetes that seem peculiarly adapted to the penetration of lignified walls and growth under conditions of relatively low moisture content and poor air supply. Certain of these are highly efficient in the assimilation of available nitrogen. All seem to use other cell wall constituents along with the cellulose.

The information as to the biochemistry of the process of cellulose decomposition is far from satisfactory. The conventional view is that by means of an exo-enzyme system hydrolysis of the cellulose to glucose is effected, and this glucose is then dissimilated intercellularly. If the structure of cellulose is represented by a simple long chain molecule the primary attack could be accomplished by a hydrolytic system that split off terminal glucose units from the long chains. Such a process would not be accompanied by the production of any considerable amount of cellulose dextrin. However, there is a strong probability of some kind of cross-linkage between chains and therefore the hydrolytic system is probably more complicated. Arguing by analogy with starch it has been often suggested that the disaccharide cellobiose may be produced in the hydrolysis of cellulose and its production has been claimed but not positively verified. To account for this, the existence of two hydrolytic enzyme systems has been suggested, one, cellulase, producing cellobiose from cellulose, and one, cellobiase, producing glucose from cellobiose. Various workers have demonstrated beyond question the presence of reducing substances in aerobic cellulose decompositions, in which the further development of the active organism has been checked either by addition of toluene or reduction in oxygen supply. Not in every case has the identification of the presence of glucose been satisfactory, and that of cellobiose is still questionable. The nature of the intermediate steps is quite unknown. Hydrolytic fragments of cellulose containing several and perhaps as many as 10-15 glucose units are soluble in water and might pass into the cell. Complete extracellular hydrolysis to glucose is not essential. It seems likely that the primary enzyme system is composite since organisms exist which can utilize the insoluble cellulose dextrin but which cannot develop on cellulose alone. The observation that certain of the more specialized cellulose organisms are inhibited by relatively low concentrations of glucose might suggest that the dissimilation does not normally proceed through this sugar. Other explanations, however, are possible.

The later stages of cellulose decomposition are not much more satisfactorily understood than the primary attack. Various products have been described but accurate balance sheets are few, and indeed the precise nature of the products formed by some of the most commonly recognized cellulose bacteria cannot be found in

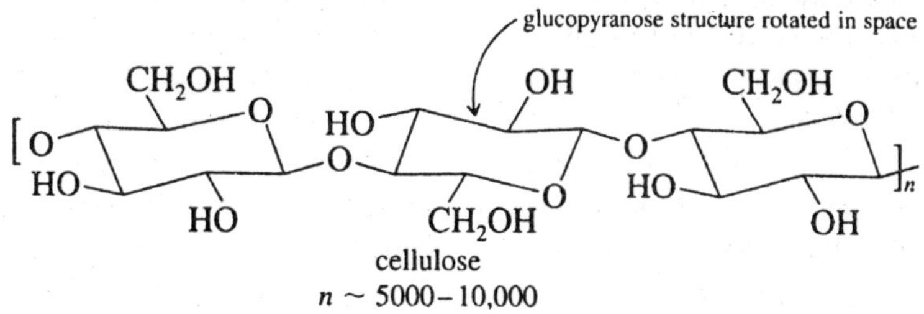

Fig. Cellulose

the literature. In general it seems that the aerobic mesophilic bacteria effect extensive oxidation, so that most of the carbon utilized can be recovered as CO_2. No more than a trace of acids, either volatile or non-volatile can be found in cytophaga cultures. In the few cases in which acid production by aerobic organisms has been found the quantity is usually small. Capsular polysaccharides containing uromc groupings are formed by some aerobic mesophilic bacteria. Anaerobic organisms on the other hand give substantial amounts of acidic products and some alcohol. Acetic and butyric acids are commonly formed, though good yields of formic acid have been reported in one case. The yields of volatile acids are usually in the neighborhood of 50-60 per cent of the cellulose fermented. Non-volatile acid and alcohol are produced only in relatively small amounts. It is unlikely that alcohol production could be increased beyond 10-15 per cent of the substrate removed. Almost nothing is known of the products of decomposition of cellulose by fungi. From sugars many fungi produce appreciable quantities of the di- and tri-basic non-volatile acids, and rarely more than traces of volatile acids. However, none of the former have been reported as being produced from cellulose despite the probability their formation.Some of the unsolved problems connected with the utilization of cellulose by micro-organisms have been a\indicated above. From a theoretical viewpoint undoubtedly the most important is that of the nature and more of action of the enzymes responsible for the initial phases of cellulose breakdown. It appears that the exo-enzymes responsible is not produced particularly freely, which is perhaps not surprising in view of the fact that the substrate is wholly insoluble. Fungi may be a better source than bacteria. It is not known whether a phosphorylation mechanism is involved, though this should be capable of ready determination. From a practical viewpoint a second problem, which had ramifications through many of the natural processes in which cellulose decomposition occurs, is the effect of the presence of other constituents on the availability and utilization of the cellulose. The cell walls of mature materials may be predominantly cellulosic but they are also incrusted and infiltered with more available polysaccharides such as the polyuronide hemicellulose and cellulosans, and less available material in the form of lignin. The relative proportions of these three major component changes with age.Increasing lignin contents is usually accompanied by icreasing resistance to decomposition in in vitro experiments, yet cellulose cannot be detected in the organic fraction of soil which is largely derived from the decomposition of partially lignified materials. Relating to this general question of mixed substrates is the effect of composite cultures. The association of organisms of entirely different characteristics, such as bacteria and fungi seems to be particularly effective in dealing with the composite substrates provided natural materials. Not far removed from the same problem is the whole subject of the biochemistry of the digestion of cellulosic materials by herbivorous animals, a process which is to a considerable extent microbial. The cellulose of roughages is ordinarily quite incompletely utilized, and great wastage in fact occurs.

Cenozoic era The era of mammals; includes the recent, pleistocene, pliocene, miocene and eocene; about 3 million years.

Centimorgan (cM) A unit of measure of the statistical probability recombination frequency between alleles. One centimorgan is equal to a 1% chance that a marker at one genetic locus will be separated from a marker at a second locus due to crossing over in a single generation. In human beings, 1 centimorgan is equivalent, on average, to 1 million base pairs.

Central dogma A phrase that refers to the concept of information flow proceeding only from DNA to RNA to protein.

Central nervous system (CNS) Pan of the body that consists of the brain and spinal cord and that controls and coordinates the activity of the entire nervous system

Centrifugal force Imaginary force that appears to act outward from the center of rotation upon a body moving in a curved path, balancing the *centripetal force.*

Centrifuge A machine that spins samples at a high speed and forces the settling of suspended material.

Centripetal force True *force* that is directed inward toward the center of rotation upon an object moving on a curved path.

Centrolecithal Having yolk accumulated at the centre of the egg (e.g. arthropods).

Centromere A specialized constricted region of a chromosome to which spindle fibers attach during cell division at which two sister chromatids (the two exact copies of each chromosome that are formed after replication) are joined, and which attach to the spindle during cell division.

Cephalin Any one of a group of phospatides found in animal and plant cells, particularly in brain and nerve tissues. It is soluble in water and most organic solvents except alcohol and acetone:

Ceramides Compounds which may be looked upon as part of cerebrosides or sphingomyelins, e.g. lignocerylsphingosine (kerasin minus galactose) found in lung and spleen.

Cerebronic acid $C_{24} H_{50}O_3$; an acid, possibly $CH_8 (CH_2)_{21}$ CHOH-COOH, forming part of the phosphatide sphingomyelin and the cerebroside phrenosin.

Cerebrosidase An enzyme which splits cerebrosides, liberating ceramides; needs activation by cysteine, glutathione or ascorbic acid.

Cerebrosides White, wax-like organic compounds found in the brain and nervous tissues with the phosphatides. They consist of one molecule of a C_{24}, or C_{25} fatty acid, with a peptide linkage to a molecule of sphingosine, which is linked to molecule of d-galactose. They are soluble in most organic solvents except ether and alcohol.

Cerebrospinal fluid A clear, colorless liquid bathing the spinal column and the brain, pH 7.4. It is a blood plasma dialysate, chemically modified by the choroid plexus epithelium.

Cerolin An alcoholic extract of brewer's yeast of indefinite composition.

Cerotic acid A saturated fatty acid, $CH_3 (CH_2)_{24}$ COOH, found in beeswax, opium wax, wool fat; m.p. 82.5°C.

Cetoleic acid An unsaturated fatty acid, $C_{22} H_{42} O_2$, with one double bond, found in marine animal and fish oils.

Cetyl alcohol Hexadecanol, m.p. 49°, which as palmitate is chief constitutent of spermaceti.

Chabrol-charonnet tests for bile salts I. Four cc. of a solution of the substance in 10 cc. 85% phosphoric acid are heated in a boiling water bath for 4 minutes, cooled, and treated with 1 cc. 0.6% aqueous vanillin solution; a rosered color indicates a positive test.

II. Four cc. of serum are treated with 36 cc. 90% alcohol and filtered after which 20 cc. are evaporated below 60°, the residue dissolved in 8 cc. phosphoric acid and heated in a water bath for 4 minutes. A rose-red color is obtained with 1 cc. 0.6% aqueous vanillin solution.

Chaperone Proteins: A series of proteins present in the endoplasmic reticulum which prevent proteins from folding prematurely and guide the proper folding of secreted proteins through a complex series of binding and release reactions.

Charles' law The volume of a given mass of a gas at constant pressure is increased by ½73 of its volume at 0° C. for each degree rise in temperature.

Chaulmoogric acid $C_{18}H_{32}O_2$; a cyclic fatty acid found in Chaulmoogra Oil which is used in the treatment of leprosy; crystallizes as leaflets from alcohol; m.p. 68°C.

Chelidonic acid A crystalline acid, $C_5H_2O_2(COOH)_2$, found in the celandine and white hellebore.

Chemical Energy A form of potential energy stored in the molecular structure.

Chemotaxis Chemiotaxis; chemotropism; reaction to chemical gradients, such as found in solutions.

Chemotherapy Chemotherapy is the science of treating infectious diseases by specific chemical compounds which exert their action systemically. The three factors involved are the infectious agent, the chemical substance and the host or organism.

The subject had its beginning with the discoveries that cinchona bark was valuable in curing malaria; ipecac in treating amebic dysentery; and mercury in the therapy of syphilis. The modern science of chemotherapy commenced with the research of Paul Ehrlich, who systematically prepared and tested a long series of organic arsenic compounds for their trypanocidal efficacy. As a guiding principle he employed the curative ratio which is the minimum lethal dose (or tolerated dose) divided by the minimum curative dose. He set 3 as the safe ratio. Ehrlich found that salvarsan (also known as 606 and now as arsphenamine) was the most effective compound against syphilis. Later he prepared neosalvarsan which has the advantage of being water soluble and neutral.

Following Ehrlich's epochal discovery many advances were made concerning the treatment of spirochetal, trypanasomal and protozoal infections.

In the treatment of syphilis the arsphenamines have retained their place of importance, but other valuable compounds have been added. Mepharsen is particularly effective. Tryparsamide is valuable in the treatment of central nervous system syphilis. Bismuth salts are important as antiluetic agents and have largely replaced mercury in the routine treatment of this important venereal disease.

Quinine remains the most important agent in the treatment of malaria, but a number of synthetic compounds, notably atabrine and plasmochin are also employed in the control of this disease. Amebiasis is combated with emetine (obtained from ipecac) and several synthetic products such as chinofon, vioform and carbarsone.

The treatment of the dreaded disease of leprosy by chaulmoogra oil offers real promise. The active principles are two unsaturated fatty acids, chaulmoogric and hydrocarpic. Leishmaniasis (kala azar and oriental sore) and bilharziasis are successfully treated with antimony. The oldest compound of this element used in therapy is tartar emetic. Several new derivatives have recently been introduced, namely antimony sodium thioglycollate, antimony thioglycollamide, and fuadin.

While the chemotherapy of diseases caused by spirochaetes, trypanosomes and other protozoal parasites steadily advanced, the treatment of bacterial diseases with chemical agents proved unsuccessful and discouraging until Domagk in 1935 discovered that prontosil, an azo dye, protected mice against streptococcal infection. The successful use of this dye against hemolytic streptococcal infection in humans was quickly established. At the Pasteur Institute in Paris it was found that in the body the dye, prontosil, splits at the azo linkage and yields p-amino-benzene sulfonamide (sulfanilamide), which proved to be the active chemotherapeutic agent. The great effectiveness and specificity of sulfanilamide against various bacterial diseases was soon established, but its limited value against certain organisms, particularly the pneumococci, soon became apparent. This led to a quest of substituted sulfanilamides. The most successful of these are sulfapyridine, sulfathiazole and sulfadiazine.

Sulfanilamide is effective against beta hemolytic streptococci, meningococci, gonococci, Cl welchii, and several other types of bacteria. Its mode of action is not known. The drug exhibits bacteriostatic action in vitro, but its much greater activity in vivo suggests that the humoral and cellular defense mechanisms of the organism contribute to its efficacy. While sulfanilamide is extremely effective against virulent bacteria, it has a low toxicity for the host. It occasionally causes mild mental symptoms, cyanosis, anemia, neutropenia, fever, and various forms of dermatitis. Rarely, however, are these of a serious nature and the important therapeutic value of the drug far outweighs all its untoward reactions.

Sulfapyridine (2-sulfanilyl aminopyridine) finds its greatest usefulness in the therapy of pneumococcal infections, which resist the action of sulfanilamide. The compound is much. more toxic than the parent drug. Nausea and vomiting are common and often severe. The low solubility of the acetylated compound formed in the body often causes the formation of crystals in the urine and these may obstruct the tract and bring about anuria.

Sulfathiazole is widely used. Like sulfapyridine, it is effective against the pneumococcus, and also against many of the group affected by sulfanilamide. The drug has also been employed successfully against staphylococci. Sulfadiazine is similar in action to sulfathiazole. Although relatively new, it is very promising since it is highly effective, and has a low toxicity.

The sulfonamide drugs have revolutionized the treatment of infectious diseases. Various infections, which prior to these drugs were invariably fatal, respond promptly and effectively to these new chemotherapeutic agents. The mortality of pneumonia has been markedly reduced and meningococcic meningitis, gonorrhea, puerperal sepsis, urinary tract infection and many other common and serious infectious diseases are effectively and promptly cured. In one important virus disease, trachoma, sulfanilamide has been employed with suc- cess.

In spite of the brilliant results already obtained, the chemotherapy of bacterial diseases is still in its infancy and it seems not too sanguine an expectation to see marked strides in this field which will dwarf in comparison all present day accomplishments.

Chenodesoxycholic acid $C_{24}H_{40}O_4$, m.p. 140°, a bile acid found in man, ox, monkeys, fowl and fish. It is 3, 7-dihyroxycholanic acid.

Chevallier-Choron test for vitamin-A in blood Twenty gm. of absolute alcohol are added to a moist-dry mixture of 3 cc. of blood and 20 gm. anhydrous sodium sulfate and the mixture allowed to stand for 1 hour. An absorption spectrum band at 325 mu indicates vitamin A.

Chiasma Place where, during meiosis, two homologous chromosomes establish close contact and where usually an exchange of homologous parts between nonsister chromatids takes place.

Chimaera A compound organism or tissue obtained by grafting or other means so that it is composed of tissues of different kinds (e.g. an andro-merogonic head on a normal diploid tail) or of different species or genera.

Chimyl alcohol Cetyl-alpha-glyceryl-ether, found in liver oils of elasmobranch fishes.

Chinese yeast Mucor Rouxii, which produces very active amylase when grown aerobically and high concentrations of alcohol on carbohydrate fermentation.

Chitin A polysaccharide that forms the hard, resistant exo-skeletons of many invertebrates, and composed of N-acetyl-ß-D-glucosamine units joined in long chains at carbon 4. On acetolysis it yields octaacetylchitobiose, and on complete hydrolysis "glucosamine" (2-amino2-desoxy-D-glucose).

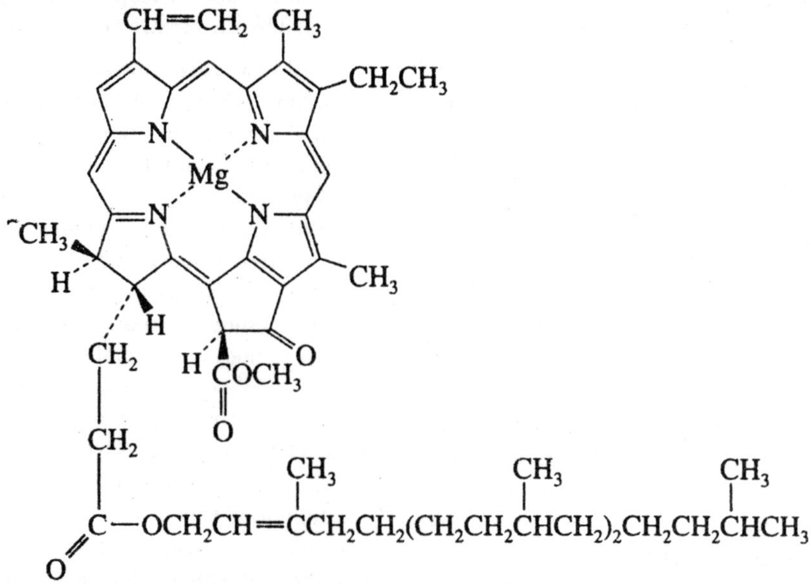

Fig. Chlorophyll a and b

Chitinase Enzyme capable of hydrolyzing chitin, found in moulds and snail digestive juice; produces glucosamine and N-acetyl glucosamine.

Chitobiose $C_{12}H_{23}O_{10}N$; an amino-sugar, 2desoxy-2-amino-4-(β-D-glucosyl)D-glucose, obtained by the partial hydrolysis of chitin in a manner analogous to the preparation of cellobiose from cellulose.

Chloral Trichloracetaldehyde, b.p. 98U+00BO; forms a hydrate which is used as soporific and anodyne.

Chloramine-T Sodium toluene-p-sulphonechloramide, a powerful antiseptic.

Chlorocruorin A dichroitic respiratory pigment of certain polychetes, e.g. Spirographis. While similar in structure to hemoglobin, the hemin group is not a red heroin as is protohemin but is a pheohemin, containing a formyl group in the 2-position instead of the vinyl group present in the corresponding position in protohemin.

Chlorocruorohaem The prosthetic group of chlorocruorin, the respiratory pigment of certain polychaete worms.

Chlorogenic acid A condensation product of quinic and caffeic acid, found in many plants

Chlorophyll The green coloring matter of plants. Two forms have been isolated, chlorophyll-a, having the empirical formula $C_{55}H_{72}O_5N_4Mg$ with ½ H_2O, m.p. 150-153U+00BO, and chlorophyll-b, having the formula C35H70O6N4Mg, m.p. 183-185U+00BO. Many decomposition products have been isolated, notably phytol which is 3,7,11,15-tetramethyl-Ä[23]-hexadecan1-ol. A common heterocycle of both forms is phorbin which is a porphin. The abstraction of Mg and phytol from the two forms of chlorophyll gives two phaeophorbids. The abstraction of phytol leaves chlorophyllids. The structures are strikingly similar to hematin. The synthesis of carbohydrates by plants under the influence of light is supposed to occur through the absorption of the light energy by chlorophyll-a after a dark reaction involving enzymatic reduction to dehydrochlorophyll. A hexose is formed in the dark reaction. The light reaction is a reoxidation with an absorption of energy.

Chlorophyllase A specific esterase found in all green plants. It hydrolyzes chlorophyll to chlorophyllide and phytyl alcohol.

Chloroplast Microscopic chlorophyll-carrying proteid body in plant cells, present in vast numbers.

Cholagogue Substance which causes emptying of gall bladder, e.g. cholecystokinin, $MgSO_4$.

Cholecystokinin A hormone of the upper intestine which causes the contraction and emptying of the gall bladder.

Cholecystotomy Operation on the gall bladder; the organ is opened and closed again without drainage.

Choleic acids Stable coordinated molecular compounds of desoxycholic acid with fatty acids, ketones, phenols, hydrocarbons and other compounds. The complexes form in the molecular ratios of 1:2, 1:3, 1:6 or 1:8. Previously confused with Desoxycholic Acid.

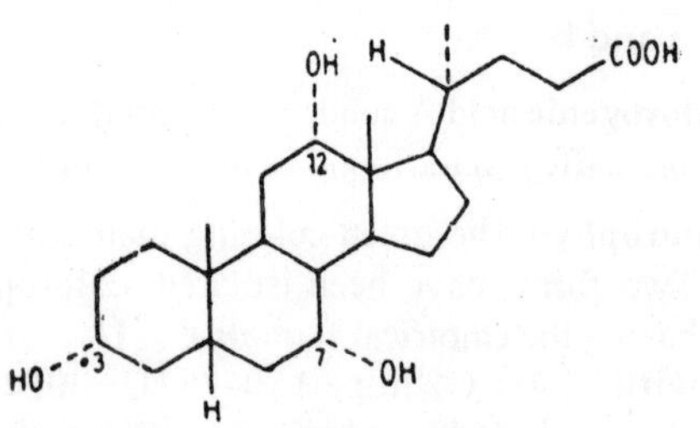

Fig. Choleic Acid

Cholesterol A sterol present in every animal cell and in more concentrated form in gallstones and brain tistue. It contains a branched side chain of 8 carbon atoms attached to C_{17} of the sterol nucleus, an OH at C_3 and a double bond between C_5 and C_6. It forms an insoluble digitonide.

Cholic acid $C_{24}H_{40}O_5$; a bile acid found in man and most animals. It is 3, 7, 12-trihydroxy-cholanic acid.

Choline dehydrogenase (liver) A cytochrome-reducing dehydrogenase which catalyzes the aerobic oxidation of choline to the corresponding aldehyde.

Choline Trimethyl Beta-hydroxyethyl ammonium hydroxide; $(CH_3)_3 \cdot NOH \cdot CH_2 \cdot CH_2OH$; a compound whose derivative acetylcholine has an important connection with nerve impulses. Choline causes a drop in blood pressure. It is formed by the breakdown of lecithin.

Choline has been shown to be "lipotropic," that is, it prevents or cures fatty livers (which may be due to depancreatization or a breakdown in fat metabolism) either in itself of as the active component of lecithin. Betaine and triethylcholine showed similar action. A factor in the effect is the promotion of the formation of phospholipids. Another factor is the transfer of methyl groups in the synthesis of methionene from homocystine and of creatine, or, in general, where labile methyl groups are required. A third factor is tire effect on the production of acetylcholine. Various deficiencies due to lack of choline might involve other factors. Rats show a falling of hair and swollen paws, along with signs of muscular weakness. Hemorrhagic kidneys, deficient vagus function, involution of the thymus, disorders of lactation, etc., have been noted. Vitamin B deficiencies shows close resemblances. Thiamin is definitely a contributing factor in the production of certain fatty livers, called "dietary fatty livers." Choline and thiamin show an antagonism in such cases. That labile methyl groups serve a useful function is shown by the fact that methionine, a source of labile methyl groups, behaves like choline in the removal of some of the above mentioned symptoms. Choline present in ester form is acted upon by the enzyme cholinesterase which has turned out to be inhibitable by thiamine, according to still fragmentary work. Choline itself may derive its methyl groups from other methylated compounds, e.g. methionine. The most plausible source of the rest of the structure is aminoethanol, As a carrier of labile methyl choline acts as a detoxifying agent, e.g. of pyridine and nicotinic acid, by trans-methylation.

Cholinergic Pertaining to nerves that liberate an acetylcholine-like substance on stimulation.

Chondroitin sulfuric acid One of the two carbohydrate complexes associated with the protein of the mucins.

Chorio-Allantoic Grafting The placing of small

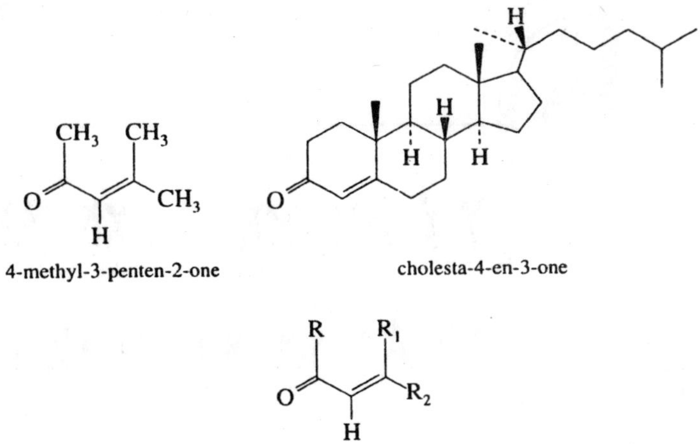

4-methyl-3-penten-2-one

cholesta-4-en-3-one

the chromophore common to both compounds

Fig. Chromophores

pieces of avian or mammalian embryos upon the chick embryo's allantois under the inner shell-membrane (the bird's egg has no chorion), in which position they become vascularised from the allantoic circulation and develop, though not in a strictly neutral medium.

Chorion The outer layer of the embryonic sac of a mammal consisting of both trophoderm and mesoderm elements and the outermost extraembryonic membrane in birds and reptiles.

Chromatid A single, continuous double stranded DNA molecule with its unique, complete grouping of genetic information, associated proteins, higher-order structures, and centromeric and telomeric regions necessary for separation and maintenance after replication.

Chromatids Two longitudinal units of a chromosome resulting from a split in early prophase, becoming daughter chromosomes later in mitosis.

Chromatids The name applied to each of the two parts of a duplicated eukaryotic chromosome for as long as the two parts remain attached at the centromere. Each chromatid consists of a DNA double helix and associated proteins, and it has the same gene sequence as its "sister" chromatids.

Chromatin A protamine nucleate; the nature of the link between the mine and the nucleic acid is not clear, but it is either a salt or a coacervate. Chromosomes are composed of chromatin.

Chromatographic adsorption The name applied to the phenomenon of selective adsorption observed when solutions of certain mixtures are passed through a column of an adsorbent. It is found that one component separates in a ring, below which is a ring with no adsorbed substance, followed by a ring of a second substance, etc. This method is useful in separating mixtures of closely related compounds.

Chromogen Any chemical group which by itself or when associated with some other group imparts color.

Chromophore A group which gives color to an organic compound but does not necessarily make it a dye. Examples: NO_2, —N=N—.

Chromoproteins In the American Classification of proteins, the chromoproteins are conjugated proteins where the prosthetic group is colored. Examples: Hemoglobin, visual purple, ovoverdin, flavoproteins.

Chromosome jumping A technique whereby one starts with a piece of DNA from one region of a

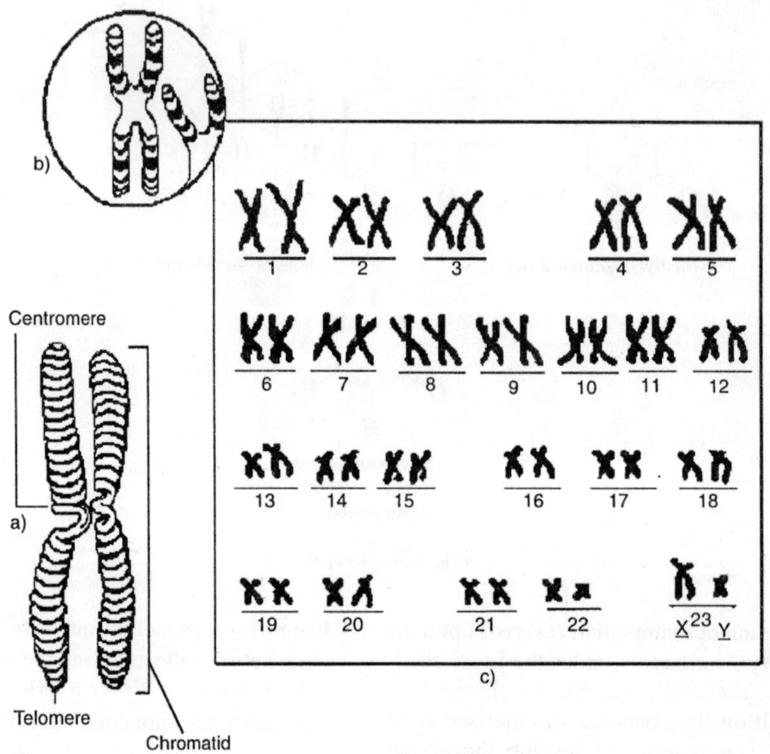

Fig. Human Chromosomes

chromosome, and obtains clones from nearby regions without cloning everything in between. One round of jumping yields new clones at distances of several tens of kb away from the starting point. In practice, this method is used when classical genetics proves that a known piece of DNA is located on the chromosome close to a gene you would like to clone (such as a human disease gene). By cloning fragments some distance away in both directions from the known fragment, one might obtain.

(1) fragments further from the desired gene (which are discarded);

(2) fragments even more closely linked to the desired gene (in which case one goes for another round of jumping); or (3) fragments from within the desired gene - the optimal result.

Chromosome walking A technique for cloning everything in the genome around a known piece of DNA (the starting probe). You screen a genomic library for all clones hybridizing with the probe, and then figure out which one extends furthest into the surrounding DNA. The most distal piece of this most distal clone is then used as a probe, so that ever more distal regions can be cloned. This has been used to move as much as 200 kb away from a given starting point (an immense undertaking). Typically used to "walk" from a starting point towards some nearby gene in order to clone that gene. Also used to obtain the remainder of a gene when you have isolated a part of it.

Chromosomes A condensed, fibrillar, self-replicating genetic structures of cells containing the cellular DNA that bears in its nucleotide sequence the linear array of genes. In prokaryotes, chromosomal DNA is circular, and the entire genome is carried on one chromosome. Eukaryotic genomes consist of a number of chromosomes whose DNA is associated with different kinds of proteins.

Chronaxie (chronaxy) Latent period between electrical stimulus and muscular response.

Chrysene A colorless aromatic polynuclear hydrocarbon, m.p. 255U+00BO, found in coal tar, usually found with a small amount of napthacene which gives the commercial chrysene its yellow color.

Chyle The milky emulsion of fat in the lymph after absorption from the intestine into the lacteals on the way to the blood stream.

Chylomicrons Fat droplets in the blood, 1 micron or less in diameter. Their number increases greatly during fat absorption after a meal.

Chyme The semi-liquid food discharged from the stomach into the duodenum, after the stomach has done its part in grinding and preliminary digestion.

Chymoinhibitor A term introduced by Tauber for substances that inhibit the milk coagulating power of proteases; found in gastric and intestinal mucosas.

Chymotrypsin A crystallizable proteinase of the pancreatic secretion effective in slightly alkaline media. It has strong milk clotting power. It exists in the pancreas in an inactive form, chymotripsinogen, which is converted to chymotrypsin by active trypsin, but not by enterokinase.

Cinchol $C_{23}H_{50}O$; a sterol with one double bond, present in cinchona bark.

Cinchona Dried bark of Cinchona, used medicinally for fevers and as a tonic, whose active constituents are alkaloids, such as quinine, cinchonidine, cinchonine and quinidine.

Cinchonidine An isomer of cinchonine, used like quinine; $C_{19}H_{23}ON_3$. CINCHONINE Alkaloid of cinchona, used like quinine; dextrorotary, $C_{19}H_{22}ON_2$. CINCHOPHEN A synthetic 2-phenylquinoline-4carboxylic acid, used for rheumatism and fever; likely to attack liver; atophan; phenoquin; quinophan.

Cinnamic acid C_6H_5·CH:CH·OOH; from storax, balsam, cinnamon oil; used as antiseptic, antitubercular.

Cis As used in molecular biology, an interaction between two sites which are located within the same molecule. However, a cis-acting protein can either be one which acts only on the molecule of DNA from which it was expressed, or a protein which acts on itself (e.g., self-proteolysis).

Citral A terpene aldehyde, $C_{10}H_{16}O$, a constituent of lemon-grass oil and others, a mixture of isomerides of which one is geranial and the other neral; has pleasant odor.

Citric acid A constituent of many juices, especially lemon (5% or more); tribasic acid $C_6H_8O_7$ or hydroxytricarballylic acid; but adds 1 molecule of water; administered for gout and as a purgative in the form of salts; formed by special fermentations.

Citric cycle Citric acid cycle; theory of Krebs and co-workers to explain catalytic promotion of oxidations in liver, kidney, muscle tissue by citrate, especially in presence of carbohydrates; the stages are: citric, cisaconitic, alpha-ketoglutaric, succinic, fumaric, malic, oxaloacetic, which reacts with pyruvic acid (the metabolite) to reform citric; isocitric acid is probably involved; insulin promotes the reaction still further in vitro, using pigeon breast muscle.

Citromycetin A yellow pigment from citromyces moulds; $C_{14}H_{10}O_7$.

Citronellal l-form found in citronella oil of Java and d-form in citronella oil of Ceylon and eucalyptus oils; $C_{10}H_{18}O$, terpene aldehyde.

Cleavage The processes of cell-division by which the original egg-cell gives rise to all the cells of the organism; if the cells are from the beginning of approximately equivalent size, the cleavage is *holoblastic* ; if some are much larger than others owing to yolk accumulation or other reasons, the cleavage is *meroblastic* ; and if cell-division occurs only at the surface of an enormous yolk-mass, the cleavage is *discoidal*. Cleavage in some species follows a definite pattern, where it is said to be *determinate* (this has permitted the tracing of "cell *lineages* "); in other species the pattern is quite lost after the first few cell-divisions.

Cleidoic Highly isolated from the environment, e.g. the eggs of birds, with walls permeable only to matter in the gaseous state.

Climacteric Period of the cessation of full sex functions, as menstruation.

Cloaca Common outlet for excretion and sexual products in lower vertebrates.

Clone An identical copy of an organism. Most plants, fungi, algae, and many other organisms naturally reproduce by making clones of themselves as a form of asexual reproduction.

Cloning vector DNA molecule originating from a virus, a plasmid, or the cell of a higher organism into which another DNA fragment of appropriate size can be integrated without loss of the vectors capacity for self- replication; vectors introduce foreign DNA into host cells, where it can be reproduced in large quantities. Examples are plasmids, cosmids, bacterial artificial chromosomes (BAC) and yeast artificial chromosomes (YAC); vectors are often recombinant molecules containing DNA sequences from several sources.

Cloning The process of generating sufficient copies of a particular piece of DNA or asexually producing a group of cells (clones), all genetically identical, from a single ancestor. In recombinant DNA technology, the use of DNA manipulation procedures to produce multiple copies of a single gene or segment of DNA is referred to as cloning DNA.

Clotting The formation of an insoluble, stringy mass of blood, tending to prevent further bleeding. One theory of clotting is: the injury causes cephalin to be liberated from the platelets. This neutralizes heparin, which then allows prothrombrin to be converted to thrombin. The latter then converts fibrinogen into insoluble fibrin.

Clupanodonic acid An 18-carbon fatty acid with four double bonds, found in Japanese sardine oil.

Coacervation The separation of microscopic liquid droplets when sols of two hydrophilic colloids of opposite electric charge are mixed. This appearance of a new phase is a special case of mutual precipitation. Biologically nucleoproteins may be an illustration of the formation and redispersion of "coacervates."

Coagulated proteins In the American Classification of proteins, those insoluble primary protein products that are formed from proteins by heat or alcohol.

Coca Dried leaves of Erythroxylum coca and Erythroxylum truxillense exerting stimulant action because of cocaine, its derivatives and truxilline alkaloids.

Cocaine $C_{17}H_{21}O_4N$; toxic alkaloid of Erythroxylon Coca, methylbenzoylecgonine, monoclinic prisms from alcohol, m.p. 98°, sl. sol. cold H_2O, giving an alkaline solution. Its hydrochloride, m.p. 200-2°, has been used in medicine as a local anesthetic, but is being replaced by new synthetics. Large doses cause death due to paralysis of the respiratory center.

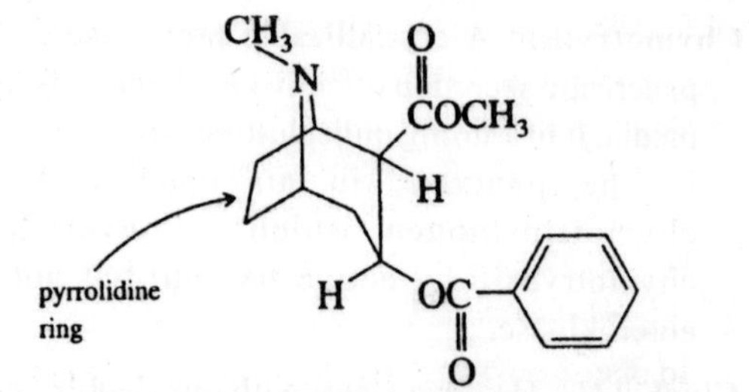

Fig. Cocaine

Cocarboxylase Diphosphothiamine; thiamine pyrophosphate; required for the activity of carboxylase, which catalyzes the decarboxylation of pyruvic acid or alpha keto acids. Except in the blood plasma it is present in tissues along with thiamin. It affects carbohydrate metabolism, carbohydrate synthesis and even synthesis of fat from carbohydrate. It is responsible for acetylations, e.g. formation of acetylcholine.

Cocceric acid $C_{31}H_{62}O_3$; a saturated monohydroxy fatty acid found in Cochineal wax.

Coccidia Protozoan parasites which often prey on experimental animals, e.g. the rabbit.

Coccyx Several vestigial vertebra of the tail in man.

Codeine $C_{18}H_{21}O_3N$; alkaloid of opium. Resembles morphine in action, but less toxic, less depressing, and slightly more stimulating, causing restlessness in large doses. Fatal dose, over 4.5 grains.

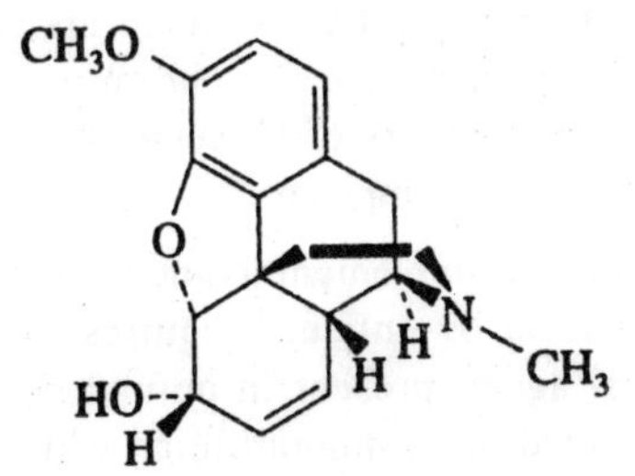

Fig. Codeine

Coding sequence The portion of a gene or an mRNA which actually codes for a protein. Introns are not coding sequences; nor are the 5' or 3' untranslated regions (or the flanking regions, for that matter - they are not even transcribed into mRNA). The coding sequence in a cDNA or mature mRNA includes everything from the AUG (or ATG) initiation codon through to the stop codon, inclusive.

Coding strand An ambiguous term intended to refer to one specific strand in a double-stranded gene.

Codon Bias The tendency for an organism or virus to use certain codons more than others to encode a particular amino acid. An important detrminant of codon bias is the guanosine-cytosine (GC) content of the genome. An organism that has a relatively low G+C content of 30% will be less likely to have a G or C at the third position of a codon (wobble position) than a A or T to specify an amino acid that can be represented by more than one codon.

Codon A group of three consecutive nucleotides within messenger RNA (mRNA) that encodes a message to initiate translation, to incorporate a specific amino acid into the growing polypeptide chain, or to stop translation. The sequence of codons in the mRNA unambiguously defines the primary structure of the final protein. Of course, the codons in the mRNA were also present in the genomic DNA, but the sequence may be interrupted by introns.

Coenzyme (Warburg and Christian) Coenzyme II; a thermostable substance, composed of 1 molecule of adenine, 1 of nicotinamide, 3 of phosphoric acid and 2 of pentose, which is reduced in the activity of the enzyme of fermentation or of the oxidation of glucose-6-phosphate.

Coenzyme A low molecular weight material which activates an enzyme; not as completely specific as a kinase.

Coenzyme-linked reactions Reactions in a mixture of two enzyme systems in which a single coenzyme undergoes a reduction by one system and an oxidation by the other, e.g. beta-hydroxybutyrate and acetoacetate and its enzyme, alcohol and acetaldehyde and its enzyme, and coenzyme I.

Colchicine $C_{22}H_{05}O_6N$; a levorotatory alkaloid of the meadow saffron; administered with belladona for gout; has shown remarkable use in affecting genetic changes; is the active constituent of the dried corms and seeds of the meadow saffron, known as colchicum.

Cole microchemical reagent for alkaloids Yellow crystalline precipitates are obtained when potassium ferrocyanide is added to solutions of the hydrochlorides of brucine, cinchonidine, cinchonine, cocaine, eucaine, heroin, hydrastine, quinoline, sparteine, stovaine, strychnine and veratrine.

Cole test for bile pigments Fifty cc. of the test solution are treated with an excess of lime or baryta water and filtered; the precipitate is treated with 5 cc. alcohol, 2 drops of concentrated sulfuric acid and 3 drops of 50% potassium chlorate solution and boiled. An emerald or bluishgreen color in the supernatant liquid indicates bile salts.

Co-linearity Refers to an exact correspondence between information encoded in DNA and the polypeptide product ultimately translated from transcripts made from the DNA. Most prokaryotic genes are co-linear with their products; eukaryotic genes often are not.

Collagens A group of albuminoids occurring only in the animal kingdom. They constitute the chief solid matter of white fibrous connective tissue and of the ground substance of cartilage and bone.

Collenchyma Thickened plant cells adapted to supporting stem.

Collodion A solution of cellulose in alcoholether. The gel resulting from the evaporation of the solvent is used as a semi-permeable membrane for diffusion, and as an ultrafilter.

Colloid A dispersion of one phase of matter in another, where the particles of the dispersed phase are coarser than molecular. The range of size of colloid particles is arbitrarily fixed between I micron and 0.1 micron.

Colocynth Dried pulp of the bitter apple, Citrullus colocynthis, which is used for purgative action.

Colorimeter An optical instrument, used in colorimetric analysis to accurately measure color differences.

Colorimetric A method of detection or determination based on a difference, qualitative or quantitative, of color.

Colostrum A secretion of the mammary glands found for 2-4 days after the birth of the offspring. Yellowish, viscous, alkaline fluid, with high protein content, acts as a purgative.

Colostrum The first secretion from the mammalian mammary giands after the birth of young; it has quite different properties from the milk which it precedes.

Competence The state of reactivity of a part of an embryo, enabling it to react to a given morphogenetic stimulus by determination and differentiation in a given direction. An embryonic cell develops as many competences as it has prospective potencies or possible morphogenetic fates. Generally speaking, determination to proceed to one morphogenetic destination cancels the other competences previously possessed by the cell. Competences appear in a succession at definite stages of development and later disappear, whether or not any of them have been implemented by the appropriate morphogenetic stimulus. Some competences exist in the adult tissues, if we may so term the state/s of reactivity to hormone action, though these effects are generally reversible

Competent Bacterial cells which are capable of accepting foreign extra-chromosomal DNA. There are a variety of processes by which cells may be made competent.

Complement, complement fixation Lysis of red cells or bacteria by antisera requires the aid of a special agent, present in normal sera and not augmented upon immunization which is called complement or alexin. It deteriorates slowly on storing and is inactivated by heating, e.g. within half to one hour at 50-60° C.

The characteristic of complement to be bound by the aggregates (precipitates, sensitized cells) formed through the interaction of antigens and antibodies is the basis for a frequently used serological test. In this "complement fixation reaction," introduced by Bordet and Gengou, antigen and antibody to be tested are mixed with fresh normal serum, containing complement, and, after incubation, hemolytic immune serum and the corresponding red cells are added as an indicator. If an immunological reaction takes place in the first stage, complement is fixed and removed from the solution and hemolysis is prevented, completely or in part, according to the intensity of the reaction

Complement Definition determined by context:

(1) The complementary sequence to a nucleic acid sequence under study.

(2) To provide a function that is missing or altered because of a mutational event.

Complementary DNA (cDNA) A DNA sequence synthesized from a messenger RNA molecule, using reverse transcriptase enzyme. cDNAs can be used experimentally to determine the sequence of messenger RNAs after their introns (non-protein-coding sections) have been spliced out. The single- stranded form is often used as a probe in physical mapping.

Complementary sequences Nucleic acid base sequences that can form a double- stranded structure by matching base pairs; the complementary sequence to G- T- A- C is C- A- T-G.

Compression The acceleration of development or extension of the period before hatching, so that the larval stages are accomplished or omitted before the organism hatches from the egg.

Concatemer Tandem arrays of monomeric DNA molecules with complementary ends. Intra-molecular reassociation of such molecules leads to circularisation while inter-molecular reaction produces concatemers.

Conformation The 3-dimensional structure of a molecule.

Conformational epitope Also called a "global epitope". An epitope comprised of contiguous but physically discontinuous components of the immunogenic molecule. In a proteinaceous antigen this could be sequences from different stretches within a polypeptide, or even from different polypeptide subunits.

Conhydrine $C_8H_{17}ON$; 2-alpha-hydroxy-propylpiperidine; odorous, basic alkaloid of hemlock; colorless leaflets, m.p. 121°; poisonous, causing paralysis of the motor nerve endings, respiratory paralysis, and death.

Coniceines A group of 6 basic amines, prepared from hemlock alkaloids. They are pyridine or piperidine derivatives.

Coniferin $C_{16}H_{22}O_8$; a glycoside of the bark of the fir tree, consisting of glucose and coniferyl alcohol; it crystallizes as needles, m.p. 185°C.

Coniine $C_8H_{17}N$; alpha-propyl-piperidine; a colorless, odorous, alkaline alkaloid of hemlock; oil, b.p. 166-7°, poisonous, causing paralysis of the motor nerve endings, resulting in respiratory paralysis and death.

Conjugation Fusion of two cells accompanied by pooling or exchange of nuclear materials and chromatin, temporarily or permanently, the cells being one male and the other female, the purpose in some cases being a rejuvenation where the sex difference may be doubted.

Conjugation Physical contact with the establishment of plasma bridges between two different bacterial cells allowing directed transfer of DNA.

Conjunctivitis Inflammation of the conjuctiva of the eye which may be

(1) simple or catarrhal, characterized by infection or associated with skin diseases, catarrhs, measles, smallpox, scarlet fever, trauma;

(2) acute contagious or "pink eye" or Koch-Weeks conjunctivitis or epidemic conjunctival catarrh, caused by the KochWeeks bacillus;

(3) Diplobacillus conjunctivitis;

(4) conjunctivitis neonatorum or ophthalmia neotatorum due to infection usually by gonococcus;

(5) adult gonococcal conjunctivitis;

(6) pneumococcal conjunctivitis;

(7) scrofulous ophthalmia or eczema of conjunctiva, occurring in tubercular cases;

(8) trachomatous or granular conjunctivitis or trachoma or Egyptian ophthalmia, which is very contagious;

(9) vernal or spring conjunctivitis characterized by flat granulations; and

(10) and many others, e.g. allergic, diphtheritic, traumatic.

Aside from causal treatment, codliver oil, iron, arsenic are administered. Atropine is used in cases of excessive irritation. Usual drugs for direct application are silver nitrate, boric acid, zinc sulfate, tannic acid, silver proteinate, ichthyol.

Connective tissue A loose tissue found in and supporting the various organs, consisting of about 60% water, 32% collagen, 0.5% inorganic matter, and the remainder largely lipids.

Consensus sequence A term that refers to sequences common to different genes within an organism, or to the same gene among different organisms, that encode a specific function. This term may be applied to either nucleic acids or proteins, since the protein sequence is completely dependent upon the nucleic acid sequence.

Conservative Substitution A nucleotide mutation which alters the amino acid sequence of the protein, but which causes the substitution of one amino acid with another which has a side chain with similar charge/polarity characteristics . The size of the side chain may also be an important consideration. Conservative mutations are generally considered unlikely to profoundly alter the structure or function of a protein, but there are many exceptions.

Constitutive expression Constantly expressed at some appreciable level.

Contig map A map depicting the relative order of a linked library of small overlapping clones representing a complete chromosomal segment.

Contig Several uses, all nouns. The term comes from a shortening of the word 'contiguous'. A 'contig' may refer to a map showing placement of a set of clones that completely, contiguously cover some segment of DNA in which you are interested. Also called the 'minimal tiling path'. More often, the term 'contig' is used to refer to the final product of a shotgun sequencing project. When individual lanes of sequence information are merged to infer the sequence of the larger DNA piece, the product consensus sequence is called a 'contig'.

Contractile vacuole A primitive organ of excretion in the cytoplasm of single-celled organisms, consisting of a liquidfilled cavity which shows periodic contraction.

Convolvulin A glycoside of jalap root, yielding on hydrolysis 4 mols. glucose, 2 mols. rhamnose and convolvulinic acid or 3:12-dihydroxyhexadecanoic acid; used as purgative.

Copaiba An oleo-resin of Copaifera which on account of its antiseptic powers has been used in the treatment of gonorrhea.

Coproporphyrin $C_{36}H_{36}O_8N_4$; a porphyrin present in urine and feces and in larger amounts in certain pathological conditions including pernicious anemia and hemolytic jaundice; an isomer is excreted in lead poisoning and forms of liver degeneration.

Coprosterol $C_{27}H_{48}O$, m.p. 102°; a saturated sterol found in feces. H. S.

Copulation Introduction of male sex cells (*sperm*) into the female reproductive tract through sexual intercourse; performed by animals in which fertilization occurs internally.

Coral Colony of tiny individual *invertebrate* marine animals called polyps that secrete external skeletons made of calcium carbonate (limestone).

Coramine Pyridine-ß-carboxylic acid diethylamide; yellow liquid; used in cardiovascular collapse, bronchial asthma as a respiratory stimulant instead of camphor; dose, 1-2 cc. of a 25% solution orally, subcutaneously or intravenously.

Corbasil β-hydroxy-β1-(3, 4-dihydroxyphenyl)-isopropylamine; a preparation related to ephedrine and epinephrine.

Cori ester Alpha glucopyranose-phosphoric acid.

Corpora striata A pair of brain ganglia serving as pathways for nervous impulses between higher and lower centers.

Corporative Differentiation due to the physiological functioning of parts, as opposed to *Individuative*, due to the action of morphogenetic fields.

Corpus callosum Bands of nerve fibres bridging both halves of the mammalian brain.

Corpus luteum The second stage of the development of the mammalian Graafian follicle of the ovary after it has ruptured and discharged an ovum. It secretes the hormone progesterone which is of prime importance in the regulation of pregnancy and its sequellae.

Cortate A commercial preparation of synthetic desoxycorticosterone acetate.

Corticosterone An active constituent of cortin, the

natural hormone of the adrenal cortex; it is $C_{21}H_{30}O_4$ or Ä4-pregnene-11,21-diol-3,20-dione; measured by its ability to counteract fatigue of adrenalectomized rats, and other effects of adrenalectomy.

Corticotropin A hormone of the anterior pituitary, to which has been ascribed the power of controlling the adrenal cortex secretion.

Cortilactin A hormone of the adrenal cortex, which apparently has some effect on the mammary glands; its nature is not known.

Cortin The hormone or hormone mixture of the adrenal cortex, the absence of which is fatal. To a large extent it controls the physical and mental sexual character. It also controls the Na/K ratio of the blood, and probably a number of other functions.

Cortipressin A hormone of the adrenal cortex which causes a rise in blood pressure; its chemical nature is not known.

Corybulbine $C_{21}H_{25}O_4N$; colorless needles, m.p. 238°; alkaloid of Corydalis; causes drop in blood pressure in mammals.

Corycavamine $C_{21}H_{21}O_5N$; rhombic columns, m.p. 149°; alkaloid of corydalis root; stereoisomeric with corycavine, and has a similar physiological activity.

Corycavine $C_{21}H_{21}O_5N$; rhombic tablets, m.p. 217-218°; alkaloid of Corydalis roots stereoisomeric with corycavamine; in frogs causes narcosis and spinal cord paralysis; in mammals stimulates tear and saliva flow and causes epileptiform convulsions; also interferes with the heart action.

Cosmid A type of artificially constructed vector used for cloning 35-45 kb of DNA. These are plasmids carrying a phage l cos site (which allows packaging into l capsids), an origin of replication and an antibiotic resistance gene. A plasmid of 40 kb is very difficult to put into bacteria, but can replicate once there. Cosmids, however, have a cos site, and thus can be packaged into l phage heads (a reaction which can be performed *in vitro*) to allow efficient introduction into bacteria.

Cotarnine $C_{12}H_{15}O_4N$, oxidized narcotine; used medicinally to check hemorrhage; trade name "stypticin" and as phthalate "styptol"; also used as ointment for shingles, eczema, etc.

Cottrell precipitation The precipitation of colloids by passing the colloid past oppositely charged plates; also called electrical precipitation, especially for dispersions of solids in gates.

Coumarin The flavor substance of the Tonka bean; odorless; the anhydride of coumaric acid made from salicyl aldehyde by closure with acetic anhydride; $C_9H_6O_2$ m.p. 70°.

Coumarine glycosides Compounds of glucose with derived coumarins, found in many plants, e.g. aesculin, daphnin, fabiatrin, fraxin and scopolin.

Coupling of oxidation processes The biological coupling of processes which enable a process requiring much energy to take place, eg. aerobic oxidations with the production of urea in the ornithine cycle which requires energy from an outside source.

Cozymase (Coenzyme I) A co-enzyme of apozymase, without which zymase preparations do not have the power to ferment. It is a mononucleotide of molecular weight about 350, containing nicotinic acid amide, pentose, adenine and phosphoric acid. It probably combines with a specific protein of zymase to become an active enzyme.

Cranium Region of vertebrate skull that covers and protects the brain

Creatine and creatinine metabolism 1. Definitions. Creatine is methyl guanidine acetic acid and creatinine is its anhydride, or methyl glycocyamidine.

Total Creatinine is the total amount of creatine and creatinine present in body tissues and fluids.

Preformed Creatinine is the difference between the total creatinine and creatine in body tissues and fluids, or if no creatine is present, the amount of creatinine, i.e., in the urine.

Creatine "As Creatinine" is the difference between the total and preformed creatinine in the urine.

Creatine is the difference between the total and preformed creatinine in the urine x 1.16.

"Apparent" Creatine and Creatinine is the respective amounts of chromogenic material in body tissues and fluids which are given by the Jaffé reaction with alkaline picrate. (This is the usual meaning of the terms creatine and creatinine expressed in the literature).

"True" Creatine and Creatinine is that amount of the Jaffé reaction with alkaline picrate which is destroyed by the specific creatinine enzyme of Miller and Dubos (i.e., 80 per cent of the chromogenic color may be due to true creatine or creatinine and 20 per cent to some other substance, such as glycocyamidine, hydantoin, acetoacetic acid, etc. The specific creatinine enzyme will destroy the 80 per cent of true creatine or creatinine but not the other 20 per cent).

"Residual" Creatine and Creatinine is that part, (i.e., the 20 per cent referred to above) of the Jaffé reaction with alkaline picrate which is not destroyed by the specific creatinine enzyme of Miller and Dubos.

Creatine Phosphate oçcurs in mammalian and heart muscle and in nerve tissue. Arginine phosphate occurs in invertebrate muscle.

Creatinemia is an increase in the creatine content of the blood above 3 mg. per 100 cc.

Creatininemia is an increase in the creatinine content of the blood above 1.5 mg. per 100 cc.

Creatinuria is the presence of extra creatine in the urine per day.

Creatininuria is the presence of extra creatinine in the urine per day.

Creatine Tolerance is the amount of creatine retained by the tissues after ingesting 1.32 gms. of creatine.

Creatinine Clearance is the number of cc. of plasma cleared of creatinine per minute after ingestion of 5 gms. of creatinine.

Creatinine Coefficient is the mg. of creatinine excreted per day per kilo body weight.

Distribution. Creatine occurs in the muscles, heart and nerve tissues of various species of animal and man. It is the chief nitrogenous constituent of muscle and its concentration will average about 0.4 per cent. About three-fourths of the creatine in muscle tissue and one-half in heart muscle is present as creatine phosphate. The creatine concentration in the blood will average about 2.5 mg./100 cc.

Creatinine occurs only in traces in the tissues. Its excretion in the urine of different individuals may vary from 0.8 to 2.5 gms. About 1.5 gm. are excreted daily in normal adult urine. Its excretion is not necessarily constant but uniform, from day to day, and is dependent upon the rate of protein metabolism rather than upon the body weight of the individual. The creatinine concentration in the blood is about 1 mg./100 cc.

Baker and Miller showed that true creatinine comprises from 80 to 100 per cent of the chromogenic material of serum and plasma and from 30 to 50 per cent in the red cells. In the urines from normal and nephritic individuals, true creatinine constituted nearly 100 per cent of the chromogenic material. Most of the chromogenic material of muscle filtrates is true creatine. In skeletal and heart muscle, testes and brain of the rat, the true creatine averages from 83 to 96 per cent of the apparent creatine, while, in intestinal muscle, pancreas, spleen, lung and liver the true creatine amounts to only 16 to 65 per cent of the apparent creatine.

Cremer reactions for phloridzin I. Phloridzin yields an amorphous difficultly solubre precipitate upon treatment with benzoyl chloride and sodium hydroxide solution. II. A solution of phloridzin in 0.1% sodium carbonate forms a red dye with benzene diazonium chloride solution, III. A red colored substance is obtained when phloridzin is treated with vanillin and hydrochloric acid.

Cretinism Stunted physical and mental development due to failure of thyroid to develop during fetal life, characterized by lowered basal metabolism.

Critical mass (physics): minimum mass of reactants required to maintain a chain of nuclear reactions involving the splitting of atomic nuclei (plural of *nucleus*)

Critical temperature, lower Level of environmental temperature below which the metabolic rate of homeotherms increases with decreasing environmental temperature and above which their metabolic rate is essentially independent of changes in environmental temperature.

Critical temperature, upper Level of environmental temperature above which an increase in environmental temperature has a thermogenic effect.

Cross-bevan reagent for cellulose Cellulose is soluble in a 30% solution of zinc chloride in concentrated (sp. gr. 1.19) hydrochloric acid.

Crossing over The breaking during meiosis of one maternal and one paternal chromosome, the exchange of corresponding sections of DNA, and the rejoining of the chromosomes. This process can result in an exchange of alleles between chromosomes. Compare recombination.

Cryoscopy Determination of the properties of substances by freezing points of mixtures.

Cryptal A terpene aldehyde found in various eucalyptus oils, $C_{10} H_{16} O$.

Cryptogams Plants that produce no true flowers or seeds; propagate by spores, e.g. ferns.

Cryptogams Plants that are not seed bearing plants, as ferns, mosses, fungi.

Crysophanic acid 4,5-dihydroxy-2-methylanthraquinone; chrysophanol; m.p. 196°, found in cascara sagrada, rhubarb, senna, etc.; used medicinally as Chrysarobin for skin diseases.

Crystals, liquid Mesomorphic state; a melt or solution of a crystalline substance which shows orientation, as by double refraction, indicating a shaping or grouping of molecules along different axes which do not show the same optical, magnetic or electrical properties in all directions.

Cube Solid with six sides of equal area and 12 edges of equal length.

Cubebs Dried unripe fruits of Piper cubeba which yield a volatile oil, used as a urinary antiseptic.

Cupreine Hydroxycinchonine; ultraquinine; alkaloid of cuprea bark, m.p. 198°, half as toxic as quinine.

Curare Extract of Strychnos containing the alkaloid curarine, South American arrow poison, used to combat tetanus in non-lethal doses.

Curine An alkaloid which lowers the blood pressure and paralyzes the heart.

Cyanide-nitroprusside reaction A test for the disulfide (-S-S-) group. The solution to be tested (sometimes saturated with ammonium sulfate) is treated with sodium cyanide and fresh nitroprusside and then made alkaline with ammonium hydroxide. An intense purple color is positive. It is caused by sulfhydryl groups, which are produced from the disulfide groups by the action of cyanide. When sulfhydryl groups are present as such, cyanide treatment is not necessary to cause a positive test (Nitroprusside Reaction).

Cyanin A pigment found in the cornflower and rose; it belongs to the group of anthocyanins, and gives on hydrolysis 2 molecules of glucose and one of cyanidin. Other cyanidin anthocyanins are: chrysanthemin or asterin of the chrysanthemum, aster and various berries and peonin from the peony.

Cyanophoric glycosides Glycosides which yield hydrocyanic acid on hydrolysis, e.g. amygdalin.

Cyclitols A group of carbohydrate-like substances consisting of polyhydroxycyclohexanes, examples of which are inositol and quercitol.

Cyclol theory A theory of protein structures based largely on mathematical probabilities, X-ray studies, and studies of unimolecular protein films. It speaks of a series of triazine hexagons, built into one another in such a way that "every triazine hexagon is surrounded by three of the triazine hexagons joined by a single diazine hexagon."

Cyclopropane Anesthetic used with oxygen, of low cardiac toxicity, but dangerous in use for lack of warning of breakdown.

Cymarin $C_{30} H_{44} O_9$; a cardiac glucoside which occurs in Canadian hemp and belongs to the digitalis group. On hydrolysis it yields strophanthidin and cymarose, m.p. 130°. Similar glycosides are: periplocymarin which hydrolyzes to periplogenin and cymarose, and sarmentocymarin which hydrolyzes to sarmentogenin and sarmentose.

Cymarose $C_7H_{14}O_4$; a sugar, 3-methyl-Ddigitoxose, obtained by the hydrolysis of cymarin, the active principle of Canadian hemp. It is a desoxy-sugar and has a m.p. 88°. Sarmentose is an isomer.

Cymenes Isopropylmethylbenzenes; $C_{10}H_{14}$; occur in many essential oils and in wood (especially p-cymene).

Cysteine HS.CH_2 CH(NH_2)COOH, an amino acid resulting from the reduction of cystine. It is presumably the carrier of the sulfhydryl groups of proteins.

Cystine metabolism Cystine, a component of most proteins, is a prominent constituent or precursor of the keratins, glutathione, tunilin, taurine, and taurocholic acid. Only plants can synthesize cystine from inorganic matter, (sulfates, etc.). However, cystine is not an essential constituent of mammalian diet although it is essential in their tissue synthesis because mammalian metabolism can produce cystine from methionine. In normal metabolism excess cystine is destroyed, being excreted chiefly as urinary sulfates.

Cystinuria An inborn error of metabolism where cystine is excreted in the urine and may be deposited in the form of kidney or bladder stones.

Cystitis Inflammation of the urinary bladder which takes on acute or chronic forms and may be of gonococcal or tubercular origin. Infections are commonly due to Escherichia coli, the staphylococcus, the streptococcus and more rarely proteus vulgaris. Among the predisposing causes are urinary stasis of any origin (calculi, faulty innervation, tumors, etc.). In cautious bladder irrigations or instillations only very dilute solutions of potassium permanganate, silver nitrate or mercuric chloride are used.

Cytase A group of enzymes which hydrolyze hemicelluloses, such as e.g. araban, mannan, xylan, galactan.

Cytidylic acid 3 - cytosine - 3 - phosphoriboside, a nucleotide found in ribonucleic acids.

Cytisine Ulexine, sophorine, baptitoxine, laburnine; from the seed of Cystissus laburnum; m.p., 152 - 153°; used as purgative, emetic, diuretic.

Cytochrome A class of hemochromogens, probably the main factor in respiration, widely distributed in aerobic cells, absent from obligate anaerobic cells. Oxidizable through mediation of Warburg's Ferment; do not combine with cyanide or carbon monoxide at physiologic pH values. Designated, according to position of absorption bands, as cytochrome a, a1, a2, a3, b, b1, b2, c, and c1. Reduced forms are reoxidized with the aid of cytochrome oxidase.

Cytochirome oxidase An enzyme which catalyzes the oxidation of ferro cytochromes by molecular oxygen; it does not oxidize p-phenylenediamine, is sensitive to HCN and H_2S and is probably identical with Warburg's respiratory enzyme.

Cytoflav A yellow water soluble pigment found in some animal tissues. It is riboflavin phosphoric acid ester. In combination with a specific protein it yields Warburg and Christian's "old" yellow enzyme.

Cytokinesis The actual splitting of a parental cell into two daughter cells; also called cytoplasmic division.

Cytosine 2-oxy-6-aminopyrimidine; a pyrimidine derivative found in the nucleic acids.

D

Dakin reagent for aldehydes and ketones Readily crystallizable, water-insoluble products are obtained when aldehydes or ketones are treated with a solution of p-nitrophenylhydrazine in 30 parts of 40% acetic acid. This is a modified Fischer reagent.

Dakin's theory of tyrosine metabolism Tyrosine is oxidized to p-hydroxyphenylpyruvic acid, which then splits open its ring, acetoacetic acid being formed; it is, therefore, strongly ketogenic.

Dalton Unit of atomic mass. One dalton corresponds to the mass of a hydrogen atom = $3.32X10^{-24}$ g.

Dambonitol Dimethyl ether of inositol, found in various latices.

Database Search Once an open reading frame or a partial amino acid sequence has been determined, the investigator compares the sequence with others in the databases using a computer and a search algorithm. This is usually done in a protein database such as PIR or Swiss-Prot. Nucleic acid sequences are in GenBank and EMBL databases. The search algorithms most commonly used are BLAST and FASTA.

De coninck reaction for phenylglycine An aqueous solution of phenylglycine is treated successively with equal volumes of concentrated sulfuric acid twice at an interval of a few moments. The heat evolved causes the formation of a violet color gradually changing to brown at the bottom of the vessel; simultaneously an odor of bitter almonds is perceived and two layers form, the upper being colorless, then turbid and finally yellow..

Deaminase Any one of a group of enzymes which remove an amine group from a compound.

Deamination Metabolic removal of amino group from amino acids and the like.

Decholin A trade name for sodium dehydrocholate, injected in diseases of the liver and bile passages.

Decidua Mucous membrane of uterus which hypertrophies toward the end of menstruation or birth.

Declination Measure of the angular distance north or south from the celestial equator to a given point on Earth's surface

Decomposition Process by which matter breaks down into its constituent parts; in biology; decomposition refers to the breakdown of dead organisms

Decose Any monosaccharide with the formula $C_{10} H_{20} O_{10}$; there are none naturally occurring.

Dedifferentiation (Einschmelzung, Catachony) Loss of differentiation, as of organelles by protozoa before cleavage, or of starving ascidians, or of the vertebrate limb stump in blastema formation.

Degeneracy In molecular biology, this term refers

to the fact that multiple different codons may encode the same amino acid. However, a given codon does not encode more than one amino acid within the nucleus of an organism.

Degrowth Decrease in spatial dimensions and weight of an organism.

Dehn-scot test for hippuric acid in the prescence of urea A few cc. of urine are treated with an excess of sodium hypobromite solution (bromine in sodium hydroxide) and the mixture heated to decompose the urea. Depending upon the concentration of hippuric acid, a smoky pale red color or an orange to brown-red precipitate is formed. **Dehydrocorydaline** $C_{22}H_{23}O_4N$; yellow crystalline powder, m.p. 112-113°; alkaloid of the roots of Corydalis; formed by oxidation of Corydaline; causes fall in blood pressure.

Dehydrogenase (1) An enzyme which activates the hydrogen of metabolites; it is aerobic, like xanthine dehydrogenase, when it can use molecular oxygen directly, or anaerobic, like lactic dehydrogenase, when it can use oxygen only through the intermediation of another system; experimentally the activity is shown as the ability to oxidize a given metabolite in the absence of air and reduce dyes of the proper potential.

(2) The complete enzyme system necessary for the activation of the substrate in cases where more than one component is involved.

Dehydrogenases (classification) 1. Aerobic, e.g. xanthine oxidase, d-amino acid dehydrogenase, tyramine dehydrogenase, uricase, which require neither coenzyme or cytochrome systems.

2. Anaerobic, which are: a. Cytochrome-linked, e.g. succinic and alpha-glycerophosphate dehydrogenases; b. Cozymase-linked, e.g. lactic, malic, beta-hydroxybutyric, citric, glucose, alcohol, triose phosphate, dihydroxyacetone, 1 (+) glutamic dehydrogenases and aldehyde mutase; c. Coenzyme II-linked, e.g. glucose and hexose monophosphate dehydrogenases.

Dehydrogenases Enzymes found widely distributed in the tissues which activate molecules of metabolites so that they can be oxidized by a hydrogen acceptor.

Dehydrogenases, classification of Green and Brosteaux classify them as:

(1) Aerobic oxidases — react with molecular oxygen, producing H_2O_2, e.g. uncase, xanthine oxidase;

(2) Cytochrome oxidase—react with oxygen indirectly through cytochrome, but not through flavoprotein, e.g. Warburg's respiratory ferment;

(3) Co-enzyme dehydrogenases—react with oxygen indirectly through flavoprotein, but not cytochrome, and require a coenzyme, e.g. lactic and malic dehydrogenases; 4. Miscellaneous, e.g. aldehydemutase, catalase.

Dehydrogenases, cytochrome reducing Enzymes which catalyze the oxidation of their substrates by hydrogen acceptors, as methylene blue, but not by molecular oxygen; especially active with the 3 cytochromes; include alphaglycerophosphoric dehydrogenase, succinic dehydrogenase, lactic dehydrogenases from yeast and Gonococcus, formic dehydrogenase (Bact. coli), choline dehydrogenase (liver).

Deletions (dna) Loss of chromosome segment, nearly always resulting in genetic disorder.

Delphinin chloride $C_{41}H_{39}O_{21}Cl.2H_2O$; a pigment belonging to the group of anthocyanins, which occur in the flowers of larkspur and many other plants. On hydrolysis it gives delphinidin chloride, 2 molecules of glucose and 2 molecules of p-hydroxybenzoic acid.

Denaturation With respect to nucleic acids, refers to the conversion from double-stranded to the single-stranded state, often achieved by heating or alkaline conditions. This is also called "melting" DNA. With respect to proteins, refers to the disruption of tertiary and secondary structure, often achieved by heat, detergents, chaotropes, and sulfhydryl-reducing agents.

Denaturing Gel An agarose or acrylamide gel run under conditions which destroy secondary or tertiary protein or RNA structure. For protein, this usually means the inclusion of 2-ME (which reduces disulfide bonds between cysteine residues) and SDS and/or urea in an acrylamide gel. For RNA, this usually means the inclusion of formaldehyde or glyoxal to destroy higher ordered RNA structures. In DNA sequencing gels, urea is included to denature dsDNA to ssDNA strands. In denaturing gels, macromolecules tend to be separated on the basis of size and (to some extent) charge, while shape and oligomerization of molecules are not important. Contrast with Native Gel.

Dendrite Thin filaments projecting from the body of a nerve cell and serving to transfer electrical signals from one cell to another.

Denigès methylglyoxal reagent 100 cc. water, 0.6 cc. bromine and 20 gm. 5% glycerine solution are heated in boiling water for 20 minutes, the excess bromine is boiled off and the solution evaporated to 100 cc. Add 20 cc. sulfuric acid to the cooled solution and distil 50 cc. The substance is mixed with 2 cc. sulfuric acid and 0.4 cc. of distillate giving characteristic colors with a number of substances.

Denigès micro-tests for creatine and creatinine

I. One drop of a saturated solution of about 1 mg. of substance in 1 drop of ammonia; a persistent orange color is obtained with creatinine. Creatine does not cause any color change. The orange color changes to yellow due to conversion of creatinine to creatine in the alkaline medium.

II. The substitution of a 1% solution of osmium nitroprusside for the picric acid gives rise to a blood-red color with creatinine and a faint rose color with creatine.

Denigès reaction for dihydroxyacetone 0.4 cc. of an aqueous solution of dihydroxyacetone (not more than 0.1%) are mixed with 0.1 cc. 4% potassium bromide, treated with 2 cc. sulfuric acid and 0.1 cc. 5% alcoholic guaiacol solution and heated on a steam bath; a blueviolet color, having an orange absorption spectrum is obtained. The use of salicylic acid in place of guaiacol gives rise to a rose-red color with a yellow and blue absorption spectrum; with gallic acid the color reaction is like that of guaiacol.

Denigès reaction for glucuronic acid 0.4 cc. of an aqueous solution of the acid is treated with 0.1 cc. of a 5% alcoholic solution of codeine and 2 cc. concentrated sulfuric acid and heated on a water bath for 5 minutes; a reddish-purple color, with an absorption band in the yellow and green spectrum, is obtained.

Denigès reaction for tyrosine 1-2 drops of tyrosine solution are added to a mixture of 3-5 drops of a 33% alcoholic solution of acetaldehyde and 2 cc. of sulfuric acid, producing a red color even in dilutions of 1.10000. 1 cc. formaldehyde in 50 cc. sulfuric acid may be used; warming the solution to 50-60° with tyrosine gives a rust-brown color gradually changing to reddish. Boiling the solution with twice its volume of glacial acetic acid causes a color change to green.

Denigès reactions for dihydroxyphenylalanine

I. A brown color is produced when dihydroxyphenylalanine is dissolved in dilute sodium hydroxide. The solution also reduces an ammoniacal solution of silver nitrate.

II. A solution of the substance is shaken with lead dioxide and filtered, yielding a filtrate which is an intense reddish brown.

III. One drop of ammonia is added to a suspension of a few particles of substance in 5 cc. of water, followed by the addition of 10 drops of hydrogen peroxide and heated to boiling; a light brown color develops.

IV. A suspension of the sample in 2 drops of water is treated with 1 drop of sulfuric acid, shaken, and 2 cc. more of acid added plus 1 drop of formaldehyde, giving rise to an intense violet color.

Denigès reactions for pyrrole I. A mixture of 5 cc. 0.01% pyrrole solution, 0.3 cc. 5% sodium nitroprusside solution and 1 cc. sodium hydroxide becomes greenish-yellow, then green. A blue color is obtained when the solution is boiled and treated with 1 cc. glacial acetic acid.

II. The addition of 2 cc. hydrochloric acid to the mixture used above gives rise to a red color on boiling.

III. A mixture of 5 cc. alcoholic pyrrole and 5 cc. concentrated hydrochloric acid, on being treated with an alcoholic vanillin solution, forms a reddish-yellow color.

Denigès reagent for aldoses and ketoses Twenty cc. of a solution of 10 gm. crystalline sodium acetate in 5 cc. glacial acetic and 100 cc. water are mixed with 3 cc. glacial acetic acid followed by the addition of 1 cc. phenylhydrazine and 1 cc. 10% sodium bisulfite solution to form a very stable reagent

Denigès test for indole 5 cc. of an 0.2% alcoholic solution of vanillin and 3 cc. concentrated hydrochloric acid are added to 5 cc. of the alcoholic test solution producing an eosin to garnet color with an absorption spectrum in the green and blue. A yellow color is obtained when cinnamaldehyde is used in place of vanillin.

Denigès test for phenolic alkaloids The following colors are produced when the substances are treated with a solution of titanium in hydrochloric or sulfuric acid: apomorphine, violet-red; cupreine, bright red; hordenine, dark orange-yellow; morphine, blood red; oxydimorphine, wine-red. No reaction is given by alkaloids without phenolic groups.

Denigès test for sulfanilids A mixture of several mg. of substance and 0.02-0.03 gm. of potassium ferrocyanide are gently heated for 20-30 seconds in a dry tube; an odor of mercaptans is a positive test.

Denigès tests for acetone I. 0.1 cc. of acetone are heated with 10 cc. bromine water for 20 minutes and the excess bromine is then boiled off. 5 cc. of the bromoacetone solution are boiled with 1:100 sodium carbonate solution and 0.4 cc. of this reaction product are treated with 0.1 cc. 1:10 potassium bromide solution, 2 cc. concentrated sulfuric acid and 0.1 cc. alcoholic salicylic acid solution giving rise to a violet to red-violet color at room temperature. Use of guaiacol for salicylic acid produces a blue color.

II. The test solution is treated with an equal volume of reagent (5 gm. mercuric oxide dissolved in 20 cc. of concentrated sulfuric acid and 80 cc. water) and heated on the water bath, a turbidity or precipitate being produced if acetone is present.

Denigès tests for mercaptans An alcoholic solution of a mercaptan is added to a few cc. of reagent (1 gm. of isatin in 100 cc. concentrated sulfuric acid) previously diluted with several volumes of concentrated sulfuric acid; a green color is produced. Aldehydes and higher alcohols are removed, if present, by shaking with sodium nitroprusside; a red color forms.

Denigiès reaction for cinchona alkaloids 0.2 gm. of alkaloid (cinchonidine, cinchonine, cupreine or quinine) are dissolved in 2 cc. glacial acetic acid and 2 cc. sulfuric acid added, producing a fluorescence. The addition of 2 cc. formaldehyde yields a bluish green fluorescence with cupreine and quinine; blue-violet fluorescence with cinchonidine and a blue fluorescence with cinchonine. The cupreine fluorescence disappears with the addition of 3-4 cc. of water.

Denitrification The removal of nitrogen from inorganic nitrates, nitrites and hyponitrites by bacterial action.

Dependent differentiation (Correlative or abhängige Differenzierung; Différentiation provoquée). The differentiation of a tissue partly as a response to factors existing outside itself, i.e. differentiation which requires the morphogenetic stimulus of an inductor from a neighbouring part of the embryo.

Desaminocoenzyme A reaction product of coenzyme I with nitrous acid, which shows some of the activity of the original coenzyme.

Desmotrypsin The inactive tryptic material left in the pancreas after glycerol extraction. It is activated to trypsin by enterokinase. There are apparently 2 fractions, an a-fraction, lyotrypsin, soluble in electrolytes a 6-fraction, soluble only in dil. HC1 or Na_2CO_3.

Desoxycholic acid $C_{24}H_{40}O_4$; m.p. 176°; a bile acid found in man, ox, goat, sheep, deer and antelope. It is 3, 12-dihydroxy-cholanic acid. Also deoxycholic acid. H. S.

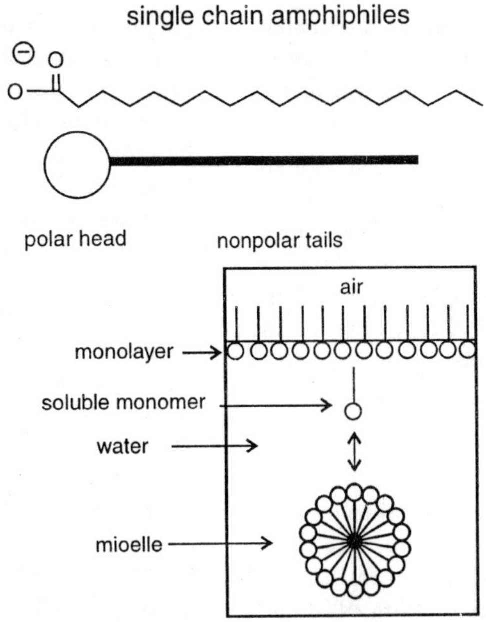

Fig. Single chain Amphiphiles

Desoxycorticosterone acetate A synthetic preparation used instead of natural adrenal cortex hormones in diseases associated with hypo-adrenal cortex function, as Addison's disease. The synthetic preparation does not always substitute completely for the natural. Even small doses may affect the circulatory system, causing hypertension and cardiac decompensation.

Desoxy-sugars Sugars in which one or more hydroxyl groups are replaced by hydrogen, e.g. 2-desoxy-d-ribose of animal nucleic acid.

Des-thio-biotin Biotin without sulfur, the place of the sulfur being taken by two hydrogen atoms; is oxidizable to pimelic acid.

Determination. The fixation of the fate of a part of an embryo, so that (if it develops further) it will perform one kind of morphoor histo-genesis, and one kind only, irrespective of the situation in which it may find itself. In using this concept one must bear in mind that all possible situations have usually not been tested; hence the word must be handled with due regard to the limitations of our knowledge. Ideally, the situations should be specified in every case, but definition will continue to be the victim of convenience until some rapid notation is devised. The term should not be used for the status of the tissue following the process, and it should have reference to an immediately subsequent developmental change. It must be remembered that determination is embryological, not genetical. This is seen with special force when ontogenetic stimulusresponse systems are considered in which the stimulator and the reactor are of genetically different tissue. The sense in which the word "determination" is here used must be distinguished from other uses, e.g.

(1) the "determination" of the original axes of the egg by external factors.

(2) the "determinate" cleavage of eggs in which the cleavage pattern is very definite and permits of accurate descriptions and predictions of cell lineage.

(3) "determinants", hypothetical units in the germ-plasm.

Detergent Term applied to materials, the solution of which aid in the removal of dirt or other foreign matter from contaminated surfaces. A common feature of detergent is that they are made up of amphiphilic molecules.

Detoxication Detoxication is the term applied to the chemical reactions employed by the organism to render toxic agents innocuous or to facilitate their elimination. In a broader sense, however, the detoxication reactions include all the chemical changes that occur in a foreign compound taken into the body. Toxins are of two types:

(1) those produced by the bacteria of the intestines and

(2) harmful foreign substances taken into the body, usually by the oral route. The main products of intestinal putrefaction are amines from the decarboxylation of amino acids; aromatic acids from the deaminization of cyclic amino acids; and phenols which are likewise the end products of bacterial action on amino acids. Specific examples are putrescine from arginine, cadaverine from lysine, tyramine from tyrosine, histamine from histidine, phenylacetic acid and benzoic acid from phenylalanine, phenol from tyrosine, and skatole and indole from tryptophane. In the detoxication the noxious agent is either destroyed or the chemical groups responsible for its physiological activity are removed or inactivated. The end products usually possess a higher acidity than the original compound. It appears likely therefore that an important factor is the conversion of a weakly acidic compound which the organism excretes with difficulty to a strong acid which the kidney eliminates with greater ease. Examples are the formation of hippuric acid, a strong acid with an ionization constant of $2.3 \times 10,_{-4}$ from benzoic acid with $CH=2.3 \times 10_{-5}$; and phenol sulfuric acid from the weakly acidic phenol. It appears likely that the organism employs the same mechanism for detoxication as it does for normal metabolic processes. The conjugation of cholic acid with glycine to form the bile acid, glycocholic acid, is very probably accomplished by the same mechanism which combines glycine and benzoic acid to form hippuric acid. Glucuronic acid which was heretofore considered almost solely as a detoxifying agent is now known to occur combined with various sex hormones. Occasionally the end products are physiologically more potent than the primary substances. Some of the acetylated derivates of sulfanilamide appear to be more toxic than the parent compounds. Oxidation is the most important reaction employed by the organism for the destruction of poisons. Many organic compounds of which ethyl alcohol is a typical example are completely oxidized to carbon dioxide and water. Phenyl-substituted fatty acids undergo ß oxidation, i.e., oxidation occurs on the ß carbon atom and the side chain is progressively reduced by 2 carbon atoms. Ultimately therefore the acids with an even number of carbon atoms yield phenylacetic acid and the odd number, benzoic acid. The aromatic ring itself is fairly resistant to oxidation. Benzene is slowly converted to phenol, polyphenols, and to a small extent to muconic acid. The introduction of a carboxyl group in the aromatic nucleus greatly reduces the toxicity. Thus, the adding of the COOH group to phenol yields the relatively non-toxic hydroxy benzoic acids.

Reduction as a detoxication reaction is unimportant. A few isolated examples can be found such as the reduction of chloral hydrate to trichloro-ethyl alcohol which is conjugated with glucuronic acid. Hydrolysis is occasionally employed in protecting the body against poison. The various glucosides including the digitalis compounds are split into glucose and the aglycone as a preliminary step in their destruction. Conjugation of compounds foreign to normal metabolic products is widely employed by the body. The coupling of aromatic acids with glycine is a reaction observed even in the lower vertebrate forms such as the frog. Since the body can synthesize glycine readily, large amounts of aromatic acids can be conjugated. The production of hippuric acid from benzoic acid is the best known example of this type of detoxication. Since the liver is an important factor in this conjugation, the rate of synthesis of hippuric acid has become a widely employed test of liver function. The presence of a group ortho to the carboxyl inhibits the glycine conjugation. This ortho blockage which is a biological type of steric hindrance offers a reasonable explanation for the commonly observed physiological potency of ortho-substituted compounds. The list of substances combined with glycine is very large and includes such interesting compounds as nicotinic acid which is now recognized as an important vitamin. Glucuronic acid is another important detoxifying agent. It can be produced in large amounts presumably from glucose and the glucogenetic amino acids. The compound can combine with aromatic acids, phenols, and alcohols. In the dogs, about two-thirds of the ingested benzoic acid is coupled with glucuronic acid. The free or uncombined acid is metabolized with difficulty and cannot be utilized for detoxication. Sulfuric acid is conjugated with phenols to form the ethereal sulfates normally found in the urine. Indican, the potassium salt of indoxyl sulfuric acid is an interesting example of this type of detoxication product. Sulfur in the form of cysteine is also employed for the detoxifying of naphthalene and various halogenated aromatic hydrocarbons such as bromobenzene. The cysteine molecule attaches itself to the benzene ring through the sulfur atom, and the resulting compound is then acetylated. The end product is known as a mercapturic acid. Acetic acid is employed by the body for the conjugation of compounds with amino groups. The acetylation of the sulfanilamide compounds is of both theoretical and practical importance. Marked variations in the acetylation reactions in different species occur. Thus p-aminobenzoic acid which is acetylated in the organism of man and the rabbit, is conjugated mainly with glucuronic acid in the dog. Certain conjugations are strictly limited to particular species. Glutamine is conjugated with phenylacetic acid only in man and in the chimpanzee. Ornithine is employed by the avian organism for the detoxication of benzoic acid. The conjugation mechanisms are no longer considered solely as detoxication reactions. The fact that various hormones have been isolated as conjugated compounds emphasizes the importance of these reactions in normal metabolism. A study of the behavior in the body of compounds not completely metabolized but leaving readily recognized end products have been exceedingly helpful in yielding important information concerning intermediary metabolism.

Development The process by which a multicellular organism is produced from a single cell.

Dextrinase An uncommon name for an enzyme responsible for the partial hydrolysis of starch.

Dextrins Intermediate hydrolysis products of starch. They are highly dextrorotatory, water soluble, and are precipitated by alcohol.

Dextro (d-) rotatory The rotation of the plane of polarized light to the right shown by solutions of certain optically active compounds.

Dhurrin $C_{14}H_{17}NO_7$; A glycoside of Sorghum Volgare; consisting of glucose and p-oxymandelonitrile.

Diabetes insipidus A relatively rare disease involving the excretion of large amounts of pale urine of low specific gravity, traceable to hypofunction of the pituitary and not to disease of the kidney. There is no sugar in the urine but there are signs of water intoxication. The blood pressure raising component, pitressin, of the posterior part and intermediate part of the pituitary, is used therapeutically by injection or as nasal spray or snuff.

Diabetes mellitus Hypo-insulinism; hyperglycemia

and glycosuria due to impairment of insulin secretion by the cells of the Islets of Langerhans of the pancreas. The onset is gradual and is supposed to be determined by both sedentary conditions of life and those of extreme emotional strain. Symptoms include polyuria (frequent urination), polydipsia (excessive thirst), polyphagia (excessive appetite) and yet loss in weight and strength. Itching (pruritus) is often present. Numerous tests have been devised for sugar in the blood and urine, for the tolerance of dextrose and the conditions of acidosis and ketosis marked by the appearance of "acetone bodies" in the urine. Treatment attends chiefly to diet. Numerous food tables, test diets, specific diabetic diets have beendeveloped, directed toward desugarization, maintenance and prevention of ketosis and acidosis. In all cases the strain on the pancreas has to be reduced. Consequently the administration of insulin is a very successful palliative, which, however, has to be watched for the prevention of insulin shock. This can be countered by the prompt administration of sugar and in extreme cases, the injection of epinephrine which causes the liberation of sugar. Sometimes both insulin and carbohydrate are administered especially if boils and carbuncles appear.

Diabetogenic principles Two diabetogenic principles, the blood-sugar-raising principle and the ketogenic principle. The former is antagonistic to insulin by possibly acting on nerve centers controlling carbohydrate metabolism, the latter increases the production of acetone bodies, especially beta-hydroxybutyric acid. The latter is also supposed to depress the metabolic rate, reduce blood lipids and increase the specific dynamic action of proteins.

Diacetyl Diketobutane; found in bay oil, butter, etc., b.p. 88°; also formed by special fermentation of glucose via methylacetyl carbinol; carrier of the aroma of butter and other foods.

Diacetyl reaction A test for proteins containing arginine. A dilute solution of the material to be tested is made alkaline and then a drop of diacetyl is added. A pink color with a green fluorescence is positive.

Diaciduria The appearance of dicarboxylic acids in the urine after feeding monoacid saturated glycerides.

Dialectric constant A specific property of substance defined by D in the equation: $f=Q_1 Q_2/rD$ f is the force of attraction or repulsion between two charges Q_1 and Q_2, separated by a thickness "r" of the substance. The dialectric constant of air is taken as unity.

Dialysis The separation of colloids from crystalloids, by utilizing the fact that the latter diffuse through a semi-permeable membrane. As the distinction between the colloidal state and the crystalloidal state is purely arbitrary, we find that borderline colloids diffuse to some extent, and that borderline crystalloids diffuse very slowly.

Diamine oxidase An enzyme of animal tissues which catalyzes the oxidation of histamine by molecular oxygen with formation of peroxide. May be a flavoprotein. Attacks putrescine, cadaverine and agmatine, but not ethylene diamine.

Diapause A state of dormancy in the normal development of numerous animals, especially insects, and distinguished from hibernation in that it must be passed through, no matter what the temperature or other external conditions may be.

Diapedesis The penetration of walls of blood vessels by cells leaving the cell walls intact, assumed to be due to a fluctuation of gel-sol condition of the cell wall.

Diastase One of a group of enzymes which convert glycogen, starch, or dextrins to reducing sugars. Also called amylase.

Dideoxy Sequencing Enzymatic determination of DNA or RNA sequence by the method of Sanger and colleagues, based on the incorporation of chain terminating dideoxynucleotides in a growing nucleic acid strand copied by DNA polymerase or reverse transcriptase from a DNA or RNA template. Separate reactions include dideoxynucleotides containing A,C, G, or T bases. The reaction products represent a

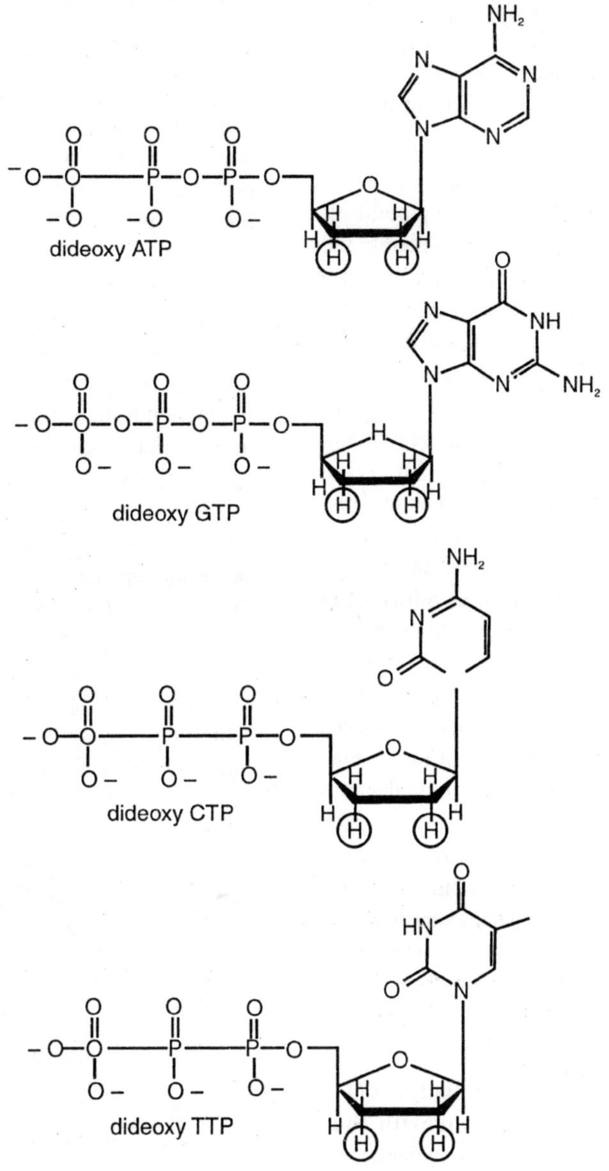

Fig. Didexoynucleotides[Dideoxy Sequencing]

collection of new. labeled DNA strands of varying lengths, all terminating with a dideoxynucleotide at the 3' end (at the site of a complementary base in the template nucleic acid), and are separated in a polyacrylamide/urea gel to generate a sequence "ladder". This method is more commonly used than "Maxam-Gilbert" (chemical) sequencing.

Differentiation centre One of the two organisation centres in the insect egg

Differentiation Increase in complexity and organisation. Increase in number of kinds of cells; invisibly, as by determination and losses of competence; visibly, as by histogenesis. Increase in morphological heterogeneity by the appearance of form and pattern.

Diffraction Apparent modification undergone by light when it passes through a narrow slit or over the edge of an opaque object; the result is a pattern of light and dark bands.

Diffusion potential When the oppositely charged ions of an electrolyte, having different mobilities, diffuse through a solution or across a membrane because of concentration differences, the mobility difference results in a separation of charge. This in turn produces electrical potential differences, known as diffusion potentials, which act to accelerate the slower and decelerate the more rapid ions so as to reduce the charge separation. Anomalous accumulation of ions in living systems (e.g., potassium in many types of cells) is sometimes attributed to this effect. However, ionic mobilities affect their distribution only during the nonstationary period of adjustment. The distribution at equilibrium, when there is no net flow, is independent of mobilities. The sole exception to this is the case of ions which participate in chemical reactions, or are adsorbed or desorbed by other components of the system. In this case the equilibrium distribution depends on the ratio of reaction rate (or adsorption rate) to the mobility. A theory of this sort is that of Theorell (Proc. Nat. Acad. Sci.,) which supposes an anion produced by cellular metabolism and diffusing out of the cell. It thus is supposed to set up a diffusion potential which acts on Na and K. These have different mobilities, and so K will be preferentially accumulated.

Digestion All living organisms essentially maintain their status quo in equilibrium with their environment. This equilibrium is dynamic and involves a continued exchange of energy and materials. The organism takes in potential energy locked in chemical molecules and, by bringing about reactions of these substances, releases and uses this energy. Some of the entering substances, or their reaction products, are also used directly to form part of the material of the body which receives them. The substances which serve both these uses are overwhelmingly the complex organic constituents of other bodies, plant or animal—the usual "foods." And these are mostly caused to react with oxygen to release energy or are partially split to simpler components which are then recombined to form body structures. Digestion has to do with the prehandling of food substances before they enter the body proper. The essential problem is this: foods are available to the organism as solid masses of substances, the large molecules of which are insoluble in water and cannot penetrate the membranes which surround all cells. Before food can really enter the body fluids or cells (in distinction from being ingested into some cavity, as a stomach, contained in the body but still connected with and essentially part of the outside world) the large masses must be broken up mechanically and the large molecules must be broken up chemically. This rendering soluble and diffusable of food material is the essence of digestion. The entire digestive apparatus of animals is organized for the mechanical and chemical splitting of foods. Independent plants, which use the sun's energy to build their body substance (the food of other organisms) from small, soluble and diffusible molecules—as water, carbon dioxide, nitrates, and other salts-neither need nor possess a digestive system. Dependent plants, (bacteria, Fungi, insectivorous plants, and the like) which use ready made food as do animals, do have and use digestive mechanisms. The number and elegance of the digestive devices of an organism vary roughly with its degree of specialization in general and with its position on the evolutionary scale. Bacteria have no special digestive organs or organelles. They simply secrete from their surface digestive substances, enzymes, which dissolve the food outside of the bacterium. The soluble digestive products then diffuse into the bacterial cell and are used. Protozoa usually sweep food particles and water into their bodies, at a fixed or a changed position; and these food vacuoles constitute temporary "stomachs" in which the digestion proceeds. Metazoa have permanent digestive organs, most elaborate in the mammals. Man's digestive system, or alimentary canal, is a continuous passage through the body, from mouth to anus, with special extensions into accessory secreting organs (glands). Mechanical digestion, depending on muscular movements, begins in the mouth with the action of the teeth and with chewing movements. The tongue and pharynx muscles then toss the chewed food into the gullet (esophagus), where other muscles pass it to the stomach. Here the muscular walls further churn and mix the food and digestive, juice for some hours, and pass the semifluid mixture along to the small intestine. Similar churning, mixing, and slow progression over houis occurs down the thirty odd foot length of the gut. By the end of this gamut, all available food has been fully digested and absorbed through the gut wall. The fluid debris then enters and slowly traverses the large intestine, or colon, from which water is reabsorbed. The remaining solid is finally evacuated as feces. Chemical digestion is far more important than mechanical digestion, it can, indeed, do the whole job alone. As muscles do mechanical digestion, so glands make and secrete the agents which carry on chemical digestion. Each gland pours into some part of the alimentary canal its special juice which contains one or more of the substances which split food molecules. These substances, enzymes, are very specific, each breaking certain chemical links in certain kinds of molecules. Saliva, discharged into the mouth by salivary glands, contains mucus to serve as lubricant and an enzyme, ptyalin, which splits starch molecules to simpler sugars. Gastric juice, formed in glands in the stomach wall, contains strong hydrochloric acid and the enzyme pepsin, which split proteins part way to their constituent amino acids. Into the upper small intestine empty the digestive juice of the pancreas and the liver's bile. The latter helps mix fat with water, and so aids in its digestion by other enzymes which are in water. Pancreatic juice is the most important and contains enzymes which digest all food types — trypsin further splits proteins, amylopsin completes starch digestion, and steapsin breaks the fats. Even the upper small intestine contributes special enzymes and enzyme activators. In all cases, in man or other animals, by simple or complex organs, insoluble food masses are comminuted mechanically and, more important, split chemically by enzymes until soluble fragment molecules are produced. Digestion is then over and absorption brings these fragments to the body cells for metabolism to begin.

Digitalis glycosides, cardio active Under the term of glycosides are included a great diversity of naturally occurring substances and of synthetic products, which are all characterized by their property of yielding various identifiable sugars, along with other chemical compounds, upon acid hydrolysis. Each sugar is, moreover, capable of forming glycosides in more than one way. The mode in linkage of the sugars with the other component or components, of the glycosides is illustrated, in its simplest expression, by the structure of some of the methyl glucosides formed by the reaction of glucose on hot methyl alcohol in presence of dry HC1 gas.

Dignin A digitanon composed of the pregnane derivative (Steroids) diginigenin, and the deoxysugar D. diginose. D. occurs together with cardial glycosiders in Digitalis, e.g. Digitalist purpurea, from which it was isolated in 1936 by Karrer.

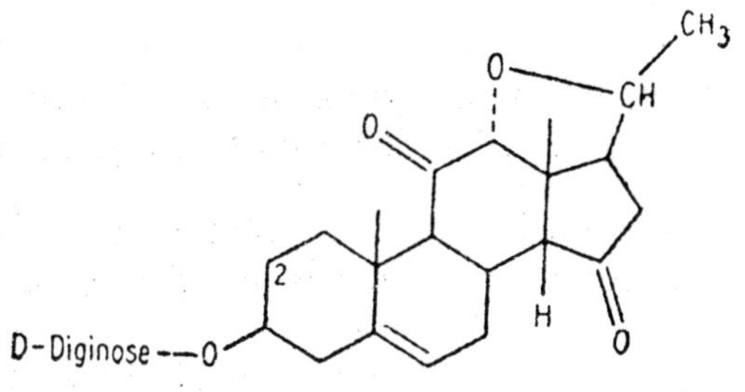

Fig. Dignin

Dihybrid Cross An experimental cross between two organisms, each of which breed true (is homozygous) for forms of two traits that are distinctly different from those displayed by the other organism. For each trait, the first generation offspring inherit a pair of nonidentical alleles.

Dihydroxyacetone m.p. 80° C., obtained in fermentations and in carbohydrate metabolism, as in the action of acetic acid bacteria on glycerol.

Dilaudid Dihydromorphinone hydrochloride, a synthetic made from morphine in order to cut down its harmful features.

Dille-kopanyi test for barbituric acid derivatives The aqueous solution of the substance (acidified with hydrochloric acid if necessary) is extracted with chloroform; 2 parts of the chloroform extract are treated with 0.1 part of a 1% solution of cobalt acetate in absolute methanol and 0.6 parts of isopropylamine in methanol. A red-violet color indicates a positive test which can be used quantitatively.

Dioscorine An alkaloid, $C_{13}H_{19}O_2N$, from the tubers of Dioscorea hirsuta, which is toxic, acting like picrotoxin.

Diphtheria A contagious disease caused by the Klebs-Löffler bacillus, characterized by the formation of false membranes on mucosae of the upper respiratory tract or the skin and a toxemia highly injurious to peripheral nerves and the heart. The Schick test for immunity by the injection of diphtheria toxin to produce an inflammatory reaction of the skin, may be of diagnostic aid. Varieties of diphtheria are pharyngeal, nasal, laryngeal (true membraneous croup), malignant diphtheria and diphtheria of various parts as the skin, genitalia, etc. There is great danger of complications, such as albuminuria, pneumonia, myocarditis, otitis media, neuropathy, secondary anemia. Various types of immunization have been developed as toxin-antitoxin, formolized toxoid, alum precipitated toxoid. The treatment leans heavily on antitoxin administration. Recovered patients frequently remain carriers of the organism.

Diploid life cycle Occurs when the only multicellular stage in an organism's life cycle is diploid.

Diploid A full set of genetic material, consisting of paired chromosomes one chromosome from each parental set. Most animal cells except the gametes have a diploid set of chromosomes. The diploid human genome has 46 chromosomes.

Direct induction The induction of a new neural axis in competent ventral ectoderm by a chemical compound acting in the same way as the naturally occurring evocator.

Direct Repeats Identical or related sequences present in two or more copies in the same orientation in the same molecule of DNA; they are not necessarily adjacent.

Disaccharides I. Reducing Disaccharides: Cellobiose is glucopyranose-4-ßglucopyranoside. Gentiobiose is glucopyranose-6ß-glucopyranoside. Isolactose is a glucose galactoside. Lactose is glucopyranose-4-ßgalactopyranoside. Maltose is glucopyranose-4-áglucopyranoside. Melibiose is glucopyranose-6-ágalactopyranoside. Turanose is ß-fructofuranose-6á-glucopyranoside.

II. Non-reducing Disaccharides: Isotrehalose is ß-glucose-ß-glucoside. Sucrose is á-glucopyranose-1-ßfructofuranoside. Trehalose is á-glucopyranose-1á-glucopyranoside.

Dismutations 1. (1) Coenzyme-linked reactions in which the same compound is both reductant and oxidant for the coenzyme, e.g. the formation of phosphoglycerate and alphaglycerophosphate from 2 molecules of triosephosphate in muscle glycolysis with the intermediation of Coenzyme I;

2. (2) In general, simultaneous oxidation and reduction, as in the Cannizzaro reaction.

Dispersal The scattering of organisms of a species, often following a major reproductive event. Spores and larvae are commonly dispersed into the environment. Pollen or gametes may also be dispersed, but in this case the intent is to target another individual so that reproduction may occur. Organisms may disperse as spores, seeds, eggs, larvae, or adults.

Dispersion Any mixture where one substance is

very intimately intermingled with another. Most frequently a dispersion refers to a colloidal suspension.

Dissociation constant For a dissociated compound at equilibrium, the product of the concentration of anion by the concentration of cation, divided by the concentration of undissociated compound, is a constant for a definite temperature. This is called the dissociation constant, or ionization constant.

Distributive operation Mathematical operation that yields the same result whether acting on a whole mathematical expression or acting on its parts individually and collecting the result.

District (Territoire) A minor morphogenetic field, possessing a general determination for production of a certain structure or organ, and undergoing progressive regional specification of detail, e.g. limb district or gill district.

disulfide bridge A covalent cross link between two polypeptide chains formed by a cystine residue (two Cys residues oxidized together).

Diversine $C_{16}H_{16}(NCH_3)(OCH_3)_2(OH)_2CO$; alkaloid; m.p. 80-93°; increases reflexes; vasoconstrictor; paralyzes peripheral motor nerve apparatus; lethal action through respiratory paralysis.

Djenkolic acid A sulfur containing dicarboxylic amino acid of velvet beans, possibly a minor constituent of other proteins.

D-lysine *a-å*-diamino caproic acid; i.p. 9.52; m.p. 224° (decomposes); an indispensable basic amino acid found in many proteins. It is a strong base and absorbs CO_2 from the air.

DNA (deoxyribonucleic acid) The molecule responsible for storing and transmitting genetic information. DNA is a double- stranded molecule held together by weak bonds between base pairs of nucleotides twisted around each other in the shape of a double helix. The four nucleotides in DNA contain the bases: adenine (A), guanine (G), cytosine (C), and thymine (T). In nature, base pairs form only between A and T and between G and C; thus the base sequence of each single strand can be deduced from that of its partner.

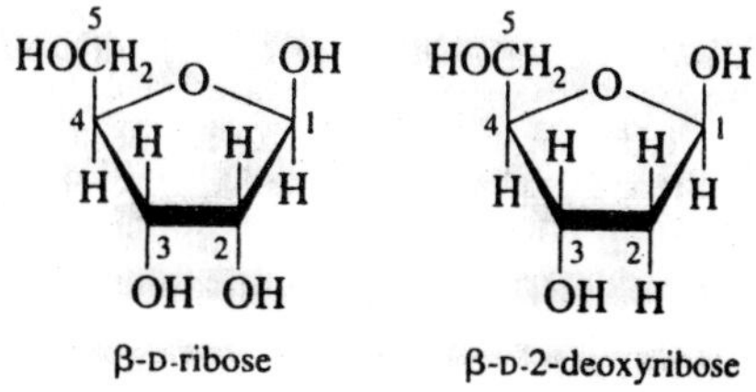

Fig. DNA

DNA cloning The production of large numbers of identical DNA molecules or cells from a single ancestral DNA molecule, nucleus or cell.

DNA library A random collection of cloned DNA fragments designed to include all or most of the genome of a given organism; also called a genomic library.

DNA ligase An enzyme that creates a phosphodiester bond between the 3' end of one DNA segment and the 5' end of another.

DNA ligase Enzymatic activity responsible for creating phosphodiester bonds between the 5' end of one strand of DNA and the 3' end of another strand. Requires the presence of a 5' phosphate on one strand, and a 3' Hydroxyl group on the second strand.

DNA polymerase Enzymatic activity responsible for catalyzing the polymerization of DNA. Is dependent upon an annealed primer from which to initiate polymerization, and a DNA template from which to copy.

DNA replicase system The entire complex of enzymes and specialized proteins required in biological DNA replication.

DNA replication The use of existing DNA as a template for the synthesis of new DNA strands. In humans and other eukaryotes, replication occurs in the cell nucleus.

DNA sequence The relative order of base pairs, whether in a fragment of DNA, a gene, a chromosome, or an entire genome.

Dnase Deoxyribonuclease, a class of enzymes which digest DNA. The most common is DNase I, an endonuclease which digests both single and double-stranded DNA.

Doca A commercial preparation of synthetic desoxycorticosterone acetate.

Domain A discrete portion of a protein with its own function. The combination of domains in a single protein determines its overall function.

Dominant allele In a diploid cell, an allele whose expression masks the expression of its partner on the homologous chromosome.

Dominant Allele that determines phenotype in a heterozygous individual carrying another recessive allele.

Donnan equilibrium If an aqueous solution of NaR (I) is separated from an aqueous solution of NaCl (II) by a membrane permeable to all ions except R—, then diffusion of Cl ions to (I) will take place. At equilibrium the product of the concentration of the diffusable ions (Na^+) (Cl^-) on one side of the membrane will be equal to the product of (Na^+) (Cl^-) on the other side. Due to the fact that the concentration of Na^+ on the side containing the non-diffusable R— is greater than the concentration of Cl—, a potential difference is set up across the membrane.

Donogany test for blood The addition of 1 cc. ammonium sulfide and 1 cc. pyridine to 10 cc. of urine produces an orange-red mixture showing a characteristic absorption spectrum. The test material may be extracted with 20% sodium hydroxide and 1 drop of extract mixed with 1 drop of pyridine on a slide. Characteristic crystals of hemochromogen form after a few hours; i.e. orange-yellow to brownish needles in star-like arrangements.

Dopa 3:4-dihydroxyphenylalanine, m.p. 282°, an amino acid of plant origin, also formed from tyrosine as it oxidizes to melanin.

Dormancy — A period of suspended growth and metabolic activity. Many plants, seeds, spores, and some invertebrates become dormant during unfavorable conditions.

Dot blot A technique for measuring the amount of one specific DNA or RNA in a complex mixture. The samples are spotted onto a hybridization membrane (such as nitrocellulose or activated nylon, etc.), fixed and hybridized with a radioactive probe. The extent of labeling (as determined by autoradiography and densitometry) is proportional to the concentration of the target molecule in the sample. Standards provide a means of calibrating the results.

Dover's powder A powder of 10% each of ipecac and opium and 80% lactose; sedative, anodyne, diaphoretic.

Drechsel test for bile acids This is a modification of the Pettenkofer test wherein syrupy phosphoric acid is used in place of sulfuric acid.

Drying oil A comparatively highly unsaturated oil which absorbs oxygen to become a stable, insoluble solid film. Linseed oil is largely used in the paint industry because of this property.

Dubsky-hrdlièka microreagent for aluminum A red, non-specific lake is obtained by the interaction of aluminum with the potassium salt of 1-aminoanthraquinone-2-carboxylic acid. Sensitivity — 1:220000. Magnesium and zinc give less sensitive.

Dulcitol The hexahydroxyhexane derived from galactose; m.p. 188° C.; optically inactive (meso).

Dune Naturally occurring wind-formed mound of sand

Duodenum Short section of the vertebrate *alimentary canal* between the stomach and the small intestine

Duplications (DNA) Type of chromosome rearrangement in which a gene sequence occurs in excess of its normal amount in a chromosome.

Dyer-baudisch reaction for cysteine and cystine The reagent (o-benzoquinone) is prepared by shaking a mixture of 1.5 gm. silver oxide, 1.5 gm. anhydrous cadmium sulfate and 0.4 gm. catechol with 10 cc. of anhydrous ether for 35 seconds. After rapid filtration, the filtrate is placed in a freezing mixture. The ether is decanted, and the crystals are immediately washed with 1-2 cc. of ether and dissolved with 8 cc. chloroform.

2. cc. of reagent are added to 2 cc. of sample and shaken vigorously for 2 minutes; a deep red color appears in the chloroform.

1. cc. hydrochloric acid and a few fragments of mossy tin are added to 1 cc. cystine solution and heated for 2 minutes. The solution is filtered, diluted with water to 15 cc. and saturated with hydrogen sulfide. 0.2 gm. charcoal are added, Shake for 1 minute and filter. Hydrogen sulfide is expelled by boiling, the solution cooled, acidified, treated with 4 cc. of reagent and shaken, giving a deep red color.

Dynamic determination Determination of a part of an embryo to carry out a certain set of morphogenetic movements.

Dysmenorrhea Painful menstruation, caused principally by excessive tension or contraction of the uterus, or by ischemia (high intrauterine pressure). These are a sign of endocrine dysfunction or structural faults or faulty sex hygiene Primary or essential dysmenorrhea is accompanied by pain only at menstrual periods. Palliatives are various analgesics and antispasmodics. Hormone therapy is adapted either to hypoovarianism or hypothyroidism as diagnosed. Progesterone is frequently administered. Intermenstrual pain may require surgical treatment or X-ray castration.

Dyspnea A condition where there is the awareness of a need of increased respiration which may be accompanied by cyanosis, a dark-bluish discoloration of the mucous membranes or the skin on account of decreased oxygen content of the blood and a simultaneous increase in carbon dioxide. Various types accompanied by cyanosis have been described, e.g. pulmonary dyspnea, cardiac dyspnea. Dyspnea without cyanosis occurs in acidosis such as due to ketosis or uremia, centric irritability (of the respiratory center). It may be hemic, i.e. due to faulty blood supply of the respiratory center, or even psychic, i.e. due to hysteria. Cyanosis without dyspnea is found in heart disease, narcotic poisoning (as in advanced stages of alcoholism, overdoses of morphine, hypnotics, etc.) and as a terminal phenomenon. Treatments are causal and involve removal of obstruction, promotion of circulation, vasodilaton, diuresis, oxygen inhalations, and judicious use of both stimulants and sedatives.

E

E 1-ecgonine $C_9H_{15}O_3NH_{20}$; tropine-2-carboxylic acid; monoclinic prisms from alcohol, m.p. 198° (dec.); an alkaloid of Erythroxylon Coca. It is the starting material for synthetic cocaine.

E. Coli (escherichia coli) A common Gram-negative bacterium present in human intestinal tract that has been studied intensively by geneticists because of its small genome size, normal lack of pathogenicity, and ease of growth in the laboratory. Useful for cloning experiments.

Echitamine Ditaine, an alkaloid, $Cn_{22}H_{28}O_4N_2 \cdot H_2O$; m.p. 206°; from bark of Alstonia scholaris; used as febrifuge or tonic in chronic diarrhea and advanced stages of dysentery.

Eclampsia Puerperal convulsions; a toxic condition of pregnancy brought on by metabolic disturbances or death of the foetus. There is a rise in blood pressure, albuminurea, increased non-protein nitrogen and other signs of kidney insufficiency. Treatments are directed toward the control of convulsions, e.g. by morphine and chloral, and relief of hypertension, e.g. by venesection.

Eclipse Movement of a celestial body across the shadow of another, or the obscuring of one celestial body by another.

Ecosystem System containing all the living and nonliving components of a particular area; refers collectively to the components and their elationships to each other

Ecotype. A local population of plants selected for the edaphic and biotic conditions of its habitat, and expressing this in its phenotype.

Ectoderm. The outermost of the three cell- or germ-layers of classical embryology. Strictly speaking, the term ectoderm should not be used until the definitive ectoderm and mesoderm have fully formed, but the loose usage avoids multiplicity of terms.

Ectoplasm The outer, stiffer portion or region of the cytoplasm of a cell which may be differentiated in texture from the inner portion or endoplasm.

Eczema An exudative skin inflammation which may be acute, subacute or chronic, produced by various exciting factors which may be chemical, mechanical, thermal, radiant (sun's rays), allergic, or endocrine. The following types are described:

(1) erythema,

(2) papules,

(3) vesicles,

(4) pustules,

(5) weeping eczema,

(6) rubrum (red),

(7) squamosum (scaly),

(8) fissum (fissures or chapping),

(9) sclerosum (chronic, leathery) and

(10) verrucosum (wart-like).

Rational treatment requires removal of cause, powdering as with zinc stearate, applying zinc pastes or boric acid lotions or starch poultices. Roentgen rays, ultraviolet rays and radium are used for obstinate chronic cases. There is also some basis for "alkalizing" and calcium therapy. Allergy must be ruled out. Special techniques are used for children's types called collectively "infantile eczema."

Edelmann test for urobilin in urine 10 cc. or urine are mixed with 5 cc. reagent A (10% alcoholic mercuric chloride solution) and extracted with amyl alcohol. The clear alcohol extract is treated with some reagent B (10% alcoholic zinc chloride solution); a rose-red color is obtained and further addition of zinc chloride produces a fluorescence.

Edema of the lungs Acute pulmonary edema; may also be chronic; a condition of accumulation of serum from the capillaries into the lung tissue or the alveoli and the bronchi. It may be provoked by poisons, inflammatory processes or abnormal osmotic relations provoked by disturbance of plasma constituents, as proteins, salts, lipoids, and sheer damages to capillary walls, as in pneumonia. Morphine, diuretics (theophylline), venesection, and oxygen administration are used to break up a vicious circle in which dyspnea sets in and further aggravates the edema. One tries to prevent death by drowning, so to speak.

Edestin A globulin from hemp seed containing all the amino acids essential for a complete protein supply.

Eefficiency ratio of useful energy delivered by a process or system to the energy supplied to it.

Egriwe test for glyceric acid A few cc. of an 0.01% solution of naphthoresorcinol in sulfuric acid are added to a drop of the test solution and heated at 90° for 3050 minutes. Dependent upon the concentration of glyceric acid, a light to dark blue color is obtained. Sensitivity—0.05 mg.

Egriwe test for glycolic acid A small amount of substance is heated with a sulfuric acid solution of 2,7-dihydroxynaphthalene; a violet-red color is obtained.

Egriwe test for lactic acid A drop of test solution is heated at 85° with 1 cc. sulfuric acid for a minute, cooled to 25° and treated with a little solid p-hydroxydiphenyl. After 30 minutes, a violet color is obtained. Sensitivity—0.1 mg.

Efficiency of food utilization This is the quotient: Change in desired effect of food divided by corresponding change in food intake.

Efficiency of food utilization, mean partial This is the quotient: (Desired effect of food plus prevention of lose which would occur without food divided by total food intake.

Efficiency of food utilization, total Quotient: desired effect of food divided by total food intake. Examples: gain in bodyweight per pound of food consumed. Calories of energy in milk per calory—or per 100 calories — of food energy consumed.

Egg (1) A large gamete without flagellae that is fertilized by a sperm cell. An egg cell is also called an ovum.

(2) A complex multicellular structure in which an animal embryo develops.

Ehrlich test for bile pigments in urine The urine is extracted with chloroform and treated with equal volumes of Ehrlich's diazo reagent and concentrated hydrochloric acid; a violet color, changing to blue-violet and finally to blue is obtained.

Ehrlich test for indican in urine 1-1.5 cc. of urine are boiled with an equal volume of reagent (0.33 gm. p-dimethylaminobenzaldehyde in 50 cc. water and 50 cc. concentrated hydrochloric acid), cooled and treated with excess ammonia or dilute sodium hydroxide. The intensity of the red color produced may be used to estimate the quantity of indican.

Ehrlich's reagent P-dimethylamino benzaldehyde, used as a specific test for indole derivatives, and therefore for proteins containing the indole nucleus. The material to be tested is treated with 20% HCl and Ehrlich's reagent (or benzaldehyde). An intense blue color is positive.

Eidogen A chemical substance which has the power of modifying the form of an embryonic organ which has been induced by other means. The term is intended to cover such substances as, may be involved, beside the primary evocator, in bringing about regional differentiation of the neural axis; but it also would be applicable to such substances as are involved in *Nachbarschaft* or *Contiguity* effects, including those which diffuse from diploid into haploid tissue, and it grades, without a very sharp dividing line, into the category of the second-grade inductor.

Einsteckung The implantation of a tissue, living or dead, or a piece of implant medium containing one or more chemical substances, into the blastocoele cavity of the gastrula, in order to test induction power.

Ekkert reactions for adrenaline and ephedrine When treated with anisealdehyde and sulfanilic acid, adrenaline and ephedrine give bright violet-rose colors; adrenaline gives a yellow to flesh color and ephedrine a dichromate red with salicylaldehyde and sulfuric acid. Adrenaline, with sulfanilic acid, 5N hydrochloric acid and 0.7% sodium nitrite solution forms deep rose-red rapidly; ephedrine takes longer. With 10% potassium hydroxide solution, adrenaline forms an evanescent red color and after evaporation a gray residue with greenish cast; ephedrine gives no color but forms an orange to dichromate red residue on evaporation.

Elastin A scleroprotein found in connective tissue, tendons, cartilage and particularly in the respiratory system. It is insoluble in water, dilute acids and alkalies and salt solutions. It is more readily hydrolyzed by trypsin than by pepsin. It has a high content of glycine and leucine.

Elaterium A product of the juice of the squirting cucumber containing "elaterin" a most powerful water producing purgative, used to relieve dropsy.

Electrode Terminal that establishes electrical contact between a conducting substance and the nonconducting part of an electrical circuit.

Electrode potential The potential developed on an inert metal, such as platinum, when it is inserted in a reversible oxidation-reduction system, or a concentration half-cell, and connected to some other half-cell, such as a calomel half-cell.

Electrodialysis A more rapid form of dialysis, which takes advantage of the fact that electrolytes are charged particles, and move more rapidly under the influence of a direct electric current to the anode and cathode. The colloidal sol is restrained by suitable membranes from moving to the anode or cathode.

Electroendosmosis The passage of liquid through a membrane or colloidal gel under the force of an applied electric current.

Electrokinetics The study of motion set up by an applied electromotive force, and of the production of an e.m.f. by motion such as the flow of liquids.

Electrolyte An acid, base, or salt, which when dissolved in water, dissociates into electrically charged particles called ions. The solution can conduct an electric current.

Electrolyte balance in muscle The facts of electrolyte balance particularly in muscle cells are admirably reviewed in the articles of Fenn (1936) and Steinbach (1940). There are several striking facts the explanations of which have given rise to much speculation. An examination of the total osmotically active constituents reveals that there exists a large excess of cation equivalents over anion equivalents in the cells of muscles. In a total of 130 there are missing 47 mille equivalents of anions per kilo of muscle.

Electrophoresis A method of separating large molecules (such as DNA fragments or proteins) from a mixture of similar molecules. An electric current is passed through a medium containing the mixture, and each kind of molecule travels through the medium at a different rate, depending on its electrical charge and size. Separation is based on these differences. Agarose and acrylamide gels are the media commonly used for electrophoresis of proteins and nucleic acids.

Electroporation A method for introducing foreign

nucleic acid into bacterial or eukaryotic cells that uses a brief, high voltage DC charge which renders the cells permeable to the nucleic acid. Also useful for introducing synthetic peptides into eucaryotic cells.

Electro-ultrafiltration A more rapid method of ultrafiltration, and one which can be used at atmospheric pressures. It utilizes the fact that both electrolytes and water will pass through semi-permeable membranes more rapidly under the influence of a direct electric current.

Elementary particle Particle smaller than an atom that is not composed of smaller parts.

Emancipation The process whereby the various organ districts acquire their boundaries in arising out of the primitive individuation field.

Emetine A dimeric isoquinoline alkaloid which is the principal alkaloid of ipecac, the ground roots of Uragoga ipecacuanha. It is very poisonous, exerting a strong simulus on mucuous membranes.

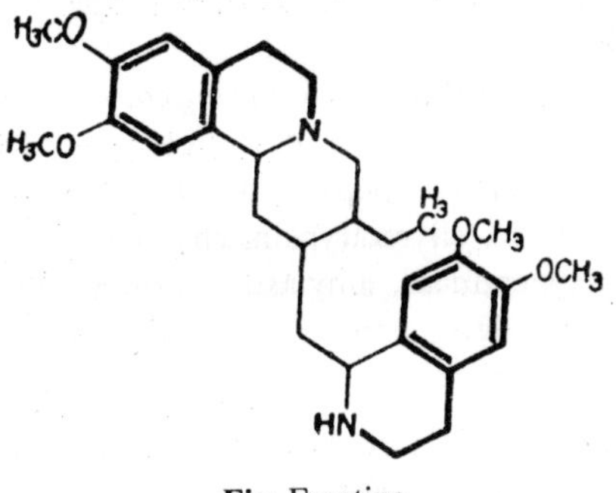

Fig. Emetine

Emodin Frangulic acid; 4,5,7-trihydroxy2-methyl-anthraquinone, obtained from rhubarb root, Cascara sagrada; m.p. 256-7°; a laxative.

Emulsin An enzyme found in bitter almonds and other places that hydrolyzes beta-glycosides into a sugar and a prosthetic group specific for the individual glycoside. Also called ß-glycosidase.

Emulsion A suspension of oil-in-water or water-in-oil. An emulsion may have dispersed particles larger than true colloids, which are stabilized by the presence of a (lyophilic) colloid, as in milk or egg yolk.

Emulsoid According to Wo. Ostwald, a colloidal suspension of a liquid in another liquid. Certain solids, such as gelatin, show all the properties of this class, so the term is not strictly accurate. Cf. Lyophilic.

Enamel The epithelium-derived outer structure of teeth, the hardest tissue of the body; has 5% water, ákeratin and inorganic matter, mostly hydroxy-apatite, and dahlite.

Encephalitis An infectious, contagious disease due to a filtrable virus, attacking the central nervous system in acute and chronic forms. Lethargy is the most frequent symptom, with eye symptoms, as diplopia, strabismus, blurred vision, irregular pupils next. Muscular twitching and paralysis (Parkinsonian syndrome) may remain after recovery. Tremors, tics, spasms, speak of extensive damage to the vegetative nervous system. Autosera are used with some success.

End labeling The technique of adding a radioactively labeled group to one end (5' or 3' end) of a DNA strand.

Endocarditis Inflammation of the endocardium. Classified clinically as simple, acute bacterial, subacute bacterial, atypical verrucous and terminal. Simple endocarditis or rheumatic endocarditis may be a complication of rheumatic fever or scarlet fever and is accompanied by damage to the heart valves. The bacterial forms are also referred to as "ulcerative" and are due to streptococcus, staphylococcus, pneumococcus, influenza bacillus, etc. They are very deadly. The atypical verrucous form (Libman-Sacks' Disease) combines many other rheumatic symptoms. A terminal form occurs following cancer, nephritis, etc. Treatments are symptomatic. So far chemotherapy or vaccine treatments or serotherapy have been failures.

Endoderm The inner skin of animal embryos, the inner primary tissue layer, or germ layer that gives rise to the inner lining of the gut and organs derived from it.

Endonuclease An enzyme that cleaves its nucleic acid substrate at internal sites (as opposed to an exonuclease, which must start at an end) in the nucleotide sequence. Examples include the restriction enzymes, DNase I and RNase A.

Endoplasmic reticulum A specialized membranous organelle within eukaryotic cells responsible for synthesis of membrane-inserted proteins, and for proteins to be exported of proteins to the cell surface or beyond.

Energy, free That portion of the total energy content of a system which is available for doing work. All forms of potential energy (mechanical, electrical, and chemical) constitute free energy in the presence of appropriate structural arrangements. Heat is convertible into work only in a fraction dependent on the difference of temperatures between which it is transferred, being completely convertible only with an infinite temperature drop.

Enhancer An enhancer is a cis-acting nucleotide sequence to which transcription factor(s) bind, and which increases the transcription of a gene. It is not part of a promoter; the basic difference being that an enhancer can be moved around anywhere in the general vicinity of the gene (within several thousand nucleotides on either side or even within an intron), and it will still function. It can even be clipped out and spliced back in backwards, and will still operate. A promoter, on the other hand, is position- and orientation-dependent. Enhancers are ususlly around 70-80 bp in length and are found, for e.g in viral DNA molecules. Some enhancers are "conditional" - in other words, they enhance transcription only under certain conditions, for example in the presence of a hormone.

Enterogastrone A hormone, postulated to be secreted by the gastric mucosa when there is fat in the intestine, and which inhibits gastric secretion and motility.

Entoderm The embryonic gut or internal cellular layer of the embryo, formed from the materials situated around the vegetative pole of the egg.

Environment Collectively, the climatic, physical, and living factors in a given area, or those factors that affect a given organism

Enzyme Complex protein which helps to speed biochemical reactions. Enzymes are important in the construction and degradation of other molecules.

Enzyme specificity The specific relation of enzyme to substrate and reaction, manifested by the fact that a given enzyme will attack only a certain compound or group of compounds, and that different enzymes produce characteristically different reactions on the same substrate.

Enzymes Definite material catalyzers of organic and usually colloid nature with specific powers of reaction, formed by living cells, but usually independent of the presence of the latter in their operation. There are a great variety of them, and they are essential to most of the functions of the living organism.

Enzymes, copper protein Copper containing enzymes which include the hemocyanins, polyphenol oxidase, monophenol oxidase (tyrosinase), laccase, hemocuprein, hepatocuprein, plant enzymes oxidizing ascorbic acid.

Enzymes, crystalline The following enzymes or their precursors have been definitely crystallized: urease, catalase, pepsin, pepsinogen, trypsin, trypsinogen, chymotrypsin, chymotrypsinogen, carboxypeptidase, amylase, yellow respiratory ferment, ficin, papain, carbonic anhydrase, ribonuclease.

Enzymes, hydrolytic May be classified as (1) esterases, which hydrolyze ester linkages, e.g. lipases, lecithinases, nucleases, phosphatases, sulfatases, choline esterase, cholesterase;

(2) carbohydrases which hydrolyze glycoside linkages falling into groups of glycosidases, like sucrase, glycuronidases, etc., and polyases, like amylase, cellulase, cytases;

(3) proteases which hydrolyze peptide linkages which fall into groups like peptidases, proteinases and amidases (hydrolytic deaminases).

Enzymes, iron porphyrin protein Oxidation catalysts containing a porphyrin structure, combined with iron, as hemoglobin, chlorocruorin, helicorubin, catalase, peroxidase, cytochrome.

Enzymes, non-proteolytic (hydrolytic) The biological catalysts of protein nature, known as enzymes, that are evolved by living cells for metabolic utilization, may be divided into two major groups: those effecting oxidations and reductions, and those producing hydrolysis. The latter group is made up of three main subdivisions representing the enzymes capable of hydrolysis, and in some cases the synthesis as well, of the three chief types of organic substances that compose living matter, i.e. proteins, carbohydrates, and lipids, and certain of their degradation products.

The present discussion is concerned with the hydrolytic enzymes that affect carbohydrates, lipids, and substances chemically related.

While enzymes may be differentiated for the most part by their specificity for particular substrates or structures, other factors offer not only additional differentiation between these enzyme types, but distinction between different enzymes acting on the same substrate. Thus the effect of pH, temperature, activators and inhibitors, the reaction kinetics, and physical separation often permit differentiation.

CARBOHYDRASES

Polysaccharidases

Amylases have been the most thoroughly studied members of this group because starch and glycogen breakdown have been of great importance commercially and of particular interest in biology and medicine.

When starch or glycogen are subjected to enzymatic action the viscosity of the substrate is reduced with an accompanying liberation of phosphate and an increase in reducing power. This action is believed by some to result from an amylophosphatase although no proof is now available that this enzyme is distinct from alpha-amylase that attacks the macromolecule to liberate dextrin units and some maltose. Beta-amylase breaks down the dextrin units to maltose and dextrins of lower molecular weight. There is evidence that the latter enzyme can also act directly on the starch molecule to whittle down the polysaccharide chain one maltose group at a time. The iodine-coloring property of starch is destroyed by alpha-amylase, but it is not entirely removed by beta-amylase. Kuhn was the first to classify amylases as alpha or beta on the basis of the mutarotation of the maltose formed, the alpha-amylase giving rise to alpha-maltose and the betaenzyme to beta-maltose.

The amylases occur in the animal body mainly in saliva, pancreatic juice, and other digestive secretions, and also in organs actively involved in carbohydrate metabolism such as the liver. Clinical tests for the function of organs secreting large amounts of an enzyme have been evolved, based on the measurement of the concentration of the particular enzyme in the blood. Thus determination of amylase in blood serum has been employed as a test for pancreatic function. However abnormalities in other organs producing the same enzyme must be considered in any interpretation of data of this nature as well as the physiological state of these organs at the time the blood is drawn. As a result the usefulness of these tests in general, and of the amylase test in particular, has been seriously limited.

In the plant organism amylases also occur particularly in those parts responsible for the metabolism of starch. This teleological trend is illustrated by the increase of amylase during the sprouting of seeds. Amylases are also found in a wide variety of fungi, bacteria, algae, and certain yeasts.

The fall in viscosity of starch solutions, the decrease in turbidity of glycogen solutions, the change in color given by iodine, and the more exact measurements of the increase in reducing substances, have all been employed to determine amylase activity. Details of these measurements, as well as procedures for the purification of amylases may be found in the standard reference works on enzyme methods. The elegant micro- and ultramicro technics of Linderstrøm-Lang and

Holter have been applied to carbohydrases as well as the esterases to be discussed later.

Pancreatic and salivary amylases have been the most thoroughly studied in the animal body, and malt amylases, because of their position of prime importance to the brewing industry, have been the most completely investigated in the plant world. The pH optimum for the action of the former is 6.8 while that for the latter falls between 4.4 and 5.2 In considerations of pH-optima in general it should be borne in mind that these values are often influenced by factors such as the nature of the buffers used, the presence or absence of salts and other compounds affecting activity, etc. Pancreatic amylase has been reported to have been produced in crystalline protein form. The amylase of the fungus, Aspergillus oryzae, has attained importance in oriental food and beverage industries and in medicine.

Cellulase and lichenase, which may be the same enzyme, and cytase are polysaccharidases that attack plant framework substances. These have been found in the lower forms of life such as thermophilic bacteria, molds, fungi, snails, and ship worms that utilize cellulose and hemicellulose as sources of nutriment. They also occur in the endosperm of germinating barley, lupines, and palm seeds. There isn't a great deal known about these enzymes but it is striking that in the thermophilic bacteria, they may exhibit activity up to 70°.

Inulinase, which breaks down inulin, the reserve carbohydrate of certain tubers and roots, to d-fructose has been found in all monocotyledonous plants, and in bacteria, yeasts and fungi. It has been claimed to occur in invertebrates, and in the spleen and placenta of some vertebrates. The optimum temperature for its action has been given as 55° and the optimum pH as 3.8.

Most of the remaining polysaccharidases, also occurring in bacteria, fungi, and germinating seeds, and having optimum activity in acid media, require more thorough investigation before their properties, nature and possible uses can be established.

Hyaluronidase deserves especial mention since this enzyme is related to the "spreading factor," an agent increasing tissue permeability and therefore concerned with the invasive ability of infective materials. Hyaluronic acid occurs in protective tissues such as skin and in bacterial capsules, and the enzyme attacking this polysaccharide lowers the viscosity of solutions of the substrate and liberates acetylhexosamine. Hyaluronidase is found in bacteria, testes, leeches, bee sting extracts, and some snake venoms. The activity of the enzyme is affected by salts, and two pH maxima for testicular preparations have been given, 4.4 and 5.8, while the single maximum, 5.8, has been reported for the enzyme from bacterial sources. At present, the question of the identity of the enzyme and the spreading factor cannot be considered completely settled. A practical application that may develop is suggested by the ability of hyaluronidase to destroy polysaccharide capsules of certain pathological streptococci. At the time of writing, no very complete review of the work on the enzyme has appeared; however a brief survey may be mentioned.

In passing, attention should be called to the work of Dubos and Avery on the interesting enzyme obtained from a soil bacillus that digests the capsular polysaccharide of Type III pneumococci. The injected enzyme can protect the mouse, rabbit, and monkey from the virulent organism.

Tetra and Trisaccharidases

Stachyase hydrolyses the tetrasaccharide, stachyose, to fructose and a mannosaccharide consisting of glucose and two molecules of galactose. This enzyme occurs in the digestive juices of crustaceans and mollusks as do the trisaccharidases, raffinase, capable of converting raffinose into fructose and melibiose, and gentianase, converting gentianose to fructose and gentiobiose. The trisaccharidases also occur in yeasts and Aspergillus niger. Raffinase, more extensively studied than the other enzymes, has been shown to have optimum activity at pH 4.5 to 5.0 at 30°. Aspergillus niger also contains melezitase that hydrolyzes melezitose to glucose and turanose. The enzyme activity in all of these cases may be measured polariscopically or by the increase in reducing action.

Disaccharidases

Maltase splits maltose to two molecules of glucose. Although the enzyme occurs in digestive juices and organs where amylase is found, as well as in resting and germinating seeds, bacteria and fungi, most of the work has been carried out on yeast maltase. This enzyme is rather unstable and loses activity even at moderate temperatures; it is injured by low concentrations of alcohol, and is inactivated when its preparations are subjected to drying. The optimal conditions for activity are pH 6.7 to 7.3 and 40°. Willstätter and coworkers, who have carried out much of the work on maltase, have shown that the course of the hydrolytic reaction is greatly influenced by the nature of the enzyme preparation.

Trehelase, a much more stable enzyme than maltase, hydrolyzes trehalose to two molecules of glucose. Little work has been done on this enzyme but it is as widely distributed in plant and animal life as maltase itself and appears to have greatest activity in slightly acid media.

Saccharase is one of the most thoroughly studied enzymes known. Its conversion of saccharose to glucose and fructose is a process of industrial importance, and the accuracy of the polarimetric method by which the course of the hydrolysis may be followed, has enabled this reaction to be studied as a model in more purely scientific investigations. The work of Nelson on the kinetics of the enzyme action is classical. The saccharase studies of Michaelis and Menten led to the theory that enzyme and substrate combined to form an enzymesubstrate compound that breaks down to yield the free enzyme and the reaction products. The velocity of the hydrolysis is determined by the concentration of the intermediate enzyme-substrate compound, and the affinity between enzyme and substrate can be expressed quantitatively in terms of the dissociation constant of this compound. The Michaelis theory was used later to explain the mode of action of enzyme inhibitors; so called "competitive" inhibition results from the inhibiting substance combining with the enzyme, in competition with the substrate, to form an enzyme-inhibitor compound thus reducing the concentration of the enzyme-substrate compound. This type of inhibition varies with the substrate concentration. "Non-competitive" inhibition results from the inhibiting substance reducing the decomposition velocity of the enzyme-substrate compound, but the inhibitor displays no affinity for the enzyme and the inhibition is independent of the substrate concentration.

Saccharase occurs in the intestines of higher animals and in the honey vesicles, saliva, and intestines of bees and other insects utilizing saccharose; but it is much more widely distributed in the plant kingdom, being particularly prevalent in those parts acting as carbohydrate stores, such as the sugar beet root. The abundance of the enzyme in yeasts has made these its chief source. The pH optimum for the yeast enzyme is 4.5, while for intestinal saccharase it falls between 5.0 and 7.0. The enzyme is inactivated by oxidizing agents, heavy metal salts, alcohol, and acetone; it is inhibited by the products of its action and is relatively insensitive to neutral salts.

Cellobiase and gentiobiase convert their respective substrates to glucose. Both occur in many seeds, grains, fungi, and bottom yeasts, and they are found together, usually in the presence of lichenase as well. The optimum pH has been given as 6.0 and the optimum temperature as 46°.

Lactase and melibiase, the galactosidases producing glucose and galactose as their hydrolytic products, are both found in bitter almonds. However only the latter occurs in bottom-yeasts while the former is prevalent in milk sugar yeasts, a number of fungi, fetal intestines, and certain bacteria. Little is known about melibiase, but it has been determined that its pH optimum is 5.5; that for almond lactase is 4.2 to 4.6, for the yeast enzyme 7.0, and for various intestinal lactases 5.0 to 6.4. The lactase of Escherichia coli has received especial consideration because it offers a classical example of the adaptive production of enzymes by bacteria. This bacterial species does not produce lactase unless the organisms are transferred repeatedly in lactose broth, whereupon the enzyme is generated rapidly. Another example of enzyme adaptation is offered by bacteria that form the enzymes that hydrolyze capsular polysaccharides of pneumococci; the enzymes adaptively developed for the hydrolysis of Type III polysaccharide have no action against the Type VIII substance.

Esterases

Lipases

Lipases hydrolyze neutral fats, glycerides of higher fatty acids, to glycerol and fatty acid. In the animal body the pancreas is the richest source of this enzyme although it occurs in rather high concentration in all of the digestive organs. It was thought earlier that stomach lipase had optimum activity between pH 4.0 and 5.0 in contrast to the enzyme in other organs that exhibit maxima from 7.0 to 8.6 depending on the source. However, Willstätter and coworkers have shown this to be due to the effects of concomitant materials naturally present, and the difference disappears on enzymatic purification. The activation of lipase action by bile salts is of physiological interest, but a variety of other surface active substances in low concentration can also produce activation. The activation effected by calcium and other salts must be considered in measurements of lipase activity.

Determinations of activity for most of the esterases are based on measurement of the acid liberated. Titration, manometric, colormetric, and stalagmometric methods have all been employed. The methodology and preparative details are available in standard reference works.

The Ricinus lipase of castor beans has been investigated more completely than the enzyme from other plant sources. The active enzyme is insoluble in aqueous media. Although the lipase from ungerminated seeds has maximum activity at pH 4.7 to 5.0, the activity at higher pH values gradually increases during germination. This change can also be produced by treating the seed enzyme with pepsin.

Lipases, as well as the simple organic esterases to be discussed next, can also synthesize esters from alcohol and acid. Sym has developed procedures for these enzymatic syntheses employing practically non-aqueous media.

Simple Organic Esterases

These esterases, liberating alcohols and simple organic acids from their esters, are found in particularly high concentration in liver. The specificity differences between these esterases and the lipases, first demonstrated by Loevenhart, have been extensively elaborated by Willstätter and his colleagues and others. Further differences between the esterases and lipases are found in their protein nature, kinetics, and influence of foreign substances. The stereochemical specificity of these enzymes is quite marked, and it appears to be a property of the enzyme itself and not the result of accompanying materials in the preparation. The inhibitors of liver esterase also exhibit stereochemical specificity. The enzyme is inhibited by high concentrations of its own substrates.

Hydrolysis by liver esterase follows a zero molecular reaction course due to the very high affinity for its substrate and the pH optimum ranging from 6.7 to 8.8 has been found to be influenced by the nature of the buffer, substrate, and enzyme source.

Cholesterolesterase

This enzyme hydrolyzes cholesterol esters to the free alcohol and organic acids, usually the latter are higher fatty acids; and it has been shown that these esters can be enzymatically synthesized by the method of Sym previously mentioned. Cholesterolesterase has been found in the liver, spleen, nerve tissue, and blood, and the chief interest in the enzyme is its relation to resorption and transport of fat in the animal body.

Azolesterases

Azolesterases are defined as the enzymes that hydrolyze esters of nitrogen—alcohols, and they have been shown to be distinct from simple organic esterases in both distribution and properties.

Cholinesterase, first demonstrated by Loewi and coworkers, has received more attention than other members of this group because of its interesting relation to the transmission of nerve impulses. The enzyme can hydrolyze, not only the physiologically important acetylcholine, but a wide variety of chemically related esters in which the alcohol component may contain a nitrogen-heterocyclic ring instead of a substituted nitrogen group. Replacement of the nitrogen in the ester by arsenic or phosphorus does not change the enzyme action, and thiocholine esters are hydrolyzed even faster than the choline compounds. The presence of a nitrogen group in the acid component of the ester prevents any enzyme action.

The richest sources of the enzyme are salivary and lachrymal glands, alimentary mucosa, and tissues of the nervous system. Optimum activity is obtained at pH 8.5 and 40°. The powerful inhibition of the enzyme by traces of physostigmine is utilified in physiological experimentation. Thiamin also inhibits the enzyme, and neutral salts and calcium ion have a definite activating effect.

Of the tropinesterases, atropinesterase was first investigated by pharmacologists interested in the destruction of atropine in the animal body. The enzyme produces tropine and tropic acid from the alkaloid. Some rabbits contain atropinesterase while others do not, evidence that this enzyme is identical with the "alkaline" phosphomonoesterase mentioned above.

Phytase hydrolyzes phytin to inositol and phosphoric acid. The enzyme is distributed in liver and blood of animals, fungi, and especially in seeds and malt. The pH optimum for the malt enzyme is 5.6 and Sawin and Glick have shown this enzyme to be an incompletely dominant hereditary factor in the rabbit. Serum, liver, and intestinal mucosa are particularly rich in the enzyme, and its optimum activity is found in pH 8.1 to 8.4 and 38°. The hydrolysis follows a zero molecular reaction course, and the enzyme is activated by neutral salts and inhibited by sulfhydryl compounds, physostigmine, cyanide and fluoride.

Tropacocainesterase and cocainesterase hydrolyze tropacocaine and cocaine respectively to tropine and benzoic acid in the first case, and probably ecgonine, benzoic acid and methyl alcohol in the second. The uncertainty about the cocaine scission derives from the fact that the products have not been isolated and the possibility remains that the enzyme may hydrolyze only one of the two ester groups in the substrate. The existence of the three tropinesterases as separate enzymes requires further evidence since the present differentiation is based only on differences in their occurrence in the blood of various species. These enzymes, in addition to cholinesterase, are contained in the alpha and beta globulin fractions of blood sera.

Morphinesterase, discovered by Wright, hydrolyzes the alcoholic acetyl group from mono- or diacetyl morphine. The enzyme is found in the sera of only some rabbits, and since it appears to parallel the occurrence of atropinesterase the possibility arises that the two enzymes may be identical.

Chlorophyllase

Chlorophyllase, discovered by Willstätter and Stoll converts chlorophyll to phytol and chlorophyllide. As would be expected, it is present in all green plant life. The enzyme retains activity in strong alcohol solutions and in the presence of acetone, and when the enzymatic action occurs in alcohol media an alcoholysis occurs by which ethylchlorophyllide is formed.

The synthetic properties of the enzyme also enable formation of chlorophyll, from the acid and phytol.

Tannase

The identity of tannase as a single enzyme is still an open question. The enzyme liberates phenolcarboxylic acids and alcohols from tannins and peptides. Methyl gallate is commonly used as the substrate in tannase investigations. Many properties of the enzyme have not yet been established. It is found chiefly in molds and fungi, and for its action it requires that the acid component of the substrate contain at least two phenolic hydroxyl groups which may not be ortho to the carboxyl radical.

Sulfatase

Sulfatase, first described by Neuberg, effects the cleavage of phenolic sulfuric acid esters. These compounds are formed by the detoxication mechanisms in the animal body for the elimination of phenols. The enzyme is found in kidney and muscle, and also in molds and fungi. An extensive review of the subject has appeared recently.

Phosphatases

Phosphomonoesterases affect diverse substrates to liberate orthophosphoric acid and alcohol components such as glycerol. These enzymes are highly important in the fat, carbohydrate, and mineral intermediate metabolisms. The measurements of serum "acid" and "alkaline" phosphatases that hydrolyze beta-glycerophosphate are used in medicine for diagnosis and prognosis of prostate cancer and bone disease. The "acid" enzyme occurring particularly in the prostate, liver, pancreas, and rice bran operates best at a pH of about 5.0, and the "alkaline" enzyme found especially in kidney, leucocytes, intestine and bone tissue is most active around a pH of 9.0. The former enzyme is not activated by magnesium ion but the latter is. A glycerophosphatase having its pH optimum at 3.0 to 4.0 is found in molds and fungi, and this enzyme is activated by magnesium ion when the substrate is alpha-glycerophosphate, but inhibited by the same ion when beta-glycerophosphate is used. Mammalian erythrocytes contain a glycerophosphatase having a pH optimum at 6.0 that is activated by magnesium ion. In contrast to the other glycerophosphatases, this enzyme hydrolyzes the alpha-substrate faster than the beta.

Nucleotidase breaks down nucleotides to a carbohydrate-purine or -pyrimidine base and phosphoric acid. It has been demonstrated in glandular tissues, nerves, and blood. There is a good deal of evidence that this enzyme is identical with the "alkaline" phosphomonoesterase mentioned above.

Phytase hydrolyzes phytin to inositol and phosphoric acid. The enzyme is distributed in liver and blood of animals, fungi, and especially in seeds and malt. The pH optimum for the malt enzyme is 5.6.

Hexosediphosphatase has not been proven to be a distinct enzyme although its preparations are devoid of activity on glycerophosphates. The enzyme plays an important role in alcoholic fermentation through its effect on fructose diphosphate, and it is also involved in the carbohydrate conversion in muscle. The particular hexosemonophosphate formed by the kidney or yeast enzyme is conditioned by the presence of arsenate.

Phosphodiesterases hydrolyze one ester linkage in diesterorthophosphates. These enzymes occur in yeasts, rice bran, snake venoms, and animal tissues. The pH optimum of the enzyme in rice bran is 5.5 and in snake venom 8.6. Since these enzymes produce monoesters, their action must be followed by phosphomonoesterase to effect the complete liberation of phosphoric acid. Lecithinases belong to this group since they hydrolyze the diglyceride esters of phosphocholine known as lecithins. These enzymes are found in many animal tissues but particularly in the kidney. A pH optimum of 7.5 has been reported. Phosphotriesters are not hydrolyzed by any known enzyme.

OTHER HYDROLYTIC ENZYMES

Pyrophosphatases

The term, pyrophosphatases, is employed in the

present instance to denote those enzymes capable of hydrolyzing the intramolecular linkages between the phosphorus units themselves in polyphosphorus compounds. This group of enzymes is of great importance in the intermediate carbohydrate metabolism of muscle.

Adenosinetriphosphatase converts adenosinetriphosphate to adenosinediphosphate. The brilliant work of Ljubimova and Engelhardt has shown that myosin, the chief protein of muscle, is probably identical with this enzyme. Confirmations have been forthcoming from Needham, Bailey, and others at Cambridge. Thus the new concept arises that an enzyme may exist, not only as a small amount of catalytic material in a bulk of tissue, but also as the main mass of a tissue. The enzyme is very unstable, losing activity at slightly elevated temperatures (40°) and when subjected to weak acidities (up to pH 4.0); it is inactivated by urea, and traces of silver ion abolish all activity. In this connection mention should be made of the crystallization of myogen, the second most prevalent protein in muscle, by Baranowski. This protein appears to be identical with an aldolase effecting decomposition of glucose.

Thiaminpyrophosphatase is of physiological interest since its substrate, thiaminpyrophosphate or cocarboxylase, is important in body oxidation mechanisms. The enzyme occurs in yeast and a pH optimum of 3.7 has been reported.

Phosphorylase

While a discussion of the many steps in carbohydrate metabolism is beyond the scope of the present article, attention should be drawn to one of the more important phosphatases, phosphorylase, involved in the reversible glucose-1-phosphate? glycogen or starch transformation. This enzyme has been investigated particularly by Cori and coworkers who have shown that, in the case of the animal enzyme, adenylic acid is essential for activity. Reducing agents enhance the activity, and the presence of some glycogen is necessary for the progress of the synthetic reaction. Change in pH produces a shift in the equilibrium finally attained. A typical glycogen is formed by the enzyme from liver, heart, or brain tissue, but the polysaccharide produced by muscle phosphorylase, in common with the enzyme from plant sources, is more like starch.

Ephedrine $C_6 H_5$ CH OH CH ($NHCH_3$) CH_3 ; An alkaloid from the Chinese drug Ma Huang; also prepared synthetically. It is dextrarotatory, but has laevorotatory salts. It raises blood pressure, strengthens heart action, and can be taken orally.

Epididymus A long coiled duct located on the top of the testicles that is important for sperm maturation and storage site.

Epigenesis The development of the organism by the new appearance of structures and functions, as against the unfolding or growth of entities already fully present in the egg at the beginning of development.

Epilepsy Falling sickness; a complex of symptoms of seizures of sudden loss of consciousness often accompanied by convulsions, attributed to injury to the central nervous system which may be congenital or acquired as in brain tumors, fractures of the skull and many diseases (cerebral arteriosclerosis, endocrine dysfunction, syphilis, heart block, etc.). Adrenalin and pituitrin predispose to attack on direct administration. Constipation, menstruation and allergy are precipitating factors. Fundamentally alkalosis and oxygen lack lead to edema and in-reased intercranial pressure affects the nerve cells so as to produce muscular spasm. The major types of attack are:

(1) grand mal, convulsions followed by coma and

(2) petit mal, a sudden helplessness and brief unconsciousness. Jacksonian epilepsy is a progressive form, beginning with the muscles on one side. Nocturnal epilepsy may occur for long times in the sleep. Pyknolepsy is a form of frequent momentary unconsciousness occurring in young children without complications till it leaves at puberty.

Epileptic equivalents are forms of uncousciousness without convulsions. Narcolepsy is a form in which drowsiness occurs with sudden sleep. Catalepsy is a sudden weakness in the legs under emotional stress.

The treatments are causal. Ketogenic diets are frequently recommended, and so are salt-poor diets. Sedatives must be used carefully, sometimes removal of bone pressure on the brain is adequate if that is causal.

Epinephrine 1-(beta-methylamino, alpha-3, 4dihydroxyphenyl)-ethanol; the active ingredient of the adrenaline glands; causes marked contraction of the blood vessels, with a corresponding rise in blood pressure.

Epitope As related to protein antigens, B-cell epitopes consist of the amino acid residues of a protein molecule which interact directly through noncovalent bonds with the amino acid residues of a particular antibody molecule (complementarity determining region). The average epitope probably involves about 15-20 contact amino acid residues, but one or two of these may be critical to the epitope's specificity and the avidity of the antibody-antigen reaction. B-cell epitopes may be either linear or conformational in nature. T-cell epitopes represent the small, processed peptides which bind to MHC class I and II molecules on the surface of T cells.

Eppinger test for glyoxylic acid A mixture of glyoxylic acid and 0.1% indole solution is floated on concentrated sulfuric acid; a red ring is formed. Skatole may be used for indole. Sensitivity - 1:20000.

Equilenin $C_{18}H_{18}O_2$; m.p. 258-259°; an estrogenic hormone found in pregnant mare urine. Its structure is that of estrone with two double bonds in ring II, and it has 1/16 of its potency.

Equilibrium Stable condition in which opposing dynamic or static elements of a system are in a balanced state.

Equilin $C_{18}H_{20}O_2$; m.p. 283-240°; an estrogenic hormone found in pregnant mare's urine. It has one alicylclic double bond more than estrone, and has 1/7th of its potency.

Ere Estrogen Response Element. A binding site in a promoter to which the activated estrogen receptor can bind. The estrogen receptor is essentially a transcription factor which is activated only in the presence of estrogens. The activated receptor will bind to an ERE, and transcription of the adjacent gene will be altered.

Ergosterol $C_{28}H_{44}O$; m.p. 163°; a sterol of ergot and yeast, having an OH at C_3, a 9 carbon chain with one double bond at C_{17} and a double bond at C_{5-6} and at C_{7-8}. On irradiation by ultraviolet light, yields successively at least five isomers; tachysterol, lumisterol, the anti-rachitic calciferol, or vitamin D_2, and suprasterol I and II.

Ergot Ergot is a fungus which grows in the heads of cereal grains and grasses. The ergot of medicinal importance, and the only one recognized in the *United States Pharmacopoeia XII*, is that which develops on rye due to the particular fungus Claviceps purpurea. The young rye flowers are infected by the ascospores of the fungus, which then develop in and replace the grain of the rye. The sclerotium thus formed is a slender, cylindrical, somewhat curved and longitudinally furrowed body, larger than the grain which it has replaced. Several such kernels or sclerotia may be found in a single head of rye. The name ergot comes from the French word for cock's spur, from the fancied resemblance of the sclerotia to this structure. Ergot shipped to the United States has come largely from Spain and Russia.

The presence of ergot in rye, as well as in other cereal and forage grains and grasses has frequently led in the past to widespread and serious epidemics of poisoning. The Saint Anthony's fire of the middle ages is now believed to have been ergotism. No serious poisoning from ergot contaminated flour has occurred in the United States for many years, although injury to stock from eating heavily infested grasses has been reported a number of times.

The powerful activity of ergot has led to extensive studies of both its chemistry and its pharmacology over many years. No other fungus, and few if any plants or plant products have been the subject of so much investigation. From ergot there have been isolated a large number of substances of biochemical interest, some of which are specific to ergot, while others are of more general distribution. The pharmacological properties of histamine were discovered largely because of an interest in ergot extracts, and the name ergamine is still sometimes applied to histamine.

Ergosterol and ergothioneine (thioneine) were named from ergot where they were first found to occur.

Of the substances peculiar to ergot the alkaloids are most important. Present day views ascribe all or nearly all of the pharmcological activity of ergot to these substances. While extracts of ergot may contain other highly active substances such as histamine, tyramine, acetylcholine, these are either destroyed or rendered ineffective in the gastrointestinal tract, so that they probably played no role either in the therapeutic action of galenical ergot preparations, such as the extract, or in the poisoning from eating ergotized rye.

Isolation of the alkaloids has been difficult. Not all of those known are present in every sample of ergot. Their relative proportions vary. Even different sclerotia from one head of rye may have different amounts of alkaloids. Crystallization of some has proved difficult. Frequently crystallization with molecules of solvent occurs, and removal of these has not been easy. Many of the chemical and pharmalogical studies have been carried out with incompletely purified materials. The terminology has been correspondingly confused. Current views recognize the existence of five pairs of optically active alkaloids, interconvertible isomerides, either existing as such in ergot or readily obtained by isomerization from the naturally occurring alkaloids. These pairs are: ergotinine and ergotoxine; ergotamine and ergotaminine; ergometrine and ergometrinine; ergosine and ergosinine; ergocristine and ergocristinine. (The alkaloid ergometrine was discovered more or less simultaneously by four different groups of workers in different parts of the world, and called also ergobasine, ergotocin, ergostetrine. The Council on Pharmacy and Chemistry of the American Medical Association adopted the name ergonovine, which term is now official in the *United States Pharmacopoeia XII*. (The form ergometrine is used in the present paragraph to emphasize the relationship to the inactive isomer ergometrinine.) Other alkaloids have also been described. Sensibamine is a complex (apparently equimolecular) of ergotamine and ergotaminine. Ergoclavine is a similar complex of ergosine and ergosinine. Further complications are added through the description of similar complexes between the active isomer of one pair, and the inactive member of another pair; e.g., ergosine and ergotaminine, ergosinine. Some fourteen such complexes have been reported, with details of crystallographic structure. While such complexes might be expected to show pharmacological activity representing the average of the two components, according to Barger such evidence as is available "imperfectly supports this view." A recently described alkaloid of smaller molecular weight, ergomonamine, $C_{19}H_{19}NO_4$, evidently belongs to a different series. It is not an indole derivative.

These substances are all alkaloids in the generally accepted sense of the term. They are soluble in organic solvents (excepting petroleum ether), and sparingly soluble or insoluble in water with the exception of ergometrine and ergometrinine. In most cases they form salts which can be crystallized. Their structure is imperfectly known. On hydrolysis the largestand most important product is a compound having the formula $C_{16}H_{16}N_2O_2$, containing an indole nucleus, a double bond, a methylated nitrogen, and a carboxyl group.

Pharmacological studies of the ergot alkaloids are most extensive with respect to ergotoxine, ergotamine, and erigonovine (ergomettine). Ergotoxine and ergotamine are absorbed slowly from the gastrointestinal tract, ergonovine (ergometrine) quite rapidly. The most important effect after absorption is that on the smooth muscle, particularly that of the pregnant uterus.

Here it increases both the tone and the rate and amplitude of the rhythmical contractions. This property is possessed most strongly by ergonovine (ergometrine). The effect on other smooth muscle mechanism resembles somewhat that of sympathetic stimulation. In the case of ergotoxine, ergotamine, and ergosine, large doses are followed by a partial or complete paralysis of the response of sympathetically innervated structures to either sympathetic nerve stimulation or administration of epinephrine, particularly in those cases where the response is motor rather than inhibitory. Ergonovine does not possess this property. Administration of ergot or of its alkaloids to roosters results in a cyanosis and bluing of the comb. Large doses or repeated administration may result in development of gangrene of the tips of the comb and the tail of the white rat has also been produced experimentally with ergotamine. Gangrene of fingers or toes was a marked symptom of ergotism in the epidemics of poisoning. It has also followed excessive dosage with the alkaloid ergotamine clinically.

Erucic acid Unsaturated fatty acid, $C_{22}H_{42}O_2$, with one double bond, found in mustard seed and rapeseed oils;

Erythema Redness of the skin; described as

(1) simplex,

(2) multiforme and

(3) nodosum. Erythema simplex may be due to drugs, sun-burn, fevers, intestinal toxemias and includes blushing. Erythema multiforme is accompanied by exudations, macules, papules, vesicles, nodules and is of toxic origin which is endogenous as well as exogenous. It is related to urticaria and purpura. Erythema nodosum is frequent upon the legs and is attributed to exposure to cold and wet. It is excited by fever conditions and the like. Therapy is usually symptomatic for all forms, but vaccines sometimes help.

Erythrocruorins Red respiratory chromoproteins of many invertebrates, which closely resemble hemoglobin in their chemical nature. Their hemin group appears to be protoheme throughout, but the protein components vary in are different from globin. The molecular weights of the erythrocruorine vary between 16,500 and several million. As a rule, erythrocruorins, locked up in red blood cells, are lowmolecular while those freely dissolved in the blood plasma are high-molecular.

Erythrocytes Red blood cells, comprising between 40-45% of the volume of blood. They are non-nucleated, circular, biconcave discs; about 8 mu in diameter, and containing about 34% by weight or $3x10^{-12}$ grams each of hemoglobin, by means of which they carry oxygen to the cells.

Erythroidine An alkaoid, $C_{16}H_{19}O_3$, isolated from certain species of erythrina, and existing in two stereoisomeric forms; both forms have curare-like action.

Esculin $C_{15}H_{16}O_9$ hydrate; aesculin; a glycoside of the bark of the horsechestnut, consisting of glucose and aesculetin; crystallizes as prisms from water or dilute alcohol; m.p. 205°.

Eserine Physostigmine, $C_{15}H_{21}O_3N_3$, alkaloid of Calabar bean; used to contract pupil of eye, also for tetanus.

Essential oils Odorous constituents of plants which steam distil. They find application in perfumery and medicine. They are hydrocarbon types, ester types, sulfides, etc. Compounds in plants, volatile with steam, extractable with hydrocarbon solvents. They include terpenes of the general formula $C_{10}H_{16}$, e.g. pinene, camphene, limonene; alcohols and ketones, e.g. camphor, menthol, borneol, menthone, cineol; a geraniol and citronellol group; benzene hydrocarbons like cymene and styrene; phenols like eugenol, thymol, carvacrol; acids and esters; aliphatic alcohols; aliphatic aldehydes; sulfides like allyl sulfide; organic bases like indol; and occasionally small amounts of paraffin hydrocarbons.

Esterases A group of enzymes that hydrolyze esters Sometimes the individual enzymes are named from the organ where they originate, as liver esterase; and sometimes the name indicates the product hydrolyzed, as sulfatase.

Estradiol $C_{18}H_{24}O_2$; m.p. 174° C; the natural female estrogenic hormone produced by the ovarian follicles. It controls the menstrual cycle, and to a large extent, the secondary sex characteristics.

Estradiol monobenzoate m.p. 195° C; The benzoyl ester of estradiol, usually at C $_3$, used clinically in treatment of menopause and disturbed estrus. Preferable to estradiol as the slow ester hydrolysis allows the use of larger and less frequent doses.

Estrogen *Hormone* secreted by the mammalian *ovary,* the organ that produces female egg cells; estrogen regulates development of female sexual characteristics and is involved in various physiological processes throughout life.

Estrogenic hormones A group of naturally occurring sterols which control the estrus cycle and secondary sex characteristics in the female. Those found in the human are estrone, estradiol and estriol, estradiol being the primary hormone secreted by the Graffian follicle; estrone being its waste product, and having only 1/10th its potency. In addition, three other weak estrogenic hormones, equilin, hippulin and equilenin are found in equine urine. Excessively large doses in rats may cause cancer.

Estrogens, synthetic The term synthetic estrogen as used in this review, includes those compounds which have estrogenic activity and yet lack the cyclopentenophenanthrene nucleus. This definition excludes derivatives of the natural estrogens from consideration, although strictly speaking, such a compound as ethinyl estradiol is a synthetic substance and does not occur as such in nature. The members of this group which have received the most attention are the following substances and certain derivatives of them: diethylstilbestrol, diethylhexestrol, and triphenyl ethylene As in the case of the natural estrogens, these substances must be assayed biologically in order to establish their potency. This may be done by determining the amount necessary to produce a cornified vaginal smear or to produce an increase in the uterine weight of immature or spayed rats or mice. Most of the values listed were obtained by the vaginal smear method on spayed rats. Any biological assay should.

(1) employ enough animals to give the results statistical significance;

(2) compare two substances of which one is a known standard;

(3) compare substances having certain pharmacological similarities such as in the rate of absorption or length of action.

Relation Between Structure and Activity

Until the work of Dodds and his collaborators appeared, it was thought that the cyclopentenophenanthrene nucleus was an indispensable structure in an estrogenic compound. Dodds and Lawson showed that the phenanthrene ring is not necessary. In fact, it is possible to obtain an estrous smear with a substance as anol containing only one benzene ring. They found that certain compounds consisting of two phenol groups joined by a carbon chain show estrogenic activity which varies with the number of carbon atoms in the chain, the position of the double bonds and of substituent groups attached to the carbon chain. Aromatic compounds lacking the hydroxyl group may perhaps be active through their ability to form such a compound in the body. For example stilbene, which requires 25 mg. to produce estrus, is oxidized in the rabbit to 4:4'-dihydroxystilbene, which requires only 5 mg. to produce estrus. It is possible (but not proved) that a similar oxidation occurs with other substances such as triphenyl ethylene. Dodds et all have pointed out that when diethylstilbestrol is depicted as shown below, it bears a resemblance to the natural estrogens. However the closer similarity produced by ring closure (dihydroxyhexahydrochrysene) is accompanied by a loss of estrogenic activity. An interesting study of the activity of an homologous series is shown in the stilbenediol series with alkyl substitutions on the ethylenic carbons. Maximum activity is obtained with diethyl substitution, chains longer or shorter than C2 having much less activity. Similarly, in the case of members of the dihydroxy diphenyl methane series, maximum activity is obtained when the intermediary carbons total six, producing a compound similar to diethylhexestrol as regards the intermediary carbons..

Biological Action in Experimental Animals

Except in a few minor details, the synthetic estrogens seem to produce to a remarkable extent the same qualitative changes in animals as do the natural estrogens. To use a convenient term introduced by Parkes, these substances are "gynecogenic" as well as estrogenic; that is, besides producing an estrous smear, they act on other tissues in a manner similar to the natural female sex hormones. Ample evidence of the essential similarity in the effect of these substances now exists. Small doses of diethylstilbestrol produce thickening and cornification of the vaginal epithelium of rodents and higher animals, growth and increase in weight of the uterus, sensitization of the uterus to the action of progesterone, and in some animals growth of the mammary gland duct system and nipples. In rats larger doses, in addition, cause cessation of growth and a decrease in body weight, atrophy of the ovaries, testes, prostate and seminal vesicles, and an increase in weight of the pituitary and in some cases of the adrenals. Diethylstilbestrol has been most studied in these respects, but such gynecogenic effects have also been produced by diethylhexestrol and the triphenyl ethylene series. The most outstanding distinction between the behavior of the synthetic and natural estrogens lies in the fact that the synthetic substances lose very little activity when they are administered orally.

Recent studies indicate that estrogens including diethylstilbestrol have certain metabolic effects which formerly received little attention. Calcium metabolism is influenced by estrogen administration. In rats, mice, and especially in birds, estrogens produce a rise in blood calcium which is apparently followed by closure of the epiphyseal lines and osteoblastic proliferation in the marrow cavity of the long bones. Diethylstilbestrol also has been found to have these effects in so far as they have been investigated. Similarly, in regard to fat metabolism, diethylstilbestrol resembles the lipemia in birds. The relation of estrogenic substances to carbohydrate metabolism is somewhat more controversial. Carbohydrate levels seem to be raised in rats by diethylstilbestrol and other estrogens. Muscle glycogen is not affected, but the blood sugar level tends to rise somewhat, while liver glycogen is considerably raised if the period of estrogen administration is prolonged. In some cases glycogen is stored in the liver to such an extent as to produce an unusual microscopic appearance of the hepatic cells. The mechanism of the glycogen storage effect does not seem to be direct. Thus estrogens also increase the insulin content of the pancreas, and since neither this effect nor the increase in liver glycogen is evident in hypophysectomized animals, it would seem that the action of estrogens in elevating liver glycogen takes place through an action on the anterior pituitary. However, the adrenal cortex may play a role in this response. Histological changes in the adrenals are well-known after estrogen administration and liver glycogen is not increased after diethylstilbestrol in adrenalectomized rats. The situation is further complicated by the observation that estrogens have a diabetogenic effect in some animals, while there are some clinical reports of diabetes mellitus being alleviated by estrogens. On the contrary, other clinical studies have shown little effect from estrogens, and diethylstilbestrol apparently does not affect experimental canine diabetes mellitus.

Toxicity

The first experimental studies on the effect of large doses of diethylstilbestrol led to the conception that the substance was potentially dangerous because it produced degenerative changes in the liver. Furthermore, the nausea and vomiting which is commonly encountered in the clinical administration of the drug demanded added, caution in its use. Recently, the trend of experimental work has indicated that the histologic changes in organs resulting from large doses of synthetic estrogens are not alarming and, in fact, are similar to those produced by natural estrogens when given in sufficiently large amounts. The administration of estrogens in the rat produces vacuolization of the hepatic cells which is due to glycogen deposition and not to a degenerative change. The adrenals are affected in some cases by a change which Cramer and Homing have caned "brown degeneration," but the significance of this change is still not clear. In the dog a marked depression of hematopoiesis occurs with estrogens; however, this effect apparently does not take place in human beings. None of the anatomical changes found in experimental animals can account for the nausea and vomiting seen clinically and it is likely that these phenomena are of central origin. Results of liver function tests on patients receiving diethylstilbestrol have usually been within the normal range.

In clinical experiences with synthetic estrogens nausea, vomiting, abdominal distress, anorexia, lassitude and vertigo are of frequent occurrence, especially when comparatively large doses are administered. Other toxic effects less commonly encountered have been diarrhea, rashes, paresthesia, and psychotic episodes.

In certain animals, estrogen administration over a long period of time will produce neoplasms. These results, first seen with the natural estrogens, have often been reproduced with diethylstilbestrol. Thus mammary adenocarcinoma can easily be produced in some strains of mice. Chromophobe adenoma of the anterior pituitary is seen in some mice and rats, and fibromata in guinea pigs have been observed to follow injections of diethylstilbestrol.

Metabolism

Little is known at the present time concerning the absorption, distribution and fate of the synthetic estrogens. Stroud injected rabbits with diethylstilbestrol, diethylhexestrol, and dihydroxydiphenylhexadiene and was able to isolate the free phenol in each case from the urine. He also obtained the free phenol after hydrolysis of the "combined" urinary fraction, indicating that conjugation of the estrogens had occurred. Of a total of 3 gm. of diethylstilbestrol injected, 25.2% was recovered in the urine, 14.2% free and 11.0% combined. These figures are probably low, since 30% of injected diethylstilbestrol has been isolated from rabbit urine in a conjugated form, identified as a monoglycuronide of diethylstilbestrol. Apparently the substance undergoes little destruction in the body, conjugation with glycuronic acid being the only known manner in which it is metabolized. This change probably takes place in the liver. The synthetic estrogens thus stand in marked contrast to the natural estrogens as regards their metabolism since the natural estrogens are largely oxidized in the body and only a small fraction is excreted in the urine.

Clinical Aspects Since its introduction, diethylstilbestrol has been subjected to extensive clinical trial as an oral substitute for the natural hormones in estrogenic therapy. Most studies have been directed to its use in the symptoms of the menopausal syn- drome. It is an effective agent in this respect. It is much cheaper than the natural estrogens and loses little potency when given by mouth, but it has the disadvantage of giving rise to unpleasant side actions which are discussed above. Diethylstilbestrol monomethyl ether, diethylhexestrol, and triphenylchlorethylene are also being subjected to clinical investigation at this time.These substances are also given in other conditions in which estrogenic therapy has been used. Diethylstilbestrol has proved to be of value in treating cases of senile vaginitis, leukoplakia vulvae, some cases of secondary amenorrhea, functional uterine bleeding and dysmenorrhea. It has apparently been successfully used in the suppression of lactation.

Ethidium Bromide Intercalates within the structure of nucleic acids in such a way that they fluoresce under UV light. Ethidium bromide staining is commonly used to visualize RNA or DNA in agarose gels placed on UV light boxes. Proper precautions are required, because the ethidium bromide is highly mutagenic and the UV light damaging to the eyes. Ethidium bromide is also included in cesium chloride gradients during ultracentrifugation, to separate supercoiled circular DNA from linear and relaxed circular DNA.

Etiolation Type of growth undergone by some plants in conditions of insufficient light; typically yields tall, weak stems, small leaves, and pale color

Eu-colloids Colloids whose colloid-particles are identical with single molecules of a substance; if many molecules coalesce the colloids are "association colloids."

Eucupine An alkaloid, isoamyl-hydrocupreine, which is used in the form of the dihydrochloride as a powerful bactericide and local anesthetic; also used internally in malaria, sepsis and influenza; externally to relieve pain in carcinoma, X-ray burns, hemmorrhoids.

Euglobulin A globulin found in the blood serum. Unlike pseudoglobulin it contains phosphorus, is insoluble in water, soluble in dilute salt solutions, and is precipitated by saturated ammonium sulfate.

Eukaryote Organism whose cells contain a membranebound nucleus and other discrete structures (*organelles*); includes all organisms other than *bacteria*

Euryhaline A term applied to animals that can withstand large changes in external salinity, e.g. shore crab Carcinus; opposite of stenohaline.

Evers reactions for ergot alkaloids A solution of ground ergot in glacial acetic acid is extracted with ether, the ether evaporated and the residue treated with an equal volume of 50% sulfuric acid. Careful mixing of the two layers gives rise to a violet-blue color.

Evipan Methyl-cyclo-hexenyl-N-methylbarbituric acid, a hypnotic causing general anaesthesia, requiring great care in use.

Evocation. That part of the morphogenetic effect of an organiser which can be referred back to the action of a single chemical substance, the Evocator.

Evolution Theory stating that new *species* emerge gradually through processes of genetic *mutation* and *natural selection.*

Ewald meal 2 slices of toast and 8 oz. water or their equivalents, used in testing power of the fasting stomach to secrete HCl, etc.

Exercise Under this head one might include the multitude of conditions relevant to muscular development. The physiologists have made their contributions concerning the complex adjustments during and followmg exercise in the respiration, circulation, and with the biochemists, in the composition of the blood and the tissues, in kidney function and in the nervous system, the endocrines, and in reaction to foreign substances which may be stimulants or depressants. The most commonly known fact is that of physiological albuminuria following violent exercise. Bicarbonate combats this condition. Occasionally there is also glycosuria after exercise. Though carbohydrates are essential they are not entirely irreplaceable as fuel for hard work. Biochemically "second wind" is increased oxygen intake along with related changes. Much has been made of the effect of hormones, which however need a great deal more of study. The effect of estrogens is very marked in catalyzing activity of mice and other animals running in cages. Sick patients are often helped by either male or female sex hormones. Special foods have been prepared for athletes and vitamins have been praised, possibly because they improve the general health. Emphasis on vitamin C1 however, is well indicated. Physostigmine is an instance of a "dope" which brings out extra activity for which the penalty may be paid later. Allergy to muscular work has been mentioned as identifiable by "hives" or asthma after exercise.

Exon Those portions of a genomic DNA sequence which will be represented in the final, mature mRNA ie. A contiguous segment of genomic DNA that codes for a polypeptide in a gene.The term "exon" can also be used for the equivalent segments in the final RNA. Exons may include coding sequences, the 5' untranslated region or the 3' untranslated region.

Exonuclease An enzyme that cleaves nucleotides sequentially from free ends of a linear nucleic acid substrate. An example is Exonuclease III, which digests only double-stranded DNA starting from the 3' end.

Explantation The development and growth of cells outside the organism from which they were derived, and in isolation from other tissue i.e. in vitro.

Expressed sequence tag (EST) A short piece of DNA sequence corresponding to a fragment of a complementary DNA (made from a cell's messenger RNA). ESTs have been used to hunt for genes, so hundreds of thousands are present in sequence databases.

Expression clone This is a clone (plasmid in a bacteria, or maybe a lambda phage in bacteria) which is designed to produce a protein from the DNA insert. Mammalian genes do not function in bacteria, so to get bacterial expression from your mammalian cDNA, you would place its coding region (i.e. no introns) immediately adjacent to bacterial

Expression Vector A plasmid or phage designed for production of a polypeptide from inserted foreign DNA under specific controls. Often an inducer is used. The vector always provides a promoter and often the transcriptional start site, ribosomal binding sequence, and initiation codon. In some cases the product is a fusion protein.

Expression To "express" a gene is to cause it to function. A gene which encodes a protein will, when expressed, be transcribed and translated to produce that protein. A gene which encodes an RNA rather than a protein (for example, a rRNA gene) will produce that RNA when expressed.

Expressivity The extent to which an animal carrying a gene and showing it is affected by it (the relative severity of the effect).

Exton reagents for albumin in urine I. A turbidity or precipitate is obtained when the urine is treated with a solution of S gm. sulfosalicylic acid and 20 gm. sodium sulfate in 100 cc.

II. One liter of a filtered solution of 50 gm. sulfosalicylic acid, 10 gm. sodium sulfate, 25 cc. of 0.4% aqueous solution of bromphenol blue may be used for the quantitative estimation of albumin.

Eye, biochemistry of One may speak of "chemical ophthalmology" as a possible division of chemical biology dealing with the chemical aspects of the entire evolution and functioning of eye tissues. The tissues of the eye resemble other tissues of the body and contain proteins much like the rest of the body with special characteristics now and then, e.g. the albuminoid of the lens substance. Keratin has not been discovered, but neurokeratin has been found. Protamines and histones have not been looked for. Many lipids have been found and there is a full quota of carbohydrates and glycogen. Many nitrogen compounds and enzymes have been reported. The inorganic constituents are distributed in an interrelated manner. Special adaptations are characteristic of the cornea, lens and vitreous humor: obtaining of oxygen partly from the air by the cornea, peripheral growth of the lens, which is without nerve and blood supply, the gel character of the vitreous humor. The role of visual purple in the rod-shaped cells of the retina is unique. The eyes are involved in diseases like xanthomatosis, xerophthalmia, night blindness and the dermatitis of avitaminosis B_1, cataract formation and glaucoma.

Divisions of the subject are: chemistry of the external secretions, chemistry of the conjunctiva, of the sclera, of the cornea, of the uveal tract (iris, ciliary body and choroid), of the retina, of the aqueous humor, of the vitreous humor, and of the lens. Chemical analyses have been made in the entire field, e.g. of tears showing many constituents including cholesterol, immunologically active proteins, enzymes, a bacteriolytic substance, lysozyme, and the like.

F

Fabinyi reaction for acetone 0.25 cc. of acetone are treated with 4 cc. N sodium hydroxide and 1 gin. of salicylaldehyde, shaken thoroughly and allowed to stand for 15 minutes. The product formed gives a red color with alkalies.

Factor V A factor necessary for the growth of certain haemophilic organisms, probably like Euler's cozymase or Warburg's coenzyme.

Fascia A sheet of band of fibrous tissue such as lies deep to the skin or invests muscles and various body organs.

Fate-map A map of an embryo in an early stage of development, e.g. the late blastula stage, indicating the various regions whose prospective significance has been established by marking methods. The regions are known as *Areas*, e.g. the area of presumptive notochord material.

Fats Esters of glycerol with three molecules of fatty acid. Fats are found in both the plant and animal kingdom, in the latter acting as supporting and protective tissue, and as a secondary source of energy. They are soluble in ether, $CHCl_3$, CS_2, etc., but sparingly soluble in ethyl alcohol.

Fat-sparing The action of sugar in preventing a flooding with fat which accomparties a ketosis or acidosis or a general carbohydrate failure of energy.

Fatty acid Monobasic aliphatic acid. Some restrict the usage to saturated acids. Cyclic acids such an chaulmoogric acid are sometimes termed fatty acids when they occur naturally as constituents of fats.

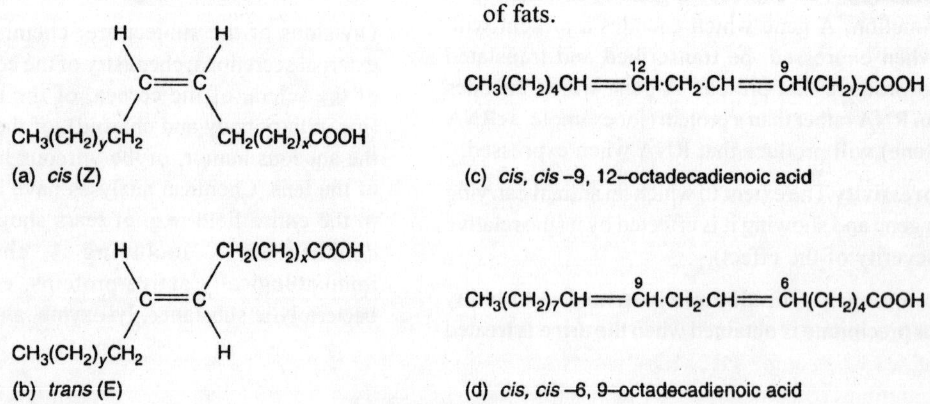

Fig. Fatty Acids

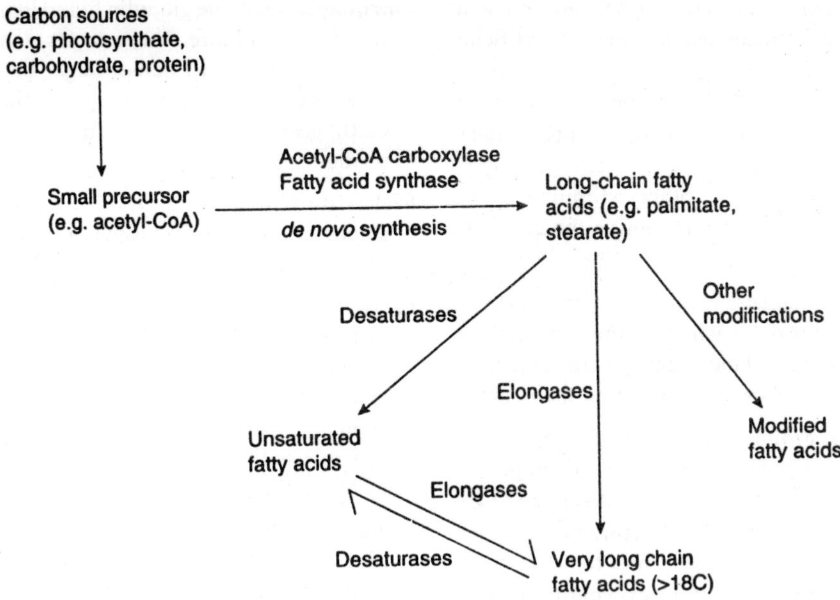

Fig.

Fig. Fatty Acid, Synthesis

Faught test for acetone 5 cc. of test solution are mixed with a solution of 0.06 gm sodium nitroprusside in 5 cc. distilled water and 5 drops of a 10% ethylenediamine hydrate solution are floated on this; acetone is indicated by a pink or red zone (no turbidity). Sensitivity—1:100000.

Fearon test for vitamin a In the presence of pyrogallol, trichloracetic acid and an oxidizing agent vitamin A gives a pink coloration.

Fechner's law The increase of intensity of sensation with the logarithm of the stimulus for moderate values above the threshold of sensation.

Fecundity Measure of the number of offspring an organism produces in a given time period

feedback inhibition Inhibition of an allosteric enzyme at the beginning of a metabolic sequence by the end product of the sequence.

Fehling reagent for glucose Solution A—34.64 gm. copper sulfate dissolved in water and diluted to 500 cc. Solution B—173 gm. Rochelle salt mixed with 150 cc. potassium hydroxide (sp. gr. 1.14) and diluted with water to 500 cc. Equal volumes of A and B are mixed previous to use. Red cuprous oxide precipitates when glucose is boiled with the solution. One cc. reagent is equivalent to 5 mg. glucose. Sensitivity—1:5000.

Fenchone $C_{10}H_{16}O$; d-form found in fennel and lavender oils, l-form in thuja oil; dicyclic ketone, m.p. 5°, b.p. 192°.

Fennel Dried ripe fruits of Foeniculum vulgare, constituent of compound licorice powder, an aperient; contains fenchone.

Fenton test for hexoses in urine A paste is formed by adding anhydrous calcium chloride to 4 cc. urine and boiled a few minutes with 10 cc. toluene and a few drops of phosphorus trichloride. The solution is cooled, after decanting the toluene, and treated with 1 cc. ethyl malonate and a little alcohol, and potassium hydroxide solution added dropwise. A rose-red color forms. A blue fluorescence is obtained when the solution is diluted with water or alcohol. Pentoses do not react.

Fermentation Energy-yielding anaerobic breakdown of a nutrient molecule, such as glucose, without net oxidation. Most commonly yields lactate in mammals or ethanol in yeast.

Fertilisation The essential stimulus to development given by the spermatozoon to the egg. In artificial *Parthenogenesis* this stimulus may be replaced by a wide variety of agents, some of which do not effect more than a few of the preliminary cytological changes preceding cleavage; this is known as *Activation* ; others stimulate more profoundly so as to put in motion the cleavage processes.

Fertilization Fecundation; union of sex cells to form a zygote; the process which stimulates cleavage and development of the ovum; generally effected by entrance of a spermatozoan into the ovum.

Fervenulin 6, 8-dimethylpyrimido 5. 4.e-1, 2, 4-triazine-5, 7-(6H, 8H) dione, Mr 193.17, m.p. 177, a pyrimidine antibiotic synthesized by Streptomyces fervens. Its structure is analogous to that of toxoflavin, but it less toxic. It has a broad spectrum of action, especially against cocci, Gram-negative and phytopathogenic bacteria and trichomonads.

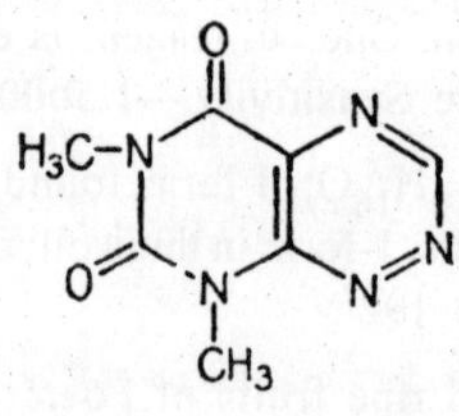

Fig. Fervenulin

Fever therapy Inductopyrexia; fever produced by electro-magnetic waves or equivalent means. The diseases treated typically are neurosyphilis, arthritis, gonorrhea, undulant fever, rheumatic fever, osteomyelitis, multiple sclerosis, allergy, optic atrophy and edema of the retina. Sulfanilamide is also used simultaneously. The outstanding successes have been in gonorrhea and neurosyphilis. The environment temperature in fever therapy is about 110° F. at a very high degree of humidity (in a cabinet). Body temperatures of 107° F. can be sustained for 3 hours. Pulse rates come up to as high as 120.

Fibrin The insoluble clotted material formed when thrombin or other material precipitates fibrinogen from the blood.

Fibrinogen A soluble globulin found in the plasma. It forms insoluble fibrin under a number of conditions, such as hemorrhage, tissue injury, or exposure to oxygen. It is essential to blood clotting and to repair of inflamed or damaged tissue.

Fichtelite $C_{19}H_{34}$; m.p. 46.5°; a highly stable saturated dimethyl isopropyl anthracene, found in peat and lignite.

Field, gradient A field in physics means a distribution of some non-material characteristic, such as force or energy, throughout portions of space unoccupied by material particles, but capable of interacting with such particles; an example is the electromagnetic field as treated by Maxwell.

The term has been broadened, especially in biological usage, to denote a spatial distribution of any characteristic, including the concentration of material particles or the rate of a chemical reaction. In this form it is virtually identical with the mathematical concept: "function of spatial coordinates."

That is: "There is a field of such-and-such in the chick embryo" means: "At every point of this embryo, the value of the characteristic 'such-and-such' is known or can be determined."

A caution is in order. The word "field" is frequently misused by using it in conjunction with a word which does not express a definite characteristic. In that case, the word is merely an elaborate circumlocution for ignorance. To say, "In the chick embryo there is a differentiation field" means simply, "At every point of this embryo a process whose nature I do not know is going on."

An associated concept is "gradient." Mathematically this is the value, at any given point of space, of the rate of change of a spatial function in the direction of its most rapid change; in a one-dimensional function it is the same as the slope.

In biological usage the term has come to have a more qualitative meaning: the existence of variation of a characteristic throughout a region of space (usually a region of a living organism), the variation being unidirectional (i.e., increase or decrease). An example is the "axial gradient" of Child. Thus, to say that a certain Planarian shows an axial gradient means that some characteristic either steadily increases or steadily decreases as one passes from head to tail.

Field A morphogenetic field is a system of order such that the positions taken up by unstable entities in one part of the system bear a definite relation to the positions taken up by other unstable entities in other parts of the system. The field effect is constituted by their several equilibrium positions. A field is bound to a material substratum from which a dynamic pattern arises. It is heteroaxial and heteropolar, has recognisably separate districts, and can, like a magnetic field, maintain its pattern when its mass is either reduced or increased. It can fuse with a similar pattern entering with new material if the axial orientation is favourable. The morphogenetic *gradient* is a special limited case of the morphogenetic field.

Filomusi—guelfi test for differentiating human and animal blood Animal blood yields characteristic crystals with a 2% sodium fluoride solution; dog blood gives acicular and rabbit blood tetrahedral crystals. Human blood does not form crystals.

Filterable virus An infectious, disease-producing agent, usually smaller in size than most bacteria and larger than most protein molecules. The introduction of a virus within the living cells of a susceptible host is followed by the production of more of the same virus. Multiplication occurs only within living cells and may be accompanied by mutation with the formation of a new virus strain. Some viruses have been isolated in the form of crystallizable high molecular weight nucleoproteins — others appear to be about as complex as bacteria.

Fischer reaction for alanine and phenylalanine An odor of phenylacetaldehyde is evolved when 0.02 gm. phenylalanine are dissovled in 2-3 cc. 25% sulfuric acid, a small crystal of potassium dichromate added and the solution heated to boiling. When a solution of alanine in 25% sulfuric acid is treated with a small crystal of potassium permanganate and boiled, acetaldehyde is given off.

Fischer-fischl reaction for glutathione in the lens of the eye When the lens is soaked in O.5M sodium or potassium cyanide for 8-18 hours, the nucleus and cortex are colored reddish-violet, the subepithelial protein layer is colorless, and the solution becomes yellow. The reddish color, due to glutathione, disappears upon acidification and is not given by glycocoll or glutamic acid.

Fischer-huppmann test for indican in urine The test sample is mixed with an equal volume of basic lead acetate solution and filtered; 10 cc. of filtrate are treated with 10 cc. of a saturated glacial acetic acid solution of acenaphthenequinone and 2 cc. concentrated hydrochloric acid. After heating to boiling, 10 cc. hexahydrotoluene are added; a red color is obtained.

FISH (fluorescence in situ hybridization) A physical mapping approach that uses fluorescein tags to detect hybridization of probes with metaphase chromosomes and with the less-condensed somatic interphase chromatin.

Fistula, biliary Arises spontaneously and as a result of injury. The spontaneous fistulae are probably always the result of inflammatory or malignant ulceration which establishes an abnormal passage between the biliary tract and some other organ or which leads to an outpouring of bile through an external opening in the abdominal wall.

Fistula, duodenal Most frequently due to the ulceration of a stone through the wall of the gall bladder. Fistulae between the common duct and duodenum also occur.

Fistula, gastric There are no particularly characteristic symptoms of this type of fistula. Occasionally, however, gall stones have been vomited, and in such cases the inference is strongly suggestive that the fistula is gastric, although it is possible that also in a high duodenal fistula there might be vomiting of calculi.

Fistula, mucous Occurs sometimes after the operation of cholecystostomy. In this condition there is a discharge of thin, stringy, almost transparent mucus without bile.

Flavianic acid 2,4-dinitronaphthol-7-sulfonic acid, used in the purification and characterization of organic bases, e.g. acetylcholine, choline, creatinine, ethanolamine, guanidine, tyramine, urea, etc.

Flavins A group of yellow pigments, including Vitamin G and Warburg's yellow respirative enzyme widely distributed in nature, showing a strong fluorescence. They all have the iso alloxazine nucleus.

Flavones The name given to the class of yellow plant pigments derived from flavone.

They are also called xanthones or anthoxanthones and may be regarded as phenyl derivatives of the 1,4-pyrone nucleus. Representatives are: quercetin, morin, brazflin, hematoxylin, gentisin.

Flavonol glycosides Include datisein, fustin, galangin, kaempferitrin, quercitrin, quercetin, quercimeritrin, serotin, incarnatrin.

Flavoprotein enzymes Flavin and protein containing enzymes which catalyze the oxidation of pyridine nucleotides and other substrates; can be resolved reversibly to lactoflavinphospbate and protein; include yeast I, yeast II, amino-acid oxidase (kidney) xanthine oxidase (milk) and fumaric hydrase (yeast), named after men or source.

Flavoprotein The flavin enzyme containing part which is cytoflav or lactoflavin-5phosphoric acid and a protein; molecular weight about 80,000; reduced form oxidized by cytochrome-c.

Fleming reaction for cysteine 0.5 cc. of a solution of 0.2 gm. of dimethyl-p-phenylene diamine in 100 cc. water are mixed with 1 cc. of test solution and treated with 1 drop 5% ferric chloride solution; a deep blue color, similar to methylene blue is obtained. The color may be salted out with sodium or zinc chloride.

Flow cytometry Analysis of biological material by detection of the light- absorbing or fluorescing properties of cells or subcellular fractions (i.e., chromosomes) passing in a narrow stream through a laser beam. An absorbance or fluorescence profile of the sample is produced. Automated sorting devices, used to fractionate samples, sort successive droplets of the analyzed stream into different fractions depending on the fluorescence emitted by each droplet.

Folic acid A vitamin-like substance from spinach and other leaves, which may be released from all animal tissues by autolysis. It promotes yeast growth and bacterial growth.

Folin and denis reaction A test for the phenol nucleus, which is also used for tyrosine. When a solution of a compound made by boiling phosphoric acid, phosphomolybdic acid and sodium tungstate together, is added to a solution of tyrosine, and then sodium carbonate is added to the point of alkalinity, a blue color results.

Folin reagents for creatinine I. Creatinine, with picric acid, forms a diffiultly soluble picrate. 11. Based on the Jaffé test, creatinine is determined colorimetricauy with the following solutions:

(1) 1.2% aqueous picric acid;

(2) 10% sodium hydroxide solution;

(3) 0.5N potassium dichromate solution for comparison.

Folin test for epinephrine (Adrenaline) Epinephrine may be detected, even at dilutions as high as 1:3000000, by the blue color produced on treatment with phosphotungstic acid solution.

Folin's reagent A test reagent for amino acids. ß-naphthoquinone sulfonic acid in the presence of strong alkalies gives a deep red color in the presence of most amino acids. It is used as a method for colorimetric estimation of an amino acid.

Folin-denis colorimetric determination of tyrosine in proteins The standard solution is prepared by solution of 0.001 gm. tyrosine in 25 cc. sodium carbonate solution and 5 cc. of the author's phenol reagent. The determination is made by heating 1 gm. of protein with 25 cc. 20% hydrochloric acid for 12 hours, diluting with water to 100 cc., adding 5 cc. of reagent to 2 cc. of the solution and mixing with 25 cc. sodium carbonate; the blue color is compared with the standard solution.

Folin-denis reagent and test for uric acid The

addition of 2 cc. of reagent and 5-10 cc. saturated sodium carbonate solution to 2 cc. uric acid solution produces a blue color. • Reagent: 10 gm. sodium tungstate are heated with 80 cc. 85% phosphoric acid and 750 cc. water for two hours, under reflux, cooled and diluted to I liter with water.

Folin-looney determination of cystine 1-5 gm. of dry protein are hydrolyzed with 25 cc. 20% sulfuric acid for 12 hours, in a kjeldahl flask, and diluted to 100 cc. 1-10 cc. of filtrate are treated with 20 cc. saturated sodium carbonate solution and 10 cc. of 20% sodium sulfite solution. 3 cc. of FolinDenis' uric acid reagent are added, after 5 minutes, and the solution compared colorimetrically with a standard solution.

Folin-looney tryptophane determination Trytophane may be determined colorimetricauy, in the same manner as tyrosine by the use of the Folin-Denis reagent. The blue color is compared with a standard.

Folin-mcellroy test for glucose Reagents—(1) 200 gm. of crystalline sodium phosphate and 50 gm. sodium thiocyanate are triturated in a procelain mortar and, after 10 minutes, mixed with 100 gm. anhydrous sodium carbonate.

(2) A 5.9% solution of copper sulfate. 0.4-0.5 gin. of (1) are added to 5 cc. of (2), heated until dissolved and about 1 cc. urine added.

Follicle stimulating hormone F.S.H., one of the hormones produced by the hypophysis. It causes an increase in the number of follicles in the female, and spermatogenesis in the male. It causes a flow of estrogens, but not androgens. The structure is apparently of a protein.

Footprinting A technique by which one identifies a protein binding site on cellular DNA. The presence of a bound protein prevents DNase from "nicking" that region, which can be detected by an appropriately designed gel.

Foreman's titration A method of determining the carboxy! group in amino acid mixtures, based on the fact that amino groups do not react basic in 85% ethyl alcohol, thus allowing direct titration of the acidic group.

Formaldehyde theory (of photosynthesis) A theory for the photosynthetic formation of carbohydrates in plants; which states that 'formaldehyde is the first product of photosynthesis and is condensed into the sugars. There are various mechanisms put forward to explain these reactions, but none have been proven.

Formic dehydrogenase A cytochrome-reducing dehydrogenase which catalyzes the oxidation of formic acid to CO2, no external coenzyme being necessary.

Formic enzyme The pyridinoprotein enzyme which catalyzes the oxidation of formate to carbon dioxide and water in presence of coenzyme I.

Formol titration A method of estimating carboxyl groups in amino acids and in proteins by binding amine groups with formaldehyde (not applicable to guanidine nucleus and hence to argnine).

Fossils Recognizable evidence of an organism that lived in the distant past. Most fossils are skeletons, shells, leaves, seeds, and tracks that are buried in rock layers before they could be decomposed.

Fouchet test for bile pigments in blood, feces and urine 10 cc. of urine and 5 cc. of 10% barium chloride solution are mixed and centrifuged. A small amount of a solution of S gm. trichloracetic acid and 2 cc. 10% ferric chloride solution in 20 cc. water is poured over the precipitate; a green color is developed in a few minutes. For blood, 5 cc. of clear serum are treated with 5 drops of reagent.

Frameshift mutation A mutation (deletion or insertion, never a simple substitution) of one or more nucleotides but never a multiple of 3 nucleotides, which shortens or lengthens a trinucleotide sequence representing a codon; the result is a shift from one reading frame to another reading frame. The amino acid sequence of the protein downstream of the mutation is completely altered, and may even be much shorter or longer due to a change in the location of the first termination (stop) codon:

Franke test for bilirubin in urine Shake 5 cc. urine

with 1-2 drops 0.2% methylene blue solution; a green color, sensitive to 1:500000, is produced. •

Fraxin $C_{16} H_{18} O_{10}$; a glycoside of the bark of the ash, consisting of glucose and fraxetin; it crystallizes as needles from alcohol, m.p. 320°.

Fredholm reaction for potassium Potassium may be detected and determined by the well defined, crystalline, sparingly soluble precipitate produced by interaction of the ion with dilituric acid (5-nitrobarbituric acid).

Freemartin A mammalian intersex due to the masculinisation of a female twin by its male partner when the foetal circulations are continuous.

Frehden-goldschmidt test for protein A small amount of sample is moistened with a saturated glacial acetic acid solution of p-dimethylaminobenzaldehyde and treated with 1 drop of concentrated hydrochloric acid, producing a violet coloration.

Frogs' spawn Concretions of a cartilaginous character formed in mucus fermentations, also known as "la fermentation visqueuse."

Fructose Levulose; fruit sugar; only hydrolytic product of inulin. Like glucose shows various ring forms and optical isomers in solution, along with mutarotation. As part of the molecule of sucrose it is a prominent constituent of the diet. The Harden-Young ester is a-fructose-1,6-diphosphoric acid and is involved in fermentation and carbohydrate metabolism, where it is in equilibrium with the Robison-Embden ester which is glucose-6-monophosphoric acid

Fructosediphosphoric acid Alpha-fructose-1, 6-diphosphoric acid; an important intermediate in the conversion of hexose to lactic acid; also formed in hydrolysis of sugar by yeast in the presence of phosphates.

Frustum Part of a geometrical solid between two parallel planes that pass through the solid parallel to its base

FSH (follicle stimulating hormone) One of the gonadotropic hormones of the anterior pituitary, which stimulates the growth and maturation of graafian follicles in the ovary, and stimulates spermatogenesis in the male.

Fucosan Polymer of 1-fucose which forms cell walls of sea weed.

Fucoxanthin Yellowish-brown pigment found in some members of the Chromista, including kelps and diatoms.

Fujita-iwatake-miyata test for ascorbic acid (vitamin c) A sky blue color is produced when 4 cc. of a solution of the substance in 2% metaphosphoric acid are treated with 1 cc. of a freshly prepared solution of 2 gm. sodium tungstate in 10 cc. N sulfuric acid, followed by the addition of 0.4 cc. 2N sodium hydroxide. 4.61 mg. ascorbic acid produces the same shade given by 1 mg. night blue.

Fumaric acid dehydrogenase An enzyme from yeast catalyzing the reaction between fumaric acid and reduced dyestuffs, such as lactoflavin.

Fumaric acid Occurs in Iceland moss and is formed by various moulds; ethylenedicarboxylic acid, m.p. 300° C.

Functional differentiation The differentiation of tissues as the result of forces arising out of the functions which they are performing.

Fundamental forces Four basic interactions that govern the behavior of matter in the universe; fundamental forces are *gravity* (acting between all particles and bodies that have mass), the *electromagnetic force.* (acting between all particles with electric charge), and the *weak* and *strong nuclear forces* (acting only on particles within the atomic nucleus)

Fusel oil A mixture of levorotatory amyl alcohol, isoamyl alcohol and some lower alcohols and their esters with the lower fatty esters formed in yeast fermentations.

Fusion protein A product of recombinant DNA in which the foreign gene product is juxtaposed ("fused") to either the carboxyl-terminal or amino-terminal portion of a polypeptide encoded by the vector itself. Use of fusion proteins often facilitates expression of otherwise lethal products and the purification of recombinant proteins.

G

Galactans Hexosans derived from galactose by the loss of water; found in gums, algae, pectins, mosses, etc.

Galactosans Polysaccharides found in plant structures. On hydrolysis they yield galactose.

Gall Abnormal growth on a plant caused by infection with a *parasite*

Gall bladder, physiologic capacity The number of hours of bile secretion the gall bladder can store and concentrate.

Gallic acid 3,4,5,-trihydrozybenzoic acid, m.p. 253° C., found in tea gall nuts, etc., as a constituent of tannins.

Gallstones Concretions usually mainly of cholesterol with small amounts of fats, phospholipids and metals, formed in the gall bladder due to an injury or the presence of a foreign body.

Gamete Reproductive cells which fuse to form a zygote. Gametes are haploid, and may be differentiated into egg and sperm.

Gametophyte The haploid stage in the life cycle of an organism undergoing alternations of generations. The gametophyte is multicellular and mitotically produces gametes. In plants, the gametophyte nourishes the zygote and young sporophyte.

Gamones Female sex determining factors, e.g. dimethyl esters of crocetin, a carotenoid dicarboxylic acid, effective for certain algae.

Ganglion Mass of nervous tissue consisting of numerous cell bodies and enclosed within a membranous sheath

Garrod test for homogentisic acid The acid is converted to the lead salt, dissolved in ether and decomposed with hydrogen sulfide, leaving the free acid in the ether; evaporation of the solvent yields crystals melting at 146-147°C This is said to be the most reliable test for the acid.

Gastrin A substance or substances found in pyloric mucous membrane which on injection causes increased flow of gastric juice. According to some, gastrin is histamine, this substance causing a copious flow of the acid-secretion of the stomach. Pilocarpine also stimulates the flow of the enzymes and other organic materials.

Gastro-enterology Gastro-enterology is that branch of internal medicine which deals with the physiology, pathology and treatment of the organs of digestion. The organs concerned with the process of digestion begin with the mouth cavity and its associated salivary glands, continuing with the pharynx, the esophagus, stomach, small intestine, colon, down to the rectum and anus. Supplementary digestive organs include the liver pancreas, and the organs comprising the bile tract.

The lips, teeth, cheeks, tongue, palate, pharynx and the salivary glands, the parotid, submaxillary and sublingual, are all concerned with the earliest steps of digestion. When these organs are diseased or missing proper mastication is interfered with and disturbance in digestion is prone to appear. The most important constituents of the saliva are a diastatic enzyme known as ptyalin, mucin, maltase, sulphocyanide of potassium, carbonate, and other inorganic ions.

One of the most important functions of the saliva is that of a solvent of various foodstuffs which facilitates the preparation of the food bolus as well as its swallowing. The diastatic ferment causes the starches to undergo their preliminary digestion. Saliva is believed by some investigators to possess also a high antiseptic action.

Esophagus

The esophagus or gullet is the beginning of the true digestive tract. It is a muscular membranous canal about nine inches long extending from the pharynx to the stomach.

After the food is mixed with saliva and well lubricated it is pressed backward by the tongue and pushed through the pharynx. Fluids or soft foods pass rapidly down the esophagus in a second or less, while it takes about four seconds for the passage of solid or thick foods.

Stenosis or obstruction of the esophagus is a common pathological finding in diseases of the esophagus.

It may be functional, due to hysteria or neurosis and is relieved in mild cases by sedatives. In more obstinate cases it is relieved by dilatation with instruments, or it may be organic due to changes in the esophageal wall.

Some of the pathological conditions responsible for organic changes are the malignancies, ulcerations either caused by the swallowing of caustic substances, tuberculosis or syphilis. Diverticuli (pouches), polypi (mucus membrane tumors attached to the wall of the organ by a stem) and varices (tortuous veins) are among the esophageal pathological conditions encountered.

Obstruction, complete or partial, and irregularities in the contour of the esophagus may be caused by pressure from external organs adjacent to the esophagus; such as an enlarged thyroid gland, heart and blood vessels, abscesses, etc.

The chief diagnostic means of abnormalities of the esophagus are fluoroscopic and roentgenographic examination with opaque mixtures, usually barium sulphate, and esophagoscopy (by use of an artificially lighted speculum inserted through the mouth into the esophagus).

The malignancies of the esophagus have at the present day a mortality of close to one hundred per cent.

The treatment is radium, deep roentgen-ray therapy, and in some early cases, surgical resection.

The Stomach

The stomach is a hollow muscular organ with a secreting surface. It consists of three layers; an outer or serous layer, an inner or muscular layer with circular and longitudinal fibers, and the third, the mucosa with its secretory glands. The stomach serves as a reservoir for the food ingested. It completes the mechanical subdivision of the food and prepares it for the small intestine. It starts the digestion of proteids and carries it to the acid albumin and proteose stage.

The stomach based upon its physiological activities divided into: cardia (upper part of stomach adjacent to the diaphragm), fundus (body of stomach), antrum (lower part of stomach adjacent to its opening), and pylorus (opening from the stomach to the small intestine).

Under fluoroscopy, shortly after ingestion of the barium meal, peristaltic waves begin at the fundus of the stomach passing down to the antrum. The antrum waves are deep constricting waves pushing the meal toward the pylorus. The pylorus opens only occasionally, not with every peristaltic wave. These waves churn up and macerate the food. The diastatic action of ptyalin on the starches begun in the mouth may continuefor a time in the fundus of the stomach and is then checked by the presence of free acid.

The gastric secretion appears shortly after the ingestion of food. It contains not only pepsin but hydrochloric acid. The acidity is normally regulated by the regurgitation of small amounts of alkaline pancreatic secretion through the pylorus.

The stomach is examined by means of a barium meal. The routine time of examination is during the meal, six hours afterward, and twenty-four hours afterward. Both the fluoroscope and x-ray films are used; the two methods supplement each other. The interior of the stomach can now be examined by direct visualization by means of the gastroscope (a flexible tube, artificially lighted, containing several lenses).

Gastric analysis is another important aid in the diagnosis of gastro-intestinal diseases. The old standard test meal is the Ewald test-breakfast, consisting of one and one-half slices of stale bread, with one and one-half glasses of water. This is taken when the stomach is empty. The stomach contents are withdrawn at the end of one hour by means of a tube passed into the stomach. The examination of the test-breakfast residue consists in testing the secretion, the quantity, presence of mucus, or blood, amount of free hydrochloric acid and total acidity.

In order to distinguish true achylia (absence of free hydrochloric acid) from false, the histamine test is used. This is done by the injection under the skin of 0.2 to 0.3 cc. of 1/1000 histamine solution.

Peptic or stomach ulcer and cancer of the stomach are of chief concern to the gastroenterologist. The cause of peptic ulcer is still debatable. Its size varies from ½ inch to 2 inches. It may be funnellike and deep, or more extensive and shallow. Peptic ulcers may heal spontaneously; burrow into a blood vessel and induce a hemorrhage; perforate through all the layers of the stomach wall and produce peritonitis; or cause an obstruction to the passage of food through the stomach by forming a scar at its opening in the process of healing.

Peptic ulcers are treated either medically or surgically.

Cancer of the stomach should be diagnosed at the earliest possible moment followed by prompt surgical removal of the malignant tissue. Cancer of the stomach is responsible for more than half of the total cancer mortality in most countries.

The Small Intestine

The small intestine is arranged anatomically into the first portion of the duodenum, which is fixed, and a movable second and third portion. The first portion surrounds the head of the pancreas. Under X-ray it appears as a smooth pointed globular cap. The second portion of the duodenum contains the openings of the bile and pancreatic ducts. The jejunum and ileum comprise the rest of the small bowel. The small intestine has four layers: an inner mucosal, a submucosal, a muscular, and a serosal layer. The mucosa contain the villi, the glands of Lieberkuhn, lymphatic tissue and large lymph nodes.

In the small intestine the main portion of the digestion of the food is carried on. The starches and sugars are broken down into simple monosaccharides and absorbed as such. The fats are split into fatty acids, glycerin and soaps. The final digestion of proteins takes place through the proteose, peptone and polypeptide stage, down to their breaking up into the amino acids.

The two common lesions seen in the small intestine are duodenal ulcer and obstruction. Lesions of the jejunum and ileum are not very common and clinical diagnosis is difficult.

The two methods of examination of the small intestine are by plain films of the abdomen in the presence of gas and after a barium meal. The progress of the barium meal is watched by repeated fluoroscopic and film examinations during the first six-hour period in order to detect obstruction or derformity.

The Large Intestine

The large intestine is about 5 feet in length. It has four layers; an inner mucosa, somewhat similar to the small bowel; a submucous layer; a muscular coat with internal circular and external longitudinal muscle layers and finally a serous coat. The cecum is the first part of the large bowel into which the small intestine is inserted by means of the ileocecal valve and from which the vermifor appendix arises. The rest of the colon is divided arbitrarily into the ascending colon, hepatic flexure, transverse colon, splenic flexure, descending colon and sigmoid.

The colon plays little part in the digestion and absorption of food stuff. It is mainly concerned with water absorption, cellulose digestion and excretion.

Peristalsis is slow and intermitant in the large bowel There is mass movement of the contents of the distal half of the colon once or twice a day. Spinal reflexes control the terminal end of the bowel, so that filling the rectum results in a reflex which causes contraction and evacuation of the bowel.

In examination of the colon, both barium meal and barium enema are used, the latter is the more important.

The outline of the colon may be changed by the filling defects of tumors or adhesions, or by kinking. The length and size of the colon may be abnormal. Spastic narrowing occurs in some cases of constipation and colitis, and organic narrowing in cancer and tuberculosis. In diarrheas the contents of the colon are seen, fluoroscopically, to rush through and in constipation the progress is slow.

The rectum filled with barium is smooth. Cancer, ulcerations, syphilis, tuberculosis or lymph granulomas produce irregular defects. All lesions in the rectum are better diagnosed by direct vision through the proctoscope (an artificially lighted tube inserted into the rectum).

The sigmoid, the part of the colon beyond the rectum, can be observed by a longer tube, about 12 inches in length, called the sigmoidoscope.

The common lesions of the colon are cancer, ulcerative colitis, tuberculosis and diverticulitis.

Cancer of the colon if diagnosed early and resected will prolong life for many years.

Ulcerative colitis, producing ulcerations in the colon, of unknown cause, is a chronic illness.

Tuberculosis of the intestine is generally secondary to pulmonary tuberculosis elsewhere.

Diverticulitis are inflamed small pouches or pockets protruding from the wall of the intestine which may become either abscessed or perforate into the peritoneal cavity.

Acute appendicitis still has a large mortality because of late treatment in the early stages. Its cause is unknown. Death from appendicitis is usually the result of peritonitis.

The infectious diarrheas are caused by various bacteria. The commonest are the amebic dysentery, acute bacillary dysentery due to the Flexner, Shiga, or Sonne organisms.

The diagnosis can be made in acute attacks by direct stool cultures or by the bacteriologic examination of the feces. Agglutination tests with the patients serum may be used in chronic cases.

The Liver

The liver, one the largest organs of the body, is more concerned with the metabolism of the body than any other organ. Bile formation and bile secretion is one of its multiple activities.

One of the main functions of bile is to assist in the digestion as well as absorption of fats. Bile checks the activity of pepsin but enhances the digestive power of the pancreatic juice. Bile salts have a mild stimulating influence on peristalsis.

Some of the diseases of the liver are: catarrhal jaundice, a mild inflammatory condition of the bile passages; acute yellow atrophy of the liver, a widespread destruction of the parenchyma of the liver; and various types of cirrhoses of the liver.

The Gall Bladder

The gall bladder is a pear shaped organ about 8 cm. in length and contains about 50 cc. of bile. It is attached to the liver. It serves as a reservoir for the storage and disposal of bile after it is formed in the liver. The gall bladded has two important functions. It stores bile during fasting and it empties this stored bile into the duodenum during digestion, especially of the fats in the food.

The gall bladder is a favorite seat for biliary calculi, mostly in persons beyond forty years of age. The theories for the causation of gall bladder disease are many. Severe inflammation of the gall bladder with suppuration, necrosis and perforation of the gall bladder wall and peritonitis may result.

X-ray is the method of choice in the diagnosis of biliary calculi. This depends on the administration of a special dye orally or intravenously, which when secreted in the bile reaches the gall bladder leaving a shadow on the X-ray film. In the presence of gall stones negative shadows will be observed. Pure cholesterol stones are not always visualized.

Biliary drainage for bile crystals will aid in the diagnosis of gall bladder calculi or cholelithiasis. The treatment in most cases is the total removal of the gall bladder.

The Pancreas

The pancreas is a fairly large gland resembling the salivary glands in its structure. It consists of head, neck, body and tail and rests on the posterior wall of the abdomen. Its anterior surface is covered by the posterior wall of the stomach.

The pancreatic duct extending through the substance of the gland passes obliquely through the wall of the duodenum terminating in an orifice common to it and the common bile duct.

The pancreatic juice digests all types of foods. It contains three important ferments: amylase which splits up the starches; lipase which splits up fats; and trypsin which digests the proteids. Chemically it contains sodium chloride, sulphate, phosphate, as well as calcium, magnesium and sulphur. The enzymes of the pancreas are activated by the intestinal juices and enhanced by the bile salts.Some of the diseases of the pancreas are acute hemorrhagic pancreatitis which comes suddenly, death often resulting in several hours; acute pancreatis, an acute infection of the pancreas occurring during the course of the disease of the stomach and duodenum; and chronic pancreatitis. This condition may follow acute pancreatitis. Calculi of the pancreatic duct, cysts and malignancy also may involve the pancreas.

Gastrula Stage of embryonic development, formed from the blastula by invagination of cells in the vegetative hemisphere and downward migration of cells of the animal hemisphere.

Gastrulation Stage of embryonic development in which cells become arranged into two or three primary tissue layers (germ layers); in humans, the layers are an inner endoderm, an intermediate mesoderm and an surface ectoderm.

Gaultherin $C_{14}H_{18}O_8.H_2O$; a glycoside of Gaultheria procumbens used in rheumatism, consisting of glucose and methylsalicylate; it crystallizes as needles or prisms from alcohol, decomposing at 200° C.

Gel shift assay (gel mobility shift assay, band shift assay) A method by which one can determine whether a particular protein preparation contains factors which bind to a particular DNA fragment. When a radiolabeled DNA fragment is run on a gel, it shows a characteristic mobility. If it is first incubated with a cellular extract of proteins (or with purified protein), any protein-DNA complexes will migrate slower than the naked DNA - a shifted band.

Gelatin A protein derivative made by treating collagens with superheated steam, or boiling in dilute acids. One of the distinguishing characteristics of gelatin is its ability to make heat-reversible gels. Often classified as an albuminoid.

Gelsemium Dried roots and rhizome of Gelsemium sempervirens containing the alkaloids gelsemine and gelseminine which are respiratory poisons; used as soporific.

Gene Discrete unit of hereditary information consisting of DNA and holding the instructions for the production of a protein

Gene Conversion The alteration of all or part of a gene by a homologous donor DNA that is itself not altered in the process.

Gene expression The process by which information coded by genes is converted into the structures present and operating in the cell. Expressed genes include those that are transcribed into mRNA and then translated into protein and those that are transcribed into RNA but not translated into protein (e.g., transfer and ribosomal RNAs).

Gene mapping Determination of the relative positions of genes on a DNA molecule (chromosome or plasmid) and of the distance, in linkage units or physical units, between them.

Gene product The biochemical material, either RNA or protein, resulting from expression of a gene. The amount of gene product is used to measure how active a gene is; abnormal amounts can be correlated with disease- causing alleles.

Gene superfamily A large group of genes related (often poorly) by sequence homologies or by the structures of their products, and by their involvement in different aspects of the same larger process (e.g. immunoglobulin gene superfamily).

Gene therapy A technique involving the use of foreign genetic material to correct a genetic defect or to modify the phenotype of an affected individual, by targeting the somatic cells.

Gene The fundamental physical and functional unit of heredity (Usually DNA, Some organisms have RNA a gene). A gene is an ordered sequence of nucleotides located in a particular position on a particular chromosome that encodes a specific functional product (i.e., a protein or RNA molecule) that contributes to or influences the phenotype of the cell. A gene usually contains coding regions, introns, untranslated regions and control regions.

Genes Any unit of the units of instruction for heritable traits. Each gene is a linear sequence of nucleotides that calls for the assembly of a sequence of specific amino acids into a polypeptide chain.

Genetic code The sequence of nucleotides, coded in triplets (codons) along the mRNA, that determines the sequence of amino acids in protein synthesis. The DNA sequence of a gene can be used to predict the mRNA sequence, and the genetic code can in turn be used to predict the amino acid sequence.

Genetic disease Heritable genetic alterations from wild type which, when expressed, result in decreased viability of the individual receiving the altered gene(s).

Genetic map The ordering of genes by the statistical determination of recombination events between them. Genes separated by greater distances are more likely to recombine.

Genetic recombination Presence of a new combination of alleles in a DNA molecule compared to the parental genotype; the result of processes such as crossing over at meiosis, chromosome rearrangements, gene mutation and recombinant DNA technology.

Genetics, biochemical aspects of As contrasted with "classical genetics," which is concerned with the mechanism by which hereditary characters are transmitted from one generation to another, biochemical genetics has to do primarily with the general question of how genes, the hereditary determiners, produce their characteristic effects. It may be said to differ from physiological and developmental genetics in that it attacks the general problem of gene action at the level of the biochemical reaction rather than at the higher levels of functional and structural organization.

The existence of genes was first inferred from the data of classical genetics. The rules by which they are transmitted indicate that they are particulate and are carried in groups (chromosomes) within which definite sequential relations obtain. The relation of the genetically inferred genes to the observed chromosomes was made possible by the techniques of cytology.

By means of biochemical methods the properties and modes of action of genes can be studied in either of two principal ways, viz.,

(1) by investigating the chemical properties of chromosomes and genes themselves, or

(2) by inferring from biochemical descriptions of hereditary characters something as to the properties of the antecedent genes.

Direct chemical analyses of chromatin from sperms indicate that chromosomes are composed mainly of thymonucleic acid (ca. 60%) combined with histones and protamines (together ca. 35%). Staining methods and the ultraviolet absorption method of Caspersson indicate that the nucleic acid is concentrated in the dark-staining bands of the giant salivary gland chromosomes of Diptera. Enzyme digestion experiments by Caspersson and by Mazia show that nuclease breaks down this nucleic acid component but does not destroy the continuity of the chromosome. Trypsin, on the other hand, does destroy the basic structure, a fact consistent with the assumption that the continuity depends on histones and protamines. As expected on this assumption, pepsin does not destroy chromosome continuity.

As described by Stanley and others, viruses show several striking similarities to genes, among them, rough similarity in size, reproduction in living cells, presence of nucleic acid, mutability, and production of specific phenotypic effects. If an essential similarity between viruses (including phages) and genes can be firmly established, much that is learned of one can be carried over to the other.

Hereditary characters in many organisms have been at least partly described in terms of their component biochemical reactions. The hereditary abnormality alcaptonuria in man, for example, is characterized by an inability to oxidize homogentisic acid (Garrod). As a consequence, this metabolic derivative of phenylalanine and tyrosine is excreted in the urine. A somewhat similar situation is found in phenylketonurics. Here there is an inherited inability to oxidize phenyl pyruvic acid to p-hydroxyphenyl pyruvic acid. A significant and interesting aspect of this disease is the fact that it is invariably associated with imbecility or idiocy (Fölling, Penrose). The usefulness of such gene-controlled disturbances as these in interpreting normal metabolism has been pointed out by Haldane and others but nevertheless has been only inadequately appreciated by biochemists, physiologists, and others.

Flower pigmentation has been studied extensively both genetically and biochemically. The substances concerned are mainly anthocyanins, anthoxanthins, carotenoids, and socalled co-pigments. The presence of these pigment components is dependent on gene activity, and they may be genetically modified either quantitatively or qualitatively. Among the qualitative variations in anthocyanins, state of oxidation, glycosidal type, and degree of methylation are determined by genetic constitution. Anthocyanins are pH indicators, and genes are known which influence pigmentation by determining the pH of cell sap. In the dahlia correlations in amounts of different pigment components have been found which indicate that the normal biosyntheses of anthocyanins, flavones, and chalkones involve a common precursor. If more of one component is produced, this tends to be at the expense of the other two.

Wright and others have made extensive genetic studies of melanin pigmentation in mammals. For the guinea pig, Wright has drawn up an ingenious detailed scheme of gene action based largely on genetic data. This scheme postulates that genes act by exerting more or less direct control over reaction rates. Because of the difficulty of working with melanin pigments chemically, unfortunately many of the biochemical predictions of the hypothesis cannot be readily tested.

By methods somewhat similar to those used by Wright in his interpretation of pigment interactions in the guinea pig, studies have been made of the development of eyepigment components in Ephestia, Drosophila, and other insects. A number of biochemical predictions made from the data of genetic and transplantation experiments have been verified. Thus tryptophane and kynurenine have been identified as precursors of the brown pigment component.

Kimball finds that mating types in Euplotes are genetically controlled and that the control involves the production of specific substances. There are six mating types dependent on three alleles of a single gene. The six types are accounted for by the three homozygotes plus the three heterozygotes. Apparently the mating types are characterized by three specific substances, produced singly by the three homozygotes and in combinations of two by the heterozygotes. The suggested one-to-one relation between gene and substance in this situation is reminiscent of the direct relation generally found between gene and antigen in genetically determined immunological characters. Irwin and Cole and co-workers, for example, find that in general in the pigeon and dove hybrids with which they have worked, there exist simple relations between genes and agglutinogens. There are, however, so-called "hybrid substances" produced in certain combinations which appear to depend on interactions of genes from the two parents.

Winge and Laustsen find that the abilities of yeasts to utilize certain disaccharides vary with different strains. In the hybrids investigated the ability to decompose a given disaccharide, presumably contingent on the presence of a specific enzyme, is dominant to the alternative inability.

On the assumption that biochemical reactions in general are controlled by genes, Beadle and Tatum have tested X-ray-treated material of the red bread mold Neurospora for loss of ability to synthesize specific growth factors. In this way a mutant strain has been found which is unable to grow in the absence of p-aminobenzoic acid. Another requires the thiazole half of the thiamin molecule and a third is unable to grow in the absence of pyridoxin. In each case the inability to carry on the synthesis results from a modification of a single gene.From facts such as those reviewed here, it seems likely that genes are specific protein molecules or groups of molecules possessing the power of reproduction under appropriate conditions in living cells. As a working hypothesis it may be supposed that their primary function is that of controlling protein specificities either by serving as models or in some other way. Since on this assumption enzyme specificities are referable to gene specificities, genes may be thought of either as synonymous with enzymes or as bearing a one-to-one relation to them. While experimental data dealing directly with this point are not abundant, there are suggestions that each enzymatically catalyzed reaction is under the direct control of one particular gene and, conversely, that each gene controls in a primary way one specific reaction. While it is probable that these considerations are based on concepts that are too simple, they may nevertheless serve to indicate that any complete view must regard genes as integral parts of the biochemical system rather than as hypothetical units conceived merely to be juggled by geneticists in the solution of intricate puzzles of heredity.

Genome projects Research and technology development efforts aimed at mapping and sequencing some or all of the genome of human beings and other organisms.

Genome The entire compliment of genetic material in the form of permanently maintained DNA for a given organism. Its size is generally given as its total number of base pairs. Mammalian genomic DNA (including that of humans) contains 6 billion base pairs of DNA per diploid cell. There are somewhere in the order of a hundred thousand genes, including coding regions, 5' and 3' untranslated regions, introns, 5' and 3' flanking DNA. Also present in the genome are structural segments such as telomeric and centromeric DNAs and replication origins, and intergenic DNA.

Genomic blot A type of Southern blot specifically used to analyze a mixture of DNA fragments derived from total genomic DNA. Because genomic DNA is very complicated, when it has been digested with restriction enzymes, it produces a complex set of fragments ranging from tens of bp to tens of thousands of bp. However, any specific gene will be reproducibly found on only one or a few specific fragments. A million identical cells will produce a million identical restriction fragments for any given gene, so probing a genomic Southern with a gene-specific probe will produce a pattern of perhaps one or just a few bands.

Genomic clone A piece of DNA taken from the genome of a cell or animal, and spliced into a bacteriophage or other cloning vector. A genomic clone may contain coding regions, exons, introns, 5' flanking regions, 5' untranslated regions, 3' flanking regions, 3' untranslated regions, or it may contain none of these...it may only contain intergenic DNA (usually not a desired outcome of a cloning experiment!).

Genomic library A collection of clones made from a set of randomly generated overlapping DNA fragments representing the entire genome of an organism. Compare library, arrayed library.

Genotype The set of genes that an individual carries; usually refers to the particular pair of alleles (alternative forms of a gene) that a person has at a given region of the genome.

Gentianose $C_{18} H_{32} O_{16}$; a non-reducing trisaccharide, a-gentiobiosyl D-fructofuranoside or, ß-D-glucopyranosylsucrose, which occurs in the rhizomes of the gentian species from which it is extracted with alcohol; on hydrolysis with invertase it gives fructose and gentiobiose; m.p. 209°.

Gentiobiose A reducing disaccharide found in many glycosides, usually prepared from the trisaccaride gentianose. It is glucopyranose-6-ß-glucopyranoside.

Gerantotherapy (benjamin) The treatment of the ageing process as an entity, or of senescence and senility as such. It is distinguished from geriatrics which limits itself to diseases of old age.

Germination The process by which a seedling emerges and develops from a seed, or by which a sporeling emerges and develops from a spore.

Gerrard test for atropine and hyoscyamine bases An alcoholic solution of the base is heated with an aqueous or alcoholic solution of mercuric chloride, forming a brick-red precipitate of mercuric oxide along with the alkaloid hydrochloride. Daturine and duboiscine, but not scopolamine, also give this reaction.

Gestation Period of mammalian development from the time a fertilized egg cell is implanted into the *uterus* wall to the time of birth

Giemsa test for quinine A solution of 10 gm. potassium iodide in 50 cc. water is mixed with a solution of 27 gm. mercuric chloride in 1500 cc. water and 2 cc. of glacial acetic acid added; an opalescence at dilutions as high as 1:100000 to 400000, is obtained with quinine solutions.

Girard's reagent A name originally applied to one compound, trimethylamino acetohydrazide hydrochloride, but since grown to include several compounds, which form water-soluble ketone derivatives, and so serve as a means of purifying them.

Glacier Large body of ice traveling gradually down a slope from a high elevation in the mountains

Gland Organ that collects substances from the blood, uses them to synthesize new materials, and secretes the new materials into the blood for a particular purpose

Glass electrode An electrode used for determining hydrogen ion concentration, which is rapidly replacing all others. It is rapid, accurate, and does not in any way contaminate the solution to be analyzed. It consists of the following concentration cell, with a glass membrane at the heavy line: Hg/HgCl, KCl/Soln. I/Soln. II/KCl, HgCl/Hg where solution I and solution II differ from each other in hydrogen ion concentration.

Glaucoma A disease of the eye in which the increased intraocular tension injures the eyeball, culminating in its hardening, opaque cornea, and cataract. The intraocular tension may reach 80 mm. of mercury and higher by tonometer. The disease may be acute or chronic and is difficult to control without surgery. Miotics, like physostigmine and pilocarpine salts, are employed widely.

Glaugine An alkaloid, $C_{21}H_{25}O_4N$, from herb Glaucium luteum, which paralyzes the heart, suspends the sensibility and influences the striated muscle while suspending the excitability.

Gliadins In the American Classification of Proteins, the gliadins are simple proteins, soluble in 70-80% alcohol, but insoluble in water, absolute alcohol and other neutral solvents. Examples: gliadin from wheat, zein from maize.

Globular protein hypothesis A theory relating to the nature of protein structure advanced by Svedberg. He finds that many proteins are globular in form. Long polypeptide chains may roll themselves up into this shape. He finds that particle weight of the proteins are multiples of 35,000.

proteins Soluble proteins with a globular (somewhat rounded) shape. Their axial to diameter ratio is less that 10:1, and usually closer to 1:1.

Globulins In the American Classification of Proteins, the globulins are simple proteins, insoluble in water, but soluble in dilute neutral solutions of salts of strong acids and strong bases. Example: serum globulin, and edestin of hemp seed.

Glomerine A quinazoline alkaloid in the defense secretion of the insect Glomeris marginate, m. p. 204 °C. It is one of the few animal alkaloids. The secretion is exuded from 8 pores arranged in a row on the 1.5 cm animal. It contains about 50 µg G. and another alkaloid homoglomerine (m.P. 149°C).

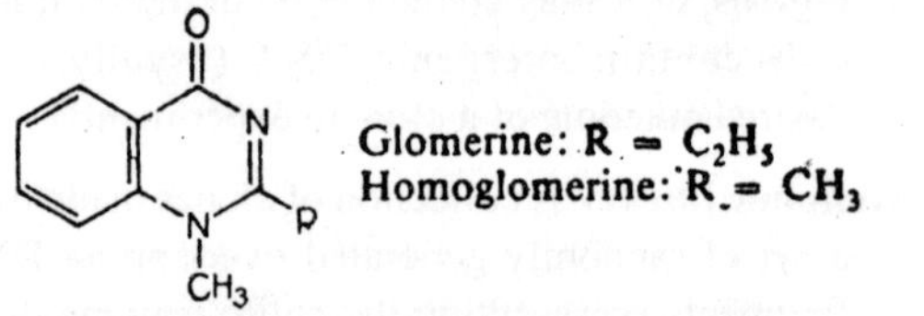

Fig. Glomerine

Glucose C6H12O6; dextrose; corn sugar; grape sugar; starch sugar; forms a monohydrate and

COMMON GLYCEROPHOSPHOLIPIDS

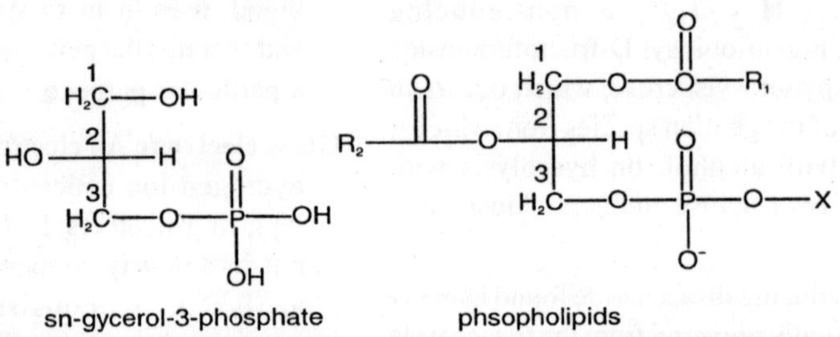

NAME OF X-OH	FORMULA OF X	NAME OF PHOSPHOLIPID
water	—H	phosphatidie acid
ethanolamine	$-CH_2CH_2NH_3^{\oplus}$	phoshatidyleth anolamine
choline	$-CH_2CH_2N(CH_3)_3^{\oplus}$	phosphatidyloholine (leoithin)
myoinositol	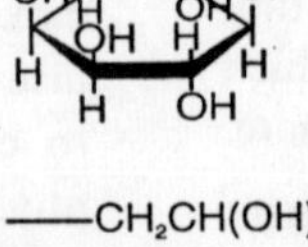	phosphatidylinositol
ghyoerol	$-CH_2CH(OH)CH_2OH$	phosphatidylglycerol

Fig. Common Glycerophospholipids

exists in numerous isomeric forms at least in solution allowing for its great variety of actions. A straight chain formula shows a free aldehyde group, oxide ring formulas make it potential but permit a space isomerism shown by á-and ß-forms of glucose. The forms á-d-glucose and ß-d-glucose are variants due to the pyranose ring. A furanose ring yields y-glucose. The alpha form has a rotation of + 110°, the beta form + 19° and the gamma form + 52.5°. The various forms achieve equilibrium in solution, which accounts for a shift toward an equilibrium value of optical rotation, known as mutarotation. There are, however, possibly eight cyclic configurations in solution besides the straight chain, the pyranose (amylene oxide) ring predominating.

Glucose is involved in photosynthesis, fermentation, blood chemistry, muscle action, diabetes, dietary balance, etc.

Glucose oxidase (aspergillus) An enzyme of the juice of Aspergillus niger which catalyzes the oxidation of d-glucose to gluconic acid, optimum pH 6, also attacks d-mannose and d-galactose.

Glutamic acid metabolism Glutamic acid may be oxidized to NH3 and alpha-ketoglutaric acid, which on ß-oxidation ultimately yields CO2 and H2O. Alternately glutamic acid plus NH3 yields glutamine. It can form glucose, thus being anti-ketogenic.

Glutamic aminophorase A transamination enzyme of animal and plant tissues which promotes the conversion of glutamic and pyruvic acids to alpha-glutaric acid and alanine and vice versa.

Glutamic enzyme (1) A pyridinoprotein enzyme which catalyzes the oxidation of 1(+)-glutamic acid to alpha-ketoglutaric acid, in presence of coenzyme I; found in kidney, liver, heart; (2) a triosephosphopyridinoprotein enzyme which catalyzes the oxidation of 1(+)glutamic acid by coenzyme II.

Glutaminase An enzyme which promotes the mutual conversion of glutamic acid and glutamine, found in beet roots and blades of rye-grass.

Glutamine $CO(NH_2)CH_2CH_2CH(NH_2)COOH$; the acid amide of glutamic acid, found in many proteins.

Glutathione $HOOCCH(NH)_2CH_2H_2CONHCH(CH_2SH)CONHCH_2COOH$; a tripeptide composed of cysteine, glycine, and glutamic acid. It occurs in plant and animal tissues both and may function as the intraorganismal transport of amino acids and biocatalytically through its sulfhydryl group is a participant in vitamin-reduction processes.

Glutelins In the American Classification of Proteins, are simple proteins, insoluble in all neutral solvents, but readily soluble in very dilute acids and alkalies. Some can be separated into two fractions by adding ammonium sulfate to an NaOH solution of the crude protein. At 3% saturation an alpha glutelin separates, at 16% saturation a beta glutelin separates. Example: glutenin from wheat.

Glutose $CH_2OH(CHOH)_2CO\text{-}CHOH\text{-}CH_2OH$; a hexose in which the keto group is on C_3; it is not metabolized, nor is it effective in treating insulin shock.

Glycerol $CH_2OHCHOHCH_2OH$; a syrupy, high boiling triatomic alcohol, which when esterified with fatty acids forms the fats, the latter being a source for glycerol. It is also prepared by yeast fermentation of carbohydrates.

á-GLYCEROPHOSPHORIC ACID A naturally occuring ester in the laevo form.

á-GLYCEROPHOSPHORIC DEHYDROGENASE A cytochrome-reducing dehydrogenase of practically every animal tissue which catalyzes the oxidation of 1(+)-alpha-glycerophosphate to glyceraldehydephosphate especially in presence of cytochromec.

á-GLYCEROPHOSPHORIC ENZYME A pyridinoprotein enzyme of heart, intestine and skeletal muscle, which catalyzes the oxidation of alpha-glycerophosphate by coenzyme I, i.e. the d(-)-isomer only.

Glycine $C_2H_5O_2N$, glycocoll, amino-acetic acid; the simplest amino-acid; i.p. 6.09, large colorless prisms, m.p. 225-230°, sol. H_2O; found in connective tissue, the keratins, and forms unabsorable compounds with many poisons in the body.

Glycine metabolism In vivo, glycine undergoes several reactions: a) oxidation to NH_3, CO_2, and H_2O, b) combination with benzoic acid analogues to form various hippuric acids, frequently for detoxification, c) combination with cholic acid to form glycocholic acid, d) formation of glucose in the diabetic. Glycine is anti-ketogenic. Under certain conditions, it is synthesized in the body.

Glycocholic acid $C_{23}H_{36}(OH)_3CONHCH_2COOH$; the peptide combination of cholic acid and glycine; a form in which the bile acid is found in human and ox bile.

Glycocyamine Guanidine acetic acid, considered to be the physiologic precursor of creatine; converted to urea and glycine by glycocyaminase (of liver).

Glycogen Animal starch; the characteristic fuel reserve carbohydrate of animals; it is a white, water dispersible powder, does not give a positive Fehlings test; hydrolyzes to glucose.

Glycogenolysis The conversion of glycogen into glucose; inhibited by insulin, caused by the enzyme glycogenase, and promoted by epinephrine.

Glycolipids Compound lipids containing carbohydrates but no phosphorus, e.g. cerebrosides or galactolipids.

Glycolytic enzyme complex The heat-labile enzyme proper and the heat stable dialyzable coenzyme of fermentation found in cellfree muscle extract.

Glycoprotein A membrane-bound protein which has attached branching carbohydrates. These may function in cell-cell recognition, such as in human blood groups and immune system response, as well as in resisting compression of cells.

Glycosides A group of carbohydrate derivatives in which some organic group is attached to the oxygen of carbon of the sugar. Glycosides are frequently found in nature accompanied by their specific hydrolysis enzymes; often are extremely poisonous. The majority are crystalline, water and alcohol soluble solids.

Glycosuria Mellituria; the appearance of glucose in the urine, may be due to food or "alimentary glycosuria",or "afternoon glycosuria"; in pathological conditions like diabetes, Addison's disease, pituitary disease, it is likely to be high and persistent.

Glycosurias, non-diabetic The most commonly used test in this country for the detection of sugar in the urine is the Benedict test. The reagent consists of an alkaline copper solution and the detection of sugar depends upon the conversion of the soluble blue cupric oxide into an insoluble cuprous oxide under the influence of heat. Since it is a reduction test, it is necessary to know that urine may contain reducing substances other than glucose. They may be normal or abnormal constituents of the urine. The normal may be phosphates.The abnormal constituents which may be found in the urine are glycuronic acid, lactose, galactose, pentose (xyloketose and arabinose) and fructose (levulose). According to Lasker and Enklewitz while doing the Benedict test, it is possible to differentiate the reducing sugar by noting the amount of reduction after certain intervals and at the end.di-nitro-salicylic acid method can also easily differentiate the various sugars. Further differentiation can be carried out with the aid of the following tests:

1. Fermentation with baker's yeast Glucose and fructose are the only fermentable sugars.

2. By polariscopic examination Glucose will be found to be dextrorotatory and fructose levorotatory.

3. Selivanoff's test Positive with fructose which can thus be differentiated from glucose.

4. Naphthoresorcinol test Determines the presence or absence of glycuronic acid or glycuronates as the reducing substance.

5. Bial's test Positive in pentosuria.

6. Mucic acid test Positive only in the presence of lactose or galactose.

7. Lead Acetate test Further identifies lactose.

8. For a more exact determination of the various

sugars characteristic osazone crystals can be prepared by treating the urine with phenyl or methyl-phenylhydrazine. The crystals can further be distinguished by their melting point.

Reducing substances in the urine which are not due to glucose are usually found in small amounts from 0.1% to 0.6%. Glycuronates come and go. Pentose is a constant finding in essential pentosuria. Fructose varies in fructosuria with the diet. Lactose and galactose vary in the amounts. Glucose may be found in small and in large amounts.

The finding of glucose in the urine does not always mean diabetes mellitus. If the blood sugar is never found to be elevated, even in the glucose tolerance test, the glycosuria is called renal glycosuria. But even the finding of an elevated blood sugar does not always mean diabetes since other conditions may cause hyperglycemia with glycosuria. The non-diabetic melliturias may thus be divided in two main groups, one without hyperglycemia and the other with hyperglycemia. The first group may show glucose or other reducing substances in the urine, while the second group shows only glucose.

NON-DIABETIC GLYCOSURIAS WITHOUT HYPERGLYCEMIA

In the discussion of the non-diabetic glycosurias without hyperglycemia, one should mention a few words about the positive Benedict test in the presence of phosphates, glycuronic acid and glycuronates. Phosphates may give a slightly positive test with Benedict solution. It will be found after a high vegetable diet. The solution is not very opaque and the precipitate is coarse, flocculent and white or grayish. It is not a constant finding and it is related to meals. Glycuronic acid may appear in the urine as a compound or in free form due to spontaneous decomposition of the compound. It is usually found in the urine following the use of coal tar or other products (aspirin, salicylates, pyramidone, antipyrin, camphor, etc.). With the omission of the preparations, the Benedict test of the urine becomes negative. The naphthoresorcinol test gives a positive reaction.

The most important non-diabetic glycosuria without hyperglycemia is renal glycosuria. It is a glycosuria with a low renal threshold. It is called by some "renal diabetes" or "diabetes innocens". The urine contains glucose. It is fermentable and gives a negative Selivanoff reaction. In treating the urine with phenylhydrazine, typical glucosazone crystals with a melting point of 205° are obtained. The blood sugar is within normal limits all the time. This means that the patient has a low renal threshold for sugar. There are no symptoms of diabetes. Renal glycosuria is inherited and therefore should be looked for in the relatives.

According to some authors patients with renal glycosuria may later develop true diabetes. Whether there is a transitory renal glycosuria there is not enough evidence except in pregnancy. The youngest case reported, that of a 20 month old infant, showed a renal glycosuria only for six weeks and then the urine became sugar-free.

Prolonged follow-up of patients with renal glycosuria by repeated glucose tolerance tests may show occasionally an elevated blood sugar in the curve. Caution should be taken to instruct the patient to follow a normal diet for at least a week preceding the glucose tolerance test since the low carbohydrate diet may effect the test.

Phlorizin glycosuria is an experimental type of diabetes which was considered by some to be similar to renal diabetes because of the glycosuria in the presence of a normal blood sugar. It is produced by the subcutaneous injection of the glycoside phlorizin. Except that the blood contains less sugar, the metabolism is as seriously affected as in pancreatic diabetes. The damage is probably in the kidney, which causes the glycosuria, and also in the liver. There is an increased destruction of protein with sugar production (hyperglyconeogenesis). This fact distinguishes this from human renal diabetes.

Glycosuria in Pregnancy. Pregnancy is the only condition in which a true renal glycosuria is found to be transitory. It may recur in the same patient with each pregnancy. It is a frequent occurrence. It usually appears during the 30th and 33rd week and disappears just before and shortly after the birth of the baby or if the pregnancy is interrupted. Williams found sugar in the urine four times as often in the second half. It has also been observed in animals. If in such animals the ovaries are removed, the glycosuria disappears, while the implantation of the ovaries of pregnant animals into non-pregnant will produce glycosuria.

The presence of sugar in the urine of a pregnant woman should be taken seriously since occasionally it may be the onset of a true diabetes. Blood sugar studies as well as the determination of the type of sugar in the urine should be carried out and the condition should be followed for some time before a definite diagnosis is made. The presence of acetonuria is believed by some to make the diagnosis of diabetes more likely. I have observed prolonged acetonuria in a pregnant woman with renal glycosuria. Some patients with renal glycosuria during pregnancy develop true diabetes later in life.

SUGAR IN THE URINE OTHER THAN GLUCOSE

Lactosuria Lactose appears in the urine in the last stages of pregnancy and during lactation, although it may occur early in pregnancy. Watkins studied intensively the lactose metabolism in women and also describes in detail a method for quantitative determination of lactose in urine. Lactose is a C_6 sugar with a slightly different configuration to glucose. It is a combination of glucose and galactose. Lactose is found in breast-fed children suffering from digestive disorders. It gives a positive Benedict test, a negative fermentation test, a positive mucic acid test and a positive lead acetate test. Lactosuria requires no treatment.

Chronic Essential Pentosuria is a harmless metabolic anomaly. The urine contains all the time small amounts of reducing substances from 0.2 to 0.6%. Only occasionally the Benedict test may be negative. A study of the urine will reveal pentose which can be a d-arabinose or a l-xyloketose. Lasker, Enklewitz and Lasker studied the urines of a large number of different cases of pentosuria reported recently and found l-xyloketose in all of them, although some of the earlier observers (Newberg, Aron , Cammidge and Howard) isolated an inactive arabinose. L-xyloketose in the urine gives a positive Benedict test in a few hours without the application of heat or in ten minutes at 55° (Lasker-Enklewitz test). Pentose is positive with Bial's reagent; it does not ferment with yeast and on treatment with phenylhydrazine it gives characteristic pentosazone crystals with a melting point from 154° to 160°. The daily excretion of pentose is uniform and is uninfluenced by diet or exercise.

Whenever a routine Benedict test is positive for sugar, the mixture of Benedict's solution and urine instead of being heated can be allowed to stand at room temperature overnight1. If it becomes positive the presence of l-xyloketose (ketopentose) and levulose should be suspected since they reduce copper solutions in cold. Fructose can only be discovered if carefully looked for. It is more easily mistaken for renal glycosuria then for diabetes since there are no symptoms of diabetes present and the blood sugar level is normal. Fructose, just like glucose, ferments baker's yeast. It is differentiated from glucose by the reaction it gives with the Selivanoff test. Fructose (levulose) is levorotatory by polariscopic examination in contrast to glucose (dextrose) which is dextrorotatory. After it has been fermented with baker's yeast the urine gives a negative Benedict test, negative Selivanoff reaction and shows no rotation. The urine is sugar-free when the patient is given a fructose free diet and shows sugar on regular diet. On treatment of the urine with methylphenyl hydrazine, fructosazone crystals are obtained. A glucose tolerance test shows a normal curve while a fructose (levulose) tolerance test gives an abnormal curve which indicates considerable delay in the removal of fructose from the blood stream.

NON-DIABETIC GLYCOSURIAS WITH HYPERGLYCEMIA

Glycosuria even in the presence of hyperglycemia does not always mean diabetes. A disturbance in the carbohydrate metabolism may be brought about by many causes and many ways. It may be of a very short duration or of a more profound and more lasting type.

Hyperglycemia and glycosuria may be due to an impairment in the function of the liver to store glycogen, as occasionally observed in liver disease, or to an increased mobilization of liver glycogen brought about by an increased adrenalin secretion resulting from numerous not infrequent causes. It is also known that stimulation of the nerves passing to the adrenal medulla may cause hyperglycemia and glycosuria. The same result is obtained following parenteral administration of adrenalin. Adrenalin causes glycogenolysis in the liver (a breakdown of the glycogen with resulting hyperglycemia and glycosuria). The following glycosurias belong in this group: those due to poisoning with chemical agents such as morphine, ether, carbon monoxide, heavy metals, etc.; in fact all which are due to severe poisonings associated with asphyxia. Emotional glycosuria (fear, fright, pain, anger) is an adrenalin glycosuria. It is produced by a stimulation of the adrenals through the splanchnic nerves.

Cerebral glycosurias. Claude Bernard by his famous puncture in the floor of the fourth ventricle was the first to produce hyperglycemia and glycosuria through stimulation of the central nervous system. Marx found the same in connection with an operation on a pituitary tumor. Brugsh, Dresel and Lewy made a more definite localization. They found a particular center of a sympathetic type located in the dorsal nucleus of the vagus (nervus periventriculary), the destruction of which causes hyperglycemia. Vonderahe also blames the cerebral glycosurias on the destruction of the nucleus paraventricularis in the hypothalamus, which he believes regulates the sugar metabolism. According to his hypothesis, the presence of sugar in the blood stimulates this center to activate the production of insulin by the islands of Langerhans. The cerebral glycosuria is probably produced by the way of the adrenals and hypersecretion of adrenalin. Transient glycosuria may follow cerebral hemorrhage, thrombosis, concussion of the brain or embolism. Glycosuria developed in many gun-shot injuries of the skull in the first World War, according to Grafe.

Alimentary glycosuria is a glycosuria observed after food intake. It should be considered as diabetes until otherwise proven since normal individuals possess a very high tolerance for carbohydrates, especially in the form of polysaccharides. It may be observed in cases with the "atypical" type of renal glycosuria whose urine shows sugar only off and on. These people show no elevation of the blood sugar and a normal glucose tolerance test. The maintenance of a normal blood sugar is the result of the rate of the removal of sugar from the blood by oxidation and storage being equal to the rate at which sugar is brought to the blood. Intravenous glucose given to a normal person at a rate faster than .85 grams per Kg. body weight per hour causes glycosuria. A normal person can ingest 100 gms. or 1.75 gms. of glucose per kg. body weight with only a slight hyperglycemia and no glycosuria.

Glycosylation The covalent addition of sugar moities to N or O atoms present in the side chains of certain amino acids of certain proteins, generally occuring within the Golgi apparatus during secretion of a protein. Glycosylation sites are only partially predictable by current computer searches for relevant motifs in protein sequence. Glycosylation may have profound but very unpredictable effects on the folding, stability, and antigenicity of secreted proteins. Glycosylation is a property of eukaryotic cells, and differs among different cell types

Glyoxalase An enzyme which catalyzes the conversion of methyl glyoxal to d-lactic acid; requires the presence of glutathione as coenzyme.

Gmelin test for bile pigments in urine Colored zones, passing from green to blue, to violet, to red and finally to yellow, are obtained when urine is floated on concentrated nitric acid.

Goiter Bronchocele; enlargement of thyroid gland or its parts with or without thyrotoxicosis. Types are:

(1) non-toxic diffuse goiter,

(2) non-toxic nodular goiter,

(3) toxic nodular goiter,

(4) toxic diffuse goiter. The non-toxic diffuse goiter, or colloid goiter or simple goiter, may be endemic due to shortage of iodine, or pubertic appearing only in puberty, or sporadic due to infection or some other cause. Depending on the type or case one uses iodides, thyroid or resorts to surgical removal. The non-toxic nodular goiter, or adenomatous goiter, is characterized by nodules and pressure symptoms. It may be aggravated by iodides. Toxic nodular goiter, or adenoma with hyperthyroidism, or Plummer's disease, may be concealed for a long time, but requires surgery. Toxic diffuse goiter includes the terms exophthalmic goiter, Graves' disease, Basedow's disease, Parry's disease, hyperthyroidism or thyrotoxicosis. It is a hyperactivity possibly brought on through excess of thyrotropic hormone of the pituitary. Psychic trauma is a prevalent "cause." Focal infection is a minor cause. Increase of at least 25% of basal metabolic rate is considered a conclusive element of diagnosis. Other symptoms are tachycardia (100 to 160 per minute) and eye symptoms. The latter are exophthalmos, or protrusion of eyeballs; von Graefe's sign, a lid-lag of the upper lid behind the movement of the eyeball; etc. Psychotherapy is frequent. Iodides may work great harm as well as help. Radium and X-ray therapy are generally not strongly recommended.

Gold number A measure of the protective efficiency of lyophilic colloids. It is that weight of the colloid in milligrams which will just fail to prevent a change in color from red to violet when 1 cc. of 10% sodium chloride solution is added to 10 cc. of a Zsigmondy (formaldehyde) red gold solution to which the colloid has been added.

Golgi Apparatus A membranous, vesicular structure which is in continuity with the endoplasmic reticulum of eukaryotic cells and generally in close proximity to the nucleus, the Golgi plays an important role in the posttranslational processing and transport of secreted proteins.

Golgi complex A complex and dynamic membranous organelle of eukaryotic cells. Golgi bodies function in the posttranslational modification (glycosylation) and secretion of proteins or insertion into membranes. Composed of cis, medial and trans compartments as well as a less well defined "trans-golgi network."

Golodetz reactions for cholesterol and oxycholesterol I. A deep blue color is obtained when cholesterol is treated with a few drops of liquid trichloracetic acid and 1 drop of formaldehyde.

II. The addition of 1-2 drops of a mixture of 5 parts concentrated sulfuric acid and 3 parts formaldehyde to solid cholesterol produces a black color.

III. Trichloracetic acid colors oxycholesterol green very rapidly; the solution has a red absorption spectrum.

Gonadotropic principles The two hormones, Prolan A and Prolan B. Prolan A stimulates male germ cells and ova and ovarian follicles to produce their hormones. Prolan B or the luteinzing hormone stimulates the conversion of the ovarian follicles to corpora lutea and their activity in secreting their hormone (progesterone).

Gout a constitutional disease with recurrent arthritic attacks on one or the other big toe and sometimes at the knee or other joints. Heredity has been blamed. Alcoholism is predisposing and often brings on attacks. Uric acid retention in the blood and reduced renal efficiency are shown. The disease may be acute, chronic or irregular. Irregular gout is really only a tendency to gout indicated by such things as itching feet, muscular pains, urinary gravel and calculus. The diet is designed to avoid uric acid formation, e.g. milk diet. Colchicine has been found of value as well as sodium iodide.

Gradient. A system of order involving progressively increasing or decreasing entities or intensities from one pole or point of a morphogenetic field to another. It is a twodimensional special case of field pattern.

Graft Portion of tissue removed from one person and transplanted either to another area of the body or another person's body; when a whole organ is removed, the process is called transplantation

Gramicidin A neutral polypeptide of a sporebearing soil bacillus (B. brevis) containing a fatty acid unit, alcohol soluble, bactericidal in vivo in very dilute solution; name also applied to a group of such substances; graminic acid; gramidinic acid.

Gravitational field strength Measure of Earth's *gravitational field* at a given point; defined as the gravitational force that acts on a 1-kilogram mass

Gre Glucocorticoid Response Element: A binding site in a promoter to which the activated glucocorticoid receptor can bind. The glucocorticoid receptor is essentially a transcription factor which is activated only in the presence of glucocorticoids. The activated receptor will bind to a GRE, and transcription of the adjacent gene will be altered.

Green & brosteaux's enzyme system An enzyme oxidation system of three components, analogous to that of Warburg & Christian. A coenzyme, from yeast or muscle, reacts with lactate in the presence of an enzyme from heart muscle to give pyruvate and reduced coenzyme. The latter, plus the third component, a dye, gives coenzyme and the reduced dye, which is reoxidized by molecular oxygen.

Grimmer test for peroxidases in milk Raw milk gives an intense brickred color when treated with 1-2 drops of 0.1% solution ethylhydroperoxide and 2 drops of a solution of guaiacol in 10 cc. alcohol and 90 cc. water. Heated milk shows no color change.

Grote test for sulfur in organic compounds Reagent — 0.5 gm. sodium nitroprusside are dissolved in 10 cc. water and 0.5 gm. hydroxylamine hydrochloride and 1 gm. sodium bicarbonate are added. After complete evolution of gas, 2 drops bromine water are added, the solution is filtered and diluted to 25 cc. 5-20 mg. of substance are dissolved in 3 cc. water and treated with 0.5 cc. reagent and excess sodium bicarbonate; a purple-red color develops within 10 minutes.

Growth Growth is a universal property of living matter by which it tends to increase in amount. However, through the years the volume of living matter on the earth remains fairly constant. The organic mass as a whole and in individuals reaches a maximum of size and efficiency of operation and then breaks down into its elementary constituents. Thus the cycle continues to revolve through the ages with immense biological and economic consequences.

Since growth is increase in size, the methods of measuring growth are: the instruments for determining weight (with due regard to specific gravity) and length of dimensions. For the latter various forms of calipers have been devised including, for microscopic organisms, graduated eye pieces and slides; for macroscopic organisms, sliding and compass calipers. By measuring (preferably the same individual) successively at time intervals, change in size and its rate are determined. Where it is desired to determine the factors involved in growth, the content of water is measured by drying; the role of various foods is determined by using experimental animals (usually rats) fed on stuffs that are complete except for one food element, e.g., a particular protein or vitamin, and measuring the subsequent growth.

Human growth of the body as a whole is determined by repeatedly weighing and by measuring the dimensions. To measure the differential growth of particular organs various landmarks and special forms of calipers have been devised, as described in works on anthropometry.

Growth may be regarded as of two types: arithmetic (additive) and logarithmic.

Arithmetic growth is shown in the increases of elongated cells or organs that are growing at their ends. Thus the growth in length of long bones in mammals is due to the proliferation chiefly in the direction of the bone axis at the end of the bone and, accordingly, from birth on to maturity the growth is an additive one, the additions absolutely diminishing after puberty.

Logarithmic growth occurs where increase is chiefly that of the organic particles upon which control of the process depends. Since the chemical processes of such particles are largely hydrolytic, oxygen is required for all growth that is due to proliferation of organic particles.

Since growth is so largely a matter of imbibition of water the work of the growth-inciting particles must be control of imbibition, i.e., they are hydrophilic. This matter of relation of water to protein has been discussed in some detail by Gortner.These relations are extensive and the role of proteins as "Bausteine" is at bottom doesn their hydrophilic properties. The evolution of the protein molecule has made possible the growth of land animals, many developing under limited aquatic conditions. Since the protein is assembled in colloidal particles, the statement of MacDougal ('20 p. 22) puts the case clearly: "It is plainly evident that growth consists of two fundamental features—hydration of the colloidal material of the plasma and the arrangement of additional material in colloidal structures with the entailed additional capacity for absorbing water." Wrinch ('41, p. 225-229) has developed the theory of protein-water systems according to which the weight of the water molecules associated with the varying numbers of units of a protein molecule may constitute 70, 80, 90, or even 99 per cent by weight of the whole. Apparently the proportion of water associated with a protein molecule may increase during development. Thus we account for the fact that the weight of water in a developing tadpole may increase from 1.03 mg. to 4.51 mg. while the weight of the dry substance increases from .80 mg. to 1.16 mg. Thus the growth of water in this case is not merely arithmetic but logarithmic, since the amount of water depends upon the increasing hydrophilic capacity of the protein molecule which stores it.

• In some cases, the cell body contains little water but the polysaccharides and proteins are closely crowded together forming polyhedrea and the number of these molecules or aggregates tends to increase logarithmically by aggregation. Animal eggs and most seeds contain comparatively little water and in these cases besides protein there are stores of carbohydrates and fats. Marvin has pointed out the law of increase in number as these polyhedra (hexoctahedra) are formed in successive layers around the central unit. The total number of units, S, in any aggregation with n layers is Sn=4n3 + 6n2+4n+1. This can be reduced to formal identity with the equation of organic growth, log W=k log (2t+1) +C in which weight is equated against time

Hence a better understanding of growth is obtained by appreciating that it follows the law of aggregation of molecules rather than that of intussusception, as held by T. H. Huxley.

Growth depends not only upon external energy (heat) and oxygen which contributes to internal energy but also upon the addition of foods and especially, in animals, the proteins and their amino acids.

Of the various amino acids taken into the body of mammals certain ones have been shown by experiments (mostly on rats) to be indispensible for growth. These are: lysine, tryptophane, histidine, phenylalanine, leucine, isoleucine, threonine, methionine, valine, and arginine of which the last can be synthesized in the animal organism, but not at a sufficiently rapid rate to meet the demands of normal growth. Just what role each plays is a matter for future research. It has been shown that lysine is essential for the estrous cycle in animals. The sulphur in methionine is essential to detoxify certain growth-inhibiting compounds that may be taken into the body. Cysteine and glutathione and other sulphydryl compounds likewise stimulate growth. Also non-sulphydryl complexes which cause sulphydryl to be released on contact with proteins may serve indirectly as growth stimulants.

Growth is also determined in large part by internal factors. The same dietary furnished to a large and to a bantam fowl results in different growth due to the hereditary, genetical factors. When a large and a small race of poultry (Rhode Island Red, and Silkie Bantam) are crossed the hybrid embryos are intermediate in size between pure bred embryos of these races. When chorio-allantoic grafts of limb buds are made from Creeper fowl (characterized by extremely short legs in the adult) the graft shows a similar slow and defective growth as is found in the embryos of this breed. Even when fed in the same institution from early childhood the Norwegain child grows rapidly in stature and size of bones while the Italian child grows more slowly. Again the humerus and the radius of a developing child are supplied by the same blood carrying the same kind of protein molecules and containing the same growth stimulating factors, yet in a particular infant the humerus grew 24 mm. in a year and the radius only 17 mm. Of the food ingested the body, or tissue, selects what its nature requires and rejects most or all of the rest.

The rate of growth shows a vast range in different organisms. Thus the human radius during infancy increases about 17 mm. in one year or 54Å in one second or about one layer of protein plus its accompanying water in two seconds. A sweet potato vine may grow 12 feet in 3 months or 3657 mm. in 8,000,000 seconds or 4570Å in one second, or nearly one thousand times as fast as the human radius. The growth processes are of course quite different in bone and in tissues of cellulose-walled plant cells, but in the growth of both water plays a prevailing part.

Many attempts have been made to express the rate of growth in a mathematical formula. In the case of dividing egg cells the number of cells at any cell generation after the individual egg (2nd, 3rd, 4th, etc.), or simply nth is 2n: thus in the 3d generation 8. But continuous compound interest, which more nearly approximates organic growth, gives at the expiration of the number of time units n, Yn=Yo(1+k)n where Yo is size at the start. Or, better still, if n be infinite, $Yn=Yoe^{kt}$ where k is the rate of growth; t the time of the growth period in convenient units, such as hours or days; e is the base 2.718. It may be added that, due to complexities in the growth process of animals and plants and to the influence of varying environmental factors upon the course of growth the actual curve rarely follows the formula for a long period.

The formulae of growth rates are based on masses; they give false ideas of growth in the individual. For the latter is characterized by varying rates during the course of development. In fact there is sometimes recognizable an irregular rhythm in the rate of growth. Thus whenever the humerus grows exceptionally rapidly the radius grows exceptionally slowly, and vice versa. Whether this is due to a rhythm in the supply of food, accelerators, and water, or to a rhythm in the utilization of that supplied has not been determined. Such alternating periods of rapid and slow growth have been observed also in tissue cultures.

Rhythmic growth may well be a consequence of the satiation of a growing part that is taking in amino acids rapidly, followed by a cessation of such feeding for a time. It has been shown experimentally that if growth is retarted by cold or starvation, when more favorable conditions ensue it is accelerated much above the normal growth rate, so that normal weight or size for the given age are attained, and may soon be surpassed.

Rhythmic growth may be due to variation in intake of accelerators of growth, which play an important role in growth processes. This hypothesis is supported by the fact of the prepubertal spurt in human growth which occurs at the time of the maturation of the gonads under the temporary influence of pituitary secretions, especially of its growth promoting factors. This spurt is preceded by some years of slow childish growth and followed by almost complete cessation of growth.

Of the accelerators of growth the most important are the hormones and vitamins. The secretions of the pituitary gland stimulate growth and the effect of these hormones is localized in the proliferating zone of cartilage. Hypophysectomy inhibits skeletal development. Whether there is a special growth hormone produced by the pituitary gland that can be separated from the other pituitary secretions is still a debated question. Prolactin alone will do much to advance growth in a dwarf mouse. The importance of thyroxin from the thyroid gland is shown in cretins where the secretion is insufficient, resulting in dwarfs. Thyroxin has an essential effect on the liver, increasing its size and its complement of protein and leading it to release sugar into the blood. It stimulates the proliferation of the cartilage cells at the growth zones of long bones. Various other stimulators of growth are known among animals such as l-cystine, dl-methionine, l-cystine disulfoxide, glutathione and SH-compounds in general. Some of these may counteract the growth inhibiting effect of iodoacetic acid.

Plant growth is stimulated by certain plant hormones. Darwin (1881) first demonstrated that the turning of plant stems toward the light was controlled by "some influence which is transmitted from the upper part of the coleoptile of a seedling to the lower part causing the latter to bend." The influence is now recognized as one of the plant hormones known collectively as the growth substance, or auxin. The existence of this auxin is demonstrated by removing the tip from a monocotyledon's (e.g. oat's) coleoptile and sticking it unilaterally on the first leaf, after pulling the leaf part way out of the coleoptile. A growth curvature that turns away from the applied tip (negative) indicates a growth promoting substance; one that turns toward the applied substance (positive) indicates a growth retarding substance in the tip. In place of the tip we may use agar blocks diffused with the growth hormone from the tip; or from other sources. The effective amount of an auxin is exceedingly small. It has been computed that one auxin molecule is active in the formation of 3x105 glucose residues for the cellulose micellae that build up the plant cell-wall. Auxin is found also in leaves and stems. Auxin usually retards the growth of the root. Auxin acts both on the water and the micellae of the cell walls.

Growth substances (auxins) have been obtained from a variety of plant and animal materials. Rhizopus (bread mold) grown on an appropriate substrate yields a growth stimulating substance named rhizopin, which is probably identical with 3-indole acetic acid (heteroauxin). Aspergillus yields a substance that reacts with tryptophane, lysine, leucine, tyrosine and phenylalanine to produce growth stimulators. Urine, after a complex treatment, yields a powerful growth stimulant called auxentriolic acid, or auxin a. Malt, maize oil and some other plant oils yield a growth stimulant called auxenolonic acid or auxin b. Yeast yields heteroauxin. A large number of other compounds have been tested as to growth activators but they are mostly less effective than the foregoing. The degree of activity is measured by applying to the cotyledon particles of lanolin to which the compounds have been added.

In testing the reactions of roots the compounds are added to the soil or injected into the stem. The relation of auxin to root formation is somewhat complicated. Strong solutions of indoleacetic acid may inhibit root growth but weak solutions stimulate it. Indol-butyric acid and naphthaleneacetic acid are also efficient. The results are more striking if sugar or Vitamin B1 are added as an after treatment; indeed the latter is regarded as a specific growth hormone for roots and is supplied by parts of the plant above ground. Nicotinic acid is an essential growth factor for pea roots. The study of the effects of auxin upon plants is opening up a new field of research into the processes involved in growth.

The various vitamins that affect growth of mammals are: thiamin (Vitamin B1), which is believed to catalyze the transformation of pyruvic acid and is thus essential in intermediary metabolism; riboflavin (Vitamin G), which catalyzes the oxidative processes in the body tissues; Vitamin D, which promotes the growth of the long bones; Vitamin C (or ascorbic acid), essential to later stages of growth of mammals; and Vitamin A essential to the growth of young rats. After exhaustion of stores already in the body, on a diet lacking in Vitamin A growth ceases.

In vertebrates, at least, certain mineral elements are necessary to the upbuilding of the mineral-containing organs of the body. Such are iron, as in the blood; calcium, as in the teeth and bones; phosphorus also required by bone.

In many of the above cases it is not general growth but that of specific essential organs that is affected.

Organic growth has this remarkable property that, after a period of logarithmic advance it reaches a point of maximum rate, whereafter it begins to decrease in a fashion that is roughly a mirror image of the former increase and eventually leads to complete cessation. The equation of this process of limiting of growth is given by Brody as dw/dt=k (A—W), W being weight at a given instant. After growth ceases the body reaches its maximum of efficiency. After a shorter or longer period, deterioration sets in and eventually death ensues. This cessation of growth is not due to some changes in the cells of older organisms, for such cells grow equally well in tissue cultures as embryonic cells. It is probably due to failure of growth stimulus.

The mechanism of growth inhibition has been inadequately studied. It has been found that high dietary cholesterol has an inhibitory effect on the growth of rodents. There are certain sulfur-containing amino acids which detoxify certain added foreign substances that would interfere with growth of rats; but a deficiency in such acids may result from the feeding of methylcholanthrene, benzpyrene and diphenyl with a consequent inhibition of growth. Varying amounts of polycyclic, particularly carcinogenic, hydrocarbons produce similar growth inhibition. This growth inhibition in immature rats may or may not be due to the same causes as inhibition at maturity.

Not only is there an inhibition of growth of the body as a whole but also a specific inhibition of growth of particular organs. Thus growth of the cranium is inhibited earlier than that of the maxilla and mandible, or that of the extremities. The epithelium of the vagina undergoes rapid growth before ovulation as soon as the corpus lutum is developed, following ovulation, this epithelial growth is inhibited, and it can be shown experimentally that the lutal secretion is the effective inhibiting agent.

Cancer is a reaction of susceptable tissue cells to growth stimulating agents, external or internal. Too often in the past there has been a failure to consider the quality of the tissue or organism. Among growth stimulating agents are cholanthrene derivatives (e.g. 4:10-dimethylene 1:2—benzanthracene);

benzpyrene; 2-methyl-3:4—benzphenathrene; O-aminoazotoluene and derivatives; — the variation in potency of derivatives of this compound, some of which are wholly inactive, is a striking fact. The estrogens, such as estrone benzoate, injected into mice cause tumors. When such estrogen is applied together with 3:4 benzpyrene the tumors appear earlier and develop more rapidly.Inhibition of tumors induced by a carcinogenic compound is effected by the same carcinogenic compound. The action is not strictly proportional to the carcinogenic power. Thus 1:2:5:6-dibenzanthracene is more inhibitory than 3:4benzpyrene. On the other hand, a series of non-carcinogenic compounds tested under the same conditions gave no inhibition of tumor growth. The carcinogenic compounds slow up the growth of the treated animals so that reduction of tumor growth may be a result of inhibition of growth of the body in general.In certain tumors (e.g. Rous sarcoma) a virus is the apparent carcinogeneous agent. If 3:4-benzpyrene is added intensified and generalized growth follows. So long as the pH of the tumor is kept below 7 virus activity remains high, but is suddenly lost at about 3 at which pH the protein molecule splits, an indication of a collaboration between virus and protein in the production of the tumor.Ultraviolet radiations accelerate the formation of tar cancer in mice; the tar products are believed to sensitize the skin to such rays, perhaps by concentration of cholesterol Yeast cells injured by ultraviolet radiation produce a substance that promotes growth. The same radiation may produce abnormal growths on the human skin. Yet X-rays are used effectively in inhibiting tumor growth. The same agent may stimulate or inhibit growth depending upon its intensity above or below the threshold of the tissue subjected to the threshold of the tissue subjected to the agent.

Growth hormone A principle produced by the anterior lobe of the pituitary gland, which has control over height and growth in general.

Gruber test for indican IN SERUM A mixture of 3 cc. concentrated hydrochloric acid, I cc. serum and a small amount of isatin is boiled for 30 seconds, cooled, shaken with 1 cc. chloroform to an emulsion, separated by centrifuging and the chloroform layer washed wtih 10 cc. 10% sodium hydroxide. Violet color indicates pathological amounts of indican.

Guaiacol Ortho-methoxyphenol, a constituent of guaiacum resin, beechwood tar; employed in lumbago, etc.

Guanase An enzyme associated with guanine (2-amino-6-oxypurine). It deaminates guanine to xanthine (2, 6,-dioxypurine).

Guanine (G) A nitrogenous base found in DNA and RNA, one member of the base pair G- C (guanine and cytosine). (2-amino-6-hydroxypurine).

Guanine gout A disease of the pig due to the absence of guanase resulting in the failure of guanine to oxidize to xanthine and uric acid.

Guanosine A nucleoside consisting of one molecule of guanine and one moleclue of d-ribose. May be considered as derived from guanylic acid through the loss of phosphoric acid. The linkage is from the C1 of the sugar to N7 of the guanine.

Guilhon test for iron pigments in liver A few drops of potassium ferrocyanide solution are added to 1 drop of fresh liver pulp which has been stirred with 0.5 cc. hydrochloric acid; the presence of hemosiderin is indicated by a Prussian blue color.

Gum arabic A mixture of several gums, the best being that obtained from Acacia Senegal; it is usually completely soluble in water.

Gums Carbohydrate derivatives consisting largely of glycoside-like polymers of hexoses and pentoses, together with complex acids.

Gunzburg test for hydrochloric acid in gastric juice Equal volumes of gastric juice and of reagent (solution of 2 gm. phloroglucinol and 1 gm. vanillin in 40 cc. alcohol) are evaporated in a porcelain dish; a red residue indicates free hydrochloric acid. Sensitivity—1:0000-20000.

Gutzeit test for arsenic One gm. of zinc is placed in a large test tube with 4-5 cc. dilute sulfuric acid and 1-2 drops copper sulfate solution and the tube loosely plugged with cotton. The cotton is moistened with a drop of lead acetate and a strip of filter paper impregnated with silver nitrate is put over the mouth of the tube. Darkening of the paper shows arsenic

Guyot biological reaction for blood The chromogen of fungus Boletus cyanescens is used as the reagent. To destroy oxidases, the cut fungus is boiled with water and macerated with 60% alcohol for 48 hours; the solution is exposed to sunlight and filtered. 2 cc. of a solution of blood in 50 cc. water is superposed with the reagent and treated with 1-2 drops hydrogen peroxide; a greenish zone turning blue is obtained. The test is only indicative, not specific. Bloods altered by carbon monoxide and leukocytes do not give this reaction.

Gymnosperms Flower bearing plants which produce naked seeds in cones or berries, usually evergreen, e.g. Coniferae as pine, spruce, larch, juniper.

Gynogenesis The fertilisation of an egg by sperm so treated that nuclear fusion cannot ensue.

H

Haas flavoprotein A bottom yeast enzyme qualitatively of the same kind as the Warburg-Christian enzyme .).

Haem The non-protein portion of the hemoglobin molecule, $C_{34}H_{32}O_4N_4Fe$, with the iron in the ferrous state; ferroprotoporphyrin; is also the base of other haemochromogens differing from hemoglobin with respect to their protein component.

Haematopoiesis Blood formation by liver, bone marrow, spleen, lymph glands; haematosis.

Haemerythrin Red respiratory pigment of certain annilid worms; contains iron.

Haemotrophe The sum total of nutritive substances supplied to the embryo from the maternal blood stream in viviparous animals.

Hagedorn-jensen reagents and method for determining blood sugar 1. To precipitate protein 0.1N sodium hydroxide and 0.45% zinc sulfate solution are mixed in the proportion of 15 shortly before use.

2. 10.6 gm. ignited sodium carbonate and 1.65 gm. potassium ferricyanide are dissolved in 1 liter of water.

3. 50 gm. sodium chloride, 5 gm. potassium iodide and 10 gm. zinc sulfate are dissolved in water and diluted to 200 cc.

4. 3 cc. of iron-free glacial acetic acid are diluted with water to 100 cc.

5. 1 gm. soluble starch dissolved in 100 cc. saturated sodium chloride solution.

6. 0.7 gm. sodium thiosulfate in 500 cc. water are standardized against a solution of 0.3566 gm. potassium iodate in 2 liters water. 0.1 cc. of blood is dealbuminized by warming with 6 cc. of (1), filtered and the residue washed 3 times with 3 cc. portions of water. Filtrate heated in boiling water bath for 15 minutes with 2 cc. of (2) and then treated with 3 cc. of (3). Mix with 2 cc. of (4), use 2 drops of (5) as indicator and titrate with (6).

Hahn oxine reagent The reagent (8-hydroxyquinoline) in 5% alcoholic solution is used to determine aluminum, zinc and magnesium; to separate aluminum or zinc from alkaline earth and alkali metals and to separate magnesium from the alkali metals.

Hair Hair is the term applied to a form of cutaneous outgrowth which normally occurs in all mammals, but in members of no other class. The "hairs" of birds, amphibians, fish and plants are so different from the hairs of mammals that there is no question of confusion nor much likelihood of close relationship. Mammalian hairs are prevailingly sub-cylindrical or flattened aggregates'of keratinized cells, probably exclusively ectodermal in derivation, although a few statements in the literature are to the effect that they also have a collagenous component derived from mesoderm. The presence of any such connective tissue derivative has not been generally confirmed.

Development. Each hair shaft, followed from tip to base, furnishes a record of a period of activity in the follicle from which it grew, and the series of successive hairs from the same follicle reveals the periodicity and changes in such activity. All hair follicles are believed to arise before or shortly after birth, the time of their appearance varying with the species and the prospective size of the follicle. In each case a strand of cells grows obliquely down from the epithelium into the corium and differentiates into the outer and inner epithelial sheaths which in turn become surrounded by a moderately firm connective tissue sheath. A specialized connective tissue papilla projects into the bulb-like dilatation at the end of the epithelial down growth. Cells of the bulb immediately above this papilla multiply and become converted into the hair shaft whose tip is pushed further and further out by basal growth. This small region is of critical importance in studying factors affecting hair growth. Sebaceous glands usually arise from the more distal part of the follicle, and in many species sweat glands do also. In nearly all follicles an arrector pili muscle runs from beneath the epithelium to the inferior wall of the follicle beneath the sebaceous gland.

Arrangement. The first follicles to appear in a prospectively hair covered area are usually rather uniformly spaced. They often appear to be in lines, but at times it is difficult to determine whether this appearance is real or only apparent. There is need for further interpretation of such geometrical arrangements as actually occur. That they have a phylogenetic significance with respect to an earlier scaly covering may be doubted. The first follicles to appear produce the largest hairs, probably throughout life. Secondary, tertiary and subsequent follicles are progressively smaller and apparently bear more or less definite spacial relation to the earlier ones already present. Since in some of the common laboratory animals many of these follicles do not appear till after birth, the possibility of their control is open to experimental test.

Aside from the groupings of follicles just referred to there are other more intimate associations. Two kinds of "hair tufts" are recognized, the term being used in a special restricted sense. In the so-called "true tufts," or compound follicles, characteristic groups appear, each consisting of one main hair and several categories of smaller hairs which arise from small papillae that have branched off from the main follicle and for the most part are enclosed in its sheath. There is usually only one set of sebaceous and sweat glands to each tuft, but occasionally one of the larger accessory hairs has a gland associated with it. "False tufts" arise when the skin around two or more independent follicles becomes depressed with the result that the hairs involved all emerge from the same pit, which may appear superficially to be the mouth of a single follicle. This latter condition can sometimes be observed in the human forearm. The two types of tufting are generally supposed to be distinct, but there seem to be some doubtful or transitional cases. A somewhat similar apperance, not to be confused with the foregoing, is produced when a hair that is about to be shed still persists in a follicle from which its successor also protrudes. In such cases the old hair is easily removed, the new one is more firmly fixed and often brighter in color. Finally, those follicles which open separately on the surface often become seconddarily pulled together in groups by surrounding masses of connective tissue fibers which gradually increase in volume during development. On the human scalp these groups, which include from one to eight hairs, increase in number during the third quarter of prenatal life and then tend to become stabilized. The groups are not altogether distinct and may be composed of different follicles at different depths in the corium.

Hair Follicles. Even small follicles frequently have large sebaceous glands and relatively strong arrector pili muscles. Larger follicles have thicker walls, and in many cases a good nerve supply. In such cases the hair and its follicle may serve as a kind of passive sense organ, especially when the follicle is made more tense by contraction of its muscle. Follicles of true tactile hairs, which occur in all animals except man, fall roughly into three categories, all of which are richly supplied with nerve terminations. In the most elaborate of these the inner connective tissue sheath is modified into an erectile body called the corpus spongiosum pili. At its upper limit there is a more or less completely circular sinus. These most highly specialized follicles are supplied by striated fibers from the neighboring musculature and are therefore under some degree of voluntary control, whence the designation active tactile hairs. Other tactile hair follicles have only smooth muscles and are called passive, some with and some without a circular sinus.

Types of Hair. Hairs themselves vary in size from the barely visible "down hair" or vellus of man to the immensely enlarged protective spines of hedgehogs and spiny anteaters. In form they range from short fusiform to greatly elongated shafts which is cross section may be circular, elliptical, prismatic or irregular in outline. Human hairs are commonly considerably flattened. Most hairs taper toward both ends but tend to be fairly uniform throughout their greater extent. Fluctuations may occur, somewhat irregular in hairs from the human scalp, or regularly as in certain hairs of lower mammals in which the soft basal part of the shaft is tipped by a thicker, firmer, but usually pointed terminal portion. The broad, scale-like tips of Ornithorhinchus hairs afford an extreme example of this type of deviation from uniformity in diameter of the shaft. In gross structure the shaft consists of cuticle, cortex and frequently medulla. The cuticle is a thin layer of translucent scales, usually without pigment. These scales may be coronal, each completely encircling the shaft or, more commonly, imbricated, their size bearing an inverse relation to the diameter of the shaft. Both types may occur in the same individual, coronal scales being limited to the smallest hairs. The cortex which constitutes the bulk of the hair shaft consists of cells generally so highly keratinized as to give the whole a translucent homogeneous appearance. In and among these cells may be a greater or less amount of pigment. In the medulla, on the other hand, cells are more loosely arranged and for this reason as well as because of the air that is likely to be present among them, they show differences in diffraction from the rest of the shaft. The medulla is usually absent at the tip and base and may be present or absent through varying extents of the shaft. In general large hairs are more likely to have medullas than small ones. Inasmuch as there are almost perfect transitions between the largest and smallest hairs, attempted classifications found in special treatises are all more or less unsatisfactory. In man the larger conspicuous, pigmented hairs are called terminal hair, the fine vellus is often referred to as down or fuzz, and intermediate forms are called transitional. Terminal hair varies in character in different parts of the body and is variously designated, e.g. capillus, cillia, tragus.

Physical and Chemical Properties. In its physical and chemical properties hair seems to share most of the characteristics common to other keratins. Analyses reported in the biochemical literature are numerous but the results, while showing good agreement on the whole, are still too diverse in some respects to justify a survey by one not a specialist in the field. The occurrence of well over a dozen amino acids in hair seem to be well established, and it would appear that at least some among them (e.g. histidine, lysine and arginine) tend to show constant ratios with reference to each other. The cystine content is relatively high in almost all hair, especially human (15+%), and probably higher in that of men than of women. It is also possibly higher in hair of adults than in that of growing children, although some authors report no differences associated with age, sex or color of hair. The nitrogen content of hair is reported as about 15% and that of sulphur is around 5%, practically all of the latter being accounted for by the cystine and methionine, except in camel's hair where there is a larger sulphur content than is needed for these amines. Other elements reported in hair include lead, arsenic and antimony. It has been maintained that these latter elements are "excreted" through the hair. This is open to doubt, particularly in the case of lead. Lipoids, in the hair of young rats, amount to about 4.5% of which 11.9% is cholesterol. In human hair there are more lipoids than in the subcutaneous tissue, but less than in the skin. The proportion of cholesterol to total lipoids varies in the hair of rat, cat, rabbit and dog, but the absolute amounts are said to be about the same, .55-.57%. The non-chemical student of this subject is likely to regret that, however excellent the analyses, the selection of the original material has often been less critical than might seem desirable. Furthermore, the difficulty in adequately separating and evaluating the pigmentary components is unfortunate because these may represent products of an essentially independent system.

The question of molecular oroganization and consequent physical properties of keratins, particularly those of hair, has been vigorously pursued for the last ten or twelve years by an active group of investigators lead by W. T. Astbury Leeds. It appears from their analysis of X-ray photographs that hair or wool can be stretched by an amount equal to about 2% of its original length without injury or permanent deformation of the keratin. In cold water it can be stretched by an amount equal to about 2% of its original length without injury or permanent deformation of the keratin. In cold water it can be stretched considerably more than that and yet return to its original condition. In steam or other softener the hair can be stretched still more, usually to twice its original length. For each medium in which it is tested the breaking point seems to be the same whether the tension is increased slowly or rapidly. Hair stretched to the limit in steam is found to undergo internal changes, which are believed to be in the nature of an extension of a normally folded polypeptide grid, made possible by the elimination of certain cross bonds resulting in conversion of the keratin from an alpha to a beta form. In this way hair or wool can be "set" in its new state; but if when first stretched to its full extent in steam it is allowed to suddenly contract in the same medium, it may then become even shorter than originally. Subsequently (the original linkages now being destroyed), stretching to twice its length in cold water is followed by complete recovery. The extensive findings in this field are of great theoretical interest and of value in the textile industry, to the practical hair dresser, and in the laboratory where it may be desirable to restore distorted hairs to their original form. The relation of keratin molecules in the different cells which go to make up the hair shaft is a matter of interest needing further study.

Shape of the cross section of a hair shaft is not closely associated with the form of the hair, as has often been supposed. The hair index (lesser diameter divided by the greater), while useful in a descriptive way, does not necessarily afford a clue to the curliness of the shaft except that hairs with a high index are often straight and those with a low index are sometimes, but not always curly. In this connection the relation of the two major axes to each other is of importance. Since hair is somewhat hygroscopic when dry, moisture tends to correct purely mechanical deformations that may have been induced by combing, by the hair's own weight, etc. In a number of abnormalities of the hair shaft such as trichorrhexis, monilithrix and pili torti, the shaft fluctuates somewhat regularly in diameter and breaks easily at the nodes. X-ray examinations of samples of the two latter anomalies are said to show no peculiarities in the molecular structure, from which it is concluded that these deviations are essentially "histological."

Pigment. Pigmentation of hair, particularly laboratory and domestic animals, has been studied with much success by geneticists. The histogenesis and chemistry of hair pigments have also been studied, but with considerably less success. Two different kinds of coloring are commonly recognized a diffuse 'stain' which is usually non-granular and in tones of yellow or red, and a clearly particulate type which occurs in tones of brown, ranging from 'black' to very light 'yellow.' The latter are all generally considered to be true melanins. There is less certainty about the former which, however, may not be so distinct as sometimes supposed. Probably all colors and patterns in mammalian hair are due to various combinations of these pigments, whose effects are often modified by structural characteristics of the hair or by other extraneous factors. The interrelation of the two possible classes of pigment has long been a matter of interest. In certain respects they behave independently in heredity but they are both influenced in the same direction, although not always to the same degree, by factors for albinism and dilution. One of the most suggestive recent attacks on their chemical relations is that of Arnow who reports that "if red human hair is extracted with boiling 0.1N hydrochloric acid, a red brown pigment is obtained in solution. This pigment has the physical properties of a pigment prepared by the mild oxidation of a dopa melanin in alkaline solution. It is suggested that the distinctive color of red hair is due to the presence in such hair of an oxidation product of melanin." Such an hypothesis has much to commend it and seems to be in accord with some of the facts established by genetics. Whether or not it will be found to agree with all of them remains tc ' seen. If granular melanin is a precursor of the red-brown pigment, it might be questioned why such genes as those for dominant white should sometimes suppress the black before they do the yellow. Results thus far obtained by the perhaps too violent methods of biochemists may ultimately require confirmation by genetic and microchemical tests.

Melanin Granules in Hair Resemble, at least superficially, those in feathers, but it has not been determined whether or not their genesis is the same. In birds, it now appears, all the melanin producing cells are derived exclusively from the neural crest of the embryo. These pigmentoblasts which migrate from their original site near the mid-dorsal line to their ultimate position in the feather germ constitute an essentially separate and independent system which determines the color and, for the most part, the pattern of the feathers. Whether or not the same is true for mammals is still uncertain, but Rawles sue of early mouse embryos transplanted into the coelom of white Leghorn chick embryos produced pigmentoblasts which migrated away from the graft into the peritoneum. Reed transplantation in young mice that pigment and possibly other cells migrate from the host into the border of the graft and enter some of the hairs. This in a measure confirms and extends the early work of Carnot et Deflandre who maintained that dark pigment migrates from graft to host skin in autoplastic transplants on the ear of guinea pigs, a finding which some later investigators have questioned. In light of the implications of these findings the graying of hair possibly becomes a little less puzzling. If the pigment producing cells of the hair are of a different nature and genesis from those that constitute the rest of the shaft, it is less surprising that the residual stock of them in the follicle should show a different and characteristic susceptibility to adverse influences such as X-rays, and perhaps even have a different rate of senescence. Koller following the earlier work of Onslow reports that in the rabbit, extracts of most colored skins give strong melanin reaction with dopa while under the same conditions extracts from dominant white skin do not form melanin, and if added to extracts of dark skin completely inhibit pigmentation. Extracts from recessive white skin form no pigment with dopa but when mixed with other extracts do not prevent them from doing so. These and other interesting oxidation phenomena parallel the genetic behavior of hair color in the same strains. Some investigators have not been able to get clear-cut results in other species with similar genetic behavior.

Differentiation of Growth Center. The various properties of the shaft must almost all be determined within a very short radius of the tip of the papilla, and in consequence special interest centers around this region. The greatest proliferative activity occurs just above the papilla which through its rich blood supply functions in conveying nutriment to the growing bulb and also possibly serves as an inductor for such growth. It has been established in other connections that destruction of the papilla terminates further hair growth, but the more difficult feat of bringing a papilla to bear on undifferentiated ectoderm has probably not been accomplished.. In hair, as to an even greater extent in feathers, epithelial cells show a high capacity for undergoing morphological and probably chemical changes at some distance (measured in cell diameters) above the basement membrane. Such changes are primarily determined by the constitution of the cells themselves but are influenced by factors brought to bear at the moment of differentiation. These latter factors seem to be predominantly physical and humoral. Free nerve endings often occur in the inner hair sheath but it is doubtful if they play any direct role in the activity of the growing bulb. It is often claimed that the shape (as shown in cross sectional view) is determined by the inclination of the papilla or the curvature of the follicle through moulding of the soft, incompletely cornified root. While these factors may very well be of some importance, it seems more likely that the form taken by the drying shaft as it emerges from the follicle is chiefly due to differentials in the character and arrangement of the cells of which it is composed. Variations between cortex and medulla are sufficient to show that almost any degree of difference in compactness and looseness of cells is possible in the hair bulb. When and how pigment comes into existence in the hair follicle is a histological problem that calls for further study. Whether intra- cellular granules are autochthonous or are deposited within the cells by migratory pigmentophores, as similar granules are deposited in feather cells, and the source of the inter-cellular granules are questions still to be answered.

Cyclic Activity of Follicles. So far as is known hair production is never continuous in any one follicle. The usual activity, as measured by the length or volume of hair produced in a unit of time, takes the form of an S-shaped curve which begins slowly, increases rapidly to the maximum which is maintained fairly constantly for the greater part of the growth period, and then slows down to a complete stop. Following cessation of growth the bulb itself cornifies and the papilla regresses. During the ensuing resting period the exposed part of the hair usually fades appreciably and is likely to suffer to some extent from attrition. Toward the end of the period the papilla shows renewed activity, a new bulb forms around it and the old bulb gradually rises to the surface and finally falls out, but often not till the new hair is projecting from the mouth of the follicle. The relative duration of growth and rest periods varies with the species and with the type of hair. In many animals the growth period is relatively short and the rest period long. There may be seasonal moults during which replacement of hair is, extensive, but even in such cases there is some replacement at other seasons. In a careful year-around study of the guinea pig, Dawson ating hairs during every month except August. Pulling out a growing hair generally results in a new hair forming somewhat sooner than it otherwise would. Commonly hair replacement is more or less continuous with little or no synchronism between follicles. This is the case in man. In the rodents certain zones or areas often moult independently of other regions, producting a pattern effect which is apparent because of the difference between the new and the faded old hairs. This tendency is often revealed following shaving or depilation of an area in the rat or mouse, when new hairs may come in quickly over part of the area but not till much later over other parts. The phenomenon, which is a normal occurence, has been mistaken for the result of experimental procedures. Although early successions of hair are quite regular in the rat (Butcher) at older ages it becomes difficult to get specimens that parallel each other and are really suitably matched to serve for experimental and control subjects.

Control of Follicular Activity. Attempts to regulate physiological processes in the follicle have rarely met with much success. That in some animals large numbers of hairs fall out and are replaced simultaneously naturally suggests that factors outside the follicles are involved, and that it should be possible to bring them under control. In other mammals, including man, hair follicles often show a high degree of autonomy, maintaining their own length of cycle and growth rate little influenced by variations in any ordinary factors. The writer has observed several follicles on one of his fingers for a period of about twenty years during which time there has been no apparent tendency for the cycles of different follicles to become synchronized, even when the average lengths differ by only a few days. Nevertheless there can be little doubt that variations in hormones, vitamins and nutritional factors exert some, although perhaps mostly an indirect influence on hair growth and pigmentation. Under such circumstances the difficulty already mentioned of securing wholly satisfactory control animals is an especially serious handicap. The often destructive effects of X-rays have long been known. Xray treatment, depending on its intensity, may produce an area of permanent alopecia, temporary alopecia followed by graying of the new hair, by complete and normal recovery, or even perhaps by slight stimulation. Numerous authors have pointed out that many hair stimulants affect the hair only as they affect the skin or general status of the individual. For example, Peck found that injection of pyrrol bodies causes increased pigmentation in the rabbit, but hastened to add that so do numerous other factors including injury to the skin. He concludes that the most important factors are general growth stimuli. It has often been assumed that cutting or shaving the hair hastens its growth, a question debated by TrotterSeymour that there is such an effect, the former explaining the apparent results as due to a statistical fallacy contingent on the fact that in any region some hairs are always approaching the end of their cycle with consequent slowing of growth rate, thus tending to depress an average covering several days as compared with one for only a single day. The need for study of individual hairs rather than groups which may be in different phases is thus emphasized. There are several factors whose abscene results in inadequate production of pigment. At least one of these, associated with the B complex, affects the pigmentation without influencing hair growth (Morgan and Simms according to these authors, an associated dermatitis and both it and the graying are corrected by extracts of adrenal cortex and thyroid, so it would appear that in the final analysis this type of graying may be due to a hormonal disturbance. Others have inferred on experimental grounds that the graying factor is distinct from any known component of the B complex. Finally Free reports that there are different kinds of graying, one of which can be produced by deficiency of iron, copper and manganese. Cures for the different types, as might be expected, are not the same.

Sex Differences. Differences in hair associated with sex have been studied rather extensively. Apparently the actual number of hair follicles in the two sexes is about the same, but the types of hair they produce may vary appreciably as in the mane of the lion or the beard of man. In many other species there seems to be no noticeable difference in the pellage of the two sexes. Zawadowsky that in some species where the sexes differ in color, as in an antelope (Portax picta), the male pattern and coloring develops only in the absence of female hormones, which is reminiscent of what occurs in some birds. But for the most part no such associations have been detected. For some purposes it is convenient to classify hair of different regions into three classes that which is uninfluenced by sex, that which is ambosexual, and that which may be considered a true secondary sexual character. It is not to be supposed that these categories are at all clearcut. In man, hair of the forehead, orbital region, and ordinarily other parts of the body is apparently uninfluenced by presence or absence of either sex gland. Follicles that produce such hair could be placed in the first class. Other hair, such as axillary and pubic, is about the same in both sexes but fails to reach full development following early gonadectomy. Traits of this sort are called ambosexual (Champy much of the hair on the limbs and trunk, including the hypogastric region (often confused with the pubic) belongs in the class of secondary sexual characters, in the sense that the degree of development is normally different in the two sexes. In what appears to be a well controlled experiment on an adult man, Whitaker marked increase in the length and weight of terminial hair over a small abdominal region to which testosterone propionate was applied, and in the dog Gardner of an inhibitory effect on hair growth by estrogens. Danforth unable to detect any effect of either testosterone propionate or theelin applied to hair follicles of the finger. Almost complete series of transitions in hair types can be found in either sex, which weakens the evidence that regulation is primarily or prevailingly hormonal. While striking effects do sometimes appear in ambosexual and secondary sexual hair of individuals suffering from tumors of the suprarenal cortex, the condition of the hair seems in general to be not so much a matter of specific stimuli as the manifestation of normal tendencies which may be more or less influenced by the milieu. This does not imply any question as to the accuracy of numerous reports in the literature which, when taken together, indicate that under certain conditions almost any hormone may have some relation to hair growth. An especially interesting bit of evidence in this connection is supplied by Hamilton observations on castrate men and boys. It has been known at least since the time of Aristotle that eunuchs seldom become bald, but Hamilton of the proper age who, judging from their male relatives, would have become bald under normal conditions, the injection of testosterone propionate is followed rather promptly by incipient, or later advanced, baldness, while those who on the same grounds would probably not have naturally become bald rarely show this tendency. When "caused" by the injection of testosterone the baldness can be relieved by its withdrawal. Hamilton's findings are in line with data from entirely different sources which indicate that baldness is a hereditary sex-limited trait. It would appear from the data at present available that baldness, if the genes for it are present, expresses itself only under normal conditions, which, in the male, involve the presence of a certain amount of male hormones. Occurrence or non-occurrence of baldness, then, is no more indicative of the hormonal condition of the individual than of his genetic constitution.

Genetic Factor. Genetic factors concerned in hair form and color have been studied extensively, and in many cases with great success. Dozens of distinct hair colors and patterns are known in the mouse, and their occurrence and frequency in various crosses can be predicted with great accuracy. Symbols indicating the genes for color (and other traits) in the mouse are now "official" so that mice with hair of a desired type can be ordered from a supply house by formula, much as chemicals can. The length and form of hair as well as temporary and permanent alopecias of many forms and laboratory mammals have been shown to be under hereditary control within closely defined limits. The genetics of hair traits is especially well known in the rat, rabbit, sheep and guinea pig, and there in much detailed information on the ox, dog, cat and horse. In human heredity these factors have not all been determined and analyzed with equal precision but the evidence indicates that they are of the same order as those of lower mammals. Among the latter, there is little evidence that perfection of new strains involving variations in hair color has been associated with much change endocrinologically.

Summary. By way of summary it may be inferred from the available evidence that the hair follicle is a highly individualized structure with an unusual degree of autonomy. During its development it acquires characters which are primarily dependent on the genetic constitution of the animal and the region of the body involved, These characters may undergo some change with age but on the whole are remarkably insensitive to extra follicular influences. Nevertheless, it has been adequately demonstrated that under suitable conditions certain follicles may respond in some measure to mechanical stimuli, to different kinds of radiation, to vitamins, to hormones and to nutritional states. These responses, however, appear to be for the most part general and non-specific. The hair that grows from the follicle serves as a record of follicular activity with great accuracy. Symbols indicating the genes for color (and other traits) in the mouse are now "official" so that mice with hair of a desired type can be ordered from a supply house by formula, much as chemicals can. The length and form of hair as well as temporary and permanent alopecias of many forms and laboratory mammals have been shown to be under hereditary control within closely defined limits. The genetics of hair traits is especially well known in the rat, rabbit, sheep and guinea pig, and there in much detailed information on the ox, dog, cat and horse. In human heredity these factors have not all been determined and analyzed with equal precision but the evidence indicates that they are of the same order as those of lower mammals. Among the latter, there is little evidence that perfection of new strains involving variations in hair color has been associated with much change endocrinologically.

Hairpin A helical (duplex) region formed by base pairing between adjacent (inverted) complementary sequences within a single strand of RNA or DNA.

Halden reaction for vitamin D A solution of vitamin D in benzene, chloroform or ether is treated with 5-10 drops of a 0.1% alcoholic pyrogallol solution and evaporated on the water bath to a few tenths of a cc. 2-4 drops of a freshly prepared 10% absolute alcohol solution of aluminum chloride are added and the solution reheated; a deep violet color appears in a few minutes.

Half-life (chemistry) time it takes for half the amount of radioactive material in a sample to decay

Hallochrome A red pigment of the skin of the annelid worm, halls; also obtained from tyrosine by action of tyrosinase.

Hamamelose Aldohexose with branched chain, obtained from a glycoside with gallic acid of the bark of Hamameli tannin.

Hammarsten reaction for cholic acid The powdered acid is shaken at room temperature with 25% hydrochloric acid in a closed flask; the color changes from yellow to yellowish-green and, after a few hours, to bluish-violet. An absorption spectrum is shown about the D line.

Hammer creatine Test A very small quantity of creatine is added to 2.5 cc. of butter culture (for acetylmethylcarbinol and diacetyl) in a test tube 0.5 in. in diameter; a volume of 40% sodium hydroxide equal to the volume of the culture is added and shaken. A red color is obtained, its rate of development and intensity at the surface are closely correlated with aroma and desirable flavor of cultures.

Hanus solution for determination of iodine number 13.2 gm. of iodine are dissolved, with gentle heating in 1 liter of glacial acetic acid, the solution cooled to 23°C and the iodine content of a 25 cc. portion determined by titration with 0.1N sodium thiosulfate. A quantity of bromine, equal to that of the iodine present, is added to the rest of the solution and the solution, protected from light, is kept in glass-stoppered bottles.

Haploid A single set of chromosomes (half the full set of genetic material), present in the egg and sperm cells of animals and in the egg and pollen cells of plants. Human beings have 23 chromosomes in their reproductive cells. Compare diploid.

Haploid-diploid life cycle — Occurs when a multicellular diploid phase, or **sporophyte,** alternates with a multicellular haploid phase, or **gametophyte**. Only plants and certain algae possess this kind of life cycle, which is also called "alternation of generations".

Haplotype A particular combination of alleles (alternative forms of genes) or sequence variations that are closely linked Ñ that is, are likely to be inherited together Ñ on the same chromosome. Originally was coined to describe MHC antigens, but now is used to describe RFLP patterns and certain other situations.

Hapten A substance which, although not antigenic, reacts in vitro with antibodies. By combination with protein hapten can often be converted into antigen.

Hardy-schultz law The generalization that the precipitating effect of a colloid by an ion is a function of its valency (approximately as the square of the concentration of the divalent, cube of the concentration of the trivalent ion, etc.).

Hart test forß-hydroxybutyric acid in urine Add 20 cc. water and a few drops of acetic acid to 20 cc. of urine and boil to a volume of 10 cc. Dilute to 20 cc. and divide into 2 portions. To tube 1 add 1 cc. hydrogen peroxide, warm for 1 minute and cool. To each tube add 0.5 cc. glacial acetic acid and a few drops of a freshly prepared aqueous solution of sodium nitroprusside. Stratify both solutions with 2 cc. concentrated ammonia and, after standing 4-5 hours, compare the two tubes. Tube 1 shows a purplishred contact ring if ß-hydroxybutyric acid were present originally while the second tube shows either no color or a faint brown contact ring due to much creatine.

Hausmann's method A method for the quantitative study of the hydrolytic products of proteins, distinguishing between ammonia nitrogen, basic nitrogen, non-basic nitrogen and humin (or melanin) nitrogen. The ammonia nitrogen is obtained on the distillate over magnesia, the basic nitrogen on a phosphotungstic acid precipitate, the non-basic nitrogen on the filtrate of the above precipitate, and humin nitrogen on the black material which separates on hydrolysis with strong acids.

Hay fever Pollinosis; an allergic catarrhal inflammation of the eyes and the upper respiratory passages due to wind-borne pollens (spring, summer and fall types). It is quite close to non-allergic types like vasometer rhinitis and allergies toward animal proteins and foods. Ephedrine by mouth and epinephrine by injection are used along with acidosis and calcium therapy.

Hazelopp determination of cysteine 2 cc. of a chloroform solution of o-benzoquinone are shaken with 2 cc. of a hydrochloric acid solution of cysteine. The chloroform layer is separated and dried over anhydrous sodium sulfate and the color compared with a solution of 1.25 gm. cobalt acetate in 10 cc. water to which has been added I cc. 9% ferric chloride solution. Color is equivalent to 2 mg. cysteine. Cystine does not give the reaction.

Hehner number The weight of the acids not volatile with steam obtained by saponification of five grams of fat.

Helenien A carotenoid found in flowers of Helenium. Its structure is ßalpha'-carotene-3-3'-dioldipalmitate.

Helicorubin A red respiratory hemochromogen found in mollusca and crayfish liver, composed of heme and a globin. It closely resembles hemoglobin.

Heller test for albumin in urine A white ring is formed, even at dilutions of 140000, when urine is carefully floated on concentrated nitric acid.

Hematin The non-protein portion of the methemoglobin molecule, $C_{34}H_{33}O_5N_4Fe$, with the iron in the ferric state; ferriprotoporphyrin; is the component of other parahaematins than methemoglobin differing from it in respect to the protein component.

Hematoblast Parent cell of red blood corpuscle found in liver or bone marrow; blood platelet; erythrocyte; microcyte.

Hematopoietic principle A substance found in normal livers, which is responsible for prevention of anemia. Its structure is not known, although it may be an enzyme, formed from an "intrinsic factor" of the gastric mucosa, and an "extrinsic factor" of certain foods.

Hematuria Blood in the urine due to a multiplicity of causes and showing both macroscopically and microscopically. The causes may be renal, ureteral, vesical, urethral, prostatic, infectious, malarial, leucemic, parasitic and drug injury. Essential hematuria is attributed to renal causes. These may be.

(1) nephrolithiasis, or stone in the kidney, (2) renal tumors (carcinoma, sarcoma, hypernephroma, embryoma),

(3) acute parenchymatous nephritis,

(4) tuberculosis of the kidney,

(5) pyelitis, pyelonephritis or pyonephrosis,

(6) hydronephrosis,

(7) polycystic kidney,

(8) general injury to the kidney from falls or blows,

(9) infarction of the kidney. Amongst the drugs which may cause hematuria are large doses of quinine, methamine, salicylates, cantharides, phenol, arsenic.Many chemical tests have been devised for the detection of blood in the urine, e.g. benzidine, Heller's, Rosenthal's, Struve's, besides microscopic and spectroscopic tests. Therapy in causal.

Hemicelluloses Polysaccharides combined with sugar acids of the uronic type, more susceptible of hydrolysis, but unrelated to celluloses.

Hemochromogen(S) The union of heme with a variety of nitrogen-containing compounds, e.g. proteins, bases (pyridine, picoline, etc.), amino acids, ammonia, cyanide, etc. While the term hemochromogen was previously restricted to the reduced form, the term parahematin being applied to the oxidized form, it is now employed for both forms. Example Pyridine ferri- and ferrohemochromogen.

Compounds of heme (haem) with proteins and bases, the combination being with two molecules of base; can combine with CO and oxidize readily to parahematins.

Hemocuprein A crystalline blue copper protein of red blood cells, m.w. about 35,000, containing 0.34% Cu which may have enzymatic activity.

Hemocyanin The respiratory chromoprotein of the cephalopods analogous to hemoglobin. The metal copper is associated with it, causing the compound to be light blue in the oxidized state and colorless in the reduced state. The weight of evidence is against the prosthetic group being a porphyrin.

Hemoerythrin A respiratory chromoprotein found in worms. It appears to contain a porphyrin which is not mammalian hematin. The red pigment of the blood that transports the oxygen to the tissues. It is composed of a porphyrin "hematin" and a protein "globin." The red pigment, also called oxyhemoglobin, gives up oxygen to the tissues and becomes purplish reduced hemoglobin. The iron of the porphyrin nucleus is in the ferrous state, and forms an oxidation-reduction system with methemoglobin.

Hemoglobin Protein complex found in the blood of most chordates and the roots of certain legumes. It binds oxgen molecules, and in chordates serves as the means by which the oxygen is supplied to the cells of the body.

Hemoglobinuria Hemoglobin in the urine. If due to an autohemolysin it is paroxysmal hemoglobinuria. Hemoglobin is found in the urine also due to malaria, hemolytic icterus, internal hemorrhage and other diseases accompanied by hemolysis, as scurvy, purpura, scarlet fever, typhoid fever, malaria, yellow fever, syphilis, septicemias, hemolytic anemias, hemolytic jaundice, etc. Various tests have been devised, e.g. Donath-Landsteiner test for hemolysis on incubation.

Hemolysis The diffusion or exudation of hemoglobin from red blood cells into the surrounding blood, sometimes due to destruction of the cell wall, sometimes to osmotic pressure.

Hemorrhagic diatheses Constitutional states predisposing to hemorrhage. Under this head comes the concept of purpura, a group of diseases in which bleeding occurs under the skin and the mucous membrane, scurvy and hemophilia, a hereditary abnormality limited to males and transmitted through the female, marked by prolonged bleeding because of the very slow coagulability of the blood. In scurvy the bleeding is due to weakness of capillary walls. The purpuras are described as thrombocytopenic and non-thrombocytopenic (angiopathic). In the former the number of blood platelets is decreased, the bleeding time is prolonged and the coagulation is normal; in the latter there is no decrease in blood platelets. The former includes essential and secondary forms. Essential purpura hemorrhagica is present in both sexes; secondary purpura is associated with tumors, anemias, leukemia, severe infections and overdoses of X-rays and radium. The non-thrombocytopenic purpuras include mild forms (purpura simplex) and anaphylactoid purpura, arthritic purpura (Schönlein's disease) and visceral purpura (Henoch's purpura). Secondary forms occur in infectious diseases (septicopyemia, scarlet fever, smallpox, black measles, cerebrospinal fever, typhus, rocky mountain fever, leukemia, syphilis); toxic purpura due to snake venom, numerous drugs or to cachexia of carcinoma, tuberculosis, Hodgkin's disease and even old age, or to neuritic complications; hepatic purpura; and purpura due to mechanical blocking of veins and bruising. Special tests have been devised to determine bleeding time, e.g. Ivy technique of blotting blood drops at intervals from a standard puncture; and coagulation time, e.g. Peterson and Mills method by observing a column of blood in a standard capillary tube. The treatments are causal.

Henderson-hasselbach equation The expression pH=pKa+log [Salt]/ [Acid] Where Ka is the ionization constant of a given acid; used in estimating the approximate pH of a solution of a weak acid in the present of one of its salts. It may also be used judiciously to estimate buffering action by replacing the salt concentration by the concentration of the anion provided the salt is well ionized and the acid is not.

Heparin A compound found in the blood, and formed largely in the liver. It maintains the fluidity of the blood by preventing prothrombin from being converted into thrombin.

Hepatocuprein A copper protein from liver containing 0.34% Cu but different from haemocuprein; may be an enzyme.

Hering reaction for ergot By extraction with petroleum benzine about 2 gm. of finely powdered ergot is freed from oil and an infusion prepared from 1 gm. of extracted ergot in 20 cc. water and 1 drop hydrochloric acid. 4 cc. of the filtered infusion are treated with I drop of ammonia and shaken vigorously with 10 cc. ether. 5 cc. of the clear ether layer are floated on about 2 cc. concentrated sulfuric acid; a cornflower-blue zone forms in a few minutes just below the interface.

Herter reaction for indoleacetic acid A reddish-violet color is produced when a solution of indoleacetic acid is treated with a mixture of hydrochloric acid and a 2% solution of p-dimethylaminobenzaldehyde.

Herzfeld test for glucose in blood An alkaline solution of methylene blue is decolorized by glucose. Solutions—

(a) 11000 methylene blue solution;

(b) 20% potassium hydroxide solution;

(c) 10% metaphosphoric acid solution.

Herzog reagent for lysine and ornithine Lysine and ornithine form addition products with phenylthiocarbamide which, on treatment with concentrated hydrochloric acid, form hydantoins crystallizable from acetone-alcohol mixture.

Heterauxesis The relation of the growth rate of a part of the developing organism (whether morphological or chemical) to the growth rate of the whole, or of another part. The part may grow more rapidly than the whole (*Tachyauxesis*) or it may grow more slowly than the whole (*Bradyauxesis*), or at the same rate (*Isauxesis*). Heterauxesis is a comparison between organisms of the same group but of different ages and hence sizes.

Heterochromatin Compact, gene-poor regions of a genome, which are enriched in simple sequence repeats. As it can be impossible to clone, heterochromatin is often ignored when calculating the percentage of a genome that has been sequenced. Heterochromatin was originally identified as regions of the genome that stained differently to euchromatin (gene-rich regions).

Heteroduplex Dna Generated by base pairing between complementary single strands derived from different parental duplex molecules; heteroduplex DNA molecules occur during genetic recombination in vivo and during hydridization of different but related DNA strands in vitro. Since the sequences of the two strands in a heteroduplex differ, the molecule is not perfectly base-paired; the melting temperature of a heteroduplex DNA is dependent upon the number of mismatched base pairs.

Heterogametic sex That sex the gametes of which are divided into two populations, one of which possesses the X-chromosome, governing sex differentiation, and the other does not.

Heteromorphosis The appearance, whether in embryonic development or in regeneration, of an organ or appendage inappropriate to its site, e.g. antenna instead of leg or eye.

Heteroplastic transplantation Transplantation between individuals of different species within the same genus.

Hexestrol Meso-hexestrol, a synthetic compound with estrogenic activity. It is not a steroid, but is used therapeutically in the same way as natural estrogens.

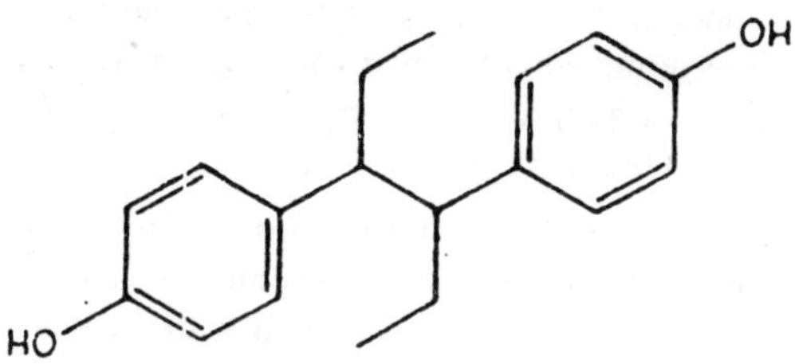

Fig. Hexestrol

Hexosans $(C_6H_{10}O_5)x$; polysaccharides derived from hexoses by loss of water. Examples are cellulose, starch, glycogen.

Hexose Any monosaccharide with the formula $C_6 H_{12} O_6$; with the pentoses make up the most important groups of the monosaccharides.

Hexosemonophosphoric acid Alpha-fructose-6-monophosphoric acid; a product associated with hexose metabolism.

Hexosemonophosphoric enzyme The pyridinoprotein enzyme containing coenzyme II which catalyzes the oxidation of glucose-6phosphate to 6-phosphogluconate. Found in red blood corpuscles and yeast.

Hibernation state characterized by low levels of metabolism that some animals enter during the cold season;

Hippomanes Packets of uterine secretion surrounded by trophoblast and invaginated into the allantoic cavity.

Hippulin $C_{18} H_{20} O_2$, an estrogenic hormone found in pregnant mare's urine; it has one alicyclic double bond more than estrone, and one-seventh its potency.

Hippuric acid Glycyl benzoic acid; an acid found in horse urine, also found as a detoxication product to remove benzoic acid from the body.

Hippuricase An enzyme that splits off the acid radical, particularly the benzoyl radical from certain peptides containing natural amino acids. The optimum pH is 6.8 to 7.0.

Hirschfeld test for pusin pathological body fluids Reagent—mixture of equal volumes of an aqueous solution of dimethyl-p-phenylene diamine hydrochloride and of 1% a-naphthol solution in 70% alcohol or 1% potassium hydroxide. When the reagent is floated on a purulent secretion, an indophenol-blue color is obtained.

Histamine ß-imidazole ethyl amine. A poisonous amine formed by bacterial action on histidine. It exerts a very powerful depressive action on the blood pressure, and stimulates gastric secretion of acid.

Histidine metabolism The nature of the metabolism is not known other than as follows. The organism is capable of utilizing large amounts of histidine, and also of synthesizing it, or of substituting imidazolacetic acid or carnosine for it. It is a precursor of uric acid and allantoin.

Histidinuria The appearance of large amounts of histidine in the urine, as in pregnancy. It is accompanied by the disappearance of histidase from the liver, and has been explained on the theory that it ensures an adequate supply of histidine for the embryo.

Histochemistry Histochemistry is a sector of biochemistry devoted exclusively to the biochemistry of tissues. Gersh defines the aims of histochemistry as.

(1) to define quantitatively the relations of chemical substances located intracellularly with those in their immediate environment and.

(2) to determine quantitatively the relations of

intracellular chemical entities to the morphological organization of the cell.

The dynamics as well as the complexity of the cell constituents, along with a lack of good quantitative analytical methods, militate against rapid progress in this field. For example, microincineration fixes the relative positions of inorganic materials, but can in no way hope to present a picture of change and diffusion unless long series of samples are taken. For most purposes, however, tissues have to be fixed to prevent diffusion due to death of the cell. Only the most common elements, e.g. calcium, magnesium, potassium and chloride, carbonate, phosphate, have been studied extensively. Amongst the heavier elements, arsenic and lead take the lead. Iron and copper claim delicate methods. Besides microincineration mentioned above there are the supplemental techniques of using X- rays and the spectroscope. In this field one might include any and all micromethods for determining vitamins, hormones and enzymes whose biological effects are great for small amounts of substances. These also depend on skillful micromanipulations of single cells or portions of cells. For some problems histochemical procedures seem to be very imperative, e.g. the site of secretion of hydrochloric acid in the stomach, bile secretion in the liver, the modus operandi of the kidney, the localization of vitamins and enzymes. Histochemists have been particularly interested in fine reactions for the cytochromes, vitamin C, the nucleic acids, cytoplasmic constituents, thyroxine, pepsin, various lipases and amylases.

Histones In the American Classification of proteins, the Histones are simple proteins, soluble in water, insoluble in very dilute ammonia, soluble in very dilute acids and in solutions of the alkali metal hydroxides. On hydrolysis the basic amino acids predominate. The yield precipitates with solutions of other proteins, which are probably coacervates.

Examples globin of hemoglobin and scombron in mackerel spermatozoa.

Histotrophe The sum total of nutritive substances supplied to the embryo in viviparous animals from sources other than the maternal blood stream, e.g. uterine glands and the like.

Hnrna (heterogeneous nuclear RNA) Heterogeneous nuclear RNA; refers collectively to the variety of RNAs found in the nucleus, including primary transcripts, partially processed RNAs and snRNA. The term hnRNA is often used just for the unprocessed primary transcripts, however.

Hoffman determination of pentoses The test substance is hydrolyzed by steam distillation with 12-20% hydrochloric acid; 1-5 cc. of distillate are neutralized, to phenolphthalein, with 50% sodium hydroxide, treated with 0.5 cc. aniline and 4 cc. acetic acid and diluted to 10 cc. After standing for 15 minutes in the dark, the furfural is determined colorimetrically against standard furfural solution. The test is positive even in concentrations of 0.00005%.

Hofmeister reaction for creatinine and kynurenic acid The addition of phosphotungstic acid to a solution of creatinine acidified with nitric acid results in the formation of a yellow crystalline precipitate.

A precipitate of rhombic plates is obtained upon the addition of mineral acid and phosphotungstic acid to kynurenic acid.

Homeobox A series of regulatory genes that control what genes are 'turned on' in their region of the growing embryo. The homeobox genes are arranged on the chromosomes in the same sequence as they are used in the body; homeobox gene 1 is responsible for the top of the head, #2 is for eye level, etc. These are present in humans, mice, fruit flies, and even as far back as flat worms.

Homeotic mutation A mutation in a homeotic gene that is responsible for controlling the activities of numerous other genes, usually during embryologic development in higher organisms.

Homocysteine $SH(CH_2)_2 \cdot CHNH_2 \cdot COOH$; an intermediate in the in vivo transformation of methionine to cysteine.

Homogentisic acid 2, 5,-dihydroxyphenylacetic acid. Found in the urine in alcaptonuria, as a product of a hereditary inability to completely oxidize tyrosine and phenylalanine, of which it is a normal metabolic intermediate.

Homoiogenetic induction The power acquired by a part of an embryo, at the Homoiogenetic induction same time as it is determined for a certain differentiation, to induce another of the same kind. Thus at the same time as the neural plate is induced by the underlying mesoderm, it acquires the power of inducing another neural plate if it is removed and implanted into the blastocoele cavity of another embryo.

Homologous chromosomes One of a pair of chromosomes that resemble each other in length, shape, and the genes they carry, and that pair with each other at meiosis. X and Y chromosomes differ in these respects but still function as homologues.

Homologous recombination The exchange of sequence between two related but different DNA (or RNA) molecules, with the result that a new "chimeric" molecule is created. Several mechanisms may result in recombination, but an essential requirement is the existence of a region of homology in the recombination partners. In DNA recombination, breakage of single strands of DNA in the two recombination partners is followed by joining of strands present in opposing molecules, and may involve specific enzymes. Recombination of RNA molecules may occur by other mechanisms.

Homologous structures Similarity in some aspect of body form or patterning between different species; evolutionary outcome of descent from a common ancestor.

Homology Indicates similarity between two different nucleotide or amino acid sequences, often with potential evolutionary significance. It is probably better to use more quantitative and descriptive terms such as nucleotide "identity" or, in the case of proteins, amino acid "identity" or "relatedness" (the latter refers to the presence of amino acids residues with similar polarity/charge characteristics at the same position within a protein).

Homozygote Individual having the same allele at the corresponding loci of two homologous chromosomes.

Hopkins reaction for indole A red color is produced when indole is treated with glyoxylic acid and concentrated sulfuric acid. Sensitivity—1 500000.

Hopkins reagent for glutathione Treatment of glutathione in dilute sulfuric acid solution with freshly prepared cuprous oxide results in the formation of a precipitate.

Hopkins test for lactic acid in stomach contents 5 cc. concentrated sulfuric acid containing a little copper sulfate are added to a few drops of stomach contents, heated for 2 minutes, cooled, and a small quantity of thiophene added. A cherry-red color indicates lactic acid.

Hopkins-cole reagent for tryptophane Tryptophane solution, made 5% with respect to sulfuric acid, is treated with the reagent (solution of 10 parts mercuric sulfate in 90 parts 5% sulfuric acid). The tryptophane, precipitated quantitatively, may be obtained pure by successive treatment with hydrogen sulfide and baryta water.

Hopkins-cole test for proteins Having found that the Adamkiewicz reactions depends upon the presence of glyoxylic acid in the glacial acetic acid used, the authors use the former acid directly. The test solution is mixed with an equal volume of glyoxylic acid and stratified with concentrated sulfuric acid; tryptophane containing proteins form a reddish color at the interface.

Hormone A chemical which is secreted by an endocrine gland in small amounts directly into the blood stream and greatly influences the functions of some specific organ and frequently of the body as a whole.

Host strain (bacterial) The bacterium used to harbor a plasmid. Typical host strains include *HB101* (general purpose *E. coli* strain), *DH5a* (ditto), *JM101* and *JM109* (suitable for growing M13 phages), XL1-Blue (general-purpose, good for blue/white *lacZ* screening). Note that the host strain is available in a form with no plasmids (hence you can put one of your own into it), or it may have plasmids present (especially if you put them there). Hundreds, perhaps thousands, of host strains are available.

Hovey-hodgins test for glycerin in the presence of ethylene glycol 3 cc. of freshly prepared 10% aqueous catechol and 6 cc. concentrated sulfuric aid are added to 3 cc. of test solution and boiled for 0.5 minutes; a blood-red color formed at about 140-145°C indicates glycerin. Ethyl alcohol, ethylene glycol and diethylene glycol give no color distinctive colors are obtained with other polyhydric alcohols.

Hübl solution for determining iodine number (1) solution of 30 gm. mercuric chloride in 500 cc. 90% alcohol.

(2) solution of 25 gm. iodine in 500 cc. 90% alcohol. The reagent is stable in the mixed form but should not be used until 48 hours after mixing.

Human Genome Initiative Collective name for several projects begun in 1986 by Department Of Energy (DOE) to.

(1) create an ordered set of DNA segments from known chromosomal locations,

(2) develop new computational methods for analyzing genetic map and DNA sequence data, and.

(3) develop new techniques and instruments for detecting and analyzing DNA. This DOE initiative is now known as the Human Genome Program. The national effort, led by DOE and NIH, is known as the Human Genome Project.

Humature Scale devised by averaging the dry bulb temperature and the percentage humidity to express the thermal demands of the environment of the organism.

Humin The dark amorphous substance formed by the reaction of tryptophane with an aldehyde group, during the acid hydrolysis of a protein.

Humus Material consisting of decomposed organic matter and constituting a carbon-rich ingredient of soil.

Hunter heactions for thymine Thymine and diazotized sulfanilic acid in sodium carbonate solution are mixed and, after a short time, treated with sodium hydroxide and hydroxylamine solution, producing an intense red color.

Hunter reactions for imidazole-acrylic acid I. The addition of siver nitrate solution to a solution of the acid, neutralized with ammonia, yields a precipitate which is soluble in excess ammonia.

II. An alkaline solution of the acid forms a red color with diazobenzenesulfonic acid.

III. The acid reduces alkaline potassium permanganate solution at room temperature.

Hunter test for ergothioneine Reagents (1) (1.5 cc. 5% sodium nitrate solution are added to 1.5 cc. of a solution of 9 gm. sulfanilic acid and 90 cc. concentrated hydrochloric acid per liter. After standing 5 minutes, 6 cc. more of sodium nitrite are added and after 5 more minutes diluted to 50 cc.

(2) 1 gm. anhydrous sodium carbonate and 10 gm. sodium acetate are dissolved in 100 cc. water. 1 cc.

(1) and 0.5 cc.

(2) are mixed and, after 15 seconds, the test solution is added. Chill for ½ minute, add 2 cc. 10% sodium hydroxide and shake; a purplishred color is obtained.

Hyaluronidase An enzyme of certain streptococcus, pneumococcus and other cultures which destroys the capsules of the organisms.

Hybridization The reaction by which the pairing of complementary strands of nucleic acid occurs. DNA is usually double-stranded, and when the strands are separated they will re-hybridize under the appropriate conditions. Hybrids can form between DNA-DNA, DNA-RNA or RNA-RNA. They can form between a short strand and a long strand containing a region complementary to the short one. Imperfect hybrids can also form, but the more imperfect they are, the less stable they will be (and the less likely to form). To "anneal" two strands is the same as to "hybridize" them.

Hybridoma A clone of plasmacytoma cells which secrete a monoclonal antibody; usually produced by fusion of peripheral or splenic plasma cells taken from an immunized mouse with an immortalized murine plasmacytoma cell line (fusion partner), followed by cloning and selection of appropriate antibody-producing cells.

Hydnocarpic acid $C_{16} H_{28} O_2$; a cyclic fatty acid found in Chaulmoogra oil which is used in the treatment of leprosy; crystallizes as leaflets from alcohol; m.p. 59-60°C.

Hydrastine Levorotatory alkaloid from Hydrastis canadensis; slows heart beat and raises blood pressure; used to contract uterus and quench hemorrhage.

Hydrastinine $C_{11} H_{13} O_3 N$; alkaloid of hydrastis Canadensis; in the eye causes dilation of the pupil; internally large doses cause paralysis of the nervous system; in smaller doses is used to prevent parturitional hemorrhages.

Hydrastis Dried rhizome and roots of Hydrastis canadensis (Golden Seal), containing the alkaloids hydrastine, berberine and canadine; used to check sweating, as tonic and as promoter of uterine contraction.

Hydrogen electrode A metallic electrode which has been coated with a thin layer of platinum black, and exposed to an atmosphere of hydrogen long enough for the platinum black to become saturated with hydrogen. It behaves as though it were an electrode made of hydrogen, and is used for conveniently and accurately measuring hydrogen ion concentration.

Hydrogenation addition of hydrogen to a carbon-containing molecule; the hydrogen atoms attach to carbon atoms, thus breaking double and triple bonds between carbon atoms in the original molecule.

Hydrolysis The addition of water to a compound causing it to split into two parts, one part chemically reacting with the H of the water and the other part with OH.

Hydronium A combination of hydrogen ion, H+, with a water molecule, giving H_3O^+.

Hydrophilic "Water loving". Hydrophilic compounds dissolve easily in water, and are usually polar.

Hydrophilicity plot A computer plot which examines the relative summed hyrophobicity/ hydrophilicity of adjacent amino acid sidechains (usually within a moving window of about 6 amino acid residues) along the primary sequence of a polypeptide chain. Generally, hydrophobic regions of proteins are considered likely to be in the interior of the native protein, while hydrophilic domains are likely to be exposed on the surface and thus possibly antigenic sites (epitopes). At best, these are crude predictions.

Hydrophobic — "water fearing". Hydrophobic compounds do not dissolve easily in water, and are usually non-polar. Oils and other long hydrocarbons are hydrophobic.

Hygrine $C_8 H_{15} ON$; l-(i-N-methylpyrrolidine) propenone-2; alkaloid of Erythroxylin Coca; alkaline liquid, b.p. 193-195°C

Hyodesoxycholic acid $C_{24} H_{40} O_4$, m.p. 197°C; a bile acid found in the hog and the hippotamus; it is 3,6-dihydroxycholanic acid.

HYOSCINE $C_{17} H_{21} O_4 N$; a syrupy alkaloid of solanaceous plants. The dl form crystallizes as prisms with H_2O, m.p. 193-194°C, has a more transient m.p. 193-194°C, has a more transcient action than atropine on the autonomic nervous system. It acts as a depressant on the central nervous system, causing fatigue and sleep.

Hyoscyamine $C_{17} H_{23} O_3 N$; long silky needles from dilute alcohol, m.p. 108.5°C; a commonly occurring alkaloid of solanaceous plants; the l-variety of atropine.

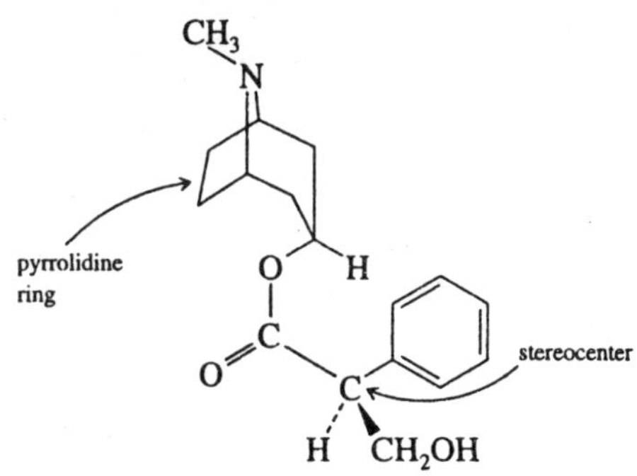

Fig. Hyoscyamine

Hyoscyamus Henbane; dried leaves and tops of Hyoscyamus niger, containing the alkaloids hyoscyamine and hyoscine.

Hyperbola Curve generated by the intersection of a plane and a right circular cone.

Hyperchlorhydria Hyperacidity; abnormal amounts of free HCl secreted by the stomach, a sign of ulcers, gall bladder disease, appendicitis.

Hyperglycemia The increased concentration of blood sugar above normal for any reason, such as diabetes.

Hyperglycemic index An estimate of rate of glucose assimilation by tissues, as the percentage obtained in dividing 2 hour level minus fasting level of blood sugar by the maximum level minus the fasting level; is zero for normal persons.

Hyperpiesia Essential hypertension; a prolonged rise in arterial pressure greater than 160 systolic or 90 diastolic without anatomic causative factor in the kidney or blood vessels, said to be familial and attributed to nervous tension. Reduction of pressure is to suit the patient's comfort rather than a therapeutic measure. Complications are often arteriosclerosis, cardiac failure, coronory thrombosis, uremia. Vasodilators (nitrites and theobromine) are used to alleviate symptoms. Cautious use of iodides, thiocyanates and hormones is advised. The most rational advice on diets restricts intake of fluids but little else outside of overeating. Psychotherapy is important.

Hyperthyroidism (Exophthalmic goitre) A constitutional disease involving excess secretion from a diffusely hypertrophied thyroid, causing any increased metabolism.

Hypoglycemia A condition of abnormally low sugar in the blood, i.e. below 80 mg. per cent. It may include hyperinsulinism and insulin shock because of the definition as a symptom and not a disease. Thus, too, liver disease and any failure in the supply of hormones, e.g. epinephrine, which inhibit insulin will bring about hypoglycemia. Addison's disease would do so by destruction of the adrenals. Pituitary destruction (Simmonds' disease) would operate by the loss of the antagonistic diabetogenic hormone. Thyroidectomy has a similar effect. Feeding carbohydrates or the missing hormone epinephrine are obvious emergency procedures. If the hyperinsulinism is due to malignant disease of the pancreas surgery is required.

Hypometabolism A depression of the basic metabolic rate in conditions of undernutrition, sleep, thyroidectomy, etc.

Hypo-ovarianism Hypogonadism in the female; deficiency of ovarian function which grades to cessation of ovarian function, or menopause, with its accompanying symptoms of nervous disturbance and amenorrhea. It may be primary when it is due to delayed or retarded development, eunochoidism or infantilism, or is due to pathological or surgical loss, or due to vitamin deficiency. Pituitary dysfunction is a more remote cause making the hypo-ovarianism secondary. Menopause may be natural (between the 40th and 50th years), premature (due to disease) or artificial (due to castration). Hypo-ovarianism shows in menstrual disturbances as amenorrhea (cessation of flow), oligomenorrhea (irregularity of flow), menorrhagia (hypermenorrhea), or metrorrhagia (polymenorrhea); eunuchism; poor development or regression of secondary sex characteristics. Nervous symptoms include "the blues," decreased memory, and insomnia. Cardiovascular symptoms are "hot flushes," palpitation and dizziness. Other symptoms attach themselves to these readily. The symptoms are even more marked in menopause. The treatment is causal and leads to substitution of estrogenic hormones to make up for their de. creases. Estriol, estrone or estradiol and their derivatives may be used.

Hypophysis The pituitary gland. A number of extremely important hormones are produced in its four parts.

Hypothalamus Part of the brain that regulates automatic body processes such as temperature and production of certain *hormones*.

Hypothalamus Of vertebrate forebrains a brain center that monitors visceral activities (such as salt-water balance, temperature control, and reproduction) and that influences related forms of behavior (as in hunger, thirst, and sex.)

Hypothyroidism A deficiency of the thyroid hormone which is shown by a decrease of basal metabolic rate. It takes on the forms of

(1) cretinism

(2) infantile myxedema

(3) adult myxedema and

(4) hypothyroidism without myxedema. The first two forms have an intra-uterine origin, leading if unchecked to dwarfism and idiocy. Myxedema of adults comes gradually. Besides low basal metabolism there is hypercholesterolemia, lowered tolerance for dextrose and diminished sensitivity to epinephrine and pilocarpine. In forms in which myxedema is absent response to thyroid medication is good. There is a general "neurasthenic syndrome" consisting of insomnia, lack of appetite (anorexia), fatiguability and vague pains.

Hypotonic (1) Marked by abnormally low tension.

(2) Less than isotonic said of solutions which are of less than isotonic concentration.

Hypoxanthine 6-oxypurine, a breakdown product of nucleoproteins via adenine and the action of adenase.

Hysteresis The dependence of the state of a physical system not only upon environmental parameters, but upon the sequence of states through which it has previously passed.

I

Icteric index A ratio used as a test for bilirubin and jaundice. Diluted blood serum is treated with acetone, filtered and compared with 110000 potassium dichromate. The ratio of the standard to the unknown, times the dilution is the icteric index, normally 4 to 6.

Idaein Cyanidin-monogalactoside; an anthocyanin of the cowberry (mountain canberry); forms a chloride, C_{21} H_{21} O_{11} $Cl._2$ ½H_2O; melting at 210°C.

Idiotope Epitopes present in the unique variable sequences of an immunoglobulin molecule. These may or may not coincide with the immunoglobulin's paratope.

Imaginary number Element of a *complex number* that contains the square root of 1

Imbibition pressure The pressure against which a lyophilic colloid will imbibe a liquid, or conversely, that pressure which will be needed to force the dispersions medium out of a gel, or similar colloid.

Immune- polysaccharides Five types of polysaccharides which usually contain uronic acids and may contain some nitrogen; isolated from bacteria and involved in the production of immunity.

Immunity Condition of being able to resist disease by protecting against the effects of disease-causing microorganisms.

Immunological phenomena Substances inciting the formation of and reacting with antibodies are named antigens. Sera that contain antibodies as the result of the injection of antigenss are called "immune sera" or "antisera," while the designation "normal" or "natural" antibodies is applied to substances found in the sera of untreated animals, which are similar in their effects to the antibodies developed by immunization. A rather detailed nomenclature has been built up around the diverse manifestations of antigen-antibody reactions, the antigens and antibodies being described with reference to the kind of reaction observed; but it is necessary to state at the outset that the same antibodies can act in various capacities, e.g., as agglutinins and precipitins.

Poisonous antigens characterized by high toxicity and their capacity of being fully neutralized, even in high multiples of the lethal dose, by their antibodies are termed toxins or exotoxins (as the toxin of the diphtheria bacillus). This name includes substances which destroy cells, e.g., the hemotoxin (or hemolysin) of tetanus bacilli, the leucocidin of staphylococci, etc.; the neutralizing antibody is called antitoxin. Toxoids are an atoxic modification of toxins which retain binding power and antigenicity. The clumping of cells is known as agglutination, and the antigens and antibodies involved are called agglutinogens and agglutinins (hemagglutinins, bacterial agglutinins) respectively. Similarly, the antibodies causing disruption of cells (lysis) are designated as lysins (bacteriolysins, hemolysins), precipitins those which cause precipitation when mixed with the inciting soluble antigens (precipitinogens), while tropins (opsonins) are antibodies that render cells susceptible to ingestion by leucocytes (phagocytosis).

Referring to the ability of antigens to induce a state of sensitivity and subsequently to produce the symptom-complex known as anaphylaxis the term anaphylactogen is in use. The anaphylactic state is induced by parenteral administration of a protein to describe a typical experiment, after 0.01 cc. of ox serum has been injected into a guinea pig (and much smaller doses may be effective, as 0.0001 cc. or less), and an interval of 10 days or longer has elapsed, a second (intravenous) injection of, say, 0.1 cc. kills the animal in a few minutes under characteristic and violent symptoms (choking, convulsions, dyspnoea); or, a rabbit previously sensitized would show intense local inflammatory and necrotic reactions on subcutaneous reinjection of the same antigen (Arthus phenomenon). The anaphylactic condition can be transferred with the serum of sensitized animals to normal ones (passive anaphylaxis) and, fundamentally, the phenomenon has the significance of an antigen-antibody reaction in vivo and can be used in place of a reaction in vitro. The opinion that these reactions generally exhibit greater specificity than in vitro tests, would require confirmation. The designation allergen connotes the ability of certain materials to induce specific allergic manifestations, denoted by Coca as atopic diseases (hay fever, asthma), and the associated special antibodies in the serum of such patients are known as reagins a name also given to some other agents (Wassermann reagins).

Antibodies produced with material taken from the animal selected for immunization, or from other individuals of the same species, are referred to as auto- or isoantibodies respectively. The expression passive immunization, in contrast to active immunization, signifies the (temporary) protection conferred upon an animal by the administration of immune sera. Reactions of an antibody with its corresponding antigen are said to be homologous, while heterologous reactions, known also as overlapping, cross or group reactions, are those taking place with (related) substances other than the inciting antigen.

Immunoprecipitation A process whereby a particular protein of interest is isolated by the addition of a specific antibody, followed by centrifugation to pellet the resulting immune complexes. Often, staphylococcal proteins A or G, bound to sepharose or some other type of macroscopic particle, is added to the reaction mix to increase the size and ease collection of the complexes. Usually, the precipitated protein is subsequently examined by SDS-PAGE.

Implantation The addition of tissue, grafts, chemical fractions in an implant medium, etc. to an embryo, without the removal of anything from it.

***In situ* hybridization** Use of a DNA or RNA probe to detect the presence of the complementary DNA sequence in cloned bacterial or cultured eukaryotic cells.

Incandescence Emission of light from an object because of its extremely high temperature.

Indian hemp Cannabis Indica; dried flowering tops of Cannabis sativa (hashish), a narcotic; habit forming.

Indican C_{14} H_{17} O_6 N3H_2 O1—A glycoside of Indigofera, consisting of glucose and indoxyl; brown rhombic crystals, m.p. 50°C.2—C_6 H_5 NH·CHCOSO_2 OK, potassium indoxyl sulfate, arising in animals from bacterial decomposition of tryptophane in the intestines.

Indicator A substance, which when added in small amounts to a chemical system, shows, by an easily discernible change of color or some other characteristic, a condition of the system or reaction; i.e. pH, etc.

Indirect induction The induction of a new neural axis in competent ventral ectoderm by a chemical compound, not acting in the same way as the naturally occurring evocator, but by liberating the masked evocator in the reacting tissue.

Individuation That part of the morphogenetic effect of the primary organiser in neural induction, especially the regional differentiation, which cannot with certainty be referred back to the action of a single chemical substance, the Evocator. Since it seems probable that forces of field type are involved in the regional differentiation of the evoked axis, the Individual field is spoken of The history of the word has already been discussed. It has, however, another embryological use which was not appreciated when it was introduced into causal morphology. Neurologists have spoken of the differentiation of local reflex systems, behaviour patterns, and the like out of the total undifferentiated diffuse response of the early embryo to stimuli, as "individuation". This is not perhaps a very good use, since it applies to parts of the organism, hence not to true *individuals*. The word "emancipation" would be preferable.

Indole CH CHC_6H_4NH m.p. 52.5°C,cyclic amine resulting from bacterial decomposition of tryptophane; partly responsible for the odor of feces and decaying protein; present in urine as indican. In large amounts it produces stupor and weak heart action.

Indophenol oxidase A component, with cytochrome, of the oxidation system of the cell; gives a blue color with paraphenylene diamine and alpha naphthol.

Indophenol test A test for the presence of oxidizing enzymes in cells and for detecting the presence of myeloblasts, etc.

Indoxyl C_8H_7ON, m.p. 85°C, excreted as a sulfate in urine as a decomposition product of tryptophane; also occurs naturally as a component of indican.

Inducer A small molecule, such as IPTG, that triggers gene transcription by binding to a regulator protein, such as LacZ.

Induction The morphogenetic effect brought about by an organiser, inductor, evocator, etc. acting on competent tissue.

Inductor A piece of living tissue, dead tissue, a chemical fraction, or a chemical substance of known constitution, which carries out an *induction* similar to that performed by an *organiser*. The word can thus be used indiscriminately for a number of types of inducing agent. To the primary and secondary organisers correspond primary and secondary inductors. (N.B. the term "secondary" here must be sharply distinguished from its use for the induced member of a twin, e.g. the "secondary" embryo as opposed to the "host" embryo.).

Inert gas any of six stable elements (helium, neon, argon, krypton, xenon, and radon) that used to be considered nonreactive, but that are now known to be able to react with certain elements in certain conditions; also called noble gases.

Informatics The study of the application of computer and statistical techniques to the management of information. In genome projects, informatics includes the development of methods to search databases quickly, to analyze DNA sequence information, and to predict protein sequence and structure from DNA sequence data.

Infrared radiation Electromagnetic radiation that is invisible to the human eye and has a wavelength just longer than the radiation at the red end of the visible spectrum.

Infusorians Obsolete term for small aquatic organisms (mostly protozoans, rotifers, one-celled algae); primarily used for protozoans.

Ingression An inward movement of cells in the amphibian blastula, especially the yolk endoderrn. (In coelenterate embryology the word has another meaning.)

Inheritance, blending The complete fusing of the paternal and maternal genetic materials.

Inheritance, particulate The nonfusion of the parental genetic material during zygote formation.

Initiation codon The codon at which translation of

a polypeptide chain is initiated. This is usually the first AUG triplet in the mRNA molecule from the 5' end, where the ribosome binds to the cap and begins to scan in a 3' direction. However, the surrounding sequence context is important and may lead to the first AUG being bypassed by the scanning ribosome in favor of an alternative, downstream AUG. Also called a "start codon". Occasionally other codons may serve as initiation codons, e.g. UUG.

Initiator trna A special tRNA responsible for annealing to the start codon, in the P site of the ribosome, to initiate polypeptide synthesis.

Inorganic compound Compound that is not and never was part of or produced by a living organism inorganic pyrophosphatase (pyrophosphatase) An enzyme that hydrolyzes a molecule of inorganic pyrophosphate to yield two molecules of phosphate. reaction

Inosine A nucleoside consisting of hypoxanthine and d-ribose, formed by deamination of adenosine; forms inosine phosphoric acid; hypoxanthosine; hypoxanthine riboside.

Inouye test for blood 10 cc. of test liquid are treated with 10 cc. alcohol and 5 cc. chloroform and shaken lightly; 10-20 drops each of a freshly prepared guaiac tincture and of ozonized turpentine oil are then added. A temporary blue-violet color is obtained in the chloroform layer.

Insert In a complete plasmid clone, there are two types of DNA - the "vector" sequences and the "insert". The vector sequences are those regions necessary for propagation, antibiotic resistance, and all those mundane functions necessary for useful cloning. In contrast, however, the insert is the piece of DNA in which you are really interested.

Insertion sequence A small bacterial transposon carrying only the genetic functions involved in transposition. There are usually inverted repeats at the ends of the insertion sequence.

Insulin Insulin, the internal secretion of the pancreas, is formed in the islet tissue of that organ and passes from there into the general circulation. Its presence is indispensable for the normal maintenance of carbohydrate metabolism in mammals. In cases of pronounced hypofunction the typical symptom complex of diabetes mellitus is observed. With excessive production of the hormone a physiological state ensues which is conveniently designated as hyperinsulinism.

Banting and Best succeeded in 1922 in obtaining from the pancreas potent hypoglycemic extracts, suitable for clinical use in the treatment of diabetes mellitus.

Isolation and Chemistry Following the successful achievement of obtaining potent extracts of the hormone from the pancreas, numerous efforts were undertaken to isolate the active principle as a crystalline chemical entity.

Insulin was first obtained in crystalline form by Abel and his associates in 1926 from highly purified commercial preparations. The method of Abel and his collaborators in obtaining crystalline preparations consists of the isoelectric precipitation of the hormone from a strongly buffered acetic acid solution by the addition of a weak base. The pH of the final solution was found to be about 5.6. Modified procedures of obtaining crystalline insulin preparations have been worked out by several investigators. Although crystalline insulin is usually prepared from beeves' pancreas, it can also be obtained from the islet tissue of certain fishes, and from pig and sheep pancreas. Insulin crystals were found to possess a physiological activity of 24 international units per milligram.

The accumulated results of the chemical investigations on crystalline insulin reveal that this hormone must be classified as a typical protein. Its complex protein-like nature affords little hope at present, for the elucidation of its exact structure and for its synthesis. The pharmacodynamic functions of the hormone are apparently connected in some way with the manner in which the component amino acids are linked, since analytical work has so far failed to disclose any especial constituent or property, which would explain the unique action of this principle in the animal body. The chemical behavior of insulin towards certain reagents seems to indicate that the physiological properties of insulin are associated with certain groupings in the molecule, e.g., the dithio (-S-S-) linkages (present as cystine), with the hydroxyl groups (tyrosine) and also apparently with a part of the free amino groups.

Standardization The evaluation of the potency of an unknown insulin preparation is made by comparing its hypoglycemic action with that of the international standard powder under carefully controlled conditions. The international unit which is now accepted is the activity contained in 0.125 mgm. of the international standard preparation. The activity of various commercial insulin preparations and of crystalline insulin is expressed in terms of this international standard.

Two procedures have been devised to determine the activity of insulin preparations by comparison with the standard and are now generally employed (a) method dependent upon the production of convulsions, and (b) method based upon the determination of the decrease in blood sugar.

Briefly stated, method (a), which usually employs mice as test animals, is based upon the comparison of the percentage of convulsions produced in white mice kept at 38°C, half of the mice being injected intraperitoneally with the standard preparation and the other half with the solution of unknown unitage. The mouse dose is defined as the quantity producing convulsions in one-half the number of mice injected.

Method (b) of assaying insulin has gained wide acceptance. Rabbits are generally used in this procedure which consists of injecting subcutaneously a suitable dose of the standard insulin preparation into one-half of a series of rabbits of 2 kilo weight and previously starved for eighteen to twenty-four hours, the other half simultaneously receiving a dose of the sample of unknown unitage. Several days later the groups are crossed over and used for the injection of the same preparations. Blood samples are usually taken at one and onehalf, at three, and at five hour intervals after the injections. From the relation between the blood sugar lowering produced by the standard insulin and by the insulin of unknown potency, the activity of the latter can be calculated.

Administration Insulin is administered to diabetic patients at present mainly by the subcutaneous route and sometimes in cases of emergency, by intravenous injection. There are associated with these generally accepted methods of administration, certain practical difficulties which investigators have for many years recognized and attempted to remedy. Chief among these difficulties are.

(1) the discomfort accompanying injection, and,

(2) the unavoidable frequency of dosage with its attendant disadvantages.

Oral administration of the hormone obviously would be the most convenient. However, since insulin activity is destroyed by proteolytic enzymes, no attempt to give the hormone by mouth can succeed unless practical measures for its protection against the action of trypsin or pepsin be perfected. In spite of various attempts, there is, unfortu- nately, at present no preparation of insulin available which can be given by mouth and which can be relied on to take the place of the insulin preparations which must now be given by injection.

Other experiments were carried out with the object of combining or mixing insulin with substances in order to decrease its rate of absorption from the tissue. Such delayed absorption would permit the use of larger doses and thus reduce the number of daily injections required. Hageaorn and his associates were the first to produce an insulin preparation possessing prolonged blood sugar lowering action and suitable for therapeutic use. These investigators demonstrated that when a solution of protamine in sodium phosphate buffer is added to an insulin solution, an insulin protamine complex is formed at pH 7.2. On subcutaneous injection of a suspension of this complex, prolonged hypoglycemia is produced. These investigators also reported successful results from the administration of protamine-insulin to diabetic patients. The importance of this finding was immediately recognized and was followed by further attempts of different investigators to obtain even more useful insulin preparations.

Shortly after Hagedorn's finding, Scott and Fisher reported that the addition of zinc to insulin prior to the addition of protamine, prolonged still further the hypoglycemic effect of the hormone. The superiority of protamine-zinc-insulin to unmodified protamine-insulin and to "regular" insulin has been substantiated by numerous clinical tests. Protamine-zinc-insulin and "regular" insulin are at present used clinically.

Various combinations of insulin with other basic compounds in the presence of certain heavy metal ions have been prepared and found to produce prolonged hypoglycemia. However as yet no preparation has been obtained which could be considered to be clinically superior to protamine-zinc-insulin.

Substitutes The clinical use of the various commercial insulin preparations for the relief of the symptoms of diabetes mellitus can be considered an entirely satisfactory and successful therapeutic measure. The fact that insulin is ineffective when administered orally has led to efforts to obtain nontoxic insulin substitutes which are active when given by mouth. At present, however, no substance fulfilling these requirements has been prepared.

The possibility of the existence of insulin-like substances in plants has been investigated at various times. Although different plant extracts have been prepared which, it was claimed, exerted a hypoglycemic effect, little has been done, either in substantiation or extension of these investigations. In many instances substances obtained from plant sources and possessing hypoglycemic action were found to be rather toxic, and have accordingly not found to be useful in medical practice, (B, C, D, G).

Frank and his associates have studied the effect of various polymethylenediguanidines on the blood sugar. Synthalin, (decamethylenediguanidine), and neosynthalin, (dodecamethy lenediguanidine), were found to be the most active of the different diguanidino derivatives studied by them. Both compounds were used for a short time in the treatment of certain diabetic cases. Because of the toxic effects of these substances their clinical use has been discontinued. Physiological Action The chief role of insulin is concerned with the regulation of important phases of metabolism. Metabolism is generally defined as the sum total of the chemical changes which occur in the various tissues of the body. Metabolism is controlled by the proper physiological coordination of various active agents in the body; as such agents we may classify enzymes, vitamins and hormones. The efficiency with which the regulatory mechanisms of the endocrine system on metabolism counteract each other, depends to a large extent on its normal balance. Any relative or absolute deficiency or preponderance of certain endocrine secretions may result in a definite abnormal shift of general metabolism. The exact role which each of the different endocrine principles plays in the process of metabolism, has not yet been fully established. The liver is in all probability the organ through which the influence of the endocrine system on metabolism is mainly exerted. The influence of certain endocrines on metabolism manifests itself by observing the main alterations in metabolism which occur in the absence of these endocrine organs.The importance of insulin with regard to metabolism becomes evident upon examination of the physiological disturbances in the body which one observes in the absence of the secretion of insulin (pancreatectomy, diabetes mellitus). The following symptoms have been found to be characteristic

1. Pronounced hyperglycemia and glycosuria.

2. Depletion of the glycogen stores in certain tissues (liver, muscle).

3. Lowering of the respiratory quotient indicating incomplete combustion of glucose.

4. Increase in the glucose-nitrogen ratio in the urine probably due to an increase in the conversion of protein into glucose.

5. Development of acidosis probably caused by imperfect fat metabolism.

Injection of insulin will relieve all these symptoms and reestablish a practically normal metabolism. Administration of insulin also affects specifically the carbohydrate metabolism of the normal organism.The following functions may be attributed to insulin.

1. Acceleration of glucose oxidation in the tissues.

2. Increase in the rate at which glucose is converted to glycogen in the muscle and in other tissues. It is still unsettled whether insulin has a direct influence on the formation of liver glycogen or whether it inhibits hepatic glycogenolysis which is caused by certain other hormones. This inhibitory effect of insulin would enable the liver of the normal animal to retain its glycogen, and would also account for the disappearance of liver glycogen in the depancreatized animal (absence of insulin).

3. Increase in glucose oxidation and in the rate of glycogen formation probably accounts for the fall in blood sugar observed after insulin injection.

4. Inhibition of carbohydrate formation in the liver from non-carbohydrate sources. Gluconeogenesis is apparently under the control of certain other endocrine principles.

Prevention of the formation of ketone bodies which are formed as a result of deranged fat metabolism.

Decrease in glucose oxidation and increase in hepatic glycogenolysis and gluconeogenesis suggests that hyperglycemia may be due to the following factors.

a. Deficient supply of insulin.

b. Liberation, either at a normal or excessive rate, of those principles which play a role in glycogenolysis and in gluconeogenesis.

c. It should be reemphasized that in any study of the physiological action of a hormone, its relationship to various other endocrine principles and also to such active agents as vitamins and enzymes has to be taken into consideration.

Insulin units The international unit is the activity of 0.125 mg. of a standard preparation of dry insulin hydrochloride. One mg. of this international standard contains 22 units as defined by the Insulin Committee of the University of Toronto, their unit being the amount needed to produce convulsions in a 2 kg. rabbit in a specified time after injection (20 minutes).

Integer (mathematics) any whole number of the set including negative whole numbers, positive whole numbers, and zero.

Integration (mathematics) process of solving for the area under a curve (a definite integral); provides a tool for performing calculations that involve varying quantities.

Interference (physics) effect of two or more waves meeting and producing a wave of either larger or smaller amplitude.

Intergenic Between two genes; e.g. intergenic DNA is the DNA found between two genes. The term is often used to mean non-functional DNA (or at least DNA with no known importance to the two genes flanking it). Alternatively, one might speak of the "intergenic distance" between two genes as the number of base pairs from the polyA site of the first gene to the cap site of the second. This usage might therefore include the promoter region of the second gene.

Intermolecular force Weak attractive force holding molecules together; also called van der Waals force.

Internal secretions The products of the adrenals, pituitary, pancreas, thyroid, parathyroid, thymus, pineal gland, sex glands.

Interphase Of cell cycles, the time interval between nuclear divisions in which a cell increases its mass, roughly doubles the number of its structures and organelles, and replicates its DNA. The interval is different for different species.

Interstitial cells Cells of testis or ovary arising from the germinal epithelium and secreting sex hormones.

Intraocular fluid The water-like fluid of the aqueous humor, pH 7.4; closely resembles cerebrospinal fluid; probably formed as a serum dialysate modified by secretions of the ciliary epithelium.

Intrinsic factor (Castle) An anti-anaemic factor obtained from desiccated stomach and duodenal mucosa.

Intron Introns are portions of genomic DNA which are transcribed (and thus present in the primary transcript) but which are later spliced out. They thus are not present in the mature mRNA. Note that although the 3' flanking region is often transcribed, it is removed by endonucleolytic cleavage and not by splicing. It is not an intron. Compare exons.

Intrusion Rock consisting of hardened magma (molten rock) or lava that has flowed into cracks in Earth's surface layers.

Inulin A white polysaccharide, found in plants, dispersible in hot water. It is hydrolyzed by inulase to fructose.

Inverse PCR Variation of PCR that makes the amplification of DNA segments of unknown sequence that flank DNA segments of known sequence possible; in brief, total DNA is digested to completion and fragments ligated under conditions that favour circularization of fragments; pair of PCR primers, designed from known sequence known sequence, are used to prime PCR from opposite strands resulting in amplification of fragment of unknown sequence.

Inversions (DNA) Type of chromosome rearrangement in which a segment that has become separated from the chromosome is reinserted at the same place but in reverse, so the position and sequence of genes are altered.

Invert sugar A mixture of glucose and fructose obtained by acid hydrolysis of sucrose, so named because the optical rotation is inverted from dextro to levo after hydrolysis.

Inverted Repeats Two copies of the same or related sequence of DNA repeated in opposite orientation on the same molecule (contrast with Direct Repeats). Adjacent inverted repeats constitute a palindrome.

Iodine number The number of grams of halogen, calculated as iodine, which 100 grams of fatty acid or other fatty substance or any unsaturated compound will take up; it is a measure of the degree of unsaturation.

Iodogorgoic acid 3,5-di-iodotyrosine; an amino acid found in the proteins of many organisms, and in marine thyroids.

Iodopsin Visual violet; a carotenoid conjugated protein with prosthetic radical which is a vitamin A derivative, found in cones of the retina; plays a similar role to rhodopsin in dark adaptation.

Iodothyroglobin The iodine containing protein of the thyroid which, at least to a large extent is the hormone of the thyroid. Thyroxine is contained in the molecule. An insufficiency results in goiter, cretinism or myxedema, an excess in an abnormally high metabolism.

Ion antagonism The mutual undoing of effects of various ions, e.g. calcium vs. sodium, magnesium vs. calcium, potassium vs. sodium. This antagonistic action has been followed with respect to surface tension effects, gelling and liquefaction of protoplasm, changes in viscosity and zeta potential.

Ion exchange Reversible replacement within a compound of one *ion* with an ion of the same charge taken from a different compound.

Ion transfer (therapeutic) Iontophoresis; the introduction of soluble salts into the tissues by means of direct current. The objective is to deposit ions in or on tissue which may exert varying effects desired, including coagulation of proteins and destruction of tissues. On the other hand drugs may be thus introduced which would have healing effects and conceivably dispersive effects. The potential insures greater effect than topical application. Various electrodes have been devised consisting of a positive metal pole in contact with a fabric saturated with the. solution to be applied. The negative pole is dispersive and is connected to the other side of the patient. The currents are of the order of 5-10 milliamperes and are applied for about 30 minutes at unulcerated places on the skin. Substances which have been applied routinely are 0.2 to 0.5 percent solutions of acetylbeta-methylcholine chloride (mecholyl chloride) and 0.1 per cent solution of the acid phosphate of histamine. The conditions treated were those of chronic rheumatoid arthritis.

Ionic bond Bond between two elements created by the exchange between them of one or more electrons and the consequential presence within them of oppositely charged *ions*.

Ionization Formation of one or more *ions* through either the exchange of electrons between atoms or the break of a *covalent bond*.

Ions The electrically charged particles that make up an electrolyte.

Ipecacuanha A dried root of Cephaelis ipecacuanha(Brazil)containing the alkaloids emetine and cephaeline; expectorant and emetic, used in colds and amoebic dysentery.

Ipuranol A sitosterol-d-glucoside obtained from the stems of Ipomoea purpurea and other plants; m.p. 290-295°C.

Isatin An oxidation product of indigo; isatinic acid anhydride; o-aminobenzoylformic anhydride; m.p. 198-203°C; used as a reagent for cuprous ions, thiophene, mercaptans and indican.

Isinglass The inner membrane from the swimming bladder of certain species of sturgeon and hake and consisting chiefly of glutin.

Isocitric enzyme The pyridinoprotein enzyme containing coenzyme II which catalyzes the oxidation of isocitrate to alpha-ketoglutarate; found in heart, kidney, adrenal gland, etc.

Isoelectric point The hydrogen ion concentration at which the undissociated residue of an amino acid is a maximum, and at which the sum of the ions is minimal. It is that concentration at which there will be a tendency for as many cations to migrate to the cathode, as anions to the anode. In terms of Zwitterions, the i.p. is that H concentration where the ampholyte exists to a maximum as a Zwitterion, and at a minimum with free acid or basic groups.

Isolating Mechanism Biological properties of individuals that prevent the interbreeding of sympatric populations.

Isomerases Enzymes which catalyze isomerizations, e.g. 3-phosphoglyceric acid to 2-isomer and phosphopyruvic acid.

Isopyrovitamin An isomer of calciferol, obtained from it by heat treatment. It contains 4 rings and 3 conjugated double bonds.

Isotope One of two or more atoms that differ in *atomic mass* because of having the same number of protons but different numbers of neutrons.

Isotopes Atoms with the same atomic number and outer electron system iving similar chemical properties, but differing in nuclear mass.

Itching "An unpleasant cutaneous sensation which provokes the desire to scratch" (Haffenreffer). It is intermediate between light tickling and diffuse burning pain. Injury to the skin provokes hyperalgesia, sensitiveness to the slightest touch. Spontaneous itches, as in eczema, prurigo, urticaria, are due both to toxic irritation and to stimulation of nerve endings. "Mnemodermia" or skin memory is a kind of sensitivity to itching on rubbing a place which had been involved in an itching skin disease previously. Chemical mediation of itching has been attributed to an H-substance, which is either histamine or a histamine-like substance, liberated from injured cells. But this may not be true in all cases. Unless some irritating foreign body or crust is removed scratching is a poor remedy. Itching may be provoked without peripheral impulses by morphine. Epinephrine may inhibit responses to itch stimuli on local injection.

J

Jacoby reagent and test for pepsin and trypsin 1 cc. of pepsin in 0.56% hydrochloric acid is mixed with 2-3 cc. of reagent (1 gm. ricin and 1.5 gm. sodium chloride in 100 cc. water) and kept at incubator temperature. Even with as little as 0.01 mg. pepsin the mixture is made perfectly clear in a few hours.

Jaffe reaction for creatinine A yellow-red to dark blood-red color is obtained when aqueous picric acid and sodium hydroxide solution are added to the test solution. Sensitivity—1200000, the intensity of the color depends upon the amount of creatinine present. Creatine does not interfere, free acids do.

Jaffe test for kynurenic acid Tetrachlorhydroxykynurin, formed by heating kynurenic acid with potassium chlorate and hydrochloric acid, forms, with ammonia, a brown color changing gradually to green and dark blue.

Jaksch test for melanin and melanogen in urine These substances, with dilute ferric chloride solution, form a black precipitate which is soluble only in potassium hydroxide and concentrated acids.

Jalapin $C_{34}H_{56}O_{16}$; scammonin; orizabin; a glucoside of Jalapa Oriza benzis; consisting of glucose and jalapinolic acid; a colorless, amorphous solid, melting between 130-150°C, used as a purgative.

Jansen test for vitamin B1 A solution of 0.001-0.02 mg. vitamin B1 is diluted with methanol to 2 cc. to which are added 1 cc. 30% sodium hydroxide and 0.1 cc. potassium ferricyanide solution; 13 cc. isobutanol are added after about 2 minutes. After shaking and centrifuging 10 cc. of isobutanol (containing thiochrome) are placed in a photoelectric fluorometer standardized for thiochrome by using quinine in sulfuric acid.

Jaundice (Hemolytic) In this group are included.

(a) primary anemias such as a pernicious anemia, splenic anemia or Banti's disease; (b) icterus neonatorum; (c) familial icterus;(d) "disassociated jaundice." It includes such conditions as those in which either the bilirubin or bile salt is retained and reabsorbed into the blood, while the other is being excreted in the normal way. Hemolytic icterus is characterized mainly by a high bilirubin content of the blood, but generally with no appearance of pigment in the urine.

Jaundice yellow staining of the skin and underlying tissues due to bile pigments. Types are distinguished as subicterus, hemolytic icterus, obstructive icterus, hepatosis (parenchymatous) icterus and icterus of the new born. The appearance of icterus (pseudo-icterus) may be brought about by yellow coloring due to other causes, e.g. carotinemia, due to excessive carotene in the food or picnic acid staining. Hemolytic citcerus shows an increase of bilirubin

A in the blood without choluria (bile acids in the urine). Obstructive jaundice may be calculous icterus, stenosis icterus or icterus due to pancreatic tumors. Bile salts are used as a choleretic to promote bile flow and possible clearing of passage. Hepatosis icterus involves liver damage by drugs, infectious diseases, etc.

In this field tests have been developed for liver function, as the galactose, azorubin, bilirubin tests in urine, the icterus (or icteric) index, van den Bergh test, and sedimentation tests.

Jelinek test for mustard gas Reagent—ammoniacal silver nitrate solution is added to a solution of isatin containing a small amount of concentrated ammonia. Mustard gas, on paper impregnated with the reagent, forms a Bordeaux red color. At room temperature characteristic color affects are produced after 1 hour, at 60-80°C in one minute.

Johnson-clapp reagent for cytosine, thymine and uracil A red color is obtained when these substances are treated with diazobenzenesulfonic acid in alkaline solution.

Jolles test for glucuronic acid in urine 200-400 cc. of urine are completely precipitated with lead acetate and filtered; the washed precipitate is suspended in water, decomposed with hydrogen sulfide and filtered. The filtrate is evaporated to 20 cc. and used for the Tollen's reaction.

Jolles test for levulose in urine Diluted urine is boiled for 1 minute with 10 drops of 20% alcoholic diphenylamine and 1 cc. concentrated hydrochloric acid; a blue color is obtained. Sensitivity-0.05%.

Jones-smith test for rotenone A mixture of 1 cc. of substance in acetone and 1 cc. of 11 nitric acid, after standing for ½-1 minute, is treated with I cc. ammonia and 9 cc. water; a blue color is produced. Sensitivity—0.1 mg.

Jordan-pryde test for fructose or other ketohexoses 8-10 mg. of hexose are heated at 40°C for ½ hour with 10 mg. recrystallized skatole in 10 cc. concentrated hydrochloric aid; an intense "permanganate" color is produced.

Juniperic acid 16-hydroxypalmitic acid $HOCH_2$-$(CH_2)_{14}$-COOH. A gymnosperms, e.g. juniper (Juniperus communis).

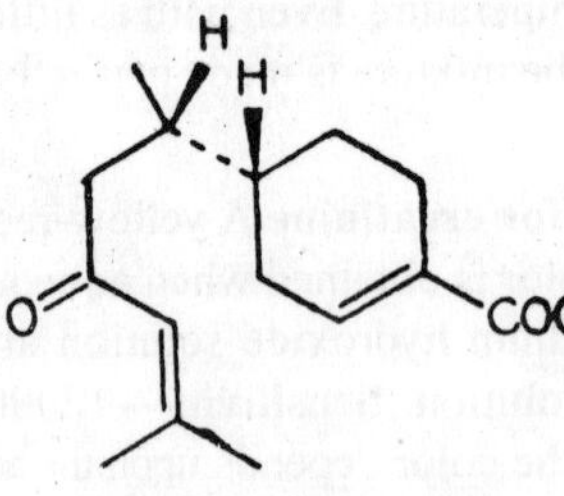

Fig. Juvabione

K

Kahlenberg reaction for cholesterol Arsenic trichloride dissolves cholesterol producing a yellow color changing to cherry-red; isocholesterol gives a cobalt-blue color changing to green; phytosterol gives a colorless solution. When cholesterol is added to a hot solution of arsenic trioxide in hydrochloric acid and heated to boiling, a red color forms in the lower part of the tube. Isocholesterol gives a violet color.

Kahn test for arsphenamine injuries Reagent—2% solution of dimethylamino-benzaldehyde in water, acidified with hydrochloric acid (Ehrlich-Kozlezkowsky solution). A red color is obtained when cold urine is treated with 2 drops of reagent, indicating injuries due to arsphenamine treatment.

Kallikrein A supposed pancreatic hormone which causes increase of pulse amplitude and peristalsis; dilates peripheral blood vessels and lowers pressure; padutin.

Kame Rocky material deposited by a stream of glacier meltwater and taking the shape of a mound or ridge.

Kamnitzer-joseph test for pregnancy In early stages of pregnancy sugar appears in the urine following intramuscular injection of 1 cc. 0.2% phlorizin solution and 1 mg. ß-eucaine. No sugar is found in non-pregnant women.

Kapeler-adler test for pregnancy A mixture of urine and a solution of 1 part of bromine in 100 parts 33% acetic acid is saturated with a mixture of 2 parts concentrated ammonia and 1 part 10% ammonium carbonate solution. A reddish to violet-red color, denoting histidine, is obtained, the shade depending upon the quantity present. A yellow to brownish-yellow color indicates a negative test.

Kapeller-adler method A specific test for histidine. The black derivative obtained on brominating histidine is dissolved in ammonia or ammonium carbonate, the former giving a purplish red color, the latter a blue-violet color.

Karst Irregular landscape of underground sinks and caverns formed by water running over limestone

Karyotype Number, shape, and size of the *chromosomes* present in an organism's cells.

Karyotype A photomicrograph of an individuals chromosomes arranged in a standard format showing the number, size, and shape of each chromosome type; used in low- resolution physical mapping to correlate gross chromosomal abnormalities with the characteristics of specific diseases.

Kawahara test for pepsin Congo red, in the digestive fluid, changes from fuchain red to violet to blue, the change toward blue depending upon the amount of pepsin present. For digestion, a boiled and filtered 0.05% solution of egg albumin or casein is used.

Kefir A preparation of milk acted upon by kefir grains or fungi which are small, irregular, yellowish granules of a gelatinous consistency, used in the Caucasus for fermenting milk.

Kelly test for ephedra alkaloids 2 gm. of air-dried drug are moistened with a 10% aqueous solution of sodium carbonate and ground with a few grains of clean, sharp sand in a mortar. The still damp material is placed in a continuous extractor and extracted with chloroform for 1 hour; the extract is shaken with 10 cc. 10% hydrochloric acid in a 25 cc. separatory funnel. The acid solution is heated with decolorizing charcoal for a few minutes and filtered. The solution is made strongly alkaline with 20% sodium hydroxide and then treated with 0.1 cc. of 2% copper sulfate solution; a typical blue or blue-violet color is obtained.

Kerasin $C_{48}H_{93}O_8N$; m.p. 180°C, levorotory; a cerebroside whose fatty acid is lignoceric acid, found in brain and nerve tissue.

Keratin Proteins of epidermal origin occurring in skin, fingernails, horns, hairs, hooves, and feathers. They are resistant to mechanical and chemical attack, insoluble in dilute acids and alkalis, water and organic solvents; not digestible by pepsin or trypsin.

Keratohyalin An albuminoid found in granules in the deepest layer of the epidermis, the stratum germinativum, probably a precursor of keratin.

Ketogenesis The production of acetone bodies, (acetone, beta hydroxybutyric acid, acetoacetic acid) that accompanies faulty fat metabolism. Found in diabetes, infections and whereever carbohydrate metabolism is disturbed. Resumption of normal carbohydrate metabolism results in cessation of acetone bodies. Amino acids, notably leucine and tyrosine can be ketogenic in severe diabetes. The keto acids drain the alkali reserve as they are neutralized by the alkali and excreted.

Ketogenic antiketogenic balance 1 gram of protein produces 2.4 millimoles, 1 gram of fat 3.43 millimoles of ketogenic material. 1 gram of glucose produces 5.56, 1 gr. of protein 3.2, 1 gr. of fat 0.57 millimoles of antiketogenic material.

Ketosis The condition produced by failure to oxidize the acetone bodies. Depletion of reserve alkali and ammonia results.

Kidney Vertebrate organ that regulates fluids, excretes waste products, and maintains the concentration of salts in the blood.

Kiliani synthesis The reaction of aldehydes and ketones with HCN to give nitriles which on hydrolysis yield an acid with one carbon more than the original compound.

Kilocalory Thousand calories. The term "large calory" is obsolescent. The term "kilogramcalory" is a misnomer.

Kinase A kinase is in general an enzyme that catalyzes the transfer of a phosphate group from ATP to something else. In molecular biology, it has acquired the more specific verbal usage for the transfer onto DNA of a radiolabeled phosphate group. This would be done in order to use the resultant "hot" DNA as a probe.

Kinesis Movement lacking in directional orientation that occurs in response to a stimulus.

Kinetic energy Energy associated with an object's motion; compare *potential energy*.

Kenetin 6-furfurylaminopurine, the model substance for cytokinins. Kinetin in conjunction with other factors, such as auxins, induces renewed cell division in resting plant tissue. It influences the nucleic acid and protein metabolism of the plant. In adaition to many other physiological effects, it prevents yellowing of isolated leves and creates an attraction centre for protein synthesis.

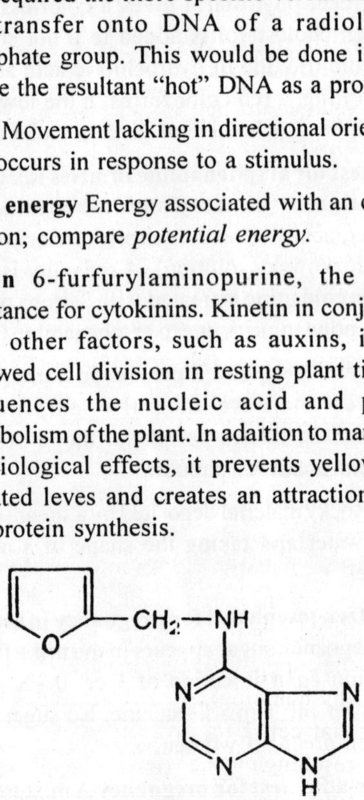

Fig. Kinetin

Kingdom one of five basic divisions that are widely used to categorize organisms; kingdoms are Animalia, Plantae, Fungi, Protista, and Monera.

Kinnersley-peters test for vitamin B 10.5 cc. diazotized sulfanilic acid are mixed with 1.25 cc. of a solution of 5.76 gm. sodium bicarbonate in 100 cc. water mixed with 100 cc. N sodium hydroxide. After one minute, 1 drop 40% formaldehyde is added, followed immediately with 0.3 cc. of vitamin B1 (pH not lower than 3.5); a pink color developing in intensity for ½-1 hour is produced. Sensitivity—1.5-2 micrograms.

Kinteochore A specialized structure found in the centromeric region of the chromosome that is responsible for attaching to the spindle during nuclear division.

Kisch test for adrenaline and tyrosine 4.5 cc. of neutral test solution are treated with 0.5 cc. of 0.02 N hydrochloric acid and 0.2 cc. 1% sodium nitrite solution and heated on a steam bath for 10 minutes; adrenaline gives a rose to rubyred color, depending upon the quantity. Tyrosine gives a yellow color changing to rose, to violet and finally to green.

Kisser reagent for calcium Soluble calcium salts in plants form characteristic crystals with a 0.264% aqueous picrolonic acid solution.

Klenow fragment The large fragment of E. coli DNA polymerase I which lacks 5' -> 3' exonuclease activity. Very useful for sequencing reactions, which proceed in a 5' -> 3' fashion (addition of nucleotides to templated free 3' ends of primers).

Knock-out experiment A technique for deleting, mutating or otherwise inactivating a gene in a mouse. This laborious method involves transfecting a crippled gene into cultured embryonic stem cells, searching through the thousands of resulting clones for one in which the crippled gene exactly replaced the normal one (by homologous recombination), and inserting that cell back into a mouse blastocyst. The resulting mouse will be chimaeric but, if you are lucky (and if you've gotten this far, you obviously are), its germ cells will carry the deleted gene. A few rounds of careful breeding can then produce progeny in which both copies of the gene are inactivated.

Knoop's deamination theory A theory to explain the degradation mechanism of the aminoacids. The a-amino acid is acted on by dehydrogenases, yielding an imine, the double bond of which adds on HOH, giving a hydroxyamine, which then loses NH3 yielding an a-keto acid.

Knoop's histidine reaction A test for histidine based on the formation of a reddish brown color when bromine is added to an acid solution of histidine and then the mixture is boiled.

Kobert reactions for saponins Ferric ferricyanide (made by mixing solutions of ferric chloride and potassium ferricyanide) turns blue on reduction by saponins. Saponin-copper is precipitated from Fehling's solution in the cold or on warming. Some saponins give a green color with Fehling's solution, even at high dilutions. Nickel and cobalt with solutions give yellow colors with saponin. Mercuric chloride is reduced to calomel when heated with saponin; all saponins form a red color with concentrated sulfuric acid in the presence of air. With Fröhde's reagent, some saponins form a yellow color; some give a blue-green color with a green fluorescence with sulfuric acid containing ferric chloride and alcohol.

Koch test for cholera bacilli The addition of sulfuric acid to a culture of cholera bacilli produces a red color, due to the action of the acid on the metabolism products,—indole and nitrite.

Koessler & hanke method A method for the determination of histidine based on the color produced by the imidazole ring of the histidine and p-diazo-benzene sulfonate.

Koessler-hanke estimation of imidazoles On the basis of the color produced by the reaction of the imidazole ring with phenyl diazonium sulfonate in alkaline medium, histidine, histamine, imidazole propionic and acetic acids and methylimidazole may be determined in quantities ranging from 0.001-0.05 mg.

Koji A culture of a micro-organism active in thehydrolysis of starch.

Kojic acid 5-hydroxy-2-hydroxymethyl-a-pyrone, produced by Aspergillus flavus-oryzae moulds on glucose.

Kola Dried cotyledons of Cola nitida containing caffeine and other material and used as tonic, stimulant, diuretic.

Kolisch reagent for creatinine The solution contains 1 gm. sodium acetate, 30 gm. mercuric chloride and 3 drops of glacial acetic acid in 125 cc. of absolute alcohol.

Kolthoff reagents and tests for sodium (1) Magnesium uranyl acetate-10 gm. uranyl acetate and 6 gm. 30% acetic acid are dissolved in water and diluted to 50 cc. 6 gm. acetic acid and 33 gm. magnesium acetate are dissolved in water and diluted to 50 cc. The solutions are mixed and, after several days, filtered. 2 cc. of test solution are treated with 10 drops of reagent and 2 cc. alcohol; a turbidity or precipitate forms within I hour, if more than 50 mg. sodium per liter are present.

(2) Zinc uranyl acetate—A hot solution of 10 gm. uranyl acetate and 6 gm. 30% acetic acid in 34 gm. water is mixed with a hot solution of 30 gm. zinc acetate and 3 gm. acetic acid in 17 gm. water and filtered after one day. When 0.5 cc. of test solution are mixed with 4 cc. of reagent, a crystalline precipitate is obtained, the time of formation depending on the quantity of sodium. With 50 mg. sodium per liter, precipitation occurs within ½ hour; alcohol sensitizes the test.

Komm-böringer reaction for tryptophane and alanine 5 cc. 15% hydrochloric acid, each 500 cc. of which contains 6 cc. 0.1% formaldehyde solution, are mixed with 5 cc. of tryptophane or albumin solution and the mixture is floated on 10 cc. 90% sulfuric acid and stirred slowly. A violet color, sensitivity 1175000, is obtained which may also be used for colorimetric estimation. Formaldehyde may be advantageously replaced by a 0.25% pdimethylaminobenzaldehyde solution in 10% hydrochloric acid; 22 cc. of test solution are treated with 2 cc. of reagent and 6 cc. 10% hydrochloric acid.

Koscis-bughi test for folliculin Even 0.001 mg. folliculin may be detected by the fluorescence obtained when 1 cc. of an aqueous or alcoholic solution is treated with concentrated sulfuric acid; the color is visible under filtered ultra-violet light. With oils, the test may be carried out after saponification with 50% sodium hydroxide; 2 layers are obtained after shaking 5-10 minutes, the test being made with the upper layer.

Kossel reaction for adenine and hypoxanthine The substance is heated on the steam bath for ½ hour with zinc and hydrochloric acid; the solution forms a red to reddishbrown color on addition of sodium chloride solution.

Kossel's "protamine nucleus" hypothesis In 1896 Kossel stated that the proteins all have protamine nuclei, i.e. are largely arginine, histidine or lysine. In 1928 Larmour found that the correlation between the total basic nitrogen and the nitrogen of these three amino acids was quite high in a number of proteins and protein fractions.

Kossel-patten test for histidine The addition of mercury salts to histidine solutions results in the formation of precipitates.

Kovacs test for indole in bacterial cultures A cherry-red color is obtained when drops of the reagent (5 gm. p-dimethy-laminobenzaldehyde in 75 gm. amyl alcohol and 25 gm. hydrochloric acid) are floated on the culture.

Krauskopp-ritter reaction for resorcinol A dark green color is obtained when a 150000 aqueous solution of resorcinol is shaken with ammortiacal cobalt solution. Phenol, pyrocatechol, hydroquinone and pyrogallol do not interfere.

Krauss test for melanin in urine Melanin-containing urine gives a deer black color with ferric ammonium sulfate solution.

Krebs cycle Final series of biochemical steps during cellular *respiration;* yields energy-providing molecules

Kreis reaction for cholesterol and phytosterol Several drops of an ether solution of the substances are evaporated and treated with 3

drops of a solution of 20 gm. benzaldehyde in 80 gm. absolute alcohol. The addition of concentrated sulfuric acid gives rise to a red-violet to dark violet color.

Kreis test A color test for rancidity in a fat due to oxidation. It depends on the formation of epihydrin aldehyde on the addition of ether, HC1 and phloroglucinol to the fat.

Kroger reaction for adenine A solution of 2 moles sodium hydroxide and 1 mole adenine are added to a solution of I mole of lead acetate. Microscopic crystals of adenine lead, $C_5H_3N_5Pb$, are obtained.

Kuhn test for bile pigments in urine Reagent—

(1) Mixture of 10 cc. of 20% ammonia and 20 cc. of 5% copper sulfate solution.

(2) Mixture of 20 cc. of 85% phosphoric acid and 20 cc. water. 2 cc. of.

(1) and 2 cc. of.

(2) are added to 20 cc. urine, followed by 6 drops of toluene. After shaking and allowing to stand a few minutes, 4 cc. of alcohol are floated on the solution; a green color in the alcohol is obtained; a pink, blue or browngray color is negative.

Kynurenin $(NH_2\ )\ C_6\ H_5C(COOH)$ =CHCHNHCOOH; a compound found in rabbit urine, resulting from feeding large amounts of tryptophane, and supposed to be a normal intermediate in its metabolism.

L

Labat reaction for hordenine and methenamine An emerald-green color is obtained when a mixture of 1 cc. of 1% hordenine solution, 1 cc. of 1% methenamine solution and 2 cc. sulfuric acid are heated to boiling. Sensitivity — hordenine 0.1 mg.; methenamine 0.001%.

Labat test for lactose in urine 100 cc. of slightly ammoniacal urine are evaporated to 10 cc. on the water bath, filtered, clarified with lead subacetate and refiltered. This is treated with an equal volume of a solution of 1 cc. phenylhydrazine, 3 cc. glacial acetic acid and 20 cc. sodium acetate soln. After heating on a steam bath for ½ hour and cooling, the characteristic lactosazone may be identified microscopically.

Laccase Enzyme of bacteria and mush. rooms responsible for converting polyphenols to their quinones.

Lactacidogen An old name for the HardenYoung ester and a little used name for the Neuberg ester.

Lactalbumin An albumin found in milk; it consists largely of lactoglobulin; m.w. about 17500.

Lactase A carbohydrase that hydrolyzes lactose to galactose and, glucose, and also ß-galactosides.

Lactation disturbances The disturbances of the period of milk formation following childbirth. These include agalactia, polygalactia, galactorrhea, fissure of nipple and mastitis. Agalactia is a deficiency or absence of milk which may be due to defective development, malnutrition, disease or emotional disturbance. Polygalactia is an abnormal increase of milk which is prolonged. Galactorrhea is a profuse flow which does not stop till a return to menstruation. Fissure threatens abcess formation. Mastitis proper is an abnormal distension which may occur in the infant as well as in the nursing woman due to infection or injury in massage. The mastitis may be parenchymatous or interstitial. Milk secretion often has to be interrupted, e.g. by belladonna administration.

Lactescence The milky appearance of serum due to large numbers of chylomicrons.

Lactic acids 2 optically active isomers of alphahydroxypropionic acid and their racemic mixture; d-lactic acid, sarcolactic acid, occurs in muscle, and is a product of carbohydrate metabolism.

Lactic dehydrogenase (Bact Coli) An enzyme of the autolysate of Bact. coli which catalyzes the oxidation of lactic acid and alphahydroxybutyric acid.

Lactic dehydrogenase (Gonococcus) A cytochrome-reducing dehydrogenase which catalyzes the oxidation of the 1-forms of lactic and alpha-hydroxybutyric acids to their corresponding keto acids.

Lactic dehydrogenase (Yeast) A cytochrome-reducing dehydrogenase which catalyzes the oxidation of lactic and alpliahydroxybutyric acids to their keto acids (10 forms), but no coenzyme is necessary.

Lactic enzyme The enzyme which catalyzes the oxidation of 1(+)-lactate to pyruvate; a pyridinoprotein, extracted from skeletal muscle of rabbit.

Lactol A term loosely used for ring modifications of sugars as well as for various other ring structures, which is yielding the field entirely to the terms pyranose and furanose.

Lactose The reducing disaccharide of milk, which is prepared commercially from milk whey. It is glucopyranose-4-ß-galactopyranoside.

Lactosuria The excretion of lactose in the urine of lactating women or of infants.

Lambrecht reaction for phloridzin One drop of a 1% alcoholic solution of a-nitroso-ß-napthol and 3 drops concentrated sulfuric acid are added to 2cc. of solution (0.5-500 mg. phloridzin per liter), heated to boiling, cooled rapidly, treated with 3 cc. of ether and 1 cc. alcohol and shaken. An unstable red coloration is obtained.

Laminarin A beta-glucan polysaccharide produced by many chromists through photosynthesis.

Landsteiner reaction for tyrosine A hydrochloric acid solution of tyrosine is treated with a sodium nitrite solution, made alkaline and treated with a-naphthol, yielding a blue-red color.

Lange's test A test for the presence of proteinglobulin in the cerebrospinal fluid, and thus for the diagnosis of cerebrospinal syphilis.

Langley-albrecht reaction for flavianic acid An orange-red color is produced when the acid is reduced with metal and acid and reoxidized by air in alkaline medium.

Lanoceric acid $C_{30}H_{60}O_4$; a saturated dihydroxy fatty acid found in lanolin; m.p. 104-1050, purity doubted.

Lanolin A crude extract of sterols from wool grease.

Lanomyristic acid $C_{14}H_{28}O_2$; an acid of wool grease.

Lanopalmitic acid $C_{16}H_{32}O_3$; a saturated monohydroxy fatty acid found in lanolin.

Lanosterol $C_{30}H_{50}O$; a pentocyclic sterol with two double bonds found in lanolin; m.p. 141°C. H. S.

Lanthionine Cystine minus one sulfur atom, a conversion product of cystine in proteins by alkali.

Larson reaction for cholesterol A 25% chloroform solution of antimony pentachloride stains cholesterol and its eaters a dark brown color.

Larva Among invertebrates, an immature stage in the life cycle which usually is much smaller than, and morphologically different from, the adult. In insects with **metamorphosis**, the larva must become a **pupa** before reaching adulthood.

L-ascorbic acid Vitamin C; the anti-scurvy vitamin found in certain greens and citrus fruits. It is unstable to heat and oxidation. It crystallizes in rectangular plates.

L-aspartic acid Amino-succinic acid; C4H7O4N; an acid amino acid present in animal and vegetable proteins largely as asparagine; crystallizes as rhombic prisms, sl. sol. cold, H_2O; m.p. 251°C.

Latent heat Heat absorbed or given off when an object or system changes state while maintaining constant temperature and pressure.

Lateral moraine Ridge of rocky material embedded in the edge of a glacier.

Lauric acid $CH_3(CH_2)_{10}COOH$; m.p. 47-48°C; a saturated fatty acid found in laurel oil.

Leaching Process by which substances in soil are dissolved in water and washed away leader A short sequence near the amino terminus of a protein or the 5' end of an RNA molecule that has a specialized targeting or regulatory function. Often the leader is cleaved after the targeting event.

Leader sequence Two usages; [1] N-terminal pre sequence of scretory proteins such as peptide hormones and membrane proteins.

[2] The untranslated sequence at the 5'-ends of mRNA molecules.

leading strand The DNA strand that, during

replication, is synthesized in the same direction in which the replication fork moves. Synthesized in a continuous fashion.

Lecithase Group of enzymes capable of splitting off fatty acids or choune phosphate group from lecithins.

Lecithin Any eater of choline, glycerol and phosphoric acid; richly distributed in nervous tissue and also essential to all cells; soluble in water with formation of colloids, and most organic solvents except acetone.

Lecithoproteins In the American classification of proteins the lecithoproteine are conjugated proteins where the additional group is lecithin or a phospholipid.

Examples proteins of the cytoplasm and tissue fibrinogen.

Legal test for acetone in urine The urine or its distillate is treated with freshly prepared sodium nitroprusside solution; a red color soon changing to yellow is obtained if acetone is present. With an excess of acetic acid, a carmine-red color is formed, changing to violet, then blue on standing.

Legumelin An albumin of the seeds of leguminous plants, such as peas or vetches.

Legumin A globulin of leguminous plant seeds, such as peas and vetches.

Lejeune test for benzoyl peroxide in bleached flour Reagent—2.5 gm. p-diaminodiphenylamine sulfate is agitated with 250 cc. alcohol for one hour and allowed to stand overnight. A mixture of 0.7 gm. flour, 2.5 cc. petroleum benzine and 1 cc. reagent is well agitated and allowed to stand. A bluish-green color develops in the supernatant liquid.

Lens Slice of transparent material with either both sides curved or one side curved and one side flat; a lens modifies the path of light rays passing through it

Lenticel Pore that serves as a site of gas exchange on plant stems and tree trunks

Leprosy Lepra; elephantiasis graecorum; a chronic, infectious, contageous disease caused by Mycobacterium leprae (Bacillus leprae) with a long period of incubation. It shows as neural leprosy, cutaneous leprosy and mixed types. Outstanding remedies are chaulmoogra and hydnocarpus oils and their derivatives taken internally or by injection.

Lepton Member of the class of lightweight particles that do not experience the *strong nuclear force* (one of the *fundamental forces*).

Leucine metabolism Leucine is oxidized to 2-keto-4methyl valeric acid, which is oxidized to isovaleric acid, then demethylated to ß-hydroxybutyric acid, which then undergoes oxidation. It is potentially ketogenic.

Leucine zipper A motif found in certain proteins in which Leu residues are evenly spaced through an a-helical region, such that they would end up on the same face of the helix. Dimers can form between two such proteins. The Leu zipper is important in the function of transcription factors such as Fos and Jun and related proteins.

Leucopterin The white pigment of butterfly wings, a dipurine derivative with guanin residues.

Leucosin An albumin of the seeds of rye, barley and wheat.

Leucotaxine A substance assumed to be liberated by injured tissue, found in inflammatory exudates; a lower polypeptide with marked capillary dilating power, attracts leucocytes injured tissue.

Leukemia Leukocythemia; a disease of the leukocyte-producing tissues showing myelogenous, lymphatic and monocytic varieties. An outstanding symptom is enormous spleen enlargement. The blood picture shows a very high leucocyte count at certain stages and changes in red cells, hemoglobin and the quality of the cells with malformations. The disease has been called "blood cancer," is very deadly, and the treatment is only for the relief of symptoms.

Leukocytes White blood cells of a number of varieties existing in blood to the extent of 5000-10000 per cu. mm. They are essential in fighting disease, in which case there is a sharp increase in their number. This bactericidal property is partly due to the proteolytic and lipolytic enzymes contained in the leukocytes.

Leukotrienes A class of peptolipids derived from arachiconic acid. These are slow reacting substances. Slow reacting substance. A (abbreviation SRS-A) is a L. involved in asthma and allergies.

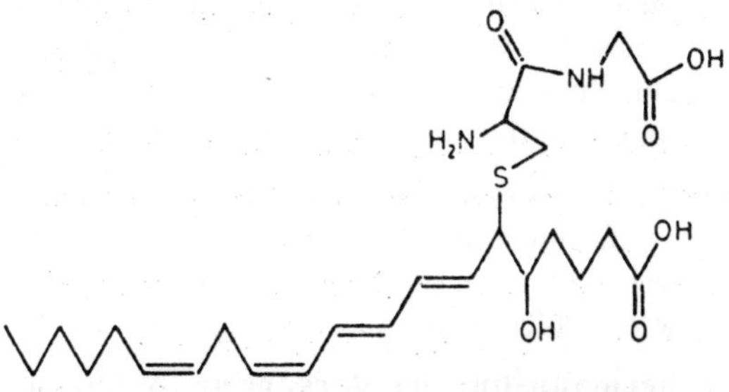

Fig. Slow reacting substance A (a Leukotriene)

Levan A water soluble fructosan of various grasses.

Levene-beatty reagent for amino acids A very concentrated solution of phosphotungstic acid (2-4 parts of acid to I part of water) precipitates concentrated solutions of alanine, aspartic acid, glycocoll, glutamic acid, leucine, etc.

Levine-bien tests for carotene I. The addition of 3 cc. of a mixture of 1 vol. 37% formaldehyde and 50 vols. sulfuric acid to 3 cc. of a chloroform solution forms a deep violet zone at the contact zone. The color goes into the acid on shaking. Sensitivity-0.01 mg. in 2.5 cc.

II. The addition of 3 drops of a mixture of 9 parts of trichloroacetic acid and I part water to 0.1 cc. chloroform solution of carotene results in the immediate formation of an intense blue color, fading on the addition of water or alcohol, but not on heating.

III. 0.5 Rm. chloral hydrate.

liquefied by heating, is treated with 1 drop hydrochloric acid and 0. 1 cc. of a chloroform, solution of substance; an immediate intense blue is given by carotene; ergosterol forms a carmine-red color changing to evanescent green and finally to a persistent blue.

Levine-mckay test for differentiating ergosterol from cholesterol Reagent-1 vol. concentrated hydrochloric acid and 3 vols. 2% aqueous sodium selenite solution. 2 cc. of reagent and 2 cc. of a chloroform solution of ergosterol are heated, in a test tube, on the water bath until the chloroform is evaporated and 2 cc. of chloroform are added. A light yellow to deep orange color is given by ergosterol.

Levine-richman reaction for cholesterol The chloroform solution of cholesterol is treated with an equal volume of concentrated sulfuric acid containing 125 mg. sodium selenite per 25 cc. acid. The top layer becomes deep purple, the bottom layer turns red-brown to very dark brown without green fluorescence. Sensitivity—0.001%.

Levo(1-)rotatory The rotation of the plane of polarized light to the left, shown by solutions of many optically active compounds.

Levulinic acid $CH_3\,COCH_2\,CH_2\,C00H$; m.p. 33.1°C; ß-acetylproprionic acid; a keto-fatty acid found resulting from carbohydrate metabolism.

Lewin reaction for proteins Proteins form a violet color with an 0.1% solution of triformoxime (trioximino-methylene) in concentrated sulfuric acid. A similar, but less sensitive, reaction is given by paraformaldehyde.

LH (luteinizing hormone) A hormone made in the anterior pituitary that promotes ovulation

L-histidine $C_6\ H_9\ O_2\ N_3$;*ß-imidazol-á-aminopropionic acid*; m.p.°C; i.p. 7.21; an indispensable basic amino acid. It dilates capillaries and increases their permeability and on injection causes a stimulation of the parietal cells of the stomach to secrete HC1.

Liberalli test for hydroxy acids Reagent—32.4 cc. 10% ferric chloride solution, 58.2 cc. 10% potassium thiocyanate solution diluted with water to 100 cc. Neutral solutions of hydroxy acids turn the reagent yellow. The same color is given by acetic and oxalic acids. The addition of 1 drop of nitric acid gives a red color with hydroxy acids, but not with others.

Library A library might be either a genomic library, or a cDNA library. In either case, the library is just a tube carrying a mixture of thousands of different clones - bacteria or l phages. Each clone carries an "insert" - the cloned DNA. A cDNA library is usually just a mixture of bacteria, where each bacteria carries a different plasmid. Inserted into the plasmids (one per plasmid) are thousands of different pieces of cDNA (each typ. 500-5000 bp) copied from some source of mRNA, for example, total liver mRNA. The basic idea is that if you have a large enough number of different liver-derived cDNAs carried in those bacteria, there is a 99% probability that a cDNA copy of any given liver mRNA exists somewhere in the tube. The real trick is to find the one you want out of that mess - a process called screening . * A genomic library is similar in concept to a cDNA library, but differs in three major ways - 1) the library carries pieces of genomic DNA (and so contains introns and flanking regions, as well as coding and untranslated); 2) you need bacteriophage l or cosmids, rather than plasmids, because... 3) the inserts are usually 5-15 kb long (in a l library) or 20-40 kb (in a cosmid library). Therefore, a genomic library is most commonly a tube containing a mixture of l phages. Enough different phages must be present in the library so that any given piece of DNA from the source genome has a 99% probability of being present.

Libration Apparent oscillation in the path of the Moon's orbit as viewed from Earth

Lichenase A polyase which splits lichenin to cellobiose; found in malt extracts, snail intestine, etc.

Lichenin Moss starch; a carbohydrate, $C_6 H_{10} O_5$ from Iceland moss, used as demulcent.

Liebermann test for ethyl sulfide in urine Reagent—8 gm. potassium nitrite are dissolved in 100 gm. concentrated sulfuric acid containing 6-7 cc. water and the liquid separated from the crystals by decantation through glass wool. Ethyl sulfide colors the reagent a transitory green.

Liebermann test for phytosterol and cholesterol Concentrated sulfuric acid is added dropwise to a cold concentrated acetic anhydride solution of the sterol giving the following color reactions rose-red to blue to green for cholesterol from gallstones; rose-red to blue to green for phytosterol from cottonseed oil; same for phytosterol from belladonna leaves and grass leaves. Onocol and onoketone give brownish-yellow to reddishbrown to green.

Liebermann test for proteins A violet-blue color is obtained when alcohol and ether washed protein is treated with hot, concentrated hydrochloric acid. Substances containing the tryptophane group give this reaction.

Liebermann's reaction A test for most proteins and specific for tryptophane or indole derivatives. Heat the protein with cone. HC1 whereupon a violet to blue-black color develops.

Liebermann-burchard reaction A test for unsaturated sterols based on the green color formed when the sterol is treated with chloroform, acetic anhydride and concentrated sulfuric acid.

Liebermann-vogt reaction for cod liver oil A cooled mixture of 20 drops chloroform, 40 drops acetic anhydride and 3 drops sulfuric acid are treated with 3 drops cod liver oil and shaken; an intense blue color, disappearing rapidly is obtained. In 20-40 seconds a permanent olive green is formed.

Liesegang phenomena If one compound, such as potassium chromate is dissolved in a gelatin gel and a solution of another substance, such as silver nitrate is allowed to diffuse into the gel, the precipitate which forms (silver chromate) forms a series of concentric rings, separated by more or less clear portions of the gel. These rings are called Liesegang rings. No completely satisfactory explanation has been offered.

Lifschotz reaction for gynocardia oil One drop of oil is dissolved in 0.5 cc. chloroform, treated with 1.5 cc. glacial acetic acid and 4-5 drops concentrated sulfuric acid added. An intense grass-green.

Color, red-violet in transmitted light, forms gradually.

Ligament Any of the strong bands of tissue that joins bones together, encloses some joints, and holds organs in place.

Ligase An enzyme, T4 DNA ligase, which can link pieces of DNA together. The pieces must have compatible ends (both of them blunt, or else mutually compatible sticky ends), and the ligation reaction requires ATP.

Ligation The process of splicing two pieces of DNA together. In practice, a pool of DNA fragments are treated with ligase in the presence of ATP, and all possible splicing products are produced, including circularized forms and end-to-end ligation of 2, 3 or more pieces. Usually, only some of these products are useful, and the investigator must have some way of selecting the desirable ones.

Lightning Discharge of electricity between either two clouds or a cloud and Earth, caused by the buildup of opposite electrical charges.

Lignin A high molecular weight polymerized compound which occurs in plants together with cellulose and lends rigidity to the cell wall. The exact structure is in doubt but the weight of evidence tends towards a polymer of 12 molecules of coniferyl alcohol, 3,4. $(OH)_2\,C_6\,H_3\,(CH)_2\,CH_2\,OH$.

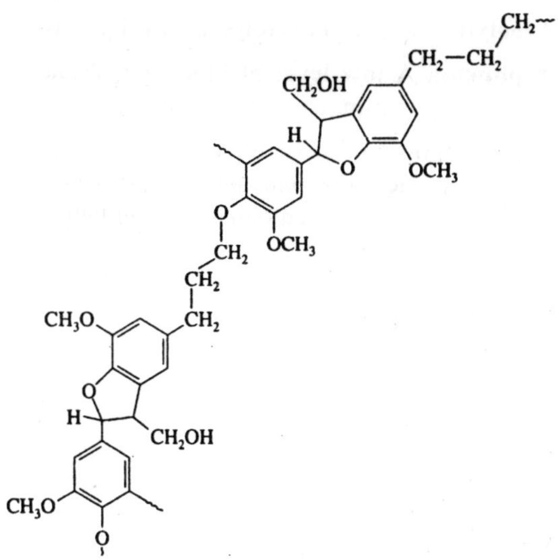

Fig. A Fragment of lignin

Lignoceric acid A saturated fatty acid, $CH_3\,(CH_2)_{22}\,COOH$, found in glucolipids and arachis oil; m.p. 81-82°C. A straight chain form is present in beechwood tar and peanut oil. A branched chain form is present in sphingomyelin and in kerasin linked as a lipin.

Limonene d-limonene, a terpene, is found in lemon, orange, etc., oils; b.p. 176°C. 1-limonene is found in pine needle, peppermint, etc., oils; b.p. 176°C.

Linamarin $C_{30}\,H_{50}\,O_{19}\,N_3$; phaseolunatin; a glycoside of flax seeds, rubber seeds and young plants, consisting of glucose and acetonecyanhydrin. It crystallizes in needles, m.p. 142°C, and has been synthesized.

Linderstrm lang's titration A method for quantitative analyis of amino-acids, based on the fact that an amino acid or protein in 90%acetone does not contain the amino acid in the zwitterion form, thus enabling direct titration to be run with alcoholic HCl or alcoholic NaOH.

Linear epitope An epitope formed by a series of amino acids which are adjacent to each other within the primary structure of the protein. Such epitopes can be successfully modelled by synthetic peptides, but comprise only a small proportion of all epitopes. The minimal epitope size is about 5 amino acid residues. Also called a sequential epitope.

Linear equation Mathematical relation between two variables that describes a straight line when plotted on *coordinate* system.

Linkage map A map of the relative positions of genetic loci on a chromosome, determined on the basis of how often the loci are inherited together. Distance is measured in centimorgans (cM).

Linkage The measure of proximity of two or more markers (e.g., genes, RFLP markers) on a chromosome determined by recombination events. The closer together the markers are, the lower the probability that they will be separated during DNA repair or replication processes (binary fission in prokaryotes, mitosis or meiosis in eukaryotes), and hence the greater the probability that they will be inherited together.

Linker A small piece of synthetic double-stranded

DNA which contains something useful, such as a restriction site. A linker might be ligated onto the end of another piece of DNA to provide a desired restriction site.

Linoleic acid Linolic acid; an unsaturated fatty acid, $CH_3CH_2)_6CHCH.CHCH(CH_2)_6COOH$, found in cottonseed oil, poppyseed and other vegetable oils. The double bonds add on oxygen from the air, forming a stable hard film, allowing it to be used as a drying oil. B.p. 229-230°C at 16 mm., m.p. 11°C.

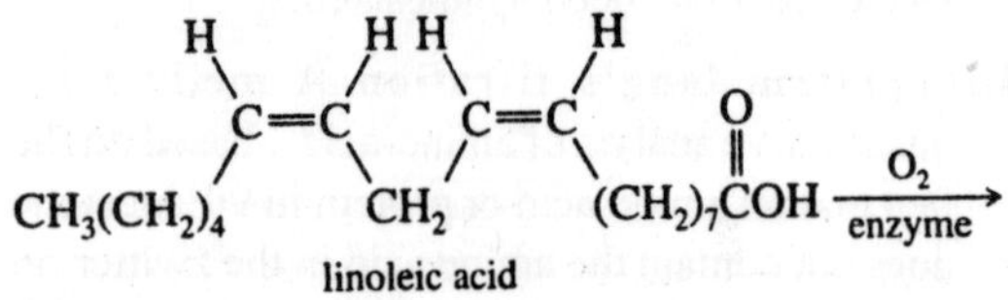

Fig. Lenoleic acid

Linolool 1-form of a terpene alcohol found in oil of linaloe; d-form found in orange and coriander oils; b.p. 198199°C.

Lipases A group of enzymes that hydrolyze the fatty acid eaters of glycerol, the fats. The individual enzymes are usually named after the source of the enzyme, as pancreatic lipase. In the presence of specific concentrations of sodium taurocholate, liver and pancreatic lipases act more rapidly.

Lipids A term used confusedly and inter. changeably to designate fats, fatty substances, fats and lipoids, and lipins.

Lipids, classification of Bloor suggests classifying lipids as follows

Lipids, derived A term applied to free fatty acids and sterols and other alcohols.

Lipins A term used like "lipids," but also restricted to nitrogen containing fatty substances.

Lipocaic A hormone of the pancreas which counteracts the production of fatty liver which, however, may be due to choline or insulin and not to special hormone action.

Lipochromes An old name for carotenoids found in animal sources.

Lipocytic coefficient The percentage of lipide phosphorus as compared to total cholesterol; tends to a constant in animal tissue metabolism.

Lipofectin A commercially marketed liposome suspension which is mixed with DNA or RNA to facilitate uptake of the nucleic acid by eukaryotic cells.

Lipoidoses Diseases marked by heavy deposition of lipids in various organs, as Niemann-Pick's disease, Gaucher disease, Christian-Schueller disease, xanthomatoses.

Lipoids Substances associated with fats; fatty compounds containing nitrogen; sometimes the term is made to include the fats.

Lipolytic Capable of hydrolysing fats; lipoclastic.

Lipopenia A low level of blood fat, found in hyperthyroidism, and certain anemias.

Lipoproteins In the American classification of proteins, the lipoproteins are conjugated proteins where the additional group is a higher fatty acid.

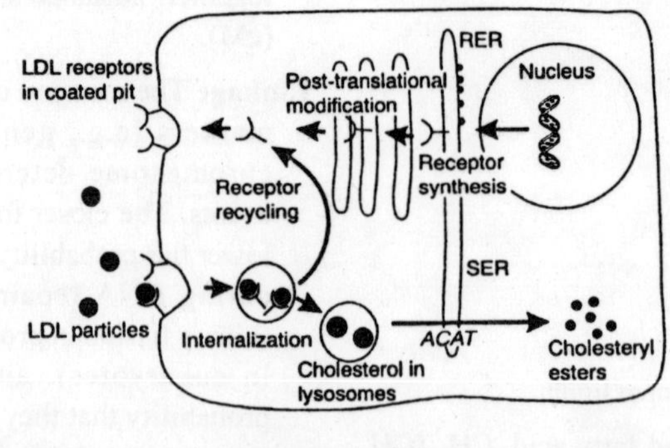

Fig. Lipoproteins

Lipp test for protein in urine Protein-containing urine is rendered turbid by powdered m-dihydroxybenzene.

Liquefaction Process by which a room-temperature gas is converted into a liquid at low temperatures and high pressures; also the process by which a soft rocky deposit is converted into a jelly by vigorous shaking

Lison reagent for peroxidases 1.5 gm. acid fuchsin or acid violet are dissolved, with the aid of heat, in a mixture of 2 cc. glacial acetic acid and 100 cc. of water. Following the addition of 5 gm. of zinc dust, the solution is cooled and treated with 2 cc. glacial acetic acid; the solution is filtered before use and 1 cc. hydrogen peroxide added to each 10 cc.

Lison test for hemoglobin Treatment with a mixture of 2.5 gm. lead acetate, 10 cc. formaldehyde and 100 cc. water converts blood hemoglobin to hematin; peroxidases in the hemoglobin are unaffected but give a characteristic blue color with benzidinehydrogen peroxide. The test is specific.

Lithification Process by which sediment is converted into solid rock.

Lithocholic acid $C_{24}H_{40}O_3$; m.p. 186°C; bile acid found in small quantities in man and in the ox. It is a 3- hydroxy cholanic acid.

Liver Vertebrate organ that performs regulatory and storage functions

Liver atrophy A condition marked by destruction of liver cells caused by toxic agents or previous disease. Jaundice develops along with cerebral symptoms and shrinkage of the liver; also known as "acute yellow atrophy of the liver" or "diffuse toxic necrosis of the liver." Intensive carbohydrate administration is used.

Liver cirrhoses Chronic diseases of the liver usually due to obstruction, showing as.

(1) portal cirrhosis or chronic interstitial hepatitis and.

(2) biliary cirrhosis or hypertophic biliary cirrhosis. The former is usually caused by alcoholism or bacterial toxins and spreads from the portal vein region; the latter shows in the bile passages and liver cells either as a primary process, icterus catarrhalis, or an obstructive type, infective cholangitis, with cholelithiasis.

Liver flavoprotein A specific aldehyde oxidase of the liver having flavinadenine dinucleotide as the prosthetic group; utilizes molecular oxygen or other hydrogen acceptors, like methylene blue, cytochrome c, nitrate, etc.

Liver The largest organ in the body, weighing about 1800 gm., and probably performing a greater volume of duties than any other organ.

Livetin A protein of the egg yolk (about 20%), perhaps identical with the serum globulin of the hen.

Lloyd reagent for alkaloids Aluminum silicate precipitates alkaloids from neutral or acid solutions; by the use of alkali and organic solvents, the alkaloid may be extracted from the precipitate.

Lobelanidine $C_{22}H_{29}O_2N$; an alkaloid of Lobelia. Prismatic scales from alcohol, m.p. 150°C. Oxidizes to dlLobeline, which it resembles in structure and physiological action.

Lobelanine $C_{22}H_{25}O_2N$; second most abundant of alkaloid of Lobelia. Needles from ether, m.p. 99°C. Reduces with sodium amalgam to lobelanidine. Resembles dl-lobeline in physiological action.

dl-LOBELINE $C_{22}H_{27}O_2N$; most active and abundant alkaloid of Lobelia. Crystallizes in prisms from alcohol, m.p. 110°C. Oxidizes to lobelanine, reduces to lobelanidine. Respiratory stimulant; excites and then paralyzes the ganglia of the involuntary nervous system.

Lobry de bruin transformation Sugar plus very dilute alkalies (0.05 N) at room temperature is converted to a series of isomers and the rotation of the solution is changed.

Localisation (Sonderung) The separation of phases ("plasms") in mosaic eggs, distinguishable cytologically, and each associated with a specific subsequent differentiation. Such phases or regions are often characterised by pigments or granules which led to the old term "organ-forming substances". Since the cytoplasm of these regions, rather than the visible elements within them, must be regarded as responsible for the morphogenetic specificity, and since evocators now deserve rather to be called "morphogenetic substances", the old term should be. abandoned.

Localize Determination of the original position (locus) of a gene or other marker on a chromosome.

Locus Set of all points located on a curve that satisfies given conditions

Locus (pl loci) The position on a chromosome of a gene or other chromosome marker; also, the DNA at that position. The use of locus is sometimes restricted to mean regions of DNA that are expressed.

Lodestar or loadstar Star that aids in navigation.

Logarithm *Exponent* of a number to a given *base.*

Lombardo test for mercury in urine A mixture of 1 drop egg albumin and 5 cc. of urine are shaken well and treated with 3 cc. freshly prepared stannous chloride solution to which some hydrochloric acid has been added. Centrifuge and observe microscopically with a magnification of 600; small globules of mercury may be observed.

Long(q) and short(p) arms The regions either side of the centromere, a compact part of a chromosome, are known as arms. As the centromere is not in the centre of the chromosome, one arm is longer than the other.

Looney test for cystine in urine The addition of phosphotungstic acid and sodium sulfite produces a blue color with the test solution; other reducing agents give the same reaction.

Lotusin $C_{28}H_{31}O_{16}N$; a glycoside of Lotus arabicus consisting of HCN, a disaccharide and lotoflavin.

Lugol test for albumin in urine Albumin-containing urine forms a precipitate when treated with a solution of iodine-potassium iodide acidified with acetic acid.

Lumichrom 67-dimethyl alloxazine, formed from riboflavin on illumination in neutral solution in methanol.

Lumiflavin 6 7 9 - trimethyl isoalloxazine, formed by the illumination of riboflavin under alkaline conditions.

Luminal Phenyl-ethyl-barbituric acid; trade name of a sedative also known as phenobarbital.

Luminescence Light emitted from a body; luminescence is produced by physiological, chemical, or electrical processes or by friction

Lumisterol $C_{28}H_{44}O$; m.p. 118°C; an inert irradiation product of ergosterol, which on further irradiation yields tachysterol.

Lung Organ that serves as the site of gas exchange in air-breathing vertebrates.

Lupinine $C_{10}H_{19}ON$; m.p. 68-71°C; alkaloid of yellow lupin seeds.

Luteinizing hormone A hormone of the anterior pituitary that causes ovulation and luteinization in the female and the production of androgens in the male. The hormone is protein in structure.

Luteose Polysaccharide obtained by hydrolysis of luteic acid, a product of the action of Penicillium luteum on a pentose or hexose.

Lycaconitine An alkaloid having a curate-like action.

Lycopene $C_{46}H_{56}$; a red carotenoid hydrocarbon pigment found in fruit of tomatoes, rose hips, berries; m.p. 175°C; soluble in chloroform and carbon sulfide.

Lycopodic acid A 16 carbon one double bond unsaturated fatty acid of lycopodium spores.

Lymph A colorless body fluid, found in the lymphatic channels, formed by permeation of blood plasma through the capillaries. Its chemical structure is similar to plasma, except for higher protein in the latter and the presence of larger numbers of lymphocytes in the former.

Lymph node Small masses of tissue that filter *lymph* and remove harmful microorganisms.

Lymphocyte Type of white blood that has a large nucleus and mediates the body's immune response by producing *antibodies.*

Lymphocyte A variety of white blood corpuscle which arises in the reticular tissue of the lymph glands and lymph nodes. The nucleus is single and is surrounded by protoplasm which is generally described as nongranular.

Lymphokines Substances secreted by *lymphocytes* to serve as chemical messengers between components of the immune system.

Lyophilic A system in which the disperse phase and the dispersions medium are mutually more or less soluble in each other.

Lyophobic A system in which the disperse phase is insoluble in the dispersions medium, and the dispersions medium does not dissolve in the disperse phase.

Lyotropic series A list of equivalent ions, in the order of their ability to flocculate a sol. Essentially the same order holds for surface tension of molten salts of these ions and for the effect of these ions in increasing the surface tension of water, probably due to the degree of hydration of the ion. Also called the Hofmeister series.

Lysocephalins The hydrolytic products of cephaline, as by the lecithinase of cobra venom.

Lysosome Membrane-bound cellular structure (*organelle*) that breaks down various biological substances.

Lysozyme A bacteriolytic, substance found in tears and also in tissues, a yellow amorphous powder soluble in water and containing no nitrogen, phosphorus or sulfur. It is a colloid and its lytic activity varies with the acidity and type of bacteria.

Lyxose $C_5 H_{10} O_5$; an aldopentose with the configuration of the ã-carbon opposite that of the a- and ß carbons. It does not occur naturally but is prepared by the degradation of galactonic acid; m.p. 106-107°C showing mutarotation (d-form).

M

M13 A bacteriophage which infects certain strains of *E. coli.* The salient feature of this phage is that it packages only a single strand of DNA into its capsid. If the investigator has inserted some heterologous DNA into the M13 genome, copious quantities of single-stranded DNA can subsequently be isolated from the phage capsids. M13 is often used to generate templates for DNA sequencing.

Macrogenesis Evolution by discontinuous change; saltational evolution.

Macrophage Type of white blood cell that engulfs and removes harmful microorganisms and other cellular debris.

Macrorestriction map Map depicting the order of and distance between sites at which restriction enzymes cleave chromosomes.

Magnetic field Area in which a magnetic or electrically charged object experiences a force because of the presence of a permanent magnet or an electric current.

Magnetic flux Measurement representing the strength of a magnetic field.

Magnetic pole Area within a magnetic field where the magnetism is the strongest; magnets have a north and south pole.

Magnetism Physical phenomenon created by the movement of electrons within either an electric current or a stationary magnet.

Magnetosphere Space surrounding a planet where objects experience the force of the planet's magnetic field.

Magnitude Measure of a celestial body's brightness.

Maillard test for indoxyl in urine 10 cc. urine are mixed with 1 cc. lead subacetate solution, filtered, and an equal volume of hydrochloric acid added to the filtrate; shaking with chloroform gives rise to a blue color. Shaking with hydrogen peroxide should be performed if no color is obtained.

Maintenance methylase An enzymatic activity responsible for maintaining the patterns of methylation on each strand of a DNA molecule after replication.

Malaquin reaction for strychnine 1 cc. strychnine solution (not more than 0.1% strychnine) is treated with 1 gm. zinc and 1 cc. hydrochloric acid for 2-4 minutes, heated to boiling rapidly, cooled and floated on sulfuric acid. A rosered ring (sensitivity-1100000) forms immediately.

Malaria A group of fevers caused by mosquito-transmitted protozoa (Plasmodia); intermittent fever, remittent fever, ague, tertian fever, quartan fever, estivo-autumnal fever, paludism. The parasite destroys the red blood cells. There are many varieties including masked and latent forms. Quinine and atabrine have been used prophylactically. Quinine, plasmochin and atabrine are administered in short or long treatments.

Male In organisms with separate sexes, the one which produces sperm.

Malic acid $C_4 H_6 O_5$; 1-hydroxysuccinic acid; 1-form occurs in acid fruits such as grapes, apples, gooseberries, and is a product of the fermentation of fumarates by Aspergillus niger; m.p. 100°C, boils and decomposes at about 140°C; antiseptic against gas bacillus.

Malic enzyme A pyridinoprotein enzyme which specifically catalyzes the oxidation of 1(-)-malate to oxaloacetate; found in heart, brain, kidney, skeletal muscle, liver, blood.

Malt Germinated barley grain, heated and dried, contains much maltose; used as a food adjunct.

Maltose A reducing disaccharide prepared by the enzymatic hydrolysis of starch. It is glucopyranose-4-*a*glucopyranoside.

Malvin A chromo-glucoside from flowers of Malva silvestris.

Mammal Member of a group of vertebrate animals that are warm-blooded and that have body hair, mammary glands (in the female), a three-bone middle ear, a two-bone lower jaw, a seven-vertebra neck, and red blood cells without nuclei (plural of *nucleus*).

Mammary gland (life science) gland that produces milk in female mammals after the birth of young

Mandelic acid Amygdalic acid; paramandelic acid; phenylglycolic acid; *a*-phenyl-hydroxy-acetic acid; 1-and dforms; m.p. 118-119°C occurs in glycosides, like amygdalin; used as urinary antiseptic in acid urines.

Mannana A mannose polysaccharide from ivory nuts consisting of mannopyranose units.

Manninotriose Glucose galactose galactoside found in ash manna; m.p. 150°C.

d-MANNITOL Manna sugar; d-mannite; a hexahydroxy hexane, obtained by reduction of fructose or mannose. It is found in many plants together with d-sorbitol which is a stereoisomer; m.p. 166-8°C; used as a mild laxative and in diabetes.

Mannosans Polysaccharides found in plant structures, e.g. ivory nuts. Unlike cellulose they yield mannose on hydrolysis.

d-MANNOSE A hexose derived from mannans; m.p. 132°C.

Mantle Region of Earth between the *crust* and the *core,* composing 82 percent of the planet's volume.

Manuiloff reagents I. 1% alcoholic solution of dahlia or methy.

l violet.

II. 1% aqueous papayotin solution.

III. 1% potassium permanganate solution. IV. 2% thiosinamine solution.

V. 40% hydrochloric acid. These are used in sex determination.

Maréchal test for bile pigments The neutral or acid urine solution is treated with 2-3 drops of tincture of iodine. An emerald-green color is obtained, changing after ½ hour to rose-red and finally to yellow.

Margaric acid n-heptadecylic acid, a synthetic acid showing no connection with margarine.

Marker Two typical usages **1] Molecular weight size marker** a piece of DNA of known size, or a mixture of pieces with known size, used on electrophoresis gels to determine the size of unknown DNA's by comparison.

2] Genetic marker An identifiable physical location on a chromosome (e.g., restriction enzyme cutting site, gene) whose inheritance can be monitored. Markers can be expressed regions of DNA (genes) or some segment of DNA with no known coding function but whose pattern of inheritance can be determined.

Marsh test for arsenic The reduction of arsenic in its compound by nascent hydrogen produces arsine, AsH3, burning with a blue flame and decomposing on heating to form a mirror of metallic arsenic.

3-5 gm. arsenic-free zinc are placed in a flask connected through a stopper to a glass tube bent horizontally and constricted to a capillary in the middle. 25 cc. of 2% sulfuric acid are placed in the flask and, after 20-30 minutes, the test sample is introduced in to the reaction flask. The tube is gently heated just before the constriction, leading to the deposition of arsenic as a black mirror in the constriction. Although antimony gives a similar mirror, it may be differentiated chemically.

Marsupial Member of a class of mammals in which females carry their offspring in a pouch for a period of time after birth

Marx-sobotka reagent and reaction for equilenin and dihydroequilenin Coupling of equilenin with diazotize p-nitrobenzene-azodimethoxyaniline gives rise to a deep blue color; dihydroequilenin gives a brownish-red precipitate on coupling which is soluble in organic solvents with the formation of a blue color.

Mass Physical property of matter that represents a measure of its *inertia.*

Mass number Sum of the number of *protons* and *neutrons* present in an *atom's* nucleus.

Material determination Determination of a part of an embryo to perform a certain type of histogenesis or morphogenesis; as opposed to Dynamic determination

Mathematical induction Mathematical demonstration proving that a law is true for a given integer if it is assumed to be true for the preceding *integer.*

Matrix Rectangular array of values that serves to solve various types of equations.

Matter Anything with detectable and measurable *mass.*

Maturation The sum total of processes whereby the egg is prepared for fertilisation, including oogenesis, the growth of the ovum, and the final nuclear changes.

May and rose method A reliable method for determining tryptophan by heating the substance to be tested with HC1 and para-dimethylamino benzaldehyde, whereupon a blue color forms which can be assayed colorimetrically.

Mayer reagent for detecting leukocyte in-crease in blood When a minimum of about 19000 leukocytes are present in blood, the characteristic guaiac reaction is obtained when the blood is treated with guaiac tincture and turpentine.

Mceration Process by which solid food substances are softened by saturation with liquid (as with saliva in the mouth).

Mean Value representing the average of more than one term.

Meander Round, nearly loop-shaped curve of a river in a flat region.

Mecholyl Mecholin, acetyl-ß-methylcholine chloride; $(CH_3)_3N{\cdot}CH_2{\cdot}CH(CH_3)O{\cdot}OCCH_5$; m.p., 171-173°C; used as parasympathetic stimulant.

Meconic acid An acid from opium, $C_7H_4O_7.3H_2O$, said to be nonpoisonous and having little physiological action.

Meconium (1) The fecal matter discharged by the new-born. It is a darkgreen substance, consisting of mucus, bile, and epithelial threads. (2) Opium.

Medes test for ascorbic acid in urine 5 cc. of urine are treated with 1 cc. of formaldehyde, 0.5 cc. of buffer solution (100 cc. 2M sodium acetate and 30 cc. 2M acetic acid) and 1 cc. of Folin's uric acid reagent; a blue color is obtained if ascorbic acid is present. The color may be used in the colorimetric determination of the acid.

Median Term within a sequence of values that has an equal number of terms before and after it; line from one corner of a triangle to the midpoint of the opposite side.

Medulla Internal portion of an organ or plant stem.

Medulla The portion of an organ or tissue that resembles bone marrow in appearance.

Megabase (Mb) Unit of length for DNA fragments equal to 1 million nucleotides and roughly equal to 1 cM.

Meiosis A two-stage type of cell division in sexually reproducing organisms. In meiosis, a diploid cell divides to produce four haploid cells, each with half the original chromosome content. For this reason, meiosis is often called a "reduction division". In organisms with a diploid life cycles, the products of meiosis are usually called gametes. In organisms with an alternation of generations, the products of meiosis are caled spores.

Melanin (1) The brownish-black pigment of hair, skin and the choroid of the eye, formed by the oxidation of tyrosine by the enzyme tyrosinase.

(2) The older, obsolescent name for humin.

Melanoprotein An intensely black chromoprotein with melanin as the prosthetic chromogen which forms the pigment of black wool.

Melezitose C_{18} $H_{32}O_{16}$; a non-reducing trisaccharide which occurs in a manna that forms on the Douglas fir and other trees as the result of the activity of a soft scale insect. On hydrolysis with invertase it gives turanose or 3-*a*-D-glucosyl-Dfructose, and glucose; m.p. of hydrate 153°C.

Melibiase A d-galactoside hydrolyzing enzyme found in bottom yeast, almonds and other plants, that splits melibiose into sucrose and galactose, optimum pH, 5.5.

Melibiose C_{12} H_{22} O_{11}; a reducing disaccharide, 6-(a-galactosyl)-D-glucose, obtained by the hydrolysis of raffinose, a trisaccharide which occurs in beet molasses, with dilute acids or invertase.

Melissic acid A saturated fatty acid, CH_3 $(CH_2)_{29}$ COOH, found in beeswax; m.p. 90-91°C.

Melting The dissociation of a duplex nucleic acid molecule into single strands, usually by increasing temperature.

Menstrual Cycle The cyclic release of oocytes and priming of the endometrium to receive a fertilized egg; the complete cycle averages 28 days in female humans.

Menthol C_{10} $H_{20}O$; hexahydrothymol; peppermint camphor; 1-form, terpene alcohol found in peppermint oils; b.p. 212°C; isomers have been reported; used as analgesic in liniments and in inhalants.

Menthone C_{10} H_{18} N; terpene aldehyde; 1-form found in peppermint, geranium, pennyroyal oils.

Mercurochrome (soluble) A sodium salt of dibromfluorescein mercury hydroxide; mild antiseptic.

Merogony The development of fragments of eggs. A fragment of an egg, containing the nucleus, may be fertilised by sperm (*Diploid merogony*). A fragment of an egg, not containing the nucleus, may be fertilised by sperm (*Andro-merogony*). A fragment of an egg, containing the nucleus, may be removed from the remainder, after fertilisation by sperm but before nuclear fusion (*Gynomerogony*), or the same experiment may be made using artificial fertilisation (*Parthenogenetic gyno-merogony*). Lastly, a fragment of an egg, not containing the egg-nucleus, can be fertilised by artificial means (*Parthenogenetic merogony*).

Merthiolate A commecial name for Na ethylmercurithiosalicylate; a powerful germicide which does not precipitate with serum proteins.

Mesembrine An alkaloid from mescal buttons exhibiting narcotic-tetanic properties.

Meso-compounds These compounds have an even number of asymmetric carbons, the members of any pair having a rotation tendency equal to each other, but opposite in sign, so that the total effect is of no optical activity.

Mesoderm In most animal embryos, a primary tissue layer between ectoderm and endoderm; gives rise to muscle, organs of circulation, reproduction, and excretion, most of the internal skeleton, and connective tissue layers of the gut and body covering.

Meson Member of a group of unstable subatomic particles (particles smaller than atoms) that are composed of two *quarks* (type of indivisible particle).

Mesozoic era Era of reptiles; includes the cretaceous, the jurassic and triassic.

Mesozoic The geological era that lasted from about 225 million years to 65 million years; the age of reptiles.

Messenger RNA (mRNA) A class of RNA molecules, each of which is complementary to one strand of DNA, characterized by a 5' 7-N-methylguanine cap and a 3' polyA tail. Its function is to carry the genetic message from the chromosome to the ribosomes.

Messenger RNA (mRNA) Proteins are not synthesized directly from genomic DNA. Instead, an RNA template (a precursor mRNA) is constructed from the sequence of the gene. This RNA is then processed in various ways, including splicing. Spliced single-stranded RNAs destined to become templates for protein synthesis are known as mRNAs, The term mRNA is used only for a mature transcript with polyA tail and with all introns removed, rather than the primary transcript in the nucleus. As such, an mRNA will have a 5' untranslated region, a coding region, a 3' untranslated region and (almost always) a poly(A) tail. Typically about 2% of the total cellular RNA is mRNA.

Metabolic body size Size of animals expressed in terms of metabolic rate, such as Calsads, similar to horsepowers as index for size of motors. Metabolic body size may also be expressed in terms of functions to which the metabolic rate is proportional such as body surface or, more recently, the ¾ power of body weight.

Metabolism Sum of all chemical changes in an organism. Includes anabolism and catabolism. Attention may be focussed on particular aspects Mineral-carbon, nitrogen, energy-metabolism.

Metabolite Any substance produced by metabolism.

Metabolizable energy Physiological fuel value of food. Energy, which is available for all metabolic processes in animal body. Heat of combustion of food minus heat of combustion of feces, urine and methane.

Metagenesis The alternation of a generation which reproduces only asexually by division or budding with a generation which reproduces only sexually by means of eggs and spermatozoa.

Metamorphism Physical changes that take place in solid rocks in conditions of increased temperature and pressure.

Metamorphosis A process of developmental change whereby a larva reaches adulthood only after a drastic change in morphology; occurs in most amphibians and insects, for some insects, this change may include another stage (pupa) before the adult stage.

Metaphase Stage of mitosis when spindle has fully formed, sister chromatids of each chromosome become attached to opposite spindle poles, and all chromosomes lie at the spindle equator.

Metaphen A commercial name for the anhydride of 4-nitro-5-hydroxymercuri-o-cresol, a very powerful germicide, used in urogenital infections.

Metaplasia The change of one kind of tissue into another; also, the production of tissue by cells which normally produce tissue of another sort.

Metaplasm Alloplasm; paraplasm; deutoplasm; lifeless material in protoplasm.

Metaproteins (1) The American classification of proteins places it in the second group of the hydrolysis products of proteins, being classified just after the proteins. They are insoluble in neutral solvents, but soluble in weak acids and alkalis.

(2) First stages in protein degradation, soluble in dilute acids and alkalies but not in water or salts.

Metastasis Spontaneous dispersion of a tumour from its original site in the body to other sites.

Metathetely The appearance of structures normally associated with an earlier stage of development, at a later stage of development, e.g. the retention of larval organs by insect pupae.

Meteor Appearance (as viewed from Earth) of a flash of light caused by space debris burning up in Earth's atmosphere.

Meteorite Space debris that falls to Earth's surface intact.

Methemoglobin A compound formed from oxyhemoglobin or red hemoglobin by mild oxidation. It has the Fe in the ferric state, unlike oxy- or reduced hemoglobin.

Methionine metabolism Methionine is an almost ubiquitous protein constituent. It is an essential component of mammalian diet because the capacity for synthesizing this amino acid from inorganic matter seems to be confined to plant life. Its methyl (CH_3) group can participate in the biosynthesis of choline, creatine, and related substances from methionine by a mechanism which is still unknown. Ingestion of excessive amounts of methionine results in excretion of urinary sulfates.

Methylcholanthrene $C_{21}H_{16}$, a cancer-producing yellow hydrocarbon prepared from cholic acid, m.p. 177°C.

Methylthiopentosides Nucleosides containing sulfur, as adenine methylthiopentoside of yeast.

Metrazol Cardiazol; $C_6H_{10}N_4$; m.p. 57-58°C; a commercial name of pentamethylenetetrazol, a vasomotor and respiratory stimulant.

Meyerhof quotient M.Q.; a measure of the resynthesis of glucose or the amount of lactic acid reconverted to glucose per unit oxygen consumption.

Micelle A particle of the dispersed phase in a colloidal system.

Michel reagent for differentiating oxyhemoglobin and co-hemoglobin Reagent prepared by dissolving 4 gm. potassium hydroxide in 75 cc. water and adding 3 gm. sodium hydrosulfite ($Na_2S_2O_4$). After solution is complete, 40 cc of 98% alcohol are added and the solution filtered; it must be kept out of contact with air.

Micro-aerotonometer An instrument for equilibrating a minute bubble of air with blood and analyzing the absorption.

Microbiology The subject of microbiology deals with microscopic or even sub-microscopic organisms, variously known as microbes, microorganisms or germs. These may be plants, animals or of an uncertain classification. Divisions of the subject follow the classifications of organisms, e.g. mycology deals with molds, protozoology with protozoa and bacteriology with bacteria. The divisions are not absolutely sharp, so that a subject like bacteriology may deal with viruses and rickettsia and selected groups of molds, yeasts and protozoa. A classification suggested by Birkeland is general bacteriology, soil bacteriology, dairy bacteriology, food bacteriology, industrial bacteriology, sanitary bacteriology, veterinary bacteriology, medical bacteriology, immunology and serology. Neighboring fields are many, e.g. epidemiology and public health science.

Bacteria and unicellular animals are classified often by biochemical activity, staining and chemical composition as well as disease-producing power, morphology and general biological characteristics. The biological phylum thallophyta (plants without roots, stems, leaves) contains the sub-phyla, algae (chlorophyll-containing) and fungi, which include yeasts, molds and bacteria. The fungi fall into the following classes (1) Schizomycetes (fission fungi, bacteria)

(2) Saccharomycetes (yeasts)

(3) Phycomycetes (forms without cross walls between cells)

(4) Ascomycetes (forming spores in a sac) and

(5) Basidiomycetes, forms in which sexual stages are unknown, and slime molds.

The Schizomycetes or bacteria are usually described as cocci (spheres), bacilli (rods) and spirilla (spirals). The streptocci appear as chain-like beads. The staphylococci appear in irregular clusters. Some spirilla look like a comma, e.g. the comma or vibrio bacillus. Many bacteria form spores and are surrounded by capsules of protein or carbohydrate nature. Classifications are far from complete. Seven or eight orders of Schizomycetes are listed.

(1) Eubacteriales, true bacteria;

(2) Actinomycetales, mold-like bacteria;

(3) Chlamydobacteriales, sheathed bacteria.

(4) Caulobacteriales, stalk bacteria;

(5) Thiobacteriales, sulfur bacteria;

(6) Myxobacteriales, slime bacteria;

(7) Spirochaetales, protozoan-like bacteria.

(8) Rickettsia. The orders are divided into numerous families, and these successively into various tribes, genera and species.

Yeasts are often described as true and false, distinguished as to whether they bud by forming endospores or not. The true yeasts include many kinds of Saccharomyces. Practical brewers use convenient designations as top yeasts and bottom yeast which differ in the vigor of fermentation. Wild yeasts are called torula, which frequently produce strong pigments. There are numerous sub-groups classified by fermentation characteristics.

Molds are fungi commonly associated with decay and food spoilage. Common types are Phycomycetes and Ascomycetes (sac-molds). The former contain the family Mucoraceae and its genera Rhizopus and Mucor, commonly referred to as bread molds. Among the Ascomycetes are the well known Aspergillus niger and the Penicillium (meaning brush-like). The flavors of cheeses are often due to Penicillia.

Microbiology is concerned with not only the morphology but the physiology of microorganisms. Bacterial metabolism, growth and participation in the cycles of nature are important fields of study. The production of enzymes by bacteria ranks as a science in itself and is used as a basis for differentiating them. The feeding characteristics of bacteria lead to their designation as autotrophic (self subsistent on simple inorganic compounds) or heterotrophic (requiring some built up organic substances as well). The metabolism is described as fermentation, with stress on incomplete carbohydrate breakdown, or putrefaction, with stress on incomplete protein breakdown. Both fermentation and putrefaction come under the term dissimilation, which is the opposite of assimilation. Microbial attack of cellulose, hemicellulose, pectin, lignin, starches, gums, sugars, waxes, etc., has been studied for a great variety of conditions, and the findings help to explain the carbon cycle in nature. Similarly the nitrogen cycle depends on its explanation at least at two points, namely nitrification and nitrogen fixation in the soil and protein decomposition and putrefaction. The autotrophic Nitrosomonas converts ammonia to nitrous acid. Nitrogen fixers are symbiotic and free living. Among the former are the genera Rhizobium and Mycobacterium, and among the latter is the aerobe, Azotobacter chroococcum, and the anaerobe, Clostridium butyricum. Very many bacteria are proteolytic. Besides carbon and nitrogen the natural history of sulfur compounds, potassium, phosphorus and calcium includes the activity of bacteria.

The genetic history and possibilities of bacteria are included in this field. Moreover, environmental influences, such as temperature, radiation, pressure, moisture, light, etc., have been studied in great detail. The effect of chemical agents has been of particular importance in the development of antiseptics (bacteriostatic substances), preservatives, disinfectants (germicides, fungicides, bacteriocides) and deodorants (putrefaction inhibitors). The so-called phenol coefficient is a measure of sterilizing power against bacteria.

The field of study of infection is intimately allied to microbiology. The manner of invasion of bodies by microorganisms and their own struggle for survival or for symbiosis constitute branches of the subject. The production of harmful or beneficial substances belongs to biochemistry rather directly. Thus bacteria may produce hemolysins, which are substances which break down red blood cells and are species specific; leucocidins, or substances which destroy white blood cells and are species specific; coagulases and fibrinolysins which cause formation of blood clots or their solution; "spreading factor," a substance which affects the permeability of tissue (also known as Duran-Reynals factor); and toxins or poisons secreted by bacteria; roughly classfied as exotoxins definitely found diffusing from the bacterial cell and endotoxins which are liberated when the bacterial cell is disrupted.

The defense of the body against the invasion of bacteria and their toxic substances leads to the entire field of immunology (q.v.). The spreading of infection when it becomes a mass phenomenon is known as epidemiology. The number of diseases is not large compared to the great variety of organisms, of which few are relatively harmful, but it runs to quite a list. Thus staphylococci might cause septicemia, pyemia, boils, pimples, abcesses impetigo, osteomyelitis. Streptococci, such as Streptococcus viridans, are credited with teeth abscesses, rheumatic fever, arthritis, sinus infections and sub-acute bacterial endocarditis. Streptococcus hemolyticus is credited with erysipelas, scarlet fever, septic sore throat, tonsillitis, bronchopneumonia, middle ear infection, puerperal sepsis. A large chapter can be written on the several types of pneumonia produced by pneumococci. For whooping cough one must study Hemophilus pertussis; for diphtheria, Corynebacterium diphtheriae or the Klebs-Loeffler bacillus; for tuberculosis, Mycobacterium tuberculosis or the tubercle bacillus; for leprosy, Mycobacterium leprae. The same general state of affairs is true for dysentery, typhoid fever, cholera, plague, tularemia, brucellosis or undulant fever, rickettsial diseases (typhus and Rocky Mountain spotted fever), syphilis, gonorrhea, etc., etc. In some cases viruses are no doubt responsible, e.g. the common cold, measles, mumps, poliomyelitis (infantile paralysis), some forms of encephalitis, rabies, psittacosis (parrot disease), smallpox, yellow fever.

Applied microbiology is not only concerned with medicine but also with food. Food spoilage, preservation and processing depend on microorganism control. Water purity and milk purity are equally vital microbial control problems. Sewage disposal is a vast field of problems involving detailed microbiological knowledge. These are more important than spectacular poisonings due to decomposing food, such as botulism due to Clostridium botulinum, and several other similar types.

In the study of microorganisms in the soil, furthermore, not only is material found of interest to agriculturalists but also to engineers, because indirectly the strength of structures, highways, airports, etc., is dependent on microbiological facts.

Microcionasterol An animal sterol of the sponge, Microciona prolifera, containing one double bond.

Microclimate Climate of a small, defined area.

Microtubule Hollow cylinder of mainly tubulin subunits; a cytoskeletal with roles in cell shape, motion, and growth and in the structure of cilia and flagella.

Migration Movement of certain animals from one location to another for breeding or feeding.

Milieu interieur Body fluids of an organism; the "internal environment."

Milk The white aqueous secretion of the mammary glands, pH 6.6-6.9, which is the sole source of nourishment of the young mammal. It contains 90% water, proteins, mostly casein, fat, mostly glycerides of the higher fatty acids, lactose, minerals, vitamins, particularly A & G, and other constitutents. Normal volume in humans varies from 1500 cc. to somewhat more than 3000 cc. at any one time.

Missense mutation A nucleotide mutation which results in a change in the amino acid sequence of the encoded protein (contrast with Silent Mutation).

Mitochondria (plural of *mitochondrion*) (life science) membrane-bound structures in cells of *eukaryotes* in which energy-providing molecules are produced

Mitochondrial matrix The aqueous portion of a mitochondrion bounded by the inner mitochondrial membrane.

Mitogenetic radiation Feeble ultraviolet radiation emitted by developing cells which can increase mitosis, or the rate of cell division; also known as Gurvitch rays; many doubt their very existence.

Mitosis Type of nuclear division that maintains the parental number of chromosomes for daughter cells. It is the basis of bodily growth and, in some cases, of asexual reproduction of eukaryotes.

Mitragynine An alkaloid, $C_{17} H_{22} N(OCH_3)(COOCH_3)_2$, from the leaves of Mitragyne speciosa, and possessing local anesthetic properties.

Modulation The reversible histological differentiations which occur as the final stages of the development of a tissue.

modulator A metabolite that, when bound to the allosteric site of an enzyme, alters its kinetic characteristics. Can be homotropic or heterotropic, inhibitor or activator.

Molar gas constant The constant R, in the General Gas law; PV=nRT, where P= pressure, V=volume, n=number of moles and T=Temp in °CA. It is equal to 0.08204 liter-atmospheres.

Molarity Number of *moles* of a particular substance dissolved in a cubic decimeter of solution; measures the concentration of the solution

Mole Unit defined by the amount of a given substance containing the same number of atoms or molecules as exist in exactly 12 grams of carbon-12.

Molecular biology The biochemical study of the genetic basis for phenotype.

Molecular cloud Enormous, cool cloud of dust and gas where stars are born.

Molecular Genetics Study of how genes function to control cellular activities.

Molecular mass *Mass* of a molecule, calculated relative to a scale on which the mass of one atom of carbon-12 equals 12.

Molecular solid Solid substance consisting of molecules joined by weak forces; melts at low temperatures compared to other solids.

Molecule Two or more atoms joined by chemical bonds and behaving as an individual unit in chemical interactions.

Molisch test A specific test for carbohydrates. a-napthol is added to the solution to be tested, and stratified over concentrated sulfuric acid. A pink to red color is positive.

Molisch test for albumin 1 cc. of test solution is treated with 2 drops of a solution of 20 gm. naphthol in 100 gm. alcohol and 5 cc. of concentrated sulfuric acid; a red or violet color is obtained if albumin or peptone is present.

Mollusca These include Lamellibranchiata — mussels, clams, scallops; Gastopoda-snails, winkles, slugs; Cephalopoda—cuttlefishes, squids, octopus.

Momentum Product of an object's *mass* and *velocity.*

Monocistronic A form of gene organization resulting in transcription of an mRNA that contains coding sequence for a single gene or gene product.

Monoclonal Antibody An antibody with very specific and often unique binding specificity which is secreted by a biologically cloned line of plasmacytoma cells in the absence of other related antibodies with different binding specificities. Differs from polyclonal antibodies, which are mixed populations of antibody molecules such as may be present in a serum specimen, within which many different individual antibodies have different binding specificities.

Monohybrid An experimental cross between two parent organisms that breed true for distinctly different forms of a single trait; heterozygous offspring results.

Mononucleotides The constituent parts of the nucleic acids, containing one molecule of either a purine or a pyrimidine derivative, a carbohydrate and a phosphoric acid molecule. The phosphoric acid is esterified with C_3 or C_5, of the sugar, which in turn is connected to a N of the purine or pyrimidine. 4 nucleotides go into making a nucleic acid molecule.

Monophenol oxidase A mushroom copper protein which catalyzes the oxidation of monophenols to their quinones, but can also attack polyphenols more slowly.

Monophyletic Of a taxon, the members of which are descendants of the nearest common ancestor.

Monose Formaldehyde. For purposes of analogy it is classified with the monosaccharides, but it is not a carbohydrate.

Monsoon Seasonal climatic phenomenon causing heavy rainfall in certain regions of Earth, particularly southern Asia.

Montanic acid A 28 carbon saturated fatty acid found in Montan wax.

Montignie reaction for sterols A red-brown color is obtained when an alcoholic solution of a sterol is gently heated with silicotungstic acid. Cholesterol, ergosterol, phytosterol and stigmasterol give the reaction; only a faint reddish-yellow color is given by á and ß-cholesterylenes and cholestenone.

Montignie test for ergosterol and its derivatives A solution of the substance in 5 cc. of alcohol is heated to 95°C and treated with 3-4 cc. of a 10% selenious oxide solution; a red precipitate developing within 2 minutes indicates ergosterol. Cholesterol, phytosterol and stigmasterol do not react.

Montignie test for indole, pyrrole and skatole The aqueous solution of the substance is heated with 8-10 drops of 10% selenious acid solution and 1 cc. nitric acid. A violet color, sensitivity—0.04 mg., is given by indole. Pyrrole forms a blue substance soluble in chloroform and a red chloroform-insoluble substance, sensitivity—0.4 mg. A red solution is given by skatole but, in the absence of nitric acid, a violet color, changing to red with nitric acid, is obtained.

Moor urine or urea reaction 5 drops concentrated ammonia and 0.1-0.2 gm. solid phosphotungstic acid are added to 5 cc. of urine which becomes an intense blue immediately. A weak reaction is obtained where metabolism is impaired or where kidney function is impaired.

Moore's test A test for sugars, depending on the decomposition of the sugar when boiled in alkalies, giving a brown color and an odor of caramel.

Morin 3,5,7,2',4',-pentahydroxyflavone; the coloring matter of the wood of old fustic, m.p. 285°C, a sensitive reagent for aluminum.

Morindin A glucoside, $C_{26}H_{28}O_{14}$, from root bark of Morinda citrifolia; m.p., 250-251°C.

Mörner test for acetoacetic acid in urine A strong odor of iodo-acetone, strongly irritating to the mucosa, is obtained when the urine is boiled with an excess of potassium iodide and ferric chloride. The odor is easily distinguished from iodine odor.

Mörner test for cysteine The addition of an alkali and sodium nitroprusside to a cysteine solution gives rise to a persistent purple-red color. The color is due to the presence of the sulfide group in the substance.

Morphactins A group of highly potent synthetic plant growth regulators. These are derivatives of flurenol (9'-hydroxy-flurene(9)-carboxylic acid, R=H), chefly by esterification of the carboxyl group and/or substitution in the ring system, e.g. chloroflurenol (R=Cl).

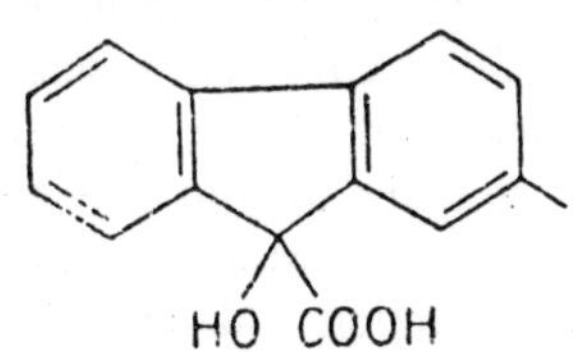

Flurenol (R=H)
Chloroflurenol (R=Cl)

Fig. Morphactins

Morphine $C_{17}H_{19}O_3N$; prisms with one H_2O; anhyd. m.p. 254°C; a strongly habit-forming analgesic alkaloid of opium. Exerts a simultaneous depressing and exciting effect on the central nervous system. Causes respiratory depression and death.

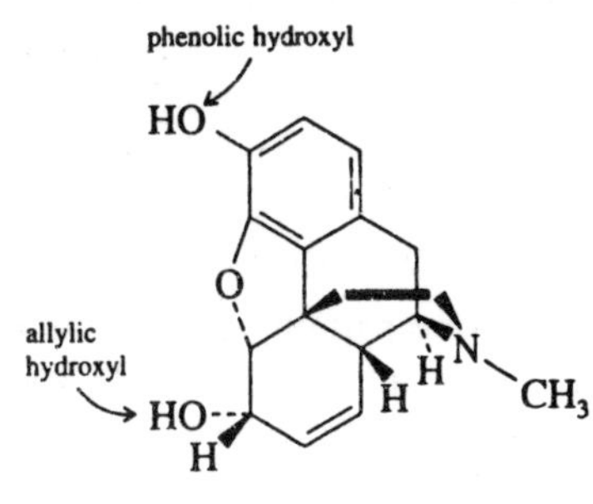

Fig. Morphine

Morphogenesis Processes by which differentiated

cells in a embryo become organized into tissues and organs, under genetic controls and environmental influences.

Morphogenetic movements The modelling movements in differentiation which change the shape of tissues, e.g. the cell streams, whereby, as in gastrulation, the eventual positions of embryonic cells are attained. They include invagination, expansion, longitudinal extension, and dorsal convergence.

Morula The segmented ovum in the mulberry stage, forming a solid mass of cells.

Mosaic (1) as opposed to *Regulative*. In mosaic development the fates of all the parts of the embryo are already fixed at or before fertilisation, so that any localised ablation or injury will later be manifested as a complete lack of the part in question. In regulative development, however, injuries can be repaired, deficiencies made up, and indifferent added tissue organised into the pattern of the body. Mosaic eggs cannot form giant embryos by egg fusion nor dwarf ones by blastomere separation, and, generally speaking, twinning is much more restricted in them.

(2) in Genetics Genetic mosaics are individuals the tissues of which contain cells of more than one genetic constitution. They may be due to somatic mutation, somatic crossing-over, or chromosome elimination.

(3) as opposed to Intermediate Organisms vary not only as to the presence or absence of organs, but also to the extent to which any one organ is formed. An organism might have all its organs complete save one, but that one entirely absent; conversely, it might have them all, but imperfectly developed. The former are, in this sense, mosaics; the latter intermediates. This sense of the word is used mainly by workers on Aphids.

Motif A recurring pattern of short sequence of DNA, RNA, or protein, that usually serves as a recognition site or active site. The same motif can be found in a variety of types of organisms.

Motile Able to move oneself about, capable of self-locomotion.

Motility Describes the ability of a cell to move independently.

Motor effect Spontaneous movement of a wire located within a magnetic field and carrying an electric current.

Mott-haliburton test for choline Choline forms a double crystalline compound with platinum chloride.

Mouse unit of estrogenic activity The least amount of estrus-producing hormone (estrone) which will cause in a spayed mouse a characteristic degeneration of the vaginal epithelium (averaged over several animals in a given colony.)

Mrna export Refers to the movement of spliced mRNA out of the nucleus to the cytoplasm.

Mrna processing Refers to the processes of polyadenylation, splicing, and addition of a 5' cap structure.

Mucic acid $COOH(CHOH)_4COOH$; tetrahydroxy-adipic acid; saccharolactic acid; an oxidation product of galactose; m.p. 255°C with decomposition.

Mucilages Salts, either with metals or organic bases, of polyeaccharide sulfuric acid esters, which gel with water. Galactose is probably at least a part of the carbohydrate part of the compound.

Examples agar-agar and carrageen.

Mucins A group of glycoproteins occuring in or made by glands, e.g. submaxillary gland and liver, containing a high percentage of carbohydrates; particularly mucoitin sulfuric acid. The mucins form very hydrophilic colloids.

Mucoids A group of conjugated glycoproteins where the protein is conjugated with chondroitin sulfuric acid. They are viscous bodies found in tendons, cartilage, etc.

Mucoitin sulfuric acid One of the two carbohydrate groups associated with the protein of the glycoproteins. It consists of two molecules of mucosin (amino glucose) with one hydroxyl esterified with sulfuric acid and the amino groups acetylated. Each of these mucosin molecules is linked to a molecule of glucuronic acid, which in turn are connected to each other.

Mucous membrane Skin that secretes mucus (a moist, protective fluid); located where animals' body cavities are in contact with the air.

Mucus A condensation product of monoses produced by bacterial action, variously described as a starch, a cellulose or an anhydride of glucose called "dextran" or an anhydride of fructose called "levan."

Multicistronic Message An mRNA transcript with more than one cistron and thus encoding more than one polypeptide. These generally do not occur in eukaryotic organisms, due to differences in the mechanism of translation initiation.

Multicopy Plasmids Present in bacteria at amounts greater than one per chromosome. Vectors for cloning DNA are usually multicopy; there are sometimes advantages in using a single copy plasmid.

Multi-gene family A group of genes that are related by sequence homologies; usually are also related by their functions or by the processes in which they participate.

Multiparous Giving birth to more than one at a time.

Multiple alternate oxidation theory A modification of the ß-oxidation theory of Knoop, advanced by Jowett and Quastel. Oxidation of fatty acids take place simultaneously along the entire molecule, at the ß, ä, æ; even-numbered carbons, preceded by desaturation of the molecule.

Multiple cloning site An artificially constructed region within a vector molecule which contains a number of closely spaced recognition sequences for restriction endonucleases. This serves as a convenient site into which foreign DNA may be inserted.

Multiplexing A sequencing approach that uses several pooled samples simultaneously, greatly increasing sequencing speed.

Mummy Dead body preserved by natural or artificial means.

Mumps Epidemic parotitis; acute contagious infection of parotid gland which may affect the gonads, the pancreas and other glands and the central nervous system, manifested by swelling and fever. Testicular atrophy or deafness may follow.

Munjistin A glycoside of 24-dihydroxyanthraquinone-3-carboxylic acid from roots of *Rubia munjista*; m.p. 229-230°C.

Muon Elementary subatomic particle more than 200 times more massive than an electron; muons are classified as *leptons*.

Murexide reaction The murexide reaction depends upon the formation of the deep violet ammonium salt of purpuric acid from uric acid and related purines. 2-3 drops of concentrated nitric acid are added to a small amount of the substance in a porcelain dish and the solution carefully evaporated to dryness. The red or yellow residue, after cooling, turns purplish-red on the addition of a drop of dilute ammonia. The use of potassium hydroxide gives rise to a purplishviolet color.

Muscarine $(CH_3)_3{\cdot}N(OH){\cdot}CH(CHO){\cdot}CHOH{\cdot}CH_2{\cdot}CH_3$; according to the older literature, an oxidation product of choline, but later shown to be a higher homologue of this oxidation product. A poison found in certain mushrooms, "fly fungus."

Muscle Animal tissue that contracts and relaxes, controlling movement.

Mustard gas Dichlorethyl sulfide; burns resemble X-ray and drug dermatitis; slow healing and periodic recurrence like the course of an infectious disease.

Mutagen An agent capable of causing mutations. Common examples are ultraviolet light, such as in sunlight, and anthracene, a material formed during the cooking of fatty meats on a barbecue grill.

Mutant A sport or variation which breeds true.

Mutarotation The change in rotation observed when a solution of an aldose or ketose is allowed to stand. The exact mechanism is not known with certainty, but it involves changes of one form to a different optical isomer.

Mutase Oxyreducase; enzyme system which

promotes the oxidation of one molecule of its substrate by the reduction of another molecule of the same substrate.

Mutation Change in the chemical makeup of *DNA* caused by cellular mistakes, natural biochemicals, and environmental influences (such as *radiation*).

Mutation Discontinuous change in chromosomal DNA, ordinarily an error in DNA replication.

Mutation A permanent, heritable change of the genetic material, either in a single gene or in the numbers or structures of the chromosomes. Mutations do not always have harmful effects. Compare polymorphism.

Myasthenia gravis A metabolic disease of the muscles possibly brought on by infection, showing as marked weakness. There is an increase of creatine and a decrease of creatinine in the urine. Prostigmine relieves symptoms. Glycocoll with ephedrine has been used on the supposition of a connection with a disturbance of creatine metabolism.

Myers-wardell test for cholesterol The blood, after drying on plaster of paris, is extracted with chloroform; the cholesterol is determined colorimetrically after adding acetic anhydride and sulfuric acid to the extract.

Myllus reaction for cholic acid A solution of 20 mg. cholic acid in 0.5 cc. alcohol is treated with 1 cc. 0.1N iodine and diluted gradually with water. Yellow crystals having a metallic luster and blue by transmitted light are precipitated. Choleic, conjugated bile and hyocholic acids do not give the reaction.

Myoalbumin An intracellular water soluble albumin, found in muscle to the extent of about 1% of the total protein; i.p. 3.3.

Myocardium, biochemistry of In 1927, Fiske and Subbarow isolated phosphocreatine in the human myocardium and with the later discovery of adenosinetriphosphate our conceptions regarding the energetics of muscular activity underwent considerable clarification.

It became apparent that compounds of phosphorus, phosphocreatine and adenosinetriphosphate in particular, are intermediaries in the transfer of chemical energy to the mechanical energy of the heart beat. These organic phosphoric acid compounds exist as potassium salts and this accounts for most of the intracellular potassium. It is now apparent that the concentration of these compounds in the myocardium serves as an index of the potential energy of the heart for work.

Creatine is present in heart muscle in about half the concentration it attains in skeletal muscle. Phosphocreatine is present in concentrations from one-tenth to one-third that of skeletal muscle. Its lower concentration in heart muscle than in skeletal muscle cannot be attributed to a difference in the amount of the intracellular phase, since there is a sharp decrease in the ratio of creatine to phosphorus. The distribution of phosphorus compounds is likewise greatly different from that of skeletal muscle. While the total phosphorus of the heart is about the same as that of skeletal muscle, a much greater part is present in the acid insoluble fraction, most of which is phospholipid. The acid-soluble fraction is accordingly less, and this difference may be traced largely to the lower phosphocreatine content of the heart.

There is good evidence, direct and circumstantial, that phosphocreatine and adenosipetriphosphate, as potassium salts, play an essential role in muscular activity, or more specifically are vital agents in the transfer of chemical to mechanical energy for muscular work.

Myoglobin (Myochrome, muscle hmoglobin) A hemoglobin found in the fibres of striated muscle. The protein component differs from blood globin, the prosthetic group is the same. The mol. wt. is only 16,500 or one quarter that of hemoglobin. It also differs from the latter with regard to the position of the absorption bands and the affinity to oxygen.

Myohemoglobin An iron containing porphyrin of the muscle closely resembling hemoglobin. Iron content is the same but molecular weight (35,000), absorption spectrum, and i.p are different.

Myokinin A betaine corresponding to arginine found in muscles.

Myopia Condition in which a person can see only nearby objects clearly; caused by a physical defect in the eye myosin. The major component of the thick filaments of muscle and other actin-myosin systems, functioning as a contractile protein.

Myosin The most abundant intracellular globulin of muscle (50 to 70% of total protein), insol. H_2O, sol. dil. salt solutions, i.p. 5.5. Hydrates to form gels. Shows double refraction.

Myosmine $C_9H_{10}N_2$; 2-(3'-pyridyl)-4,5dihydropyrrol; an alkaloid of tobacco smoke, m.p. 42-43°C.

Myrcene $C_{10}H_{16}$; an acyclic monoterpene, found in oil of hops, verbena, etc.; b.p. 166-168°C.

Myricyl alcohol $CH_3(CH_2)_{29}CH_2OH$, m.p. 87°C, present in beeswax as the palmitate.

Myristic acid A saturated fatty acid $CH_3(CH_2)_{12}COOH$, found in nutmeg butter; m.p. 57-58°C C.

Myristoleic acid Unsaturated fatty acid, $C_{16}H_{30}O_2$, containing one double bond, found in cod liver and whale oils.

Myrosin A sulfatase present in mustard seeds which hydrolyses its glycosides; also present in animal tissues.

Myrtenol An alcohol derived from pinene found in myrtle oil.

Mytilitol C7H14O6; probably a methyl ether of inositol found in valve muscles of *Mytilus edulis*.

N

Nachbarschaft (Contiguity effects) The morphogenetic effects produced by neighbouring tissues or structures on a developing embryonic organ.

Nadi reagent Mixture of dimethyl-p-phenylene diamine and alpha naphthol used as a test for indophenol oxidase.

Nadir Point on the *celestial sphere* directly under an observer; compare *zenith*

Naiman reagent for vitamin B1 Vitamin B_1 gives an orange-red precipitate with potassium-bismuth iodide; the weight of the precipitate is proportional to the vitamin content.

Naphthacene A yellow hydrocarbon found in small amounts with chrysene. It is a linear benzologue of anthracene.

Narcosis A condition of profound stupor brought about by narcotic drugs.

Narcotine $C_{22}H_{23}O_7N$; needles, m.p. 176°C; opianin; opian; narcosine; alkaloid of opium, 3-18%, a benzyl isoquinoline derivative. Slightly poisonous. Paralyses smooth muscles. Simultaneously excites and depresses central nervous system; antineuralgic, antispasmodic and antiperiodic; has been used for malaria and migraine.

Nasse reaction A specific test for tyrosine. Heat the material to be tested with an aqueous solution of mercuric acetate or sulfate and a few drops of a 1% solution of sodium nitrate. Tyrosine gives a cherry-red color, tryptophane gives no color.

Nastic movement Movement of a plant caused by an environmental stimulus but independent of the direction of the stimulus.

Native Gel An electrophoresis gel run under conditions which do not denature proteins (i.e., in the absence of SDS, urea, 2-mercaptoethanol, etc.).

Natural frequency Frequency at which a mechanical system vibrates spontaneously

Natural radioactivity *Radioactivity* produced by naturally radioactive elements in Earth's crust

Natural selection Process by which some individuals—as a result of genetic *mutation*—are better adapted to surviving and reproducing in their environment. This differential reproductive success gradually causes the traits of these individuals to become more frequent within a population. Natural selection is a key element of the theory of *evolution*

Natural Selection Differential success in reproduction of different phenotypes resulting from interaction of organisms with their environment. Evolution occurs when natural selection causes changes in relative frequencies of alleles in the gene pool.

Neighbourwise (Ortsgemäss) Behaving after transplantation in accordance with the new situation in which the graft finds itself, i.e. plastic, pluripotent, undetermined. "Falling into step"

with the new neighbour cells and going along with them to their morphogenetic destination.

Nembutal Pentobarbital-sodium; Na salt of ethyl-alpha-methylbutyl barbituric acid; a popular hypnotic.

Nencki test for mercaptan in urine The gas evolved from the distillation of urine with oxalic acid is led through mercuric cyanide solution. The resulting precipitate is distilled with hydrochloric acid and the distillate passed into lead acetate solution forming a yellow crystalline precipitate.

Nencki-sieber test for urobilin in urine 20 cc. of urine are shaken with 10 cc. amyl alcohol and treated with a few drops of 1% ammoniacal alcoholic solution of zinc chloride. The presence of urobilin is indicated by a green fluorescence and a characteristic absorption spectrum.

Neo-Darwinism Darwin's theory of evolution, but rejecting any inheritance of acquired characters.

Neopine $C_{18}H_{21}O_3$, needles, m.p. 127°C; an alkaloid of opium, isomeric with codeine. The hydrobromide crystallizes as prisms, m.p. 282-283°C.

Neoplasm, neoplastic A tumour, tumorous.

Neoteny Sexual maturity occurring in the larval period.

Nephelometry The measurement of the concentration of dispersed particles, based on the measurement of the intensity of the Tyndall cone.

Nephron Any of the thousands of microscopic units of the vertebrate kidney in which urine is formed.

Nerol A terpene alcohol present in neroli, petit-grain and bergamot oils, resembles geraniol.

Nerolidol A sesquiterpene alcohol found in neroli oil and Peru balsam.

Nerve Band of *nervous system* tissue that transmits electrical impulses between parts of the nervous system and the rest of the body.

Nervone m.p. 180°C; a cerebroside, whose fatty acid is nervonic acid.

Nervonic acid $C_{24}H_{46}O_2$; cis-14-tetracosanic acid; an acid found associated with the sphingomyelins and in the cerebroside nervon, $CH_3(CH_2)_7CH{=}CH(CH_2)_{13}COOH$.

Nervous system System in animals that receives, processes, and interprets external stimuli and transmits signals to the rest of the body; in vertebrates, the nervous system consists of the brain, spinal cord, and *nerves.*

Nervous system Living organisms maintain themselves in an equilibrium state which slowly shifts with time (growth, etc.). But from moment to moment they are called upon to react to environmental changes which act upon them. The external event constitutes a stimulus, usually one involving but little energy, which evokes from the organism some active response, commonly involving much more energy than was in the stimulus. This sequence involves, therefore, amplification or "trigger action" by the organism; and, further, the energy released is not squandered but is directed to certain uses. The response is such as to counteract the stimulus, to maintain the equilibrium status, and so to adapt the organism. Such "adaptive amplification" is a universal property of organisms.

Protoplasm itself is irritable, it responds to stimuli, but inefficiently and sluggishly. Relatively strong stimuli are required to disturb it, the excitation finally set up does not spread far or fast from the stimulated region, and the response is slow and inexpert. The slow formation of a pseudopod by an amoeba, on the surface opposite a portion poked with a needle, is a fair example; the drooping of a burned leaf of the sensitive plant is another. Sensitivity, speed, and delicacy of behavior are greatly enhanced in the more evolved animals when special organs for adaptive amplification have appeared.

Sense organs, or receptors -special cells or cell groups each particularly sensitive to one type of stimulus—enable the animal to become aware of and to respond to more and feebler environmental changes. Minute amounts of light excite the retina of the eye, only intense light affects the amoeba. And the eye discriminates light intensities, colors, and directions as the undifferentiated amoeba does not. Similar relations hold for the other sense receptors, for sound, touch, temperature, chemicals, etc. In like fashion, other cell groups, the effectors, have become specialized for response or action. The muscles produce movement which is more rapid, more powerful, more graded, and more finely controlled than are the amoeba's movements. The glands make more specific and often more complex chemicals and release them at particular times and places. And other special effectors have been invented—electric shockers, light flashes, and the like.

Most important of these new organs for enhancing the behavioral capacities of animals is the nervous system. This serves to conduct, with great rapidity, the excitation set up at receptors and ultimately to deliver it to effectors; and it serves the more difficult function of sorting and directing and relaying these exciting messages so that the appropriate effectors are appropriately activated for each of the various stimuli which demands attention. The nervous system, then, serves as conductor, mainly via its long nerve strands and paths which connect to all parts of the body; and as coordinator or integrator, mainly via the nerve cells and fibers and their interconnections which constitute the mass of the central nervous system itself. Nearly all of the ten billion odd nerve cells, or neurones, of the body are gathered in the long spinal cord and its swollen upper end — the central nervous system which lies in the backbone and fills the cavity of the skull. Only a few neurones, gathered into small knots, or ganglia, are scattered among the other body tissues as parts of a subordinate, largely self-regulating (autonomic) nervous system, which helps control visceral activity—as the size of blood vessels, the movements of the digestive tube, secretion of saliva, the diameter of the pupil of the eye, the speed of the heart, and the like.

Each neurone has long extensions —branching dendrites which carry nerve impulses to the cell body and a mostly unbranched axon which carries away impulses, often for half the length of the body. These branches connect various neurones, an axon of one, (or more often of many) meeting, synapsing with, the dendrite of another (or many others). Receptors connect with the dendrites of sensory neurones, which run in the peripheral nerves; their axons connect with other neurones, nearby or in far portions of the central nervous system; soon or late, connections are made with motor neurones whose axons are again peripheral nerves and run to one or another muscle. This simple pathway—from receptor, through sensory, intermediate, and motor neurones, to effector—is the unit of neural function, the reflex arc.

Since impulses travel from neurone to neurone mostly along these anatomical tracks, the connections from each nerve are unique to it. The whole nervous system does not function as a single unit but as a coordinated group of units each with its particular duties. Thus simple reflexes —as pulling the foot or finger away from an injurious stimulus — are completed at whatever level of the spinal cord the relevant nerves connect. A dog with the lower cord cut from the remaining parts of the nervous system will still execute leg reflexes of pulling from an injury, scratching an irritation, even of stepping. But it will have no sensory experience of pain with the injury nor be able to move the legs voluntarily. The paths carrying sensory messages to regions of the upper brain, where they are interpreted, or motor messages from other regions of the brain, where "voluntary" motion is initiated, have been interrupted.

Anatomical, physiological (such as the above), electrical, and other studies have elucidated in great detail the myriad interconnections within the nervous system and the special functions of each portion, as well as the general role of the whole. Thus, simple limb and body reflexes depend mostly on the spinal cord. Head reflexes and the control of many visceral actions (breathing, salivation, vomiting, regulation of blood pressure and heart rate, etc.) depend on the medulla oblongata, an oblong expansion of the upper cord. Above this, the midbrain is especially concerned with the control of posture and balance; injury here may make a man or animal unable to stand or walk smoothly. Still higher, the top of the older nervous system, is the thalamus, which largely controls the autonomic nervous system and is concerned wtih emotional expression, as of rage, fear, and pain.

The newer brain regions are the cerebellum, above the midbrain, and the cerebrum, above the thalamus. These have their neurone cell-bodies (which constitute the grey matter; the fibers make the white matter) on or near the surface, in an outer layer or cortex. And these cortexes have increased in relative and absolute size in the "higher" animals from mammals through man. The cerebellum coordinates voluntary movements and is larger as motor skills increase. The human cerebral cortex is as large as the remainder of the nervous system, and it is further especially subdivided into many regions with special functions —seeing, hearing, talking, etc. Yet somehow it continues to act as a single whole, for on its functioning depends man's mind—his reasoning, remembering, and abstracting. It enables his to conceive a future and to direct his efforts towards one of his preference.

Nessler reagent for ammonia Reagent—a solution of 10 gm. of mercuric iodide and 5 gin. potassium iodide in 50 cc. of water to which is added a solution of 20 gm. potassium hydroxide in 50 cc. water. A yellow to brown-red precipitate is obtained with ammonia or ammonium salt solutions.

Nested pcr A very sensitive method for amplfication of DNA, which takes part of the product of a single PCR reaction (after 30-35 cycles), and subjects it to a new round of PCR using a different set of PCR primers which are nested within the region flanked by the original primer.

Net energy That part of the food energy which appears in the animal's products (milk, meat, eggs, wool, work). Metabolizable energy minus calorigenic effect of food.

Neubauer's theory of tyrosine metabolism Tyrosine metabolism takes two paths; in one p-hydroxyphenyl pyruvic acid is formed, and in the other 2,5-dihydroxyphenyl alanine; both then form 2,5-dihydroxyphenyl pyruvic acid and then homogentisic acid; which then is decomposed to acetoacetic acid. Strongly ketogenic.

Neuberg ester Hexosemonophosphoric acid; fructofuranose-6-monophosphate; present in muscle.

Neuberg reaction for pyruvic acid, succinic acid and alanine, arginine, betaine, etc A pine shaving moistened with hydrochloric acid turns red when heated with ammonium pyruvate, due to evolution of pyrrole. Ammonium succinate, heated with zinc dust, also gives the pyrrole reaction. The same reaction is given by the following on heating Alanine, asparagine, betaine, cystine, diamonosuccinic acid, glycocoll, histidine, lipsine, leucine, sarcosine, taurine and tyrosine.

Neuberg reagent for aliphatic alcohols and amino acids Alcohols and amino acids form crystalline compounds of high molecular weight with á-naphthylisocyanate. The amino acids may be estimated quantitatively.

Neuberg test for glyoxalic acid A little naphthoresorcinol is added to 2 cc. of test solution and, after the addition of 5 cc. concentrated hydrochloric acid, the solution is heated to boiling. After cooling and diluting with water, the solution is shaken with ether; a violet or deep red color indicates glyoxalic acid.

Neufeld phenomenon The bactericidal and lytic effect of bile salts for certain streptococci, pneumococci, staphylococci, etc.

Neurine Trimethylvinyl ammonium hydroxide; present in brain and other products; product of putrefaction of lecithin; forms a trihydrate and is a strong base.

Neurogen A term which may be used for the primary evocator and all other chemical substances which bring about neural induction in vertebrates (cf. Carcinogen, Oestrogen, etc.). Not to be confused with adjectives for muscular contractions, said to be neurogenic or aneurogenic (myogenic) according as to whether they require nervous stimulus or not.

Neurohumors Local nerve hormones, e.g. sympathin.

Neurophysiology Neurophysiology is that branch of physiology concerned with the study of the biological functions of the nervous system. Although a tremendous amount of detailed and experimentally proven factual material has been brought to light (most of it in the last half-century) our exact knowledge regarding neural organization and functioning must be considered inadequate and relatively meagre. The following account will be concerned with some of the more general aspects of this branch of physiology.

The transmission of excitation or the conduction of nerve impulses forms the basis of all neural activity. Consequently, the nature of impulses in single nerve fibers has been the subject of much investigation. The membrane hypothesis, which is now generally accepted, supposes that depolarization of the fiber surface is, in fact, the nerve impulse. According to this theory, the polarized membrane breaks down locally under the action of the stimulus and so establishes a potential difference between the active and inactive regions. Local currents thus set up, serve as the stimulus which makes neighboring segments active. The depolarization is instantaneous while restoration of the surface requires a certain time. During the first part of the period of restoration, the surface is unable to break down again, it is absolutely refractory, and thus the depolarization spreads down the fiber. Among the many aspects of neural activity which are of primary interest in this connection may be listed the action potential of nerve; the refractory period; the all-ornone nature of conduction; antidromic conduction; velocity of conduction; metabolism of the nerve fiber; the respiration of nerve; chemical changes in nerve; the transmission of excitation at the neuromyal junction.

The simplest reaction of the nervous system is the spinal reflex, a reaction evoked by afferent impulses entering the spinal cord at the level from which the motor impulses emerge. Because of the enormous complexity of the central pathways, analysis demands that central conduction be studied in as simple a form as possible. For this reason, much attention has been devoted to the spinal reflex arcs. Each functional unit of the central nervous system is a cell which has no direct anatomical continuity with any other functionally related cell, action of one cell upon another being communicated solely through discontinuous interfaces. Foster and Sherrington introduced the term "synapse" (from the Greek, to clasp) to describe the relationship between contiguous neurons. In the reflex arc the passage of neural impulses is in one direction only, from the afferent or sensory neurons to the efferent or motor neuron. This is because the synapse possesses a valve-like action which imposes a polarity on the direction of conduction. The reflex arc has been utilized in the study of the temporal and spatial summation of inadequate stimuli, the overlapping central distribution of afferent pathways resulting in the phenomena of convergence and occlusion, the principle of reciprocal innervation, reflex discharge and reflex tetanus, the nature of central excitation and inhibition, synaptic transmission and graded synaptic resistance.

The early phylogenetic development of the nervous system led to a central organ arranged in segments. In the lower worms each segmental ganglion exhibits a high degree of autonomy, so that most of its segmental activities are undisturbed by separating it from adjacent ganglia. The process of encephalization, which took place as the invertebrates developed and which occurred in an even more striking manner during vertebrate evolution, resulted in an ever increasing dependence of the aboral segments on the head ganglion or brain. With the increase in size and complexity of the brain there has been a tendency of this organ to take over, elaborate, and control functions which in lower forms are governed almost entirely by the individual segments. But even the spinal segments of man retain some small portion of the primitive autonomy and when the spinal cord of a carnivore is completely separated from the brain it remains capable of mediating simple items of behavior which are but slightly removed in character from similar movements evocable in the intact animal. When the lower parts of the brain stem are left connected with the cord, more complex activity is possible. If only the cerebral cortex, the most recent and most complex product of encephalization, is removed in a cat or dog, the behavior deficit is even less. Study of these behavioral activities indicates that each level is organized to govern specific functions rather than to control anatomical units. In other words, the central nervous system is organized on the basis of what may be termed "levels of function."

Hughlings Jackson, in 1884, suggested that inhibitory impulses stemming from the higher or more rostral (and phylogenetically more recent) parts of the nervous system prevent the lower and more primitive levels from dominating behavior. Removal of the higher, more discriminative levels results in the release of certain more primitive and less integrated functions controlled by the more archaic portions of the cerebrospinal axis. In working with decerebrate animals, Sherrington found that this concept of Jackson's was eminently sound. In the case of the decerebrate preparation the most striking release phenomenon is the greatly exaggerated tone which is present in the extensor muscles due to the release of a vestibulo-spinal mechanism from higher extrapyramidal control. Because reflex responses can be predicted and easily evoked in the unanesthetized decerebrate animal, this preparation in the acute state enabled Sherrington to formulate many basic concepts of reflex action and integration. Recently Macht and Bard have prepared decerebrate cats which they were able to maintain and study for periods ranging up to five months after operation.

Surgical ablation is one of the most extensively used experimental methods in neurophysiology. Analysis of the capacities and abnormalities of different animals in the spinal, decerebrate and decorticate states has been responsible for much of our knowledge concerning the integration and working of different functional levels.

Postural coordination and its central control have received much attention since Sherrington's early work on decerebrate rigidity, Liddell and Sherrington's investigations of the stretch or myotatic reflex, and the studies of Magnus and De Kleyn on attitudinal and righting reactions. All phases of the postural mechanism are influenced by the cerebral cortex although many basic reactions are integrated in the medulla oblongata. The righting reactions are for the most part dependent upon the integrity of the midbrain, although recent experimental work indicates that some of these reflexes are elaborated by cats from which all neural tissue rostral to the pons has been extirpated. Tonic labyrinthine reflexes include the acceleratory reflexes and the positional reflexes. The former arise in the semicircular canals and are responsible for a group of special reactions involving the eye (vestibular nystagmus) and the skeletal muscles. The positional reflexes arise from the otolithic maculae of the labyrinth, and have to do principally with the righting reactions.

Studies of sensation or the mechanism involved in the transmission of afferent impulses from specific receptors of the body to certain way stations in the brain where these impulses are translated into the "awareness" of sensory experience, constitute an important field of neurophysiology. The great sensory ganglion of the brain stem is the dorsal thalamus; all somatic sensory impulses must pass through this portion of the diencephalon before being forwarded to the cerebral cortex. Extensive clinical and experimental investigations of the role played by the thalamus in sensation have been carried out by Head and Holmes, Dusser De Barenne and others. Impulses from the dorsal thalamus pass to the somatic sensory area of the cortex. The outstanding feature of cortical sensibility is the recognition of fine detail in sensation. Various methods have been employed in attempts to determine the position and extent of the somesthetic area of the cortex. The effects of cortical lesions, electrical stimulation of the intact brain in unanesthetized patients, and additional clinical studies by Cushing, Foerster and others indicate the existence of a well-defined segmental localization of sensory function in the parietal lobes. A far more detailed picture of cortical sensory representation has been obtained as a result of the discovery by Marshall, Woolsey and Bard that the application of a discrete tactile stimulus to a cutaneous area produces in the cortex of the anesthetized animal a well-localized potential wave. These authors have been able to demonstrate very accurately the representation of tactile sensibility in the monkey's cortex as revealed by the cortical potentials set up on stimulation of touch receptors in different parts of the contralateral half of the body.

The autonomic nervous system is the efferent pathway to the visceral effectors. The distinction between the autonomic and the cerebrospinal systems is a functional rather than an anatomical one. Anatomically, the two are continuous rather than distinct; the autonomic being a visceral extension of the cerebrospinal. Functionally, they are different in that the cerebrospinal controls skeletal muscle, while the autonomic innervates visceral effectors (glands, smooth muscle, and cardiac muscle). Skeletal muscle fibers are connected with the central nervous system by motor neurons which have their cell bodies in the brain or spinal cord, and there is only one nerve cell between the cerebrospinal axis and the effector. The innervation of the viscera, on the other hand, is characterized by the fact that the path from the central axis to the effector is broken by a synapse and the axon directly in contact with the effector is part of a neuron which lies entirely outside the central nervous system. The cranial and sacral outflows of the autonomic comprise the parasympathetic system, while the thoraco-lumbar division is usually called the sympathetic system. Most of the visceral tissues possess a double nerve supply, parasympathetic and sympathetic. Wherever this occurs, the two systems, as a rule, exert opposed or antagonistic effects.

Cannon and his coworkers have emphasized the fact that the sympathetic system tends to go into action as a whole in situations which require an emergency response. The action of the sympathetic leads to an increase of the available potential energy by increasing the amount of sugar in the blood. Unnecessary activity of the digestive system is stopped. Acceleration of the heart and peripheral vasoconstriction produces a more rapid circulation through active tissue. Respiratory processes are aided by dilation of the bronchioles. In other words, there is an integrated body of activity directed toward the mobilization of energy, protection of the organism, and maintenance of the milieu intérieur in the face of an emergency situation. Conditions under which the sympathetic acts include pain, asphyxia, emotional excitement, exposure to cold, hypoglycemia, and vigorous muscular exercise. Neural stimulation of the adrenal medulla causes the secretion of a chemical agent, adrenin, capable of inducing practically all of the changes which occur when the entire thoraco-lumbar outflow is active. Cannon and others have demonstrated that a sympathico-mimetic substance is given off into the blood stream when impulses in postganglionic sympathetic fibers induce a change in effector cells. This substance, to which has been given the name, sympathin, has properties similar to adrenin and acts in conjunction with it.

The parasympathetic division of the autonomic system does not go into action as a whole. Its restricted distribution enables it to exert effects upon single organs without affecting the other viscera. Likewise, there is an absence of any parasympathetic hormone analogous to the adrenin of the sympathetic system which is capable of acting extensively upon organs innervated by this system. While there is good evidence that a chemical agent, acetylcholine, is the transmitter of excitation and inhibition at parasympathetic postganglionic endings, this substance is not capable of exerting widespread effects because of its very rapid hydrolysis through the agency of cholinesterase.

In this brief account, no mention has been made of the functions of the cerebellum, basal ganglia, special senses, the motor areas of the cerebral cortex, the pyramidal and extrapyramidal motor system, the hypothalamus or the electroencephalogram. For a detailed account of these and other subjects within the province of neurophysiology, the reader is referred to the sources listed below.The most challenging problem presented by the nervous system is the explanation and elucidation of mental phenomena. However, it must be admitted that the neurophysiologist has made little progress in this direction. Analysis of mental experience in terms of reflex action and neural organization is still completely unsatisfactory. The principal advances in this connection have been made by the physiological psychologist through the use of conditioned reflex experiments and other special methods.

Neurotoxin Poison which interferes with nerve function, usually by affecting the flow of ions through the cell membrane.

Neurotransmitter Biochemical that transmits signals from a nerve cell to another nerve cell or to other parts of the body.

Neurula The stage in development at which the neural axis is fully formed and histogenesis rapidly proceeding.

Neutral medium A situation for the culture of young undetermined parts of embryos such that they are not exposed to the action of unknown inductors (e.g. Holtfreter salt solution, as against chorioallantoic grafting, yolk-sac implantations, or culture in eye- or body-cavity).

Neutralization Elimination of a substance's basicity or acidity by the addition of an acid or base.

Neutrino Any of three subatomic particles (tau neutrino, muon neutrino, and electron neutrino) that have no electric charge and a minuscule amount of mass.

Neutron Subatomic particle with no electric charge located in an atom's nucleus; and it is made up of three *quarks*.

Neutron rays, biological effects of The neutron of mass 1 but possessing no charge, can be ejected with varying amounts of energy in cyclotron disintegrations of matter. It has biological effects comparable but also different from those produced by X-rays. It operates chiefly by the production of ionization in tissue. In the last decade techniques of generation and the measurement of the quantity of radiation have been developed so that one might estimate tissue doses. Retardation of growth of normal and malignant tissues have been noted. Bacteria have been killed and bacteriophage has been inactivated. In every case a quantitative difference exists, principally in favor of dosing with neutrons. Therapeutic results in cancer cases are encouraging when fast neutrons are used. Slow neutrons exercise an enhanced effect in the presence of boron compounds which have been introduced into neoplastic tissue.

Neutron star Small, dense star made up primarily of *neutrons*

Newton Unit of *force* defined by the amount of force required to accelerate a 1-kilogram object by 1 meter per second per second

Niche The multidimensional resource space of a species; its ecological requirements.

Nick translation A method for incorporating radioactive isotopes (typically 32P) into a piece of DNA. The DNA is randomly nicked by DNase I, and then starting from those nicks DNA polymerase I digests and then replaces a stretch of DNA. Radiolabeled precursor nucleotide triphosphates can thus be incorporated.

Nick In duplex DNA, this refers to the absence of a phosphodiester bond between two adjacent nucleotides on one strand.

Nicol prism A specially constructed piece of calcite or Iceland spar ($CaCO_3$), which when ordinary light is passed through, causes the emitted light to vibrate in only one plane; i.e. to be plane polarized.

Nicotimine $C_{10} H_{14} N_2$; an alkaloid found with nicotine. Colorless alkaline oil, b.p. 250-255°C. Aurichloride forms yellow leaflets, m.p. 182-185°C.

Nicotine $C_{10} H_{10} N_2$; 1-methyl-2-(3'-pyridyl)pyrrolidine, or pyridyl-N-methylpyrrolidine; chief alkaloid of tobacco; colorless oil, b.p. 246.1°C 730.5 mm. Causes transient stimulation, then paralysis of autonomic ganglia. Largest use is as an insecticide. Fatal dose 1-4 drops, effect usually rapid; chemical antidotes, tannic acid, hydrogen peroxide, intracardiac, injection of epinephrine.

Nicotinic acid $C_6 H_{502} N$, m-pyridine carboxylic acid; precursor of pellagra preventing factor of B complex and cures dog disease, "black tongue."

Nicotoine An alkaloid, $C_8 H_{11} N$, from Turkish tobacco; b.p., 208°C.

Nicotyrine $C_{10} H_{10} N_2$; alkaloid of tobacco. Oil b.p. 280-281°C. On dehydration yields nicotine, which it resembles physiologically.

Nidation The first stage (8 to 10 days) of the establishment of a fertilized ovum in the walls of the uterus.

Nidicolous Birds which hatch in a relatively undeveloped stage and hence have to remain in the nest highly dependent on their parents for some considerable time.

Nidifugous Birds which hatch in a relatively highly developed stage, and can therefore leave the nest at once or almost so.

Ninhydrin reaction Alpha amino acids or compounds containing the á-amino group give a color ranging from deep blue to violet or even to red when treated with ninhydrin solution. In a 110000 solution, glycocoll and alanine give a blue color; tyrosine gives the color at 15000.

Nirvanol 5, 5 - phenylethyl - hydantoin, m.p. 199°C, a hypnotic and sedative.

Nisinic acid A 24 carbon unsaturated fatty acid with 6 double bonds, found in cod liver, herring, shark liver oils.

Nitratase (bact Coli) nitrate reductase An enzyme of the extracts of Bact. coli which catalyzes the reduction of nitrate to nitrite; not well known.

Nitrogen balance The ratio of nitrogen (protein) intake and output. Positive N balance indicates retention of protein, and tissue building; negative N balance indicates loss of protein, and tissue destruction.

Nitrogen cycle The cycling of various forms of biologically available nitrogen (N_2, nitrate, nitrite, ammonia, etc.) through the plant, animal, and microbial worlds, and through the atmosphere and geosphere.

Nitrogen equilibrium The condition of having equal nitrogen (protein) intake and output. Indicates equilibrium of tissue building and destruction.

Nitrogen fixation Conversion of atmospheric nitrogen (N_2) into a reduced, biologically available form by nitrogen-fixing organisms, often in symbiosis with legumes.

Nitrogenase complex A system of enzymes capable of anaerobic reduction of atmospheric nitrogen (N_2) to ammonia in the presence of ATP.

Nitrogenous base A nitrogen- containing molecule having the chemical properties of a base.

Nonconservative Substitution A mutation which results in the substitution of one amino acid within a polypeptide chain with an amino acid belonging to a different polarity/charge group.

Nondisjunction Failure of the two homologous chromosomes of a pair to go to opposite poles at the first meiotic division; as a result one daughter cell has both chromosomes and the other neither.

Nonose Any monosaccharide with the formula $C_9 H_{18} O_9$. There are none naturally occurring.

Nonsense Mutation A change in the sequence of a nucleic acid that causes a nonsense (stop or termination) codon to replace a codon representing an amino acid.

Nontranslated RNA (NTR) The segments located at the 5' and 3' ends of a mRNA molecule which do not encode any part of the polyprotein; may contain important translational control elements.

Norgestrel A synthetic gestragen. It is used inoral contracceptives, and it is the most ptent of the orraly active gestagens.

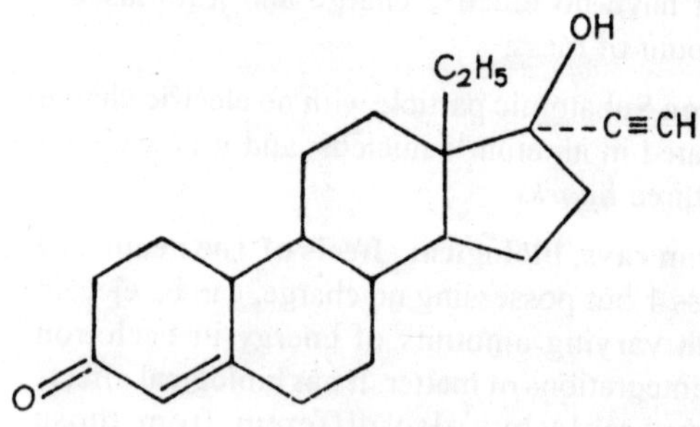

Fig. Norgestrel

Normal (1) A measure of concentration of a solution, equal to one gram equivalent of solute per liter of solution.

(2) In organic chemistry the designation applied to straight carbon chain aliphatic compounds.

Nornicotine $C_9 H_{12} N_2$; 2 - (3 - pyridyl) - pyrrolidine; an alkaloid of tobacco. Properties vary depending on origin; b.p. about 139-140°C. Physiologically similar to nicotine.

Northern blot A technique for analyzing mixtures of RNA by transfer of size-separated RNA fragments to a synthetic membrane, whereby the presence and rough size of one particular type of RNA (usually an mRNA) can be ascertained. After Dr. E. M. Southern invented the Southern blot, it was adapted to RNA and named the "Northern" blot.

Nuclear Envelope A double membrane that is the

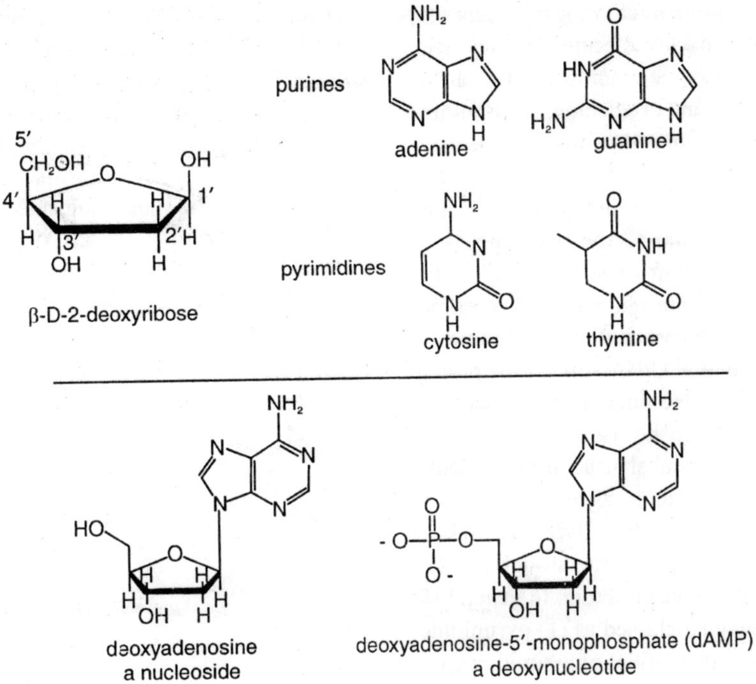

Fig. Nucleic acids monomers

outermost portion of a cell nucleus.

Nuclear inductor A morphogenetic stimulating substance given off by the nuclear material bearing the factors of heredity (genes and chromosomes), usually operating only within the cell in question, but in some cases more diffusible.

Nuclear run-on A method used to estimate the relative rate of transcription of a given gene, as opposed to the steady-state level of the mRNA transcript (which is influenced not just by transcription rates, but by the stability of the RNA). This technique is based on the assumption that a highly-transcribed gene should have more molecules of RNA polymerase bound to it than will the same gene in a less-active state. If properly prepared, isolated nuclei will continue to transcribe genes and incorporate 32P into RNA, but only in those transcripts that were in progress at the time the nuclei were isolated. Once the polymerase molecules complete the transcript they have in progress, they should not be able to re-initiate transcription. If that is true, then the amount of radiolabel incorporated into a specific type of mRNA is theoretically proportional to the number of RNA polymerase complexes present on that gene at the time of isolation. A very difficult technique, rarely applied appropriately from what I understand.

Nuclease An enzyme which degrades nucleic acids. A nuclease can be DNA-specific (a DNase), RNA-specific (RNase) or non-specific. It may act only on single stranded nucleic acids, or only on double-stranded nucleic acids, or it may be non-specific with respect to strandedness. A nuclease may degrade only from an end (an exonuclease), or may be able to start in the middle of a strand (an endonuclease). To further complicate matters, many enzymes have multiple functions; for example, Bal31 has a 3'-exonuclease activity on double-stranded DNA, and an endonuclease activity specific for single-stranded DNA or RNA.

Nucleases Enzymes which hydrolyze nucleic acids.

Nucleic acids The prosthetic group of the

nucleoproteins. Four nucleotides (which see) combine to make up one molecule of nucleic acid. On hydrolysis all nucleic acids yield 2 purine bases, usually guanine or adenine, 2 pyrimidine bases, usually cytosine, thymine, uracil, or 5-methyl cytosine; phosphoric acid, and a carbohydrate, ribose or desoxyribose. The nucleic acids are vital to the life process. A decrease in the amount of nucleic acid in the body is associated with old age. Chromosomes are nucleic acids plus protamin, the combination existing as chromatin. On dissolving the protein from chromosomes the nucleic acid remains in bands that appear to be associated with the genes. The exact chemical structure is in doubt, but probably the pyrimidines are connected to the carbohydrates through a glycosidal linkage at N3, and the purines similarly through N7, The sugars have the ä-oxide structure. In general the nucleic acids may be classed as (1) pyrimidines or amino, and oxy derivatives of pyrimidines, e.g. cytosine, alloxan; (2) purines or amino and oxy derivatives of purine, e.g. adenosine, xanthine, caffeine, pterins; (3) nucleosides or purine or pyrimidine N-glycosides, e.g. adenosine, uridine, vicine; (4) nucleotides or phosphoric acid esters of nucleosides, e.g. adenylic acids, cozymase, partly depolymerized nucleic acids and nucleic acids in increasing order of molecular complexity.

Nuclein A combination of protein and nucleic acid, the protein being in most cases of somewhat smaller molecular weight than the protein in nucleoprotein, from which nuclein may be derived by partial hydrolysis.

Nucleoproteins In the American classification of proteins, the nucleoproteins consist of one or more protein molecules with nucleic acid. Example nuclein.

Nucleoside A term refering to the combination of adenine, cytosine, guanine, or thymine with a ribose or 2-deoxyribose sugar moiety. A nucleoside is not phosphorylated.

Nucleotide A building block of DNA or RNA consisting of a nitrogenous base (adenine, guanine, thymine, or cytosine in DNA; adenine, guanine, uracil, or cytosine in RNA), a phosphate molecule, and a sugar molecule (deoxyribose in DNA and ribose in RNA). Thousands of nucleotides are linked to form a DNA or RNA molecule.

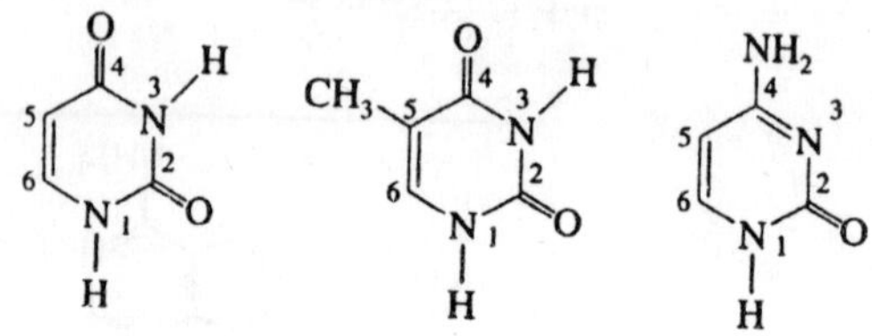

uracil *found in RNA* — thymine *found in DNA* — cytosine *found in both RNA and DNA*

pyrimidine bases

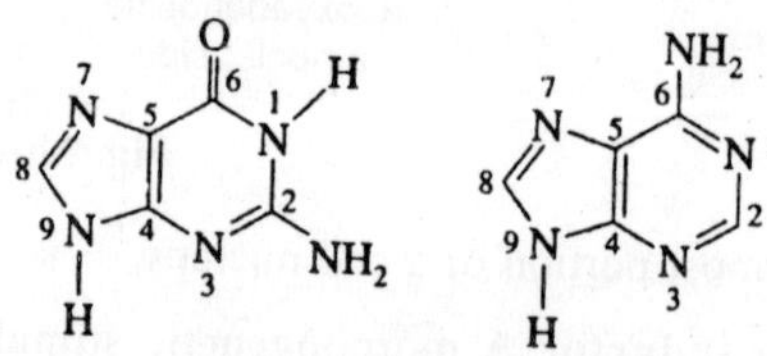

Fig. Nucleotide

Nucleus membrane-bound cellular structure that holds the *chromosomes*, positively charged core of an atom, containing the *neutrons* and *protons* nucleus In eukaryotes, a membrane-bounded organelle that contains chromosomes.

Nutriacholic acid $C_{24}H_{40}O_5$, m.p. 198°C; a bile acid found in the beaver.

Nux vomica Dried ripe seeds of Strychnos nux vomica containing alkaloids strychnine and brucine; a heart tonic and is also used in dyspepsia.

Nylander-almen test A test for reducing compounds, whereby alkaline bismuth subnitrate is reduced to metallic bismuth.

O

Occluded front Region of weather in which a cold front overtakes a warm front, pushing the warm air upward and causing clouds and rain.

Occultation Temporary disappearance of a star from sight because of the passage of a celestial body across its path.

Ocean ridge Mountain range on the floor of the ocean formed by two tectonic *plates* moving apart and magma rising to the surface.

Ocean trench Deep gorge in the ocean floor formed by the movement of one tectonic *plate* beneath another.

Ochratoxins Mycotoxins produced by Aspergillus ochraceus during food spoilage. In rats and mice these cause pronounced liver damage. The three known O, are designated A, B and C.

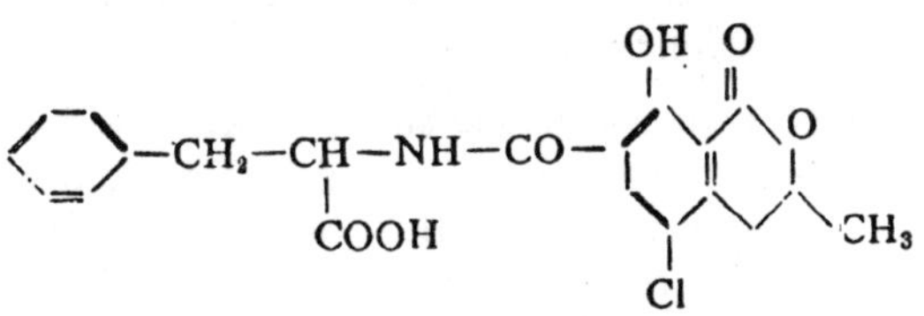

Fig. Ochratoxin

Oconhydrine $C_8H_{17}ON$; 5-hydroxy-2-propylpyridine; an alkaloid of hemlock; needles from ether, mp. 105-6°C; poisonous, causing paralysis of the motor nerve endings, respiratory paralysis, and death.

Octose Any monosaccharide with the formula. $C_8H_{16}O_8$. There are none naturally occurring.

Oenin A glycoside of the skin of purple grapes, consisting of glucose and oenidin.

Oestrogen A chemical substance with activity analogous to that of the female sex hormones in the oestrous cycle.

Offer test for uric acid Uric acid reduces an alkaline solution of phosphomolybdic acid, even in the cold, forming a blue color. A deep blue precipitate of hexagonal prisms is obtained if 0.05% or more uric acid is present.

Okuda-nishijima reaction for cysteine A small quantity of cysteine solution is treated with a few drops of faintly acid zinc chloride solution and a freshly prepared aqeous sodium nitroprusside solution. Ammonia or sodium hydroxide is added until zinc hydroxide precipitates; a ruby-red color, stable for several hours, is formed.

Oleandrin Glycoside composed of digitalose and digitaligenin.

Oleic acid An unsaturated liquid fatty acid, $CH_3(CH_2)_7CH=CH(CH_2)_7COOH$; found in most fats and oils; m.p. 14°C C.

Oleoresins The sticky exudate of pine or fir trees, consisting of a solution of resin acids in essential oils. On standing the essential oils evaporate, leaving a hard glass-like resin.

Oligodeoxyribonucleotide A short, single-stranded DNA molecule, generally 15-50 nucleotides in length, which may be used as a primer or a hybridization probe. Oligodeoxyribonucleotides are synthesized chemically under automated conditions.

Oligomenorrhea Abnormal prolongation of time intervals between menstruations. Etiology same as for Amenorrhea (which see).

Oncogene A gene in a tumor virus or in cancerous cells which, when transferred into other cells, can cause transformation (note that only certain cells are susceptible to transformation by any one oncogene). Functional oncogenes are not present in normal cells. A normal cell has many "proto-oncogenes" which serve normal functions, and which under the right circumstances can be activated to become oncogenes. The prefix "v-" indicates that a gene is derived from a virus, and is generally an oncogene (like *v-src*, *v-ras, v-myb*, etc).

Oncogenic transformation A change in the behavioral phenotype of a cell to one lacking in normal proliferative control, and often involving invasive characteristics.

Oncotic pressure The osmotic pressure of colloids, e.g. blood plasma has a pressure of 28 mm. Hg. One can also clone more complex organisms, with considerable difficulty. The much-publicized Scottish research that resulted in the sheep 'Dolly' exemplifies this approach.

Oocyte undeveloped female sex cell.

Oogenesis The origin of the female germ cell or egg.

Open cluster Aggregation of young stars.

Open reading frame That segment of a nucleic acid sequence that lies between two stop codons, when translated in a given reading frame. The presence of an open reading frame is necessary to encode a polypeptide sequence (or an exon thereof), but the presence of an open reading frame is not proof that a polypeptide sequence is encoded in that sequence.

Operon Genes, which are grouped together for coordinate regulation by the same regulator.

Opium The dried latex of unripe capsules of Papaver somniferum, contains numerous alkaloids, as morphine, codeine, narcotine; used as a narcotic, in cough medicines and to control diarrhea; laudanum is a tincture of 1% morphine content; paregoric in a twentieth the strength of laudanum.

Optical activity The ability possessed by certain chemicals to rotate the plane of polarized light when the latter is passed through the substance.

Optical rotation The number of degrees of rotation that an optically active substance produces in the Plane of polarized light for a given length of passage.

Optochin Ethylhydrocupreine, made by hydrogenating quinine, then demethylating and ethylating; pneumococcicidal even in dilutions of I to 400000 in serum.

Orbit Elliptical path of one celestial body around another.

Orbital, atomic One of variously shaped regions surrounding the *nucleus* of an atom within which *electrons* are found.

Orcinol Orcin, 3,5-dihydroxytoluene, 5methyl-resorcinol; mp., 58°C (107°C when anhydr.); b.p. 290°C; occurs in many species of lichens and is used as reagent in test for sugars.

Orcinol-hydrochloric acid test A test for pentoses depending on, the formation of blue green or violet colors when a sugar is heated with HCl and orcinol.

Organ Structure of the body that has a specific function or set of functions.

Organelle Membrane-bound structure within the cell of a *eukaryote* having a specialized set of functions organelles Membrane-bounded structures found in eukaryotic cells which contain enzymes and other components required for specialized cell functions.

Organic compound *Compound* that contains the element carbon.

Organiser (*cont.*) pendent upon the prior function of the primary organiser, which therefore *seems* to be more manifold in its inductive effects. The chemical substance emitted by an organiser is called an *Evocator*.

Organiser A living part of an embryo which exerts a morphogenetic stimulus upon another part or parts, bringing about their determination and the following histological and morphological differentiation. The organiser which acts first in development is known as the *Primary* or *First-Grade* Organiser; those acting at successive stages are known as *Secondary (Second-grade)* and *Tertiary (Third-grade)* Organisers. The appearance of the latter is generally de.

Organizer (spemann's) A chemical substance, probably of the sterol type, in the gastrula, which can organize primitive cells to develop according to the development pattern of its source.

Organogenesis The development of morphological differentiation.

Organogenesis Formation of a female gamete, from a germ cell to a mature haploid ovum.

Origin of replication A site at which DNA replication is initiated. There is only one in bacterial chromosomes, but numerous origins in eukaryotic chromosomal DNA. (Abbr. "ori").

Origin recognition complex An organization of protein factors assembled at an origin of replication for the purpose of initiating replication.

origin The nucleotide sequence or site in DNA where DNA replication is initiated.

Orlow reaction for lecithin A mixture of alcoholic solutions of lecithin and alloxan gives a pink to red color and a red precipitate is produced.

Ornithine cycle The formation of urea in a cycle involving the addition of 2 molecules of ammonia and 1 of carbon dioxide and the abstraction of water from ornithine to produce citrulline, the addition of ammonia and the abstraction of water to citrulline to produce arginine, and the decomposition of arginine by arginase to form urea and to reform ornithine for another cycle; this amounts to the synthesis of urea from ammonia and carbon dioxide; the site of action is the liver.

Ornithine $NH_2CH_2(CH_2)_2CHNH_2COOH$; a basic amino acid; diamino valerianic acid; obtained from arginine by splitting off urea; found in the urine and excrement of fowls. On decomposition it gives rise to putrescine.

Ornithuric acid Ornithine dibenzoate; the detoxification product of benzoic acid by birds (analogous to the formation of hippuric acid by mammals).

Orosin A term suggested for associated coagulable serum protein systems.

Orotic acid 4-uracilcarboxylic acid, a pyrimidine found free in milk.

Orthocellulose A little used name for pure cellulose.

Orthogenesis Hypothesis that a rectilinear trend in evolution is due to a built-in finalistic principle.

Orthostatic proteinuria The appearance of protein in the urine of certain few individuals when they stand erect.

Orthotopic transplantation Transplantation between corresponding (identical) sites in such a way that the graft maintains its normal orientation.

Oryzanin A name originally applied to the crude thiamine chloride preparations, now sometimes used for the pure substance.

Osazone The reaction product of two molecules of phenylhydrazine with one molecule of sugar. The two molecules of phenylhydrazine add on to the carbonyl carbon and the one adjacent to it. Useful in the identification of the sugars.

Oscillation One cycle (to-and-fro movement) of a vibrating or swinging system.

Osmol or osm The amount of electrolyte equivalent in osmotic pressure to that of a molar or molal solution of nonionized solute.

Osmosis Movement of a fluid through a semipermeable membrane so as to equalize the *concentrations* of the solutions on each side of the membrane.

Osmotic pressure The pressure observed between solutions of different concentrations separated by a membrane permeable to the solvent but not the solute, due to the greater diffusion of solvent molecules from the dilute to the concentrated solution. The osmotic pressure of one mole of solute per liter is 22.4 atmospheres.

Osone A 1,2,-dicarbonyl derivative of sugars. Obtained by heating an osazone with fuming HCl, thus splitting off the two phenylhydrazine groups.

Ossein An albuminoid of the matrix of bone, apparently identical with collagen.

Osseomucoid A mucoid of the matrix of bone.

Osteasterol A sterol, $C_{29}H_{48}O$, isomeric with stigmasterol and possessing two double bonds; present in oysters and clams and probably replaces cholesterol in all the Lamellibranchiata.

Osteoclasts Bone-destroying cells which make way for rearrangement and growth.

Ouabain Acocantherin; $C_{29}H_{44}O_{12}$; m.p. 187°C; Gratus (G) Strophanthin; glycoside from Strophanthus gratus, composed of rhamnose and ouabagenin; used as cardiac stimulant.

Ovalbumin Egg albumin; m.w. about 40,000; obtained from egg white.

Ovary In female animals, the primary reproductive organ in which eggs form. In seed bearing plants, the portion of the carpal where eggs develop, fertilization takes place, and seeds mature. A mature ovary and sometimes other plant parts is a fruit.

Overhang A terminus of a duplex DNA molecule which has one or more unpaired nucleotides in one of the two strands (hence either a 3' or 5' overhang). Cleavage of DNA with many restriction endonucleases leaves such overhangs.

Oviduct Duct through which eggs travel from the ovary to the uterus. Formerly called the fallopian tubes.

Oviparity The bringing forth of young from eggs developing outside the maternal body.

Ovoviviparity The bringing forth of young from eggs developing within the maternal body, but having little or no connection with the maternal tissues.

Ovulation Release of a female sex cell from the *ovary*.

Ovulation During each turn of the menstrual cycle, the release of a secondary oocyte from the ovary.

Ovule Any of one or more structures that form on the inner wall of the ovary of seed bearing plants and that, at maturity, are seeds. An ovule contains the female gametophyte with its egg, surrounded by nutritive and protective tissues.

Oxalic acid m.p. 189.5°C., occurs in many plant cells free and as K acid salt or Ca salt in lichens and moulds.

Oxaluric acid The form in which part of oxalic acid is excreted in the urine.

Oxbow lake U-shaped lake located along a river; it is typically formed by flood water being separated from the river.

Oxidation chemical process by which an atom or molecule loses electrons, gains oxygen atoms, or loses hydrogen atoms, chemical reaction between the iron in rocks and the oxygen in water that causes weathering and yields a reddish deposit oxidation, The loss of electrons.

Oxidation-reduction potential The e.m.f. developable by a system capable of mutual oxidation and reduction. It is due to the flow of electrons from the oxidant to the reductant. Usually the potential is referred to an external standard electrode, as a standard hydrogen or calomel electrode.

Oxocyclo-desmotropy Tollen's formulation of sugar structure showing an oxygen ring to explain the masking of the ketonic oxygen.

Oxyacanthine An alkaloid, $C_{37}H_{40}N_2O_6$; m.p., 216-217°C; from root of berberis vulgaris; causes stomach pains and salivation; affects respiration and is said to cause mydriasis.

Oxycellulose The product of progressive action of oxidizing agents on a cellulose, like cotton.

Oxygen debt The partial asphyxia due to violent exercise or the like.

Oxygen poisoning A body fall in temperature of

rats produced by exposure to oxygen at 4 atmospheres pressure.

Oxygenase (bach) A system capable of converting oxygen into a peroxide (Bach).

Oxyhemoglobin The loose combination of hemoglobin with oxygen, which takes place in the lungs. The compound loses the oxygen to the tissues becoming reduced hemoglobin. The iron is in the ferrous state.

Oxyproteic acids A term sometimes used for sulfur-containing peptides.

Oxyreducase A component of the fermentation group of enzymes, that, in conjunction with cozymase causes a Cannizzaro reaction to take place with aldehydes.

Oxyrenin Some recent studies have shown that one of the causative factors in the genesis of "renal hypertension" might be a deviation of the metabolism of the aminophenols and phenolic amino acids. Schroeder, and Schroeder and Adams furnished experimental and clinical evidence showing that the alpha phenolase, "tyrosinase," inactivates the pressor amines, tyrosine, tyramine, adrenaline, and also to some extent renin under definite conditions, and is capable of reducing the blood pressure when injected in patients with "essential hypertension" due to hydronephrosis or to renal ischemia of the Goldblatt type. However the experiments of Bing and of Holtz and co-workers would rather suggest that the enzymatic destruction of the kidney pressor factor proceeds in the organism by way of a deamination reaction. In fact, angiotonin and renin are inactivated by the deaminase used in the experiments of the Croxattos. This enzyme differs from tyrosinase in a number of biochemical properties which preclude the possibility of the two possible reactions proceeding together on the same molecule. That both types of inactivation mechanism may play a role in the organism is suggested by the experiments of Holtz showing that in the presence of kidney tissue the phenolic amino acids (l-dopa, tyrosine, etc.) are decarboxylated under anaerobic conditions (as in kidney ischemia probably) to the state of hydroxytyramine or like hypertensor bodies, while their deamination to the state of dihydroxyphenyl acetic acid or like bodies is an aerobic reaction. Although Mason, Evers and Blalock found the oxygen consumption identical in the normal and in the ischemic kidney, Werle, Madlener and Herman found the deamination process Increased and the decarboxylation process reduced after ischemia of the kidney in situ. Contrary to this, however, the injection of l-dopa into an ischemic kidney in situ is followed by a strong hypertensive reaction similar to that produced by the presence of hydroxytyramine in the organism (Bing).3A possible relationship of the above deamination-decarboxylation reactions to the presence of blood pressure active substances in the mammalian kidney is suggested by the experiments of Hermann and Bacq showing that the enzymatic destruction oxidation (by a vegetable phenolase) of pressor bodies like adrenaline and tyramine proceeds, under suitable conditions, through the intermediate state of a strongly hypotensive substance (adrenoxine).That a similar behavior is exhibited also by renin has been recently demonstrated by Jonnard. The progressive oxidation of pure "renin E" by a vegetable phenolase has been carried out to the state of a heretofore unidentified hypotensive substance which has a prolonged effect upon the arterial blood pressure of dogs and cats. The reaction is somewhat accelerated by the presence of diphenols, and particularly resorcinol (whose oxidation products are known to be without harmful effect upon the mammalian organism).Awaiting further information concerning the chemical nature of the new hypotensive substance resulting from the oxidation of renin E, it is suggested to name it "OXYRENIN."The experiments illustrated here bear a remarkable similarity to those of Hermann and Bacq, only the time scale being considerably lengthened. They are also interesting in regard to the observation of Martin, Ichniowsky, Wisansky and Ansbacher who have shown that the enzymatic inactivation of some pressor amines can be enhanced by a number of phenolic substances among which is l-dopa, whose anaerobic decarboxylation in the ischemic kidney yields a strongly pressor substance, probably hydroxytyramine.

Oxytocin A polypeptide hormone of the posterior lobe of the pituitary. Causes contraction of smooth muscle fibres.

Oxytocin A hormone made in the hypothalamus and released from the posterior pituitary that stimulates muscle contraction in the mammary glands and the uterus.

Ozonide The product formed by the addition of ozone (O_3) to a double bond. It has the structure which on hydrolysis breaks between the two carbons, thus affording a means of determining where the original double bond was.

P

Pace-maker The substance whose rate of reaction in a series of linked reactions sets the pace for the series; lactic acid may be considered as the pace-maker in fermentation.

Pachytene State during the prophase of meiosis during which the homologous chromosomes are completely paired.

Pacini-taras test for vitamin a in oils One drop of oil, treated with 1 drop of guaiacol, 2 drops phenol and 1 cc. perchloric acid in 5 cc. chloroform, forms a purple color, developing into bright red. The color appears specific for vitamin A.

Package In recombinant DNA procedures, refers to the step of incorporation of cosmid or other lambda vector DNA with an insert into a phage head for transduction of DNA into host.

Paeonin Peonin; anthocyan from deep violet-red peonies, the diglucoside of cyanidinmonomethylether.

Paget reaction for adrenaline and adrenalone An adrenaline solution gives a reddish-brown or amber color with 10% ammonium molybdate solution, which changes to a greenish fluorescence after the addition of sodium hydroxide. Adrenalone forms a yellow color or yelloworange precipitate which is not changed sodium hydroxide.

Paleontology The study of fossil organisms.

Paleozoic era Era of invertebrates; includes the permian, carboniferous (amphibians), devonian (fishes), silurian (invertebrates), cambrian.

Palindromic sequence A nucleotide sequence which is the same when read in either direction, usually consisting of adjacent inverted repeats. Restriction endonuclease recognition sites are palindromes 5'—>3' GAATTC EcoRI recognition site CTTAAG 3'<—5'.

Palisade induction A neural induction in which the stimulus has not been sufficiently strong, or the mechanical conditions not sufficiently favourable, to allow of the formation of a well-shaped neural tube. Instead of this, the histologically differentiated neural cells are arranged in palisade fashion around the implant.

Palliative Medical treatment intended to alleviate symptoms of a disease.

Palmitic acid A saturated fatty acid, $CH_3(CH_2)_{14}COOH$; found in animal and vegetable fats; m.p. 63-4°CC.

Palmitoleic acid A 16 carbon single double bond unsaturated fatty acid of cod liver and herring oils.

Pancreas Gland of the vertebrate digestive system that releases enzymes that break down starches, proteins, and fats.

Pancreatic enzymes These include pancreatic amylase (amylopsin), maltases, lactases, sucrases; lipase (steapsin); proteolytic enzymes, as trypsin, chymotrypsin, heterotrypsin, polypeptidases, dipeptidases; enterokinase activator.

Pancreatic juice The external secretion of the pancreas; aids digestion; has a pH of 7.5-8.0; is 98.7% water, the remainder being various enzymes, organic and inorganic compounds.

Pancreatic prolipase An inactive thermolabile protein of pancreatic extracts, which in the presence of tissue extracts becomes an active fat hydrolyzing enzyme.

Pancreatropic principle The hormone of the anterior pituitary which controls the islets of Langerhans and their insulin production and action.

Pangenesis Hypothesis that all parts of the body contribute genetic material to the reproductive organs, and particularly to the gametes.

Pantothenic acid Probably the most important constituent of "bios" or "yeast growth substance," now known to have the constitution alphagamma - dihydroxy - beta - beta methyl-butyroalanine or CHOHC (CH_3), CHOHCONHCH, CH_2 COOH, and to have vitamin properties. It is freed from a bound condition by enzyme action. Its deficiency is shown by gray hair, adrenal necrosis, kidney complications, ulcers in the intestinal tract in hogs, dogs, mice, rats and chicks. Feeding extra pantothenic acid increases hatchability of eggs. Pantothenic acid is low in the blood of patients with riboflavin deficiency, beriberi and pellagra.

Papain Papayotin; vegetable pepsin; an intra-cellular proteinase of the pan-pan tree, Carica Papaya; activated by glutathione and other SH containing compounds, by HCN, and by photokinase; shows an optimum pH between 4 and 7.

Papainases A group of intracellular proteinases in plants with an optimum pH of 4-7. Papain is a typical example. They are inactivated by copper salts and iodoacetate and activated by SH compounds and cyanide.

Papaverine $C_{20}H_{21}O_4N$; prisms or needles, m.p. 147°C; opium alkaloid of the benzyl isoquinoline group; weakly poisonous, large doses causing mild tetany; used as antispasmodic since it paralyzes smooth muscle.

Parabiosis The grafting together of two or more embryos, usually side by side, in order to study their later mutual morphogenetic and endocrine effects.

Parabola Curve formed by the intersection of a cone and a plane that cuts the sloping side of the cone.

Paracasein The product of the action of the enzyme rennin on casein.

Parahematins Haematin combined with certain denatured proteins and bases, the combination being with 2 molecules of base, and the iron being in the ferric state.

Parallax Apparent change in the position or direction of a celestial body against its background when observed from a new location.

Parallel lines and parallel planes Lines or planes that are always the same distance apart and thus never intersect.

Parallelogram Four-sided geometrical figure having opposite pairs of equal-length parallel sides and opposite angles of equal degree.

Paranuclein 1. The essential substance of true nucleoli.
2. A nuclein derived from cytoplasm.

Parapatric Referring to two species having contiguous geographic ranges but no (or only minimal) interbreeding in the zone of contact.

Parasite Organism that lives on and in a host organism and obtains its nutrition from the host, typically at the host's expense.

Parathormone The protein hormone of the parathyroid glands which regulates blood calcium, preventing convulsions.

Parathyroid hormone units One U.S.P. unit is one-hundredth of the amount of material necessary to raise the Ca level of 100 cc. of blood serum of normal dogs 1 mg. within 16-18 hours.

Parathyrotropic principle A hormone of the

anterior pituitary which regulates the proliferation of the parathyroid gland cells.

Paratope That region within the antigen-binding site of an immunoglobulin molecule responsible for recognition of epitope structure.

Parenchyma Ground tissue; protoplasm-rich tissue.

Parental Cross a monohybrid cross between two parent organisms that breed true for distinctly different forms of a single trait; heterozygous offspring results.

Parietal Pertaining to a wall or lining.

Parillin $C_{26}H_{44}O_{10}.H_2O$ a glycoside of sarsaparilla root consisting of 2 molecules of glucose, rhamnose and parigenin. It is a hemolytic poison even in 10 parts in a million.

Parotid Cheek gland secreting saliva.

Parthenogenesis The stimulation to development, more or less complete, of an egg by artificial means.

Pasteur enzyme An intracellular catalyst, responsible for the inhibiting action of molecular oxygen on fermentation (Pasteur reaction). According to the photochemical studies of K. G. Stern and associates, the enzyme is a pheohemin protein and closely related to but not identical with Warburg's respiratory ferment. Like the latter, the Pasteur Enzyme reacts with cyanide and with carbon monoxide, besides oxygen. The combination with CO is reversible and dissociated by light.

Pasteur reaction Pasteur-Meyerhof reaction; the cessation of fermentative processes when anaerobic conditions are removed with replacement by oxidative processes.

Pasteurization Heat treatment of milk short of complete sterilization; "holder" method involves 30 minutes at 150°C F.

Pathogen Microorganism (such as infectious *bacteria*) that causes disease.

Pathology The study of the nature of diseases, and their causes and symptoms.

Patten test for cystine The action of phenyl cyanate on cystine results in the formation of hydantoic acid.

Pauly reaction (1904) A not too specific test for tyrosine. A sodium carbonate solution of tyrosine and diazobenzene sulphonic acid gives a red color. Histidine gives a very similar color.

Pauly reaction for histidine and imidazole The test solution is made alkaline with sodium carbonate and then treated with diazobenzenesulfonic acid. A dye, pure orange in acid and cherry-red in alkaline solution, is produced. Sensitivity - 1100000.

Pavolini test for sugars 5 cc. of test solution are shaken with 0.5-1 cc. of alkaline mercuric potassium iodide solution. A yellow or yellow-green precipitate is given by dextrin, fructose, lactose and glucose; sucrose does not give it.

Pectase An enzyme that demethoxylates pectin, yielding pectic acid.

Pectic acid A member of the group of pectins, derived from pectin by demethoxylation.

Pectin A group of substances which consists entirely of a long chain of galacturonic acid units joined 1,4, some of which are esterified with methyl alcohol; water dispersible. In the presence of acids and sugar pectin forms the basis of household jellies.

Pectinase An enzyme that hydrolyzes pectin to sugars and galacturonic acid.

Pectinoxanthin The brilliant red animal xanthophyll of the ovaries of the scallop, Pecten.

Pectins Complex high molecular weight carbohydrates containing galacturonic acid, galactose, xylose, methanol, acetic acid and possibly arabinose. There are three recognized members protopectin, pectic acid and pectin, which see.

Pectization An obsolescent word describing the coagulation of sols, and their transformation to gels.

Pectose Obsolete name for protopectin.

Pelargonin Salvinin; monardin; a glycoside

anthocyan of the genarium, consisting of two molecules of glucose and one of pelargonidin. The chloride, $C_{27} H_{31} O_{15} Cl$ forms scarlet-red needles, water and alcohol soluble.

Pellagra-preventive factor Nicotinic Acid Amide; the vitamin necessary to prevent pellagra in humans and black tongue in dogs. It is found in buttermilk, beef, yeast, etc. It forms an oxidationreduction system with its reduced form.

Pellerin test for adrenaline in urine A mixture of 10 cc. urine, 5 cc. saturated aqueous mercuric acetate solution and 1 gm. sodium acetate is vigorously shaken for a minute. The vessel is immersed in a boiling water bath for 10-15 seconds, shaken again, cooled and filtered. A transitory reddish or pink color in the filtrate is positive.

Pelletierine $C_8 H_{15} N$; punicine; alkaloid of pomegranate bark; oil, b.p. 195°C; anthelmintic and local anesthetic.

Pellotine An alkaloid, $C_{13} H_{19} O_3 N$; m.p., 110-112°C; from mescal buttons of Anholoniun Williamaii.

Peltier effect Change in temperature at the interface of two distinct metals when an electric current flows from one metal to the other.

Penetrance Refers to the proportion of individuals heterozygous for a given dominant allele that express the phenotype of that dominant allele.

Penis A male organ which deposits sperm into the female reproductive tract.

Pentosans Carbohydrate constituents of gums and mucilages. Have the general formula $(C_5 H_8 O_4)x$.

Pentose Any monosaccharide with the formula $C_5 H_{10} O_5$; with the hexoses make up the most important group of the monosaccharides.

Penumbra Area of Earth from which a viewer is able to see only a partial eclipse of the Sun or Moon; compare *umbra*

Pepsin A proteolytic and milk-clotting enzyme of the stomach which hydrolyzes all classes of proteins except keratins to proteoses and peptones. With casein its optimum pH is 1.8. It exists in an inactive pre-stage called pepsinogen, formerly thought to be converted to pepsin by HC1, but since shown to be inactive only because of H^+ concentration. On increasing H^+ concentration activity is brought about. i.p.=2.75.

Peptidases Enzymes which attack peptides but not proteins. Their point of attack is at the end of the peptide chain, i.e. at the CONH group, so that they are also called exopeptidases.

Peptide Bond A covalent bond between two amino acids, in which the carboxyl group of one amino acid (X_1—COOH) and the amino group of an adjacent amino acid (NH_2—X_2) react to form X_1-CO-NH-X_2 plus H_2O.

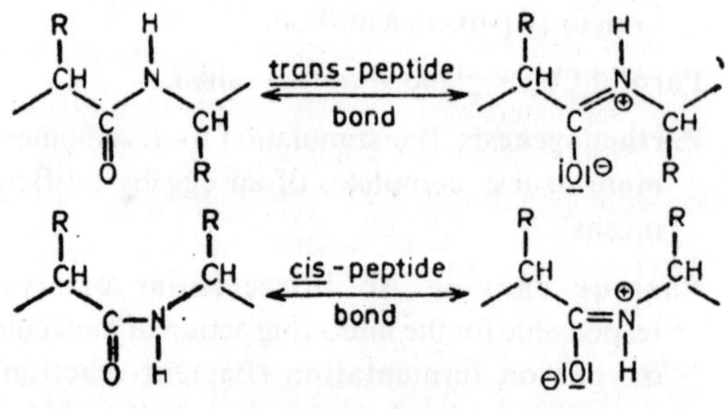

Fig. Peptide Bond

Peptide linkage The linkage between the amino group of one molecule of an amino acid with the carboxyl group of another molecule of the same or a different amino acid. This type of linkage builds up proteins.

Peptide A molecule formed by peptide bonds covalently linking two or more amino acids. Short peptides (generally less than 60 amino acid residues, and usually only half that length) can be chemically synthesized by one of several different methods; larger peptides (more correctly, polypeptides) are usually expressed from recombinant DNA.

Peptide-binding groove That region of a Class I or II MHC molecule that is responsible for binding processed antigen peptides for presentation.

Peptidoglycan Carbohydrate polymer cross-linked by proteins. It is found in the cell wall of Gram positive bacteria, where it stains with the dye crystal-violet.

Peptidyl transferase An enzymatic activity of the ribosome responsible for formation of the peptide bond between the nascent polypeptide chain and the amino acid carried by the charged tRNA in the A site of the ribosome. In so doing, the ribosome is moved along the mRNA by one codon.

Peptine ring hypothesis A theory of protein structure based on the postulation of the existence in the protein molecule of piperazine rings with hydrocarbon side chains, and cyclic or heterocyclic rings attached to the side chains.

Peptization The dispersion of a substance to a sol, usually by some added foreign substance.

Peptones In the American classification of proteins, those secondary protein derivatives characterized by being soluble in water, not coagulated by heat, and not being precipitated from solution by saturating their solution with ammonium sulfate.

Percentage Representation of a number as a fraction of 100.

Percorten A trade name for synthetic desoxycorticosterone acetate.

Pereirine An alkaloid, $C_{20}H_{26}ON_2 \cdot ½H_2O$; m.p., 135°C with decomposition; from bark of geissospermum vellosii; used in certain fevers.

Perennial plant Plant that lives throughout the year, persisting for more than two years.

Pericardium Membrane around the heart.

Perigee Point at which an object traveling on an elliptical orbit is closest to the body it is orbiting (often referring specifically to objects orbiting Earth).

Perihelion Point at which an object orbiting the Sun is closest to the Sun.

Perillaldehyde Monocyclic terpene aldehyde; l form present in perilla oil.

Periplocin $C_{30}H_{48}O_{12}$; m.p. 205°C; a cardiac glycoside of the bark of Periploca graeca, yielding, on hydrolysis, periplogenin, hexose and cymarose; used like digitalis in functional cardiac arrhythmias and pseudo-anginas.

Periplogenin $C_{23}H_{34}O_5$; m.p. 185°C; the 3,5,14trihydroxy aglycone of the cardiac glycoside periplocin.

Peritonitis Inflammation of the peritoneum which may be caused by bacterial infection. It may take the forms of.

(1) acute peritonitis or.

(2) tuberculous peritonitis. In the first case primary, secondary and postoperative forms are distinguished, in the latter, acute and chronic forms. The organism may reach the peritoneum via the blood, lymphatics, by extension of other inflammations, perforation of appendix, stomach, bowel ulcers, hernia, perforations of gall bladder, etc., and postoperative complications. Prophylaxis plays an important role. Parenteral administration of fluid (dextrose and sodium chloride) is used in treatment where nothing can be taken orally.

Permeability Permeability may be defined qualitatively as the readiness with which a dissolved substance penetrates the surface layer or membrane of a phase. In the biological case, with which this article is concerned, the phase is generally a cell. It might, however, be a cell-part-e.g., the nucleus. The readiness of penetration through macroscopic surface layers, such as intestinal epithelium, capillary epithelium, frog skin, etc., is generally considered under this head; but the problem of penetration in those cases is more complex, and will not receive extensive consideration in this place.

The intensity of speed of permeation of any substance is measured by the quantity (in grams, moles, number of molecules, or other units, according to the unit of concentration chosen) which crosses unit area of the surface in unit time. This quantity may be termed the surface current or surface flux. It is generally related linearly to the concentrations of the substance just inside and just outside the membrane. The form of this relationship will be more fully discussed below. Methods.

Many different experimental procedures, and empirical "constants" derived from them, have been used in investigating permeability. These may be grouped roughly into direct and indirect methods.

Direct methods are those in which there is, or appears to be, some direct (usually visible) index of concentration changes inside the cell. The use of vital dyes is prominent among such, the intensity of the dye supposedly affording a measure of concentration. Similar procedures, usually applicable only in special cases, are changes in the optical properties (transparency) of the cell cytoplasm which are known to be correlated with changes in content of the substance being investigated. There are two principal and related objections to optical methods. One is that the number of complicating factors-adsorption by heterogeneities in the cytoplasm, effect of pH, diffusion coefficient in the cytoplasm, secondary changes due to specific physiological effects, etc.—makes any clean-cut interpretation of the results in terms of membrane permeability fairly difficult. Only by means of a complete theoretical treatment, involving many difficulties to carry out, could the various side effects be separated from the aim of the investigation—or by a set of additional experiments equally if not more laborious. The second objection is that the relation between the concentration to be determined and the optical effect which is measured is unknown, not necessarily linear, perhaps not even monotonic. Thus, even if permeability could be reliably studied qualitatively by such procedures, they would not, without much further labor, furnish a quantitative scale.

Indirect methods are principally osmotic. They rely on the fact that cells show visible changes of volume (plasmolysis, etc.) in nonisotonic solutions of substances which do not penetrate them with infinite ease. Thus plasmolysis and its reversal, studied in a time series of observations, may give a measure of rates of penetration. These methods are more usable than the optical ones, and for a long time have been more popular. They are by no means completely free from objection, however. The principal difficulty is that the procedure is likely to produce permeability changes of unknown and perhaps irreversible character. Even more objectionable are methods which frankly capitalize on the production of drastic and irreversible changes in the cell, an example being the use of time required to produce hemolysis of erythrocytes as a measure of permeability. The number of unknown complicating factors here is so excessive as to render this common technique, at least in the reviewer's opinion, a fertile source of error rather than information.

Purely physiological methods, such as those which employ contracture of isolated muscle or the beat of the isolated heart as indices of the penetration of physiologically active substances, are quantitatively useless, and, in view of the fact that most such substances (electrolytes) may conceivably or are known to affect physiological activity without penetrating (e.g., effect of potassium on muscle and nerve), are likely to be qualitatively useless and even misleading as well.

What might be termed indirect physico-chemical methods have centered around measurements of conductivity (of single cells or of cell suspensions). These measurements, if carried out and interpreted with great caution, can be very useful. They are, however, limited to studying electrolytes. They are also open to the possible objection that the method of measurement may induce secondary permeability changes, or damage the cell under investigation.

Undoubtedly the best method theoretically is the direct method of chemical analysis. This is open to the obvious objection that it is quite difficult to carry out. However, a fuller exploitation of the resources of microchemistry than has been customary, and the analysis of large numbers of sufficiently similar cells, should make this method at least feasible. It would seem that the greater expenditure of time and labor would be sufficiently rewarded by reliable results with some quantitative significance.

Naturally, the problem of interpreting such analyses still remains, and much caution is needed. In many cases there is reason to believe that an appreciable part of the substance being studied is not present in the cell free in solution, but bound in some fashion to permanent cell components (e.g., proteins), and probably in some sort of equilibrium with the amount in solution. In such cases analytical results giving total content of the substance in the cell would not be sufficient. The same may hold if the substance is not uniformly distributed through the cell, either by virtue of being involved in cellular metabolism or by reason of heterogeneity of the cell in regard to solubility of the substance. In the latter case, precipitation methods, which permit distribution of the substance to be studied, would be a valuable auxiliary to straight analysis. In case of binding by the cell matrix, conductivity measurements might reveal the true case, though in some cases they may give fallacious results, (e.g., if the time required to take the measurement is as great as the time required for the measuring current to change the state of binding itself). Obviously an adequate theoretical treatment is needed to assist experimental investigation in such cases.

A brief word may be said regarding the material used, which is by and large extraordinarily stereotyped. Among the most popular types of cells for use in permeability studies are the algae Valonia and Nitella, because of their convenient size and the rather large amounts of cell sap available for analysis, and the erythrocytes of various species, apparently on account of their ready availability. The highly specialized plant cells are objectionable insofar as they are little capable of furnishing useful information regarding the permeability behavior of most animal cells. The erythrocyte is in the highest degree objectionable because it is so degenerate a cell as scarcely to deserve the name at all. Its metabolism is so faint that phenomena connected with the metabolic activity of the cell are likely to go undetected. The erythrocyte is important enough to be worth studying for itself; but it is highly unlikely that the results will shed any light on the permeability behavior of other more active animal cells. Among the best cells so far used are the ova of various species (e.g., frog, fish, echinoderm). Steinbach has made ingenious use of the rather plentiful protoplasm in the giant axon of the squid. Results.

To give a complete account of the various findings on the permeability of different substances for different cells, as well as an adequate critique of their often very dubious reliability, would greatly exceed the scope of this brief review. We content ourselves here with a rough summary.

Experiments on artificial "model" membranes sugggest that watersoluble substances in the main are permeable according to the relation between their molecular volume and the pore size of the membrane. Substances soluble in lipids or in lipid solvents, on the other hand, seem to permeate lipid-containing membranes according to their lipid solubility. Lipid solubility, of course, is an exceedingly vague term; and in the literature of permeability it has been employed to cover a multitude of sins. It is indisputable that many cells have surface layers rich in phospholipids, although the evidence here is most convincing for the erythrocyte. But it is not so clear that the speed of penetration of a phospholipid surface necessarily bears a simple relation to solubility in olive oil or ether. Some investigators have attempted to account for the various permeability phenomena of cells by postulating mosaic membranes consisting partially of lipid areas; others, such as Davson and Danielli, have argued for multiple-layer membranes consisting of alternate protein and phospholipid layers.

The studies on artificial membranes are more or less, in the same not too precise sense, borne out by studies on actual cells. Numerous classes of substances, however, show special properties. Thus many cells show electrolyte permeability which may be regarded as due to the charge of the membrane. Moreover, most animal cells are relatively quite permeable to anions, and relatively impermeable to cations. Cation impermeability is more relative than has been supposed, and much doubt has been thrown on past experiments and hasty conclusions by the careful studies of Steinbach. Still not entirely accounted for, but most probably related to selective binding by cell proteins, is the selective retention of K by some cells, and by Na by others.

Mineral acids and alkalis penetrate far less freely than weak organic acids and bases, and NH_3 and CO_2. It seems possible that the undissociated forms of these compounds may penetrate, while the ions are unable to do so.

Special chemical effects are found among electrolytes. Thus series of cations and anions may be established in the order of permeability, which cannot clearly be related to valence, or even to mass or molecular volume. Attempts have been made to account for these series in terms of the hydration of the ions, but unfortunately no complete quantitative treatment has been carried out. Special ion effects have also been attributed to the influence of the ions on the colloidal state (also, unfortunately, a vague concept) of the surface layer; thus the effects of Ca ions are attributed to the coagulation (and increased density) of the membrane by the ions. Decision on these hypotheses awaits a more satisfactory theory of the behavior of biocolloide.

Among the external influences which alter permeability, the chief is temperature. It may be said that in general increase of temperature increases permeability. But the temperature coefficients which have been measured vary rather widely, and their interpretation even more widely. Further and much more systematic investigation is required here.

Permeability, glycerol type The uneven distribution of a substance in the body on ingestion, as in the case of glycerol, sucrose, sulfate, thiocyanate, chloride.

Permeability, physiological The permeability of a membrane while physiological processes go on, such as active oxidation.

Permeability, urea type The uniform distribution of a substance in the fluids and the cells of a body, as in the case of ingestion of urea or sulfanilamide.

Peronnet-truhaut reactions for amino acids and uric acid A solution of of 1 gm. of amino acid in 2 cc. 10% sodium hydroxide solution is stratified with 2 cc. of 0.1% alcoholic m-dinitrobenzene solution; an intense violet ring is formed. The reaction is given by glycine, lysine and phenylalanine. Uric acid gives a violet color when treated with 0.1% alcoholic m-dinitrobenzene solution and 1 cc. 10% sodium hydroxide.

Peroxidases A group of enzymes widely distributed in plant and animal tissues; usually prepared from horseradish; transfer peroxide oxygen to oxidizable substances. The pH range of activity is from 4 to 10, optimum between 5-6. There probably is a reduced heme group in the structure.

Perpendicular Refers to a right-angle relationship between two lines, planes, or other geometrical entities.

Perrin reaction for inositol 2 drops of the concentrated test solution are evaporated on platinum foil with 1 drop of silver nitrate and ignited. The hot residue has a pale pink color which disappears on cooling and reappears on heating.

Perspiration Process of excreting water and waste salts from the skin (in animals) through sweat *glands*.

Pervaporation The evaporation of the dispersions medium from a colloidal system by suspending the liquid in a collodion bag, and heating. If the liquid also contains crystalloids these deposit on the outside of the bag.

Petal Flower structure that attracts pollinators (birds and insects that transfer *pollen* among flowers) with bright color or fragrance.

Pettenkofer test for bile acids in urine The addition of sucrose and concentrated sulfuric acid to urine gives rise to an intense red to violet color in the presence of bile acids.

Pfiffner-myers colorimetric determination of methylguanidine 6 gm. sodium nitroprusside and 8.5 gm. sodium ferrocyanide are dissolved in 100 cc. water; 15-20 minutes before use 1 part of the solution is mixed with 10% sodium hydroxide and 2 parts 3% hydrogen peroxide. 1 cc. of this solution is treated with 4 cc. of the unknown guanidine solution and compared with the color of a standard solution.

Ph A direct measure of acidity or an indirect measure of alkalinity.

Ph measurement The most wide-spread method is the glass electrode, which relies on empirical correlation with the hydrogen electrode, good to 0.01 pH; the quinhydrone electrode is subject to many errors, as salt and protein effects; and colorimetric methods are limited by the sensitivity of the color reaction used.

Phagemid A type of plasmid which carries within its sequence a bacteriophage replication origin (ori). When the host bacterium is infected with "helper" phage, the phagemid is replicated along with the phage DNA and packaged into phage capsids.

Phagocyte Cell that engulfs and digests foreign cells and substances.

Pharbitin Pharbitisin; isoconvolvulin; $C_{54}H_{96}O_{27}$; a glycoside of seeds of Pharbitis nil, composed of glucose, rhamnose and ipurolic acid (311dihydroxymyristic); violent cathartic.

Pharmacology Pharmacology (Gr. pharmakon, drug; logos, science) is the science of drugs or medicinal agents. The subject includes the nature, properties, preparation, action and uses of all materials used as medicine. Because of the wide scope of pharmacology it is subdivided into Pharmacy, which deals with the preparing, compounding and dispensing of drugs; Pharmacognosy, which deals with the physical properties and origin of crude drugs; Materia Medica, which is the study of the source, composition, properties, preparation and dose of all medicinal agents; Toxicology, which is the science of poisons; Pharmacodynamics, which is the study of the action of drugs on the living organism; and Therapeutics, which is the application of drugs in disease.

In the development of pharmacology, more and more emphasis has been given to the action of drugs in the organism so that today the term pharmacology is often considered synonymous with pharmacodynamics. Therapeutics has steadily gained ground and has markedly influenced the trend of pharmacology. In the early period of the science crude plant products and minerals such as calomel, tartar emetic and magnesium sulfate were the important materials of medicine. Then came the isolation of the active ingredients from potent drugs—morphine from opium, atropine from bella donna, strychnine from nux vomica and quinine from cinchona bark. This was followed by a period in which the introduction of synthetic drugs into therapy was the dominant activity. Chloral hydrate as a soporific, the salicylates as analgesics, procaine as a local anesthetic, cinchophen as an antipodaric are examples. At the turn of the century came the first chemotherapy era with Ehrlich's salvarsan the most outstanding achievement. After a lapse of several decades another intensive period of activity in this field was initiated by the discovery of the effectiveness of prontosil and of sulfanilamide against streptococcal infection. In recent years the hormones and the vitamins have come into the limelight and the study of their pharmacological action is well under way.

The key objective of pharmacology is the study of the action of drugs on the organism. Much information has been acquired by empirical observations, by accidental discoveries, and by planned research. Animal experimentation has been and is still of utmost importance in advancing the science. The correlation of chemical structure and physiological activity is responsible for much of the rapid progress. As soon as it was discovered that sulfanilamide was chemotherapeutically active, its structure became the basis for the synthesis of related compounds such as sulfathiazol, sulfadiazine and sulfapyridine, which are of even greater therapeutic value.

It is common practice to classify drugs on the basis of the systems on which they act. The central nervous system depressants form a large class, since it includes all the agents capable of producing general anesthesia. Likewise it covers the hypnotics and sedatives as well as the analgesics and the antipyretics. Morphine and the opium alkaloids receive special attention because of their great clinical and theoretical importance. The local anesthetics constitute a separate group. Cocaine and procaine are the most representative members of this class. The stimulants of the central nervous system include the xanthines, strychnine, picrotoxin, metrazol, and coramine.

The drugs acting on the peripheral nervous system constitute an important group which is subdivided into the parasympathetic and the sympathetic. The most important stimulants of the parasympathetic system are acetylcholine, acetyl-ß-methylcholine, pilocarpine, physostigmine, and prostigmine; the depressants are atrophine, scopolamine, hyoscyamus and their derivatives. The stimulants of the sympathetic system are epinephrine, ephedrine and related compounds.

The cardiovascular drugs because of their great value in therapy occupy an important place in pharmacology. The digitalis group, quinidine and the nitrites are the most important agents under the above heading. The diuretics and the antidiuretics constitute another group. To understand their action a knowledge of water and mineral balance is required. • A group of drugs acting specifically on the gastrointestinal tract is treated as a class and subdivided as follows emetics, carminatives, bitters, hydrochloric acid, anacids, digestants, cathartics, and anthelmintics.

The oxytocic drugs constitute a small but important group. The principal agents are the ergot alkaloids, quinine and the posterior pituitary hormone. Another group are the drugs acting on the blood forming organs. Iron salts and the antipernicious anemia principles in the form of liver extract and desiccated stomach are the most effective agents. It is to be noted that no successful drugs for the control of the production of leukocytes have been found. Of utmost importance are the drugs used to combat pathogenic microorganisms. The antiseptics and germicides are usually general rather than specific in their action. Among the important specifics are quinine and atabrine in malaria; chaulmoogra. oil in leprosy; the arsenicals, bismuth, iodides and mercurials in syphilis; emetin in amebiasis; and the sulfonamides, particularly in streptococcic and other bacterial infections.

The newest chapter in pharmacology deals with vitamins and hormones. Some of the hormones, such as epinephrine and thyroid have, however, occupied a prominent place in pharmacology for many years.

Pharmacology is not an independent science. It draws freely upon physiological chemistry, physiology and the other related branches of science. Its aims are practical rather than theoretical and it is therefore the backbone of therapeutics.

Phase One of the physical states of matter (such as gas, liquid, or solid). (*physics*) state of an oscillating system; for example, two pendulums are in phase if their swings occur in the same direction at the same frequency.

Phase variation Alternation in the type of flagellum produced by a bacterium.

Phellandral A terpene aldehyde of water fennel oil, etc.

Phellandrene, alpha and beta Monocyclic terpenes; d-alpha occurs in bitter fennel, ginger-grass oils; l-alpha in eucalyptus oils; dbeta in water fennel and lemon oil; l-beta in some peppermint oils.

Pheromone A chemical substance which an animals releases that influences the behaviour of another creature of the same type, for example by attracting it sexually.

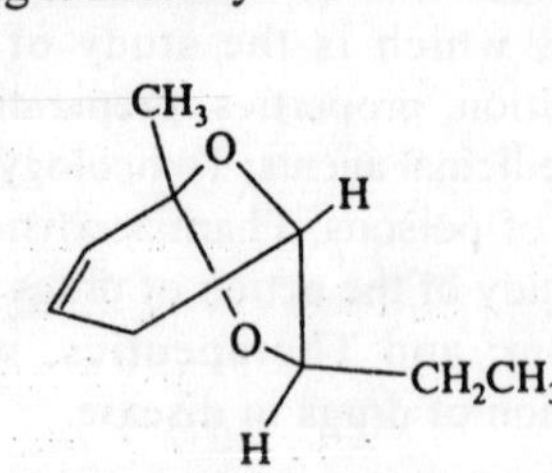

Fig. Pheromones

Phenacetin Aceto-p-phenetidine, an antipyretic and analgesic.

Phenobarbital $C_6H_5(C_2H_5)$ C $(CONH)_2CO$; phenyl ethyl barbituric acid or phenylethylamalonylurea; used as a hypnotic in nervous insomnia and states of nervous excitement, and as a sedative in epilepsy.

Phenocopy The imitation of a morphological or physiological character due to one genotype in an organism possessing a different genotype, by the action of external physical and chemical factors, not leading to any inheritable change.

Phenocritical period A period in the development of an organism during which an effect produced by a gene can most readily be influenced or initiated by externally applied factors.

Phenolphthalin Phthalin; dihydroxy-triphenylmethane-2-carboxylic acid, m.p. 225°C, a reagent for oxidases, blood, HCN, peroxide.

Phenolsulphonphthalein test A test of renal function. On a given amount of the phenolsulphonphthalein the amount excreted in two hours is determined.

Phenotype Visible characteristics determined by an organism's genetic makeup phenotype The observable characteristics of an organism.

L-phenylalanine ß-phenyl-a-amino propionic acid; C9H11O2N; needles or leaves, v.s. hot water, sl. sol. cold, m.p. 283°C. i.p. 5.36; an indispensable amino acid found to a large extent in proteins of seeds and young plants.

Phenylalanine metabolism Phenylalanine is mostly converted to tyrosine, whence it is further catabolized. To some extent, particularly in phenylketonuria, it is converted to phenylpyruvic acid. Strongly ketogenic.

Phenylketonuria The occurrence of phenylpyruvic acid in the urine, found in certain mentally deficient individuals. The disease is hereditary, traceable to a single recessive Mendelian character.

Phenylpyruvic acid $C_6H_5CH_2COCOOH$; an intermediate of phenylalanine metabolism, found particularly in phenylketonuria.

Philothione An old name for glutathione.

Phlobaphenes Inert, insoluble, reddish or brownish, amorphous anhydrides of tannins.

Phlorizin Phloridzin; 2-phloretin-ß-glucoside; asebotin; a glycoside of rosaceae (apple, cherry, pear and plum) consisting of glucose and phloretin. Causes glycosuria, (phlorizin diabetes) on subcutaneous injection by preventing kidneys from reabsorbing glucose and by inhibiting phosphorylation; needles, m.p. 108-9°C; used as test for renal insufficiency.

Phlorizinize To bring about an artificial diabetic condition by the administration of phlorizin.

Phloroglucinol 135-trihydroxybenzene; found in many glycosides; m.p. 219°C anhydrous.

Phloroglucinol-hcl test A test for pentoses depending on the formation of a red color when the sugar is heated with phloroglucinol and HC1.

Phosphagen Phosphocreatine; the compound, which on breaking down to creatine and phosphoric acid, provides the energy of muscle contraction. It is resynthesized. The term is sometimes confused with arginine phosphate in the discussion of invertebrate muscle.

Phosphatases A group of widely distributed esterases, important in bone formation, muscle activity, and lactation. They hydrolyze nucleic acids and phosphoric acid compounds of fats, proteins, carbohydrates, amino acids, etc. Rickets is accompanied by a high serum phosphatase.

Phosphate bond energy A general interrelationship between phosphate metabolism and energy supply has become increasingly apparent through the growing understanding of muscle chemistry. Therewith, phosphorylation of organic compounds which occurs during metabolic processes became of new significance. First, it was found that the phospho-creatine bond served as a reservoir of con- traction energy. This discovery was followed soon by the isolation of a number of unusual compounds adenosinepolyphosphate, phosphopyruvate and lately of the carboxylphosphates. These compounds, like creatinephosphate, contain energy condensed in phosphate bonds which although differing in chemistry appear to be equivalent in energy content. The average derivable from hydrolytic decomposition of these energy-rich phosphate bonds is 11,000 cals.

This group of compounds has to be distinguished clearly from the ordinary phosphate esters glycero-, hexose-phosphate, phosphoglycerates and the other compounds where, in ordinary ester linkage, phosphate is linked to an alcoholic hydroxyl. The ordinary ester linkage has an average energy of only 3000 cals.

The enzymatic mechanism of phosphate transfer resembles very closely that of hydrogen transfer. Generally, phosphorylation-dephosphorylation systems are comparable to hydrogenation-dehydrogenation systems. Such resemblance is not merely superficial. As an O/R potential is assigned to a hydrogenation system, a phosphate group potential may be assigned to a phosphorylation-dephosphorylation system. The phosphate group potential is measured adequately through the energy liberated by dephosphorylation.

When the biological important phosphate derivatives are grouped according to bond energies, there is an aggregation around two levels of potential (1) the 11,000 cals. level of the energy-rich bond and (2) the 3000 cals. level of the ordinary ester bond. The free inorganic phosphate is at zero level on the base line of our scale.

Phosphodiester Bond The covalent bond between the 3' hydroxyl in the sugar ring of one nucleotide and the 5' phosphate group of the sugar ring of the adjacent nucleotide residue within a nucleic acid 5'-Ribose- 3' - O - $P(O)_2$ - O - 5' -Ribose - 3' - etc.

Phospholipids Compound lipids containing phosphoric acid and nitrogen; as lecithin, cephalin, sphingomyelin.

Phosphoproteins In the American classification of proteins, the phosphoproteins are conjugated proteins, in which ophosphoric acid is esterified with the hydroxy group of a hydroxy amino acid, especially serine. Example Casein.

Phosphorylation The addition of a phosphate monoester to a macromolecule, catalyzed by a specific kinase enzyme. With respect to proteins, certain amino acid side chains (serine, threonine, tyrosine) are subject to phosphorylation catalyzed by protein kinases; altering the phosphorylation status of a protein may have dramatic effects on its biologic properties, and is a common cellular control mechanism. With respect to DNA, 5' ends must be phosphorylated for ligation.

Phosphotidic acids Diglyceride phosphoric acids.

Photolysis Chemical reaction or decomposition that is fueled by light or *ultraviolet radiation.*

Photometer An instrument which measures the intensity of light.

Photon Particle of energy by which light is carried.

Photosynthesis The term photosynthesis might be applied to any synthetic process dependent upon radiant energy. Common usage of the term by plant scientists to denote the assimilation of carbon dioxide by green plants under the influence of light, has resulted in fairly general acceptance of this more limited meaning. Assimilation of carbon dioxide by organisms other than green plants, such as algae of various colors, and certain classes of bacteria, is also called photosynthesis, provided the energy required for assimilation is supplied by absorbed radiation, whether visible or not. Certain groups of organisms assimilate carbon dioxide by mechanisms which derive the necessary energy from chemical reactions, rather than from absorbed radiation. This type of metabolism is called chemosynthesis, and is not included under photosynthesis. For example, the colorless sulfur bacteria (Thiobacteria) oxidize compounds of sulfur or elementary sulfur, and utilize the energy set free by this oxidation to assimilate carbon dioxide. No irradiation is required, apart from the maintainance of physiological temperatures. These sulfur bacteria are classed as chemosynthetic. Certain other sulfur bacteria

(Thiorhodaceae) also oxidize compounds of sulfur or elementary sulfur in connection with carbon dioxide assimilation, but the assimilatory process is dependent upon the absorption of radiation, either visible or infrared. These organisms are classed as photosynthetic.

Most of the organisms capable of photosynthesis are classified in the plant kingdom, and the subject has therefore been regarded as belonging to the science of botany. But a number of outstanding contributions to our knowledge of photosynthesis have come from investigators primarily interested in fields other than botany, notably chemistry and physics. The photosynthetic process presents certain problems which have frequently been overlooked by botanists and plant psysiologists, but which have excited the interest of outstanding chemists and physicists.

Prior to the beginning of the nineteenth century, it was believed that the carbon content of plants in general was derived from the humus of the soil. Indeed this belief is to be found in text books of plant physiology up to 1839, although more than a century earlier the Dutch iatro-chemist van Helmont had reported an experiment which should have sufficed to refute the humus theory of plant nutrition. van Helmont planted a willow cutting in a tub of soil, and watered it with rain water for a period of about 2 years. During this time the dry weight of the soil decreased only 2 ounces, while the willow gained in weight about 160 pounds. Knowledge of chemistry and the nature of matter was so limited in van Helmont's time that correct interpretation of this remarkable experiment had to wait until a later date. An understanding of the nature and significance of photosynthesis developed in the latter part of the eighteenth and early nineteenth centuries, coincident with the rejection of the phlogiston theory of combustion as a result of the work of Priestly, Scheele, Lavoisier, and others. Early studies of combustion stimulated an interest in the gas exchange of plants and animals, and it was established by Ingen-Housz that in light the green parts of plants produced a gas (oxygen) which would support combustion, as well as the life of small animals. He recognized that in darkness plants absorbed oxygen, and that the non-green parts absorbed oxygen in both light and darkness. de Saussure carried out more precise experiments on the gas exchange of plants, and gave an account of the process of photosynthesis which was later published as one of Ostwald's *Klassiker der Exakten Wissenschaften*. This account includes most of the fundamental facts concerning photosynthesis which are recognized today.

The botanists of the nineteenth century made but few advances beyond the achievements of de Saussure although the dependence of photosynthesis on various external factors was given a good deal of attention. It was observed that the photosynthetic activity of various organisms depended upon the light intensity, the temperature, and the carbon dioxide concentration, but the results of different workers were often in wide disagreement, and no systematic understanding of the significance of these factors was developed until after 1900, when F. F. Blackman showed that the influence of any one factor on rate of photosynthesis depended on the concentration or intensity of the other factors. Blackman found that temperature changes were almost without effect on the rate of photosynthesis if the light intensity was low, but that high light intensities resulted in a marked dependence of the rate on temperature. He formulated a general statement of the results of his work, known as the theory of optima and limiting factors, and this theory exerted a strong directive influence upon subsequent work, because it was clear that the study of the dependence of photosynthesis upon any one factor was significant only if other factors also were controlled. But the most important result of Blackman's work was the recognition that two types of reaction are involved in the process of photosynthesis, and that the characteristics of these different reactions may be studied separately by proper adjustment of the experimental conditions. Thus at low light intensities, the rate of the process is limited by the photochemical steps, which are not dependent on temperature, while at high light intensities the photochemical steps are completed so rapidly that they are no longer rate-limiting, and the rate of the process is limited by the chemical steps, which have a high temperature coefficient. The chemical steps are called collectively the Blackman reaction.

Warburg and Negelein made a careful study of the characteristics of the Blackman and photochemical reactions, by working at high and low light intensities, with the unicellular green alga, Chlorella pyrenoidosa. Their work has become a classic in this field. Among other things, they noted that at high light intensities, when the rate was limited by the Blackman reaction, intermittent periods of light and darkness gave a higher yield of photosynthesis than the same duration of illumination given continuously.

Later work with intermittent light, with more elaborate methods, has shown that it is a valuable means of separating the Blackman and photochemical reactions. The photochemical steps take place only during the light flashes, which may be made so short in comparison with the dark periods that the slower processes of the Blackman reaction proceed only to a negligible extent during the light flashes. The Blackman reaction runs its course during the intervening dark periods, and some of its properties have been revealed by measurements in intermittent light. In this way the average time required for completion of the dark processes has been estimated to be between 0.01 and 0.02 seconds, at a temperature of 25°C C. From measurements of the maximum rate of photosynthesis in continuous light, it has been calculated that per molecule of chlorophyll present, one molecule of carbon dioxide is reduced every 20 seconds. At maximum photosynthesis, the rate is supposed to be limited by the Blackman reaction, so it may seem surprising that the time for this process calculated from the flashing light experiments should be so different from that calculated on the basis of the measurements in continuous light. Very likely several processes are involved in the Blackman reaction, and they may require different lengths of time for completion. But in any case the above comparison is not strictly admissible, because in the flashing light we estimated the time of the Blackman reaction from the maximum number of carbon dioxide molecules reduced per flash, and in the continuous light we based the calculation on the maximum number of carbon dioxide molecules reduced per chlorophyll molecule.

We do not know how many chlorophyll molecules may take part in the reduction of a single molecule of carbon dioxide, but calculations have been made from the flashing light experiments which show that with light flashes so bright that increased intensity causes no increase in photosynthesis, and with dark intervals long enough so that the Blackman reaction can run to completion between each pair of light flashes, the maximum number of carbon dioxide molecules which can be reduced per flash is only about ½500th of the number of chlorophyll molecules present.

From thermodynamic considerations we are sure that more than one quantum of light must be absorbed per molecule of carbon dioxide reduced. Measurements indicate that under conditions of maximum photosynthetic efficiency, about 10 quanta of light must be absorbed per molecule of carbon dioxide reduced. We may be quite sure that all these 10 quanta need not be absorbed by the same chlorophyll molecule in order to reduce a particular carbon dioxide molecule, for if this were so the rate of photosynthesis at low light intensities would vary as a higher power of the light intensity. Actually, the rate of photosynthesis at low intensities varies as the first power of light intensity. But no fully satisfactory explanation has yet been given of the manner in which the absorption of the 10 quanta of light are distributed among the 2500-odd chlorophyll molecules which appear to be present per molecule of carbon dioxide reduced per flash. Since the 10 quanta do not provide much of a margin of energy for the reduction of carbon dioxide to carbohydrate, there is not room for any large proportion of ineffective absorptions, so it appears that the assimilatory mechanism must have some provision for bringing the energy absorbed by widely separated chlorophyll molecules to act in the reduction of a single carbon dioxide molecule. Probably the interpretation of these basic observations will require the best ingenuity of physical chemists.

The attention devoted to the kinetics of photosynthesis is partly the result of failure to develop a more constructive angle of attack. While much has been learned about the chemistry of other physiological processes from experiments with ground cells and cell extracts, and from the introduction of supposed intermediate substances, such lines of attack have so far been quite unsuccessful in the study of photosynthesis. The least disturbance in the cell, such as pricking the membrane with a needle, causes complete cessation of photosynthetic activity. None of the organic substances tested have shown evidence of acting as intermediates in photosynthesis. Substances have either proved to be toxic, or have been first oxidized by the cell to carbon dioxide and water before being built into carbohydrates. We do not know the first substance with which carbon dioxide combines after entering the cell, nor the immediate precursor of carbohydrate. It is also uncertain whether starch or sugar is the first product of photosynthesis, or if there is any single first product of photosynthesis.

Regarding the general nature of the photosynthetic process, our views are probably narrowed by the tendency to regard the process as essentially a synthesis of carbohydrate material. This is because the over-all ratio of exchange of oxygen and carbon dioxide is equal to unity, indicating carbohydrate synthesis.

Since most of the material of average plants is of the same level of reduction as carbohydrate, this ratio is to be expected, but it is quite possible that products both more and less reduced than carbohydrate are also produced in photosynthesis, and that the average exchange misleads us into regarding the entire process as carbohydrate synthesis.

Chemists have long recognized the probability that so profound a change as the synthesis of carbohydrate from carbon dioxide and water must take place in small intermediate steps. There has been much speculation concerning substances of 2 or 3 carbon atoms, as possible intermediates in the process, but there has been no good experimental evidence to support any of the theories advanced. Recent experiments with radioactive isotopes of carbon have indicated that carbon dioxide is probably added as a carboxyl group to pre-existing compounds of high molecular weight (about 1000), and that reduction of the carboxyl group follows. The present view is that this is a more probable mechanism for the synthesis of compounds of high molecular weight than the polymerization of such reduction products as formaldehyde or formic acid, or other compounds of small molecular weight.

However, it has been found that a large number of organisms have the ability to fix carbon dioxide as carboxyl groups on certain carbon compounds, and it is not yet clear that this characteristic of photosynthetic organisms is a part of the photosynthetic mechanism proper.

Experiments with purple bacteria have provided evidence for regarding carbon dioxide assimilation as primarily a process of hydrogen transfer from some hydrogen donor to carbon dioxide, rather than as a removal of oxygen from carbon dioxide and the addition of water. In the case of green plants, the hydrogen donor is water, and oxygen is evolved. In the case of the purple sulfur bacteria, hydrogen sulfide may be the hydrogen donor, and sulfur instead of oxygen is produced. Certain purple bacteria can also use fatty acids or molecular hydrogen as hydrogen donor. Ordinarily green plants appear to be incapable of obtaining hydrogen from other sources than water, and oxygen is always produced. But it has been shown that certain unicellular algae can use molecular hydrogen for carbon dioxide reduction, provided they have first been kept for some time under anaerobic conditions.

Experiments with a heavy isotope of oxygen support the opinion that the oxygen evolved in photosynthesis comes from water, and not from carbon dioxide.

In higher green plants, the seat of photosynthetic activity is the plastid. The bacteria, and certain algae, do not have plastids but have their light absorbing pigments distributed throughout the cells. The need for a light absorbing pigment is obvious, since neither carbon dioxide nor water absorb visible light or the near infrared. The activity of the chlorophyll in green plants seems to be limited to the spectral region from about 400 to about 700 mì. Photosynthetic bacteria carry on carbon dioxide assimilation in the near infrared, as well as in the visible spectrum.

Chlorophyll is apparently responsible for most of the photosynthetic activity of green plants, but there is evidence that the carotenoid pigments which always accompany chlorophyll may play some photochemical part in carbon dioxide assimilation. The red pigment of certain purple bacteria, evidently a carotenoid type of pigment, shows definite photosynthetic activity. The purple bacteria, and also the green bacteria, have a green pigment akin to chlorophyll, bacteriochlorophyll.Evidence concerning the activity of accessory pigments is drawn largely from measurements of photosynthetic efficiency in different wave lengths. These measurements are compared with the estimated absorption of the different pigments, based on measurements with extracted pigments. The absorption spectra of the pigments are altered more or less by extraction, so it is impossible to be sure just how the absorption is divided among the various pigments in the living cells. This introduces an uncertainty in the reasoning concerning the activity of accessory pigments, but the evidence is pretty clear in some cases at least.In recent years the fluorescence of chlorophyll has been studied in order to obtain evidence concerning the mechanism of photosynthesis. Since chlorophyll always fluoresces red, whether it is excited by blue, green, or yellow light, it is believed that the emission of fluorescence indicates it has been raised to an energy level of a certain stability, and that this energy level is the one required for photosynthetic activity. It is not believed that the higher excitation produced by shorter wave lengths can make any more energy available for photosynthesis. Many attempts have been made to correlate the fluctuations in chlorophyll fluorescence with the rise and fall of photosynthetic activity. No certain conclusions can be drawn from this work as yet.There have been a number of reports of artificial photosynthesis, but these claims have not been substantiated, and it appears that green plants are unique in their ability to reduce carbon dioxide with visible light. The energy provided by a single quantum of red light is too small for the reduction of carbon dioxide to carbohydrate. The process must be carried out by elementary steps of small enough energy requirements so that a single quantum of red light suffices for each elementary step. At the present time this is the feature of photosynthesis which especially attracts the interest of physicists and physical chemists.The more important photosynthetic organisms are of course the crop plants and timber trees upon which our civilization depends. These organisms are not adapted to precise quantitative experiments on the nature of the photosynthetic process. Most of our precise knowledge has come from studies on unicellular algae and bacteria. It is to be hoped that when our understanding of the fundamental processes is sufficiently advanced, it will prove to be of value in connection with larger human problems. There has been much study of photosynthesis under natural conditions, and some efforts have been made to correlate the knowledge gained from more precise studies with the behavior of plants under field conditions.

Phototropism Reaction of a living form to light.

Phrenosin (E) Cerebron; $C_{48}H_{93}O_9N$; m.p. 212°C; a cerebroside whose fatty acid is cerebronic acid.

Phthiocerol $C_{34}H_{67}(OH)_2OCH_3$; an optically active alcohol typical for human tuberculosis bacteria.

Phthiocol 2-methyl-3-hydroxy-1, 4-naphthoquinone; a yellow pigment of human tubercle bacilli that has strong anti-hemorrhagic powers.

Phycocyan A chromoprotein fond in sea algae.

Phycocyanin Blue, water-soluble pigment found in the cyanobacteria and the red algae.

Phycoerythrin Red, water-soluble pigment found in the cyanobacteria and red algae.

Phycoerythrin A chromoprotein found in sea algae.

Phycophaein Pigment of brown sea weeds.

Phycoxanthin Yellow pigment of diatoms.

Phyllite Rock formed under gradually increasing temperature and pressure; phyllite's internal crystal structure causes it to break along smooth, shiny edges

Phylogeny The study of the ancestral history of organisms.

Phylum One of the 36 groups, below the level of *kingdom*, into which organisms are classified.

Physalien Physalin; zeaxanthin dipalmitate; $C_{72}H_{116}O_4$; m.p. 99°C; a carotenoid found in fruits of Physalis alkekengi, and other members of the species. Its structure is ß-ß'-carotene-3-3'-dioldipalmitate.

Physetoleic acid A 16 carbon, one double bond, unsaturated fatty acid, found in Caspian seal oil.

Physical map A map of the locations of identifiable landmarks on DNA (e.g., restriction enzyme cutting sites, genes), regardless of inheritance. Distance is measured in base pairs. For the human genome, the lowest- resolution physical map is the banding patterns on the 24 different chromosomes; the highest- resolution map would be the complete nucleotide sequence of the chromosomes.

Physiological antagonists Physiological antidotes or antidotes which merely mask the symptoms, e.g. atropine vs. morphine, chloroform vs. strychnine, barbiturates vs. cocaine, picrotoxin vs. hypnotics.

Physiology The study of the functions of organisms.

Physiology, developmental A branch of physiology dealing with experimental embryology and chemical embryology.

Phytin Calcium magnesium salt of phytic acid (inositol hexaphosphoric acid); derived from numerous seeds and grains; used in neurasthenia, anemia, tuberculosis.

Phytohormones Plant growth hormones; include the auxins (cyclopentene derivatives), heteroauxin (indole-3-acetic acid) and more or less unidentified factors, traumatic acid, caulocaline (stems), phyllocaline (leaves) and rhizocaline (roots).

Phytokinase A papain activator found in pappain preparations. It is a polypeptide composed largely of glutamic acid and cystine.

Phytol $C_{20}H_{40}O$; b.p. 202.5-204°C at 10 mm.; diterpenic alcohol produced by the action of alkali on chlorophyll.

Phytosterols Sterols found in plants, e.g. sitosterol, stigmasterol, white solids, optically active.

Picramic acid 4, 6-dinitro-2-aminophenol, m.p. 169°C, a reagent for albumin.

Picroaconitine An alkaloid, $C_{32}H_{45}O_{10}N$, from the tuber of Aconitum napellus; m.p., 150°C; used as cardiac sedative.

Picrocrocin Saffron-bitter; $C_{16}H_{26}O_7$; m.p. 154156°C; glucoside of safranal, a carotenoid aldehyde, which is a female determining factor in certain algae, or a gamone used as an abortive.

Picrolonic acid 1-(p-nitrophenyl) -3-methyl - 4 - nitropyrazolon, m.p. 116-117°C, a reagent for alkaloids, tryptophan, phenylalanin and calcium; also used in forming crystalline derivatives of amino acids for purification.

Picrotoxin Cocculin, an amaroid, $C_{30}H_{34}O_{13}$; an equimolecular compound of picrotin and picrotoxinin; m.p., 200°C; from berries of various species of cocculus; used as antihidrotic.

Pilocarpine $C_{11}H_{16}O_2N_2$, m.p. 34°C, alkaloid of Jaborandi species, causes profuse perspiration, slows heart beat; antidote to belladonna, stimulant for hair growth.

Pineal principle A hormone of the pineal gland which causes tadpoles to become transparent and produces sexually precocious dwarfs on injection in successive generations.

Pinitol Matezite; sennite; methyl ether of inositol found in senna, coniferin.

Pinnaglobulin Brown respiratory pigment containing manganese, in blood of lamellibranch, Pinna squamosa.

Pinnate leaf Leaf arrangement in which rows of many small leaflets grow on either side of a *central rib.*

Pinocampheol Secondary alcohol derived from pinene; l-form in oil of hyssop.

Pinocamphone A dicyclic ketone; l-form found in oil of hyssop.

Pinocarveol A dicyclic terpene alcohol of Eucalyptus globulus.

Pinocytosis The incupping of surfaces of single cells due to disturbances of permeability.

Pion Subatomic particle that decays into *muons* and *neutrinos;* both neutral and positively charged forms of pions exist

Piperazine $C_4 H_{10} N_2$; piperazidine; arthriticin; dispermin; hexahydropyrazine; diethylene diamine; use for gout and rheumatism, and to dissolve urinary stones; m.p. 105-106°C, forms hydrate.

Piperidine An alkaloid, hexahydropyridine, $C_5 H_{11} N$; m.p., 9°C; b.p., 105-106°C; from black pepper; is a strong base.

Piperine $C_{17}H_{19}O_3N$; piperoylpiperide, monoclinic needles from alc. m.p. 128-129.5°C. Pungent mildly irritative alkaloid of Piper Nigrum, has pepper-like taste. Synthesized from piperoyl chloride and piperidine. Forms periodide; $B_2 HI.I_2$, steel-blue needles m.p. 145°C.

Piperitol A secondary terpene alcohol; lform found in eucalyptus oils; dform in a species of Andropogon.

Piperitone A terpene ketone; l-form found in eucalyptus oils; d-form in Japanese peppermint; camphor-like odor.

Piperonal Heliotropin; piperonylaldehyde; m.p. 37°C, $C_8 H_6 O_3$, found with vanillin; a perfume also used as a pediculicide.

Pistil Female reproductive structure of a flower, consisting of either a single *carpel* or more than one carpel fused together

Pitocin An aq. soln. containing the oxytocic principle from the pituitary gland, posterior lobe; used to stimulate uterine contractions in labor.

Pitressin An aq. soln. containing the pressor and diuretic-antidiuretic principle of the pituitary gland, posterior lobe; raises blood pressure, promotes muscular tone; used in post-operative ileus, shock, enuresis, diabetes insipidus.

Pittarelli reaction for creatinine and urea 5 cc. of the strongly alkaline test solution are treated with 0.05 gm. sodium nitroprusside and 0.1 gm. potassium persulfate. A red color develops within 20-30 seconds, sensitivity for creatinine—150000; for uric acid 1500.

Pittarelli test for ascorbic acid (vitamin c) I. A mixture of the vitamin solution is slowly treated with dilute ammonium thiocyanate forming a white precipitate turning to a green color on further addition of ammonium thiocyanate.

II. Mercurous chloride is precipitated from mercuric chloride solution by the vitamin.

Pituitary gland Of vertebrate endocrine systems, a gland that interacts with the hypothalamus to coordinate and control many physiological functions, including the activity of many other endocrine glans. Its posterior lobe store and secretes hypothalamic hormones, the anterior lobe produces and secretes its own hormones.

Pituitrin Aqueous extract of the posterior lobe of the pituitary.

Pk A convenient method of expressing and plotting ionization constants. It is defined as the negative logarithm of the ionization constant.

Placebo Substance that is administered to a medical patient but that has no medically active ingredients; patients receiving placebos are generally unaware of whether they are receiving a true drug or a placebo.

Placenta The organ formed at the junction between foetal and maternal tissues and blood streams. Its various types have been classified morphologically and physiologically.

Placental barrier The inability of many substances to cross from the maternal to the fetal system.

Plane polarized light Light vibrating in one plane along the path of propagation. It is usually obtained by passing ordinary light through polarizing crystals.

Plankton Small organisms (animals and plants) that float in the water; particularly algae and crustaceans.

Plant growth hormones The term hormone means a chemical messenger, i.e., a substance which is produced in one part of the organism and is then transported to other parts where it exerts its effect. It is restricted further to substances active in minute amounts. The substances controlling growth in plants are of many types water, the nutrient salts (K, Ca, Mg, Fe, Mn, Zn, Cu, Mo, NO_3, PO_4, SO_4, BO_3), carbohydrates, amino acids, and purines all influence growth. These are not hormones, however. The most important growth hormones are the auxins, a group of ethersoluble, unsaturated organic acids whose structure is discussed below.

Assay Methods The auxins were discovered through the experiment of Páal in 1918 in which the tip of the oat coleoptile was cut off and then replaced asymmetrically, i.e., to one side . The side below the replaced tip grew more than the opposite side, causing curvature. Went then showed that a similar result was obtained if the tips were placed first on agar and the agar then applied to one side of decapitated coleoptiles. The curvature so obtained is proportional to the concentration of the hormone in the agar, up to a limiting maximum angle. Skoog has increased the sensitivity of the test some five or six times by removing the seed 24 hours beforehand.

If the agar be placed symmetrically on top of the coleoptile, the increased straight growth is also proportional to concentration of hormone, and this straight growth could be used for assay. Other methods of assay have been developed. If stems of etiolated pea plants (Went), flower stalks of dandelion (Jost), or coleoptiles of oat or maize are slit lengthwise and immersed in a solution of the hormone the two halves curve inwards, while in water alone they curve outwards. A modification of this test (Thimann and Schneider) gives measurable response to 0.003 mg. active substance per liter of solution.

The essence of the assay methods is, of course, to deprive the test plant as far as possible of its own auxin supply, just as animals on vitamin-deficient diets are used for vitamin away. Since auxins are produced in light, etiolated plants are desirable, the buds, tips, or other producing zones being removed. However, some workers use green plants in the light but the procedure is obviously open to objection, though usable results have been so obtained.

Chemistry

Auxins are produced in the coleoptile tip, in growing buds, embryos, and young leaves of all plants studied, and to some extent in all green tissue in the light. They are also formed by many fungi and bacteria. On account of their presence in plant material, auxins are found in blood, animal tissues, and in considerable amounts in urine. From urine and from corn germ oil Kögl, Haagen Smit and Erxleben isolated auxentriolic acid (auxin a) and auxenolonic acid (auxin b); from urine was also isolated indole3-acetic acid. Indole-3-acetic acid was isolated from yeast by Kögl and Kostermans, from the fungus Rhizopus suinus by Thimann, and recently from corn meal by HaagenSmit, Leech, and Bergren. A large number of synthetic substances are active as auxins. These all comprise an unsaturated or an aromatic ring, with an acid sidechain having at least one methylene group between the ring and the carboxyl . Salts of the acids are active, but there is some reason to believe that at the pH of plant sap (about 6.0) it is the undissociated acid that is active. Application of inorganic acid to a cut surface (the cut allows entry) will often accelerate growth, probably by formation of the free auxin acid from its salts. Esters and amides are also active, though not in all cases; whether they are converted to the free acid in the plant is not certain. Activity varies in an alternating manner with length of side chain; it varies widely with the nature of the ring, indole, indene, and naphthalene giving much more activity than benzene, cyclohexene, pyrrole, or cyclopentene rings; it is affected by substitution in the ring and is destroyed by the saturation of the ring with unsaturated side chains (as far as studied) the cis-form is active, the trans not. From a consideration of all recorded active substances, Koepfli, Thimann and Went concluded that "the specificity of physiological activity does not necessarily depend on the nucleus of a substance but upon a particular molecular configuration. The minimum structural requirements for cell elongation activity in higher plants . . . are (a) a ring system as nucleus, (b) a double bond in this ring, (c) a side chain, (d) a carboxyl group (or a structure readily converted to a carboxyl) on this side chain at least one carbon atom removed from the ring, and (e) a particular space relationship between the ring and the carboxyl group. Recently Zimmerman and Hitchcock have found an apparent exception, activity (on green plants in the light) being exerted by 2bromo-3-nitro benzoic acid, in which the COOH group is not separated from the ring. At the time of writing, this has not been tested under critical conditions.

Tropisms

Shoots curve toward a source of light and away from gravity. These curvatures are due to unequal auxin distribution, the lower side in the case of the geotropism, or the shaded side in phototropism, receiving more auxin than the opposite side. The mechanism of perception is not understood; in the case of light the spectral distribution of response corresponds to that of the carotenoids present and the extreme sensitivity of the tip of the coleoptile, as compared with the base, agrees with the presence of carotenoids almost wholly in the tip (Bünning). The connection between perception by the carotenoids and the effect on transport of auxin is unknown. The action of light is complicated by two further factors inactivation of some of the auxin by light, and reduction in the responsiveness of the tissue in presence of light. It is the latter which is largely responsible for the fact that green plants in light do not elongate as much as in the dark.

Root Formation

Evidence that formation of roots on stems or cuttings is controlled by a hormone was first given by Bouillenne and Went in Java; they ascribed the action to a special substance, rhizocaline. In attempts to purify this, Thimann and Went found that the activity of various preparations during purification went parallel with the auxin activity. Finally it was proven (Thimann and Koepfli) that pure indole-3acetic acid causes root formation in the standard test on etiolated pea cuttings, and Kögl found the same thing with Tradescantia internodes. It is now clear that virtually all auxins cause root formation. The most active are indole-3-acetic and indole-3-butyric acids and naphthalene-1-acetic acid and its amide. These are marketed in alcoholic solution, or dissolved in lanoline, or as powders mixed with talc under various trade names. They are usually applied to the base of the cutting in concentrations (aqueous solution) of from 25 to 200 mg. per liter, according to the sensitivity of the plant material, for 24 hours. In talc or lanoline the concentrations are from 10 to 100 times as high. The cuttings are then kept in the propagation bench for periods between 10 days and 4 months. Auxins accelerate rooting, increase the number of roots per cutting, and cause cuttings to root which would not do so otherwise. Certain plants, however, including many forest trees, resist even optimal auxin treatment. Seeds may also be treated before planting, to obtain increased root systems, with a subsequent corresponding increase in growth rate, plant size, or even fruit set.

Inhibition of Growth

Although root formation is induced by auxins,

the elongation of roots is inhibited. Extremely low concentrations, however, (about 10-9 molal) somewhat accelerate root growth. The significance of this for normal growth is unknown since the root tip produces only moderate amounts of auxin.

The inhibition of lateral buds by auxins is of great importance. The terminal bud on a shoot inhibits development of laterals, and when it is removed the laterals develop. This is the basis of pruning. The inhibition was shown by Thimann and Skoog to be due to the auxin formed by the terminal bud, and it can be imitated by supplying a pure auxin to the cut surface after pruning. A few of the known auxins, particularly phenylacetic acid, will not cause this inhibition, but most do. Dormant buds do not produce appreciable amounts of auxin, and consequently do not inhibit other buds. In dwarf plants the auxin formed in the terminal bud is destroyed more rapidly than normal (van Overbeek); this accounts not only for the reduced growth but also for the greater branching or bushiness which commonly accompanies dwarfing, because insufficient auxin is produced to keep the lateral buds inhibited.

Allied to bud inhibition is the inhibition of the abscission layer. Mature leaves and fruits fall from the plant due to formation of a special layer of cells whose walls pull apart from one another and thus virtually sever the organ from the stem. In some way this is inhibited by auxins. The phenomenon has found important application by apple orchardists, who use an auxin spray (usually naphthalene-1-acetic acid) towards the end of the season to prevent early fruit drop.

Other Functions of Auxins

Auxins are present in algae, and it is probable that they control growth in them as in higher plants. Although widely formed by fungi and bacteria, their role in the growth of these organisms is uncertain.

The activation of cambium in trees in the spring, leading to rapid cell division and the laying down of xylem and phloem, is apparently under the control of auxin. The developing young buds produce considerable amounts of auxin and this is transported downwards into the cambium below them. The transit of the auxin down the stem roughly parallels the downward spread of cambial activity. Snow, Söding and others have shown that introduction of indole-acetic acid into the stem will cause great activity of the cambium, many layers of wood being laid down. With such artificial auxin, however, the effect does not spread so far downwards as when caused by buds.

The growth of the ovary, after fertilization, into a fruit, is started, apparently, by the auxin present in pollen. By removing the anthers and, instead, applying auxin in lanolin, or as spray, directly to the ovary, Gustafson and other workers have caused fruit to develop without seed (parthenocarpy). Reedless fruits of tomato, squash, pear, holly, and many other plants have been so produced.

High concentrations of auxins applied locally to young stems cause swellings, tumors, etc. Some of the phenomena accompanying bacterial and other infections in plants (e.g., crown gall, root nodules, etc.) are believed to be due to the local liberation of large amounts of auxins. In nodules it is clear that large quantities of free auxin are formed. There is good evidence that a large reservoir of auxin in an inactive bound form is present in most plant tissues, and this may be the source from which the invading organism produces free auxin. The auxin may also, however, consist of indole-acetic acid produced by oxidative deamination of tryptophane, which is, of course, present in plant proteins.

Transport and mode of action

The outstanding peculiarity of the auxins is that in the plant they are produced always at apical or terminal points, and are transported thence towards the base. This transport is strictly polar, i.e., it moves only from apex toward base, even in cut, isolated sections. The polar transport is not influenced by inverting the section, in other words the auxin continues to move from the morphologically apical to the morphologically basal end. However, turning the section through

90°C causes partial diversion of auxin to one side (cf. geotropism above). Also, as was shown strikingly by van der Weij, auxin will continue to be transported polarly even against an externally applied gradient of auxin concentration. If applied to the roots, or to the base of cuttings, auxins are drawn up in the transpiration stream just as are salts and other solutes, but as soon as they diffuse out of the xylem into living tissue they are re-exported polarly downward.

A supply of carbohydrate is needed for auxin to cause growth. In part this is required for cell-wall formation, since the wall consists of cellulose, which is built of hexose units. In part, however, it is required for respiration. Growth goes on only in O_2 and is poisoned by HCN and by ICH_2COOH just as is respiration. There is good evidence that the growth process is linked to one part of the total respiration, a part which involves also malic or fumaric acid. The inhibition of growth by ICH_2COOH is removed by adding malate. Starved coleoptiles show little response to auxin unless malate is added as well. The combination of auxin and malate then causes an increase in respiration which closely parallels the increase in growth. It has also been shown by Sweeney and Thimann that application of auxin to cut coleoptiles causes an immediate increase in the rate of protoplasmic streaming in the cells. This increased rate requires carbohydrate for its maintenance, is prevented by iodoacetate, and promoted by malate. It requires oxygen. There is thus a very close parallel between the action of auxin on growth, respiration, and streaming. The small part of the respiration involved represents less than 10 per cent of the total O_2 consumption of the tissue, but apparently controls all of the growth. Whether these deductions hold true for other tissues than the coleoptile remains to be proven.

Other Growth Hormones

For the growth of isolated roots in culture solution thiamin is essential. Some roots also require nicotinic acid or pyridoxin (Bonner). Since these materials are formed in the leaves and transported to the roots, they must be considered plant hormones. In the rooting of cuttings, auxin causes initiation of roots, but a few species need application of thiamin for the roots to grow out. Biotin and adenine also contribute to root formation in some instances. If plants produce unusually small amounts of thiamin, their growth may be accelerated by supplying thiamin to the whole plant by watering with a dilute solution (about 0.5 mg. per liter). This is true for camellias but for most other plants has been disputed and conflicting results have been obtained.

The growth of isolated embryos in culture provides an excellent test for the necessity of growth substances. They require in addition to thiamin both biotin and pantothenic acid. Growth is in some cases also stimulated by estrogens. If very young, still another factor, as yet unidentified but present in coconut milk, has been shown to be essential by van Overbeek, Conklin, and Blakeslee. Coconut milk is essentially a liquid endosperm; this factor is normally supplied to the embryo by the endosperm. Apparently it is needed only for the earliest stages of development.

When tissues are injured, the cells near the wound elongate or divide. This effect is due to a "wound hormone" set free by the injury. Extracts of ground or wounded tissues are active in causing this reaction, and the cells of the inner pericarp of the bean pod, which are especially responsive, have been used as assay material. When a drop of the active solution is placed on this tissue, a small wart or nodule is formed, the height of which is linearly proportional, within limits, to the concentration of the active substance. This wound tissue may involve cell division or it may not, according to the variety of bean employed. In this way English and Bonner isolated traumatic acid, or 1-decene-1-10-dicarboxylic acid, which is responsible for at least a large part of the wound hormone effect. There are also small nonspecific effects, exerted by water, neutral salts, auxins, and citric acid, but these never approach the magnitude of the traumatic acid response. The saturated acids, decane-1-10 dicarboxylic and sebacic ($C_{10}H_{18}O_4$) possess about half the activity of traumatic acid.

When leaves of Mimosa pudica, the "sensitive plant," are touched, burned, or otherwise stimulated, a stimulus travels down the petiole causing the leaflets to fold and the petiole to droop. Other leaves above or below may be stimulated. There is good evidence that this stimulus is due to a rapidly transported substance of hormonal nature. A hyand certain anthraquinone glucodroxyacid of unknown constitution, sides, have been implicated.

In order to account for the growth of leaf blades and other organs, Went has postulated the existence of a number of organforming substances, phyllocaline controlling leaf growth, caulocaline controlling stem growth, and rhizocaline controlling root growth. These are considered to act in addition to auxin. The evidence for them is as yet indirect and not entirely conclusive.The phenomena of flower formation, particularly the influence of the photo-period on flowering, has given rise to the concept of a flowering hormone or florigen. There is indeed evidence that such a substance is produced in leaves and transported to the buds, causing therein the change from the vegetative to the flowering condition, but there is no clear proof of its hormonal nature.

Plasm A region of cytoplasmic localisation in mosaic eggs which gives rise later to a definite organ or organ system.

Plasma Blood serum from which the fibrinogen has not been removed.

Plasma membrane Cell membrane.

Plasmalogens Acetal phospholipides which are present in small quantities in tissues and contain palmitaldehyde and stearaldehyde in place of fatty acids.

Plasmaphoresis The reduction of serum protein to a very low level by bleeding, removal of the cells from the exuded blood and reinjection of the cells in a protein-free solution.

Plasmid A class of circular, autonomously replicating, extrachromosomal DNA elements found in many bacteria. Contain origins of replication to ensure their maintenance. Some plasmids are capable of integrating into the host genome. A number of artificially constructed plasmids are used as cloning vectors or to alter the characteristics of the bacteria. Common plasmids are pBR322, pGEM, pUC18.

Plasmochin N-diethylamino-isopentyl-8-amino6-methoxyquinoline, light yellow; used in malaria to destroy gametes of parasite.

Plasmogamy A process of fusion of the cytoplasm of two cells; the first step in syngamy.

Plasmolysis The exosmosis of water from a cell, as when the cell is placed in a hypertonic solution. In a plasmolyzed cell the protoplasm shrinks away from the cell wall.

Plasmolytic method A method of determining the osmotic pressure of cells by placing the cell in solutions of varying concentrations. By microscopically observing the resulting turgidity or plasmolysis of the cell. one can determine the concentration of a solution which is isotonic to the cell and from this its osmotic pressure.

Plasteins The name given to the group of proteins synthesized in the presence of pepsin.

Plasticity The resistance that a solid or semi-solid offers to deformation of shape by shear or flow. In many respects analogous to viscosity in a liquid.

Plasticity Equivalent to *Pluripotence* ; the readiness of a part of the embryo for a large number of possible fates.

Plastid Structure in plant cells that is bound by a double-walled membrane and contains *DNA;* the *chloroplast,* which contains the light-absorbing substance chlorophyll, is one type of plastid.

Plate One of several slabs of Earth's crust that make up the ocean beds and continents and that float on and move around above the *mantle*; also called tectonic plate.

Plate tectonics Geological theory according to which the earth's crust consists of moving continental plates.

Plateau Region of fairly flat land located at a high elevation; or a series of mountain peaks of the same height.

Platelets Cellular constituents of blood, ordinarily numbering between 250,000-400,000/ccm. They contain the cephalin, etc., which is necessary to blood clotting.

Platyhelminthes Class of flat worms and numerous parasitic worms.

Pleiotropic A gene that affects several characteristics of the phenotype.

Pleiotropism The causation of multiple effects by one genetic factor, probably not directly, but because of an influence on metabolism which manifests itself in many different ways.

Plisade cell Plant cell rich in *chloroplasts,* located beneath the outer cell layer of the top of a leaf; site of *photosynthesis.*

Plummer-mendenhall reaction for theophylline A mixture of equal volumes of methyl alcohol solutions of saturated cupric acetate and theophylline gives a pale blue-green precipitate. Sensitivity—0.2 mg. per cc.

Pluripotent In a condition of having a large number of possible fates; undetermined.

Pneumonia Inflamation of the lungs due to infection by bacteria or specific viruses after a period of incubation followed by chills and fever, pain in chest and many systemic symptoms. The sputum is typed for the different kinds of pneumococci, and so is the blood, and even lung sections may be taken. Complications involve pleurisy, pericarditis and pulmonary edema. Serum therapy in order of frequency employs antisera of types I, II, V, VIII, VII, IV and XIV. Sulfapyridine therapy is used for all types. In emergency the sodium salt of sulfapyridine is injected intravenously. The caution in the use of sulfa drugs is to watch out for the development of acute hemolytic anemia and anemia, agranulocytosis. Oxygen therapy to prevent cyanosis is frequently indicated especially in the case of children and infants. Sulfathiazole has also been used with success.

Podoyhyllin A mixture of resins of the American mandrake; a purgative and liver stimulant.

Poh A term, analogous to pH, sometimes used to express directly the degree of alkalinity of a solution.

Point mutation A single nucleotide substitution within a gene; there may be several point mutations within a single gene. Point mutations do not lead to a shift in reading frames, thus at most cause only a single amino acid substitution.

Poise The unit of viscosity. It is the force (one dyne) which, when exerted on a unit area between two parallel planes one square cm. in area and one cm. apart, produces a difference in streaming between the two planes of one cm. velocity per second.

Poison ivy Rhus toxicodendron; poison vine; poison oak; contains toxicodenrol, toxicodendric acid and a resin; has been used as a rubefacient and irritant and internally in rheumatism.

Polar Bodies Small cells generated during during gamete meiosis, enabling much cytoplasm to be partitioned into just one of the four meiotic products, the ovum.

Polar Molecules Which contain elements differing in electronegativity, making them soluble in water.

Polarimeter An instrument used to measure the number of degrees an optically active substance rotates the plane of polarized light. It consists of two prisms so arranged that polarized light from the first prism can pass through the second when the instrument is set at zero. The rotation of a substance is determined by placing it between the two prisms and rotating the second to compensate for the change that the substance has caused in the plane of the light emerging from the first.

Polarized light Light carried by *electromagnetic waves* that are traveling along a single plane.

Pole Either of the north or south points on Earth's surface that sit on the axis of Earth's rotation.

Polenske number The number of cc. of 0.1N KOH needed to neutralize the non-volatile acids obtained by the saponification of 5 grams of fat or oil.

Poliomyelitis Infantile spinal paralysis;

HeineMedin disease; an acute infectious disease caused by a filtrable virus after an incubation period and showing involvement of the central nervous system in a meningitis stage and in an acute paralytic stage of many varieties, followed by atrophies in a prolonged postinflammatory stage. The chief mainstay in its treatment is convalescents' serum.

Pollen Grains that contain the male sex cells of a seed plant.

Pollination Process by which *pollen* travels from one plant to another, via wind, water, insects, birds, bats, or small mammals.

Polya tail After an mRNA is transcribed from a gene, the cell adds a stretch of A residues (typically 50-200) to its 3' end. It is thought that the presence of this "polyA tail" increases the stability of the mRNA (possibly by protecting it from nucleases). Note that not all mRNAs have a polyA tail; the histone mRNAs in particular do not.

Polyacrylamide gel (PAGE) Used to separate proteins and smaller DNA fragments and oligonucleotides by electrophoresis. When run under conditions which denature proteins (i.e., in the presence of 2-mercaptoethanol, SDS, and possibly urea), molecules are separated primarily on the basis of size.

Polycistronic Refers to a form of gene organization resulting in transcription of an mRNA that contains the coding sequences for multiple gene products, each of which is independently translated from the mRNA.

Polygalitol 15-anhydromannitol, found in Polygala species; sweet, m.p. 142°C.

Polygenic disorders Genetic disorders resulting from the combined action of alleles of more than one gene (e.g., heart disease, diabetes, and some cancers). Although such disorders are inherited, they depend on the simultaneous presence of several alleles; thus the hereditary patterns are usually more complex than those of single- gene disorders. Compare single- gene disorders.

Polygeny The determination of a phenotypic character by several genes.

Polymenorrhea Metrorrhagia (old term); too frequent menstrual flow.

Polymer— a large molecule constructed from many smaller identical units. These include proteins, nucleic acids, and starches.

Polymerase chain reaction A technique for replicating a specific piece of DNA *in vitro*, even in the presence of excess non-specific DNA. Primers are added (which initiate the copying of each strand) along with nucleotides and heat stable Taq polymerase. By cycling the temperature, the target DNA is repetitively denatured and copied. Because the newly synthesized DNA strands can subsequently serve as additional templates for the same primer sequences, successive rounds of primer annealing, strand elongation, and dissociation produce rapid and highly specific amplification of the desired sequence. PCR also can be used to detect the existence of the defined sequence in a DNA sample. A single copy of the target DNA, even if mixed in with other undesirable DNA, can be amplified to obtain billions of replicates. PCR can be used to amplify RNA sequences if they are first converted to DNA via reverse transcriptase. This two-phase procedure is known as 'RT-PCR'. * Polymerase Chain Reaction (PCR) is the basis for a number of extremely important methods in molecular biology. It can be used to detect and measure vanishingly small amounts of DNA and to create customized pieces of DNA. It has been applied to clinical diagnosis and therapy, to forensics and to vast numbers of research applications. It would be difficult to overstate the importance of PCR to science.

Polymerase An enzyme which links individual nucleotides together into a long strand, using another strand as a template. There are two general types of polymerase - DNA polymerases (which synthesize DNA) and RNA polymerase (which makes RNA). Within these two classes, there are numerous sub-types of polymerase, depending on what type of nucleic acid can function as template and what type of nucleic acid is formed. A DNA-dependant DNA polymerase will copy one DNA strand starting from a primer, and the product will be the complementary DNA strand. A DNA-dependant RNA polymerase will use DNA as a template to synthesize an RNA strand.

Polymerization Formation of a distinct compound by joining two or more molecules of the same kind.

Polymorphism Difference in DNA sequence among individuals. To be called a polymorphism, a variant should be present in a significant number of people in the population. Genetic variations occurring in more than 1% of a population would be considered useful polymorphisms for genetic linkage analysis. Compare mutation.

Polynomial Mathematical expression having one or more *variables.*

Polynucleotidase A heat resistant enzyme which becomes inactive at 85-95°C but regains its activity on cooling to 60°C (Dubos).

Polynucleotide Kinase Enzyme which catalyzes the transfer of the terminal phosphate of ATP to 5' hydroxyl termini of polynucleotides, either DNAor RNA. Usually derived from T4 bacteriophage.

Polypeptidases Peptidases which split peptides more complex than two amino acid groups, distinguished as amino polypeptidases and carboxypolypeptidases.

Polypeptides The combination of two or more amino acids, the amino group of one molecule combining with the carboxyl group of the other amino acid, with a molecule of water being split off in the process.

Polyphenol oxidase (1) Potato—4 copper protein enzyme which catalyzes the oxidation of polyphenols (catechol, adrenaline, dihydroxyphenylalanine) to quinones; can absorb CO.

(2) Agaricus Campestris — of mushroom origin, qualitatively as above.

Polyploid Having more than two sets of haploid chromosomes.

Polyploidogen A chemical substance which, by inhibiting certain phases of nuclear division, gives rise to polyploid cells and tissues.

Polyprotein A giant polypeptide that contains multiple individual protein sequences embedded within it and which must be proteolytically cleaved to yield the individual proteins.

Polysaccharides Carbohydrates formed by the condensation of two or more monosaccharides. Usually refers to the condensation product of an indefinite but very large number of monosaccharides, such as in starch and cellulose. They generally form hydrophilic colloids rather than true solutions.

Polysaccharides, classification of A convenient classification (Gortner) is.

I. Starches or Hexosans Glucosans, fructosans, mannosans, galactosans, amyloids.

II. Gums or Mucilages Pentosans, natural gums, pectins, plant mucilages.

III. Polymerized Glucosamine Chitin.

IV. Celluloses Hemicelluloses, true celluloses, compound celluloses.

One must remember that noncarbohydrate groups are frequently also present in the natural compounds.

Polysome A complex of an mRNA molecule and many ribosomes.

Polysome Complex of mRNA and several ribosomes.

Polyzime A preparation of the enzymes of Aspergillus oryzae.

Pons reagent for albumin A 11000 aqueous solution of the sodium salt of sulfochondroitic acid is a specific reagent for albumin, forming precipitates in dilute acetic acid solutions. Limit of detection—5 mg. albumin per liter.

Population Any group of interbreeding organisms.

Populin $C_{20}H_{22}O_8$; 6-benzoylsalicin; a glycoside of the bark of Populus, consisting of glucose, saligenin and benzoic acid; needles, m.p. 180°C; used as antipyretic in urinary affections.

Porphyrins A group of pigments, part of the

hemoglobins and chlorophyll, possessing the porphyn structure. The more important porphyrins derived from hematin are.

Hemin—$C_{34}H_{32}O_4N_4FeC_l$

Protoporphyrin—$C_{34}H_{34}O_4N_4$

Hematoporphyrin—$C_{34}H_{38}O_6N_4$

Mesoporphyrin—$C_{34}H_{38}O_4N_4$

Aetioporphyrin—$C_{32}H_{38}N_4$

Deuteroporpyhrin—$C_{30}H_{30}O_4N_4$

Deuterohemin—$C_{30}H_{28}O_4N_4FeC_l$ Which differ in the nature of the side chains which are methyl, ethyl, -CH_2CH_2COOH and CH_2CH-.

Porphyropsin A compound of the retinae of fresh water fishes, analogous to rhodopsin, consisting of a retinene component and vitamin A_2.

Porphyroxine $C_{19}H_{23}O_4N$; prisms, m.p. 134-135°C; opin; an alkaloid of opium. The hydrochloride crystallizes as prismatic needles, m.p. 155°C.

Position effect A change in the phenotypic effect of a gene owing to a change of its position on the chromosome.

Post absorptive condition State of an animal in which last food intake does not essentially affect its metabolic rate. (Twelve hours after last food for humans, dogs and rats, four days for ruminants.)

Post-transcriptional regulation Any process occurring after transcription which affects the amount of protein a gene produces. Includes RNA processing efficiency, RNA stability, translation efficiency, protein stability. For example, the rapid degradation of an mRNA will reduce the amount of protein arising from it. Increasing the rate at which an mRNA is translated will increase the amount of protein product.

Post-translational processing The reactions which alter a protein's covalent structure, such as phosphorylation, glycosylation or proteolytic cleavage.

Post-translational regulation Any process which affects the amount of protein produced from a gene, and which occurs AFTER translation in the grand scheme of genetic expression. Actually, this is often just a buzz-word for regulation of the stability of the protein. The more stable a protein is, the more it will accumulate.

Potential difference Between two points, the amount of electrical energy required to move a single unit of electric charge from one point to the other.

Potential energy stored energy that results for the specific location or arrangement of matter.

Potentials, bioelectric It is a century since the work of Du Bois Reymond clearly demonstrated the existence of potentials in tissues and changes in these potentials with physiological state. Although the energy involved in bioelectric phenomena is negligible, they contribute the most important mechanisms for the release and control of action. For, probably, all living cells have voltages across their outer membrane (of the order of 50 mV); these voltages and the membrane resistance (some hundreds of ohms per square centimeter) across which they act drop sharply during excitation and response; and the resulting currents flow to other cells and cell regions and so help to pass on activity. Such bioelectric changes are, indeed, at the basis of the propagated nerve impulse, itself.

The very elongated cells or fibres of nerve and striated muscle and the large cells of certain algae (Nitella, Valonia, Halicystis, Chara) have especially lent themselves to the analysis of bioelectric phenomena; but potential, resistance and capacity measurements on the heart, skin, eye, gut, glands, brain, plant structures, etc., have been of great use as a means of studying these organs and tissues. The very special electric organs evolved by certain rays, skates, eels, etc., as a weapon of attack or defense and able to deliver shocks at several hundred volts—have also been useful objects for study, since they are only modified muscle elements built in series instead of in parallel.

When a microelectrode is thrust into a single cell or fibre, it registers about 50 mV negative to a similar electrode placed outside the unit. With both outside or both inside, essentially no potential is recorded when the system is at rest —indicating that the potential is generated across the membrane. (Some potentials exist from pole to pole of intact elongated cells, but these are as yet little studied.) The membrane potential is maintained, despite leakage of current, by the cell metabolism. When cell oxidations are stopped by deprivation of oxygen or by respiration poisons (cyanide, anaesthetics; or by blocking fermentation) the potential falls. The cell thus acts as a self-charging storage battery.

The exact mechanism for transforming chemical into electrical energy is not known, but the potassium ion is probably involved. Cells regularly contain potassium salts in their protoplasm (or inner sap) in many times higher concentration than these salts are present in surrounding fluids; and the membrane potential is about what it should be for such concentration differences. Further, when metabolism is interfered with the internal potassium leaks out; also, when outside potassium is raised the membrane potential falls. Other ions may well play a role, and other substances, especially acetylcholine, are also able to influence strongly the membrane potential.

In nerve or muscle, with units several centimeters long arranged parallel to each other, it is possible to follow electric phenomena on the gross structure. A narrow flat muscle (the frog's sartorius) has been much used. When ordinary electrodes are placed on the two ends of an isolated but essentially uninjured specimen they record little or no potential. When one end is cut off, or the fibres there otherwise injured, that electrode becomes negative (some 15 mV) to the other. This injury or resting potential (or current, as the case may be) is really a portion of the membrane potential, the injured region being in effect a partial lead from the inside. If, now, the muscle is made to contract the potential difference becomes momentarily less. Since the injured end does not change, this means that the electrode outside the intact portion of the muscle fibres has become temporarily less positive. This is the action potential (or current)—a decrease of the voltage across the cell membrane when the cell is active. It can be demonstrated with a pair of microelectrodes across the membrane, as in the case of the resting membrane potential; and, indeed, the change during activity may be so great that the resting potential is actually reversed in direction.

The action potential is not simultaneously present over an entire nerve or muscle fibre but travels from the region of stimulation. In human nerves, the nerve impulse and the attendant action potential travel at rates up to 150 meters a second; in muscle, propagation is under 5 meters a second. In either case, the change spreads away from the point, stimulated, equally well in either direction. If, therefore, one end of the muscle is stimulated and electrodes are placed at two positions along the muscle's length, the action potential will reach the near one first, and make it negative to the far one, while shortly after the far one will become active and negative to the near one. Such a diphasic response is obtained from intact tissue. If the far end has been previously injured so that it cannot become active, then only the first negative wave occurs and the response is monophasic. A monophasic action potential can also be obtained with microelectrodes across a single portion of the membrane.

Besides the simple resting and action potentials, other pre- and after-potentials are related to excitation and response, especially in nerve. The latter may be positive and negative alternately and may last many seconds or even minutes beyond the main action potential, or spike, which is ordinarily complete within a thousandth of a second. They, however, rarely amount to more than a few per cent of the spike in voltage. The prepotentials are associated with excitation and occur before the spike or, when stimuli are too weak to evoke a full response, without it. They may also oscillate, as do the after-potentials, but more rapidly.

Membrane potential and resistance are profoundly related to the spread of the nerve or muscle (or heart, etc.) impulse. The region initially stimulated loses its polarization, the potential across its membrane; and this loss of polarization leads to the cell response. Thus the negative pole of a current, applied to the outside of a muscle, neutralizes the original membrane potential (which is positive outside), and the positive current pole increases it. The cell is depolarized at the negative pole and it is here, not at the positive pole or anywhere between, that the response begins. As a region of membrane is depolarized some critical physicochemical reaction occurs in the membrane—the full action potential develops and the membrane's electrical resistance falls greatly. This causes currents to flow between the active part of the membrane and adjacent regions which are still at rest. But these currents depolarize, and so stimulate, these new regions which, in turn, give their explosive action and cause currents to flow through yet further resting regions. This propagated excitation of each resting region by the oncoming activity is the mechanism of transmission in the body—of the nerve and muscle impulse.

Whether electric currents are similarly responsible for excitation from one cell to another is not so certainly established. Receptors do develop potentials when stimulated which might excite the sensory nerve fibers. Currents do flow across nerve cells and fibers at synapses, and between nerve fibers and muscle (with the aid of special connecting end-plates and their potentials), and might easily excite across the cell junctions. It is also possible, however, that acetylcholine and other substances, which are freed in connection with the activity of nerves and their endings and which can excite effectors and depolarize them, are involved. Probably electrical and chemical mechanisms cooperate in transmission across junctions as they do in transmission along nerve or muscle fibers.

Potentiometric titration Titration whose course is followed by measuring the changes in the e.m.f., due to changes in pH, which occur when one reagent is added to the other.

Pozzi-escot micro-test for aminobenzoic acid The ester-containing urine distillate is saponified in a closed tube with sodium hydroxide solution, neutralized and treated with copper sulfate solution. Blue crystals of copper anthranilate are formed if o-aminobenzoic acid is present.

Pozzi-escot reaction for rotenone When sulfuric acid solutions of mercuric oxide and rotenone are mixed, a white precipitate insoluble in the cold is formed. The precipitate dissolves on heating, yielding an intensely colored solution.

Pre Progesterone Response Element A binding site in a promoter to which the activated progesterone receptor can bind. The progesterone receptor is essentially a transcription factor which is activated only in the presence of progesterone . The activated receptor will bind to a PRE, and transcription of the adjacent gene will be altered.

Prebula-mccollum reaction for vitamin b_1 Vitamin B_1 gives a characteristic purple-red compound with a diazotized solution of either p-aminoacetanilid or methyl-p-aminophenylketone. The compound is stable, very insoluble in water and completely soluble in xylene.

Precipitation Solid substance that separates from a *solution* as a consequence of a physical or chemical change occurring in the solution. (*Earth science*) deposition of hail, mist, rain, sleet, or snow onto a planet's surface.

Precipitation Curve, Heidelberger Curve A Plot of the quantity of precipitate formed during the titration of antibody with an antigen, or vice versa. The antibody must be at least bivalent. The precipitation curve reaches a maximum at the Equivalence point, then decreases as the concentration of one component exceeds its optimum (Fig.)

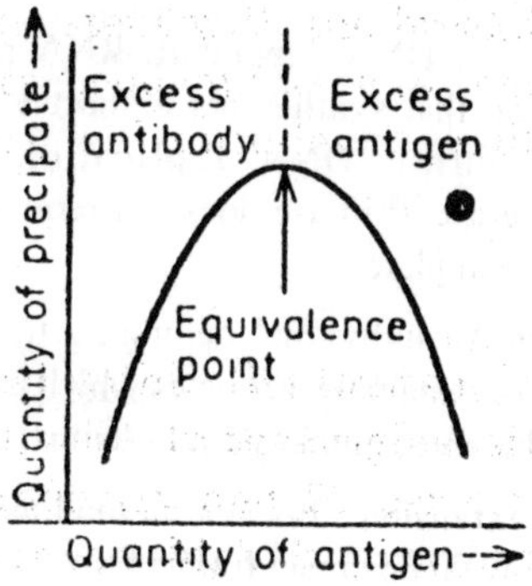

Fig. Preceipitation Curve

Preformation The theory that all the structures of an organism are present in one of the gametes.

Pregnandiol $C_{21} H_{36} O_2$; m.p. 235°C; an inactive sterol found with progesterone in the corpus luteum and in pregnancy urine. Its structure is that of a saturated progesterone with the two carbonyl groups reduced to alcoholic groups.

Pre-mrna An RNA molecule which is transcribed from chromosomal DNA in the nucleus of eukaryotic cells, and subsequently processed through splicing reactions to generate the mRNA which directs protein synthesis in the cytoplasm.

Pressure (physics) force exerted by a fluid on each square surface unit of an object immersed in the fluid

Presumptive Having regard to the normal fate of the part in question, e.g. presumptive skin or presumptive ear-vesicle. Primary structure The sequence of amino acids and any interchain and intrachain disulfide bonds of a protein.

Primary structure Refers to the sequence of amino acid residues or nucleotides within protein or nucleic acid molecules, respectively.

Primary transcript When a gene is transcribed in the nucleus, the initial product is the primary transcript, an RNA containing copies of all exons and introns. This primary transcript is then processed by the cell to remove the introns, to cleave off unwanted 3' sequence, and to polyadenylate the 5' end. The mature message thus formed is then exported to the cytoplasm for translation.

Primase An DNA-dependent RNA polymerase function that synthesizes a short RNA primer used during DNA replication by DNA polymerase. Primase does not require a primer, only a template.

Primates An individual belonging to the highest order of mammals, Primates which includes man and the apes, monkeys, and lemurs.

Primer extension This is a method used to figure out how far upstream from a fixed site the start of an mRNA is. For example, perhaps you have isolated a cDNA clone, but you don't think that the clone has all of the 5' untranslated region. To find out how much is missing, you would first sequence the part you have, and figure out which strand is coding strand (usually the coding strand will have a large open reading frame). Next, you ask the DNA Synthesis Facility to make an oligonucleotide complementary to the 5'-most region of the coding strand (and thus complementary to the mRNA). This "primer" is hybridized to mRNA (say, a mixture of mRNA containing the one in which you are interested), and reverse transcriptase is added to copy the mRNA from the primer out to the 5' end. The size of the resulting DNA fragment shows how far away from the 5' end your primer is.

Primer A small oligonucleotide (anywhere from 6 to 50 nt long) used to prime DNA synthesis. The DNA polymerases are only able to extend a pre-existing strand along a template; they are not able to take a naked single strand and produce a complementary copy of it de-novo. A primer which sticks to the template is therefore used to initiate the replication. Primers are necessary for DNA sequencing and PCR.

Primeverose Glucose-6-beta-d-xyloside, a constituent of the glycosides, primeverin, gaultherin, rhamnicosin, genticaulin; m.p. 208°C.

Principle of segregation Diploid organisms inherit a pair of genes for each trait (on homologous chromosomes).

Prion (acronym for *proteinaceous infectious particle*) (life science) ultrasmall (more than 100 times smaller than a *virus*) particle composed of protein, believed to cause certain diseases of the nervous system in some mammals; it may be considered a microorganism, although it lacks genetic material.

Probe Single- stranded DNA or RNA fragment of specific base sequence, labeled either radioactively (often incorporating 32P or 35S) or immunologically, that are used to detect the complementary base sequence by hybridization. For example, if you want to quantitate the levels of alpha subunit mRNA in a preparation of pituitary RNA, you might make a radiolabeled

RNA in-vitro which is complementary to the mRNA, and then use it to probe a Northern blot of the pit RNA. A probe can be radiolabeled, or tagged with another functional group such as biotin. A probe can be cloned DNA, or might be a synthetic DNA strand. As an example of the latter, perhaps you have isolated a protein for which you wish to obtain a cDNA or genomic clone. You might (pay to) microsequence a portion of the protein, deduce the nucleic acid sequence, (pay to) synthesize an oligonucleotide carrying that sequence, radiolabel it and use it as a probe to screen a cDNA library or genomic library. A better way is to call up someone who already has the clone.

Proboscideans Relatives of the elephant, including the extinct mammoth and mastodon.

Procaine Planocaine; novocaine; ethocaine; diethylamino-ethyl-p-aminobenzoate; m.p. 59-61°C; used as hydrochloride for local anaesthesia; less toxic than cocaine.

Processing The reactions occurring in the nucleus which convert the primary RNA transcript to a mature mRNA. Processing reactions include capping, splicing and polyadenylation. The term can also refer to the processing of the protein product, including proteolytic cleavages, glycosylation, etc.

Progesterone $C_{21} H_{30} O_2$; needles, mp. 121°C, prisms m.p. 128°C; hormone of the corpus luteum. It inhibits the release of ova during pregnancy, and is apparently necessary for the maintainance of the pregnancy. Given subcutaneously in oil solution. One gram=1000 International Units.

Progesterone A hormone produced by the ovaries that controls secretion patterns of other hormones involved in reproduction.

Progynon A trade name for the female sex hormone.

Prokaryote Cell or organism lacking a membrane-bound, structurally discrete nucleus and other subcellular compartments. Bacteria are prokaryotes. Compare eukaryote.

Prolactin Galactin; mammotropin. The protein hormone of the anterior pituitary that stimulates the mammary glands to secretion.

Prolactin units The assay of the activity of a preparation in increasing the weight of the crop sac of the pigeon dove in terms of 0.1 gram of an International standard preparation.

Prolidase A proteolytic enzyme that hydrolyzes peptide linkages involving the imino group of proline.

Prolinase The only proteolytic enzyme that splits a peptide linkage where the carboxy group of proline is involved.

Proline metabolism Proline is oxidized to glutamic acid, which then may be further oxidized to ammonia and ketoglutaric acid, which on ß-oxidation ultimately yields $C0_2$ and H_2 0. Alternately, the glutamic acid may combine with ammonia to yield glutamine. It is anti-ketogenic as it can form glucose.

Prolycopene A naturally occurring stereoisomer of lycopene, especially in the "tangerine tomato." It is a C_{40} carotenoid and is one of 72 possible isomers of lycopene.

Prolysine An amino-acid of dubious constitution, believed to be alpha-aminoä-hydantoin valeric acid; m.p. 222°C.

Promoter site Region of a DNA molecule 5' to a coding sequence that is responsible for assembly of an RNA polymerase complex and the initiation of transcription.

Promoter The first few hundred nucleotides of DNA "upstream" (on the 5' side) of a gene, which control the transcription of that gene. The promoter is part of the 5' flanking DNA, i.e. it is not transcribed into RNA, but without the promoter, the gene is not functional. Note that the definition is a bit hazy as far as the size of the region encompassed, but the "promoter" of a gene starts with the nucleotide immediately upstream from the cap site, and includes binding sites for one or more transcription factors which can not work if moved farther away from the gene.

Prontosil $C_{12} H_{13} O_2 N_5$ HCI; m.p., 247-251°C; used for treatment of streptococcal and staphylococcal infection, similar to sulfanilamide.

Proof-reading Mechanism for correction of errors made during synthesis of nucleic acids or polypeptides by scrutiny of the products after the nucleotides or amino acids have already been incorporated.

Prophase The first stage of cell division, when chromosomes condense and become visible. During prophase of meiosis 1, synapses and crossing over occur.

Prosecretin An inactive precursor of secretin, into which it is transformed by the action of acid.

Prospective potency (prospective Potenz; Potentialité totale) The sum total of possible fates of any part of the embryo at the beginning of development. In regulative eggs prospective potency is much wider than prospective significance; in mosaic eggs the two are identical. "Potency" should carry the undertone of "possibility", not "power". Potency differs from competence in that it characterises the whole period of development, including the time before any competences have arisen. It makes no reference to any causal process.

Prospective significance The actual normal fate of any part of the embryo at the beginning of development.

Prostag Landins A member of Lipid Compounds that are derived from fatty acid and have important function in animal body. They are mediators and have a variety of strong physoilogical effects.

Prostaglandin Any of a family of chemicals that serve to transport messages between cells.

Prostate gland Gland in male mammals that produces a fluid present in *semen*

Prostate A male gland that produces a milky white, alkaline fluid that activates sperm.

Protagon An extract of brain, probably impure lecithin.

Protalbinic acid A sol containing the sodium salts of the degradation products of albumin by alkali.

Protamines In the American classification of proteins, the protamines are simple proteins, soluble in water and other neutral solvents. They are large polypeptides, containing only about 15 peptide linkages, with the preponderance of amino acids being diamines. Therefore the protamines are basic. They precipitate other proteins from aqueous solutions, probably as coacervates. Highly toxic on inintravenous injection. Example clupeine from herring, scombrine from mackerel.

Proteans In the American classification of proteins, the proteans are the first products of hydrolysis of proteins. The closely resemble the globulins and glutelins.

Proteasome A specialized organelle within the cytoplasm of a cell that is responsible for degradation of cytoplasmically situated proteins. The proteasome plays a key role in normal protein turnover and in peptide presentation by MHC Class I antigen.

Protective colloid A lyophilic colloid adsorbed by a lyophobic colloid, and which acts to stabilize the latter, or vice versa.

Protein Class of biochemical compounds constructed from amino acids. Proteins may be structural, such as those that make up hair and cartilage, or they may be reactive, such as the enzymes.

Protein Any of the many substances composed of *amino acids* that are built by cells according to instructions encoded in the genetic material; proteins perform a wide variety of essential functions in living organisms protein A macromolecule composed of one or more polypeptide chains. Each polypeptide chains has a characteristic sequence of amino acids linked by peptide bonds. The molar mass is usually above 100,000.

Protein derivatives, primary In the American classification of Proteins, those protein derivatives arrived at by hydrolysis, which involve only slight changes in the protein molecule. Insoluble in water and neutral salts, coagulated by heat. Consist of proteans, metaproteins and coagulated proteins.

Protein derivatives, secondary In the American Classification of Proteins, those hydrolysis products of proteins which represent further cleavage of the protein molecule than in the primary protein derivatives. Soluble in water, not coagulated by heat. Consists of the proteoses, peptones, and peptides.

Protein A large molecule composed of one or more chains of amino acids in a specific order; the order is determined by the base sequence of nucleotides in the gene coding for the protein. Proteins are required for the structure, function, and regulation of the bodys cells, tissues, and organs, and each protein has unique functions. Examples are hormones, enzymes, and antibodies.

Proteinaceous Describes any structure which is composed of protein.

Proteinases Proteolytic enzymes that attack proteins and peptides of suitable structure. Their point of attack is at the middle of the chain, i.e. at -C·N=, so that they are called Ö endopeptidases. Their action is better in the absence of nearby free amino or carboxyl groups.

Proteins Extremely complex, nitrogen containing organic compounds (or mixtures), found in every living organism and essential to its life. Used largely as the building material for the body, but in times of necessity can be utilized for energy. On hydrolysis with acid, base or enzyme, yield mostly áamino acids.

Proteins, acid and base binding by Since proteins form colloidal solutions this factor has always made it difficult to study the binding of acid and of base by proteins. Nevertheless, demonstrations have been made of their amphoteric nature in several ways.

(1) protein compounds may be precipitated out, e.g. calcium caseinates.

(2) proteins may be titrated by dissolving water insoluble substances, e.g. edestin dissolving strychnine;

(3) potentiometric titration;

(4) compound formation with gaseous bases or acids out of solution; and.

(5) other physical methods, as measuring electrical conductivity, depression of freezing point which, however, are not clear cut.

Proteins, american classification A classification recommended by committees of the American Physiological Society and the American Society of Biological Chemists is.

I. The Simple Proteins.

The Albumins.

The Globulins.

The Glutelins.

The Prolamines or Alcohol-Soluble Proteins.

The Albuminoids.

The Histones.

The Protamines.

II. Conjugated Proteins.

Nucleoproteins.

Glycoproteins or Glucoproteins.

Phosphoproteins.

Chromoproteins.

Lecithoproteins.

Lipoproteins.

III. Derived Proteins.

A. Primary Protein Derivatives.

Coagulated proteins.

Proteans.

Metaproteins.

B. Secondary Protein Derivatives.

Proteoses.

Peptones.

Peptides.

Diketopiperazines.

For definitions.

Proteins, conjugated In the American Classification of proteins a combination in a form other than a salt of a simple protein with a non-protein group.

Proteins, derived The various products obtained by the action of acids, enzymes, bases, or heat on proteins, including the hydrolysis products, and the synthetic proteins and polypeptides.

Proteins, primary derivatives In the American Classification of proteins, the primary protein derivatives are those derivatives produced by hydrolysis, which only involve minor changes in the protein molecule.

Proteinuria The presence of excess protein, particularly albumin and globulin, in the urine. Found in Bright's disease, nephrosis and other febrile conditions, and temporarily after violent exercise.

Proteoses In the American classification of Proteins, those secondary protein derivatives characterized by noncoagulability by heat solubility in water, and precipitation from solution by saturating half with zinc sulfate; albumoses.

Prothetely The appearance of structures normally associated with a later stage of development, at an earlier stage of development, e.g. pupal organs in larval insects.

Prothrombin The inactive precursor of thrombin. It is a protein, possibly a globulin.

Protist Single-celled *eukaryote* belonging to the *kingdom* Protista.

Protocatechuic acid 3,4-dihydroxy-benzoic acid, a constituent of tannins and occurring free in the onion.

Protochordata Trasition forms of chordata which have a notochord but no true backbone.

Protocotoin Piperonyl-phloroglucinol-dimethyl esater; $C_{16} H_{14} O_6$; crystalline principle from paracoto bark.

Protocurarine An alkaloid, $C_{19}H_{24} O_2$,N; from "pot curare" from Strychnos castelnaea Wedd; a powerful poison.

Protoferriheme The non-protein portion of catalase.

Protones Intermediate products of the hydrolysis of protamines; give strong biuret reaction, more soluble than protamines which they resemble.

Proto-oncogene A gene present in a normal cell which carries out a normal cellular function, but which can become an oncogene under certain circumstances. The prefix "c-" indicates a cellular gene, and is generally used for proto-oncogenes (examples *c-myb* , *c-myc* , *c-fos* , *c-jun* , etc).

Protopectin The member of the group of pectins which is the precursor of the other two members.

Protopectinase An enzyme that hydrolyzes protopectin away from the cell walls, probably yielding pectin.

Protopine $C_{20} H_{19} O_5$ N; monoclinic crystals, m.p. 207°C, aurichloride is yellow crystalline powder, m.p. 198°C; fumarine; alkaloid of a large variety of plants, notably Dicentra spectabilis, and of opium. In frogs is a narcotic. In mammals causes epileptiform convulsions and circulatory damage. Is also a local anesthetic.

Prunitrin Glucoside,$C_{22} H_{24} O_{11} \cdot 4H_2O$, from the bark of a species of Prunus.

Prussian blue $Fe_4 [Fe(CN)_6]_3$; ferric ferrocyanide used in making and grading ultrafilters.

Pseudoephedrine An alkaloid; stereoisomer of ephedrine from leaves of ephedra vulgaris; m.p., 118-119^{0}C.

Pseudogene A region of DNA that shows extensive similarity to a known gene, but which cannot itself function, either because it has lost the signal required for transcription (the promoter sequence) or because it carries mutations that prevent it from being translated into protein. It is generally assumed that pseudogenes are copies of mRNA. Pseudogenes are known in the globin system but also in many other multigene families.

Pseudoglobulin A globulin found in the blood serum. Unlike euglobulin it contains no phosphorus, is soluble in water and is precipitated by ½ saturated ammonium sulfate.

Pseudohyoscyamine An alkaloid, $C_{16} H_{32}O_1$ N; has been suggested as sedative and hypnotic to control excitability.

Pseudoknot A feature of RNA tertiary structure; best visualized as two overlapping stem-loops in which the loop of the first stem-loop participates as half of the stem in the second stem-loop.

Pseudo-peroxidases Enzymes, such as iron porphyrins, which catalyze the same kinds of reactions as peroxidase, but at a much lower rate; differ also by being thermostable.

Pseudorevertant A mutant virus or organism which has recovered a wildtype phenotype due to a second-site mutation (potentially located in a different region of the genome, or involving a different polypeptide) which has eliminated the effect of the initial mutation.

Pseudostrophanthin A glucoside from seed of Strophanthus Kombe; is commonly called H-strophanthin and is said to be twice as powerful as Kstrophanthin.

Pseudoyohimbine An alkaloid, $C_{21}H_{26}O_3N_2$; from bark of Coryanthe Yohimbe and melting at 264-265^{0}C.

Psoriasis A chronic skin disease of unknown cause showing as papules and scales which are shed. Purine-low diets relieve symptoms, but the disease is incurable and is the stamping ground for any new remedy. Temporary results have been reported for sodium cacodylate, thyroxin, chrysarobin (an irritant), radiation (X-rays, sunlight, ultraviolet light).

Psychiatry, biochemistry in relation to Recent reviews of the literature, on biochemical investigations in psychiatry present a confusing array of conflicting reports which leave the reader with the impression that biochemical methods have contributed little to the understanding of mental disorders. Cameron in the recent revision of his book, *'Objective and Experimental Psychiatry'* has, however, made a beginning in relieving this confusion. It is the purpose of the present communication to demonstrate that biochemistry has already contributed a great deal to our understanding of mental disorders and to indicate the direction of present day investigations.

Biochemistry, the chemistry of living organism, is such a broad subject and has undergone such rapid development in so many fields that few people realize the extent of this science and its import for medicine and psychiatry.

In fact, every kind and degree of adjustment that a person makes in his daily life involves complex biochemical changes in his body. For example, social situations which produce fear or rage or other emotions constitute one of the most effective ways of bringing about biochemical changes in man and animals. The recognition of the fact that mental activity influences bodily processes in general has led many people to postulate that most diseases were "caused" by mental and emotional disorders. The great contributions of Pavlow, Freud, and Cannon have served to support such a conception; it should be emphasized however that these authorities themselves, being aware of the limitations of their work, made no such one-sided generalization.

Dramatic impetus has been given to an interest in the methods of physics and biochemistry by the extraordinary, if transient, relief of symptoms in mentally ill patients induced by insulin or metrazol, or electric shock. Unfortunately this interest tends to be as transient as the periods of symptomatic relief in the patients. The fact, however, that chemical and physical methods can produce these striking changes, should lead to a thorough exploration of biochemistry and physics in relation to mental processes in an effort to find more effective methods of treatment.

Of the numerous ramifications of biochemistry the development of our knowledge of energy exchange in man and animals may be chosen as an excellent example of the intimate relations between behavior and the chemistry of the body. Originally this process was conceived of as being relatively simple; food as calories when burned in the body produced an output of energy which could be measured in units of heat and of work. The elucidation of the steps involved in this energy exchange has opened whole new fields in chemistry and in medicine. Yet many gaps remain to be filled. Numerous contributions from these fields are important for psychiatry and will probably be even more important in the future.

The recognition of the importance of thyroid disorders in producing states of excitement or lethargy is one of the most brilliant contributions which may be cited. Not many years have elapsed since the time when patients with myxedema and hyperthyroidism could be found immured in mental hospitals along with the stuporous and excited states of unknown etiology. Although methods for the diagnosis and treatment of thyroid diseases have been remarkably advanced, much remains to be learned. Many patients in states of lethargy and with low basal metabolic rates do not respond to treatment with thyroid hormone and presumably do not have thyroid deficiency. Again, many excited and agitated patients appear to have hyperthyroidism with basal metabolism rates above normal, yet removal of their thyroid glands does not relieve their symptoms. These patients are frequently mistreated because of the lack of methods for determining the amount of thyroid activity.

While it has long been known that iodine constituted a unique part of the thyroid hormone, the determination of blood serum iodine as an index to the amount of hormone has only recently been possible due to the technical difficulties of microchemistry. Man and Riggs, , , have devised a practical method for determining the protein-bound iodine in serum and have shown that this fraction can be employed as a measure of the amount of active thyroid hormone. This method has proved effective in differentiating excited and depressed patients with hyperthyroidism from those with non thyroid disorders.

In connection with the biochemistry of thyroid disorders it should also be emphasized that these conditions are often precipitated by emotional conflicts. Consequently one has the opportunity to observe the interplay of constitutional predisposition, environmental difficulties, both psychological and dietary, and biochemical factors which enter into the process of making the person ill.

Only one aspect however, of the problem of energy exchanges in man is illuminated by the advances in biochemistry of the thyroid. Each endocrine organ produces one or more chemical substance; the hormones. The amount each gland secretes is controlled by the nervous system, which itself produces acetyl choline, and by the interaction of the hormones themselves. While it is known that animal drives and energy output are dependent on these delicate chemical intergrations, our knowledge is too incomplete to make it possible to understand and control the behavior of the whole animal.

One of the most common complaints which physicians are asked to relieve is lack of energy. While it is well known that any disturbance of the delicately balanced chemical and physiological arrangements of the body, may produce varying degrees of diminution of energy output, a large proportion of these patients who lack energy do not appear to suffer from any of the known organic diseases. Many of these patients are eventually referred to a psychiatrist. As a considerable number of them are seriously incapacitated for months or years, no matter what psychotherapeutic methods may be employed, the psychiatrists have had ample opportunity to observe the course of their conditions.

We have learned that the ability or lack of ability to convert food into a large amount of work or play tends to be characteristic of certain families and that in members of such families high levels of activity may not be sustained but are interrupted by periods of inactivity. We have also observed that these cyclic disorders are more common in women and in people of stocky body-build. Such observations have led some psychiatrists to lay stress on genetic and constitutional factors as determining a person's potential energy and susceptibility to disorders in energy output.

In addition to these constitutional factors, many others take part in determining a person's potential capacity for work or play. Only a few can be mentioned here.

The brain must have a continual supply of oxygen, carbohydrates and enzymes. Biochemical and physiological investigations, , , have demonstrated that a few moments of oxygen and carbohydrate deprivation result in interruption of the nervous system activity. If, the period of deprivation is prolonged more than these few moments death occurs; or if restoration occurs shortly before death, some cells of the brain may be destroyed, thereby permanently altering the person's potentialities for certain kinds of activity. This knowledge has led to a revision in the psychiatric interpretation of the significance of birth injuries and of the effects of carbon monoxide poisoning, of convulsive seizures and of other conditions which impair the oxygen supply to the brain. Particularly relevant is the work of the neurologists, showing that certain brain lesions may produce queer forms of overactivity while others will result in underactivity.

The successful correlation of physiological, electrophysical and chemical methods in the investigation of patients with epilepsy, by Lennox, Gibbs, Gibbs, and Nims, , constitutes an important advance. They were thus able to demonstrate defects in the control of respiratory metabolism in epileptic patients which would have been overlooked by a one-sided approach.

The development of the field of carbohydrate metabolism has made it possible to recognize and treat previously mysterious states of weakness, fainting and convulsions due to low blood sugar. The effects of the pancreatic adenomas should be mentioned in this connection. The demonstration of the sensitivity of the brain to lack of carbohydrate and Cannon's work on the effects of emotions have led psychiatrists all over the world to investigate carbohydrate metabolism in excited and depressed patients. As has so frequently happened in other fields of psychiatry, a great variety of conflicting discoveries have been reported. On the basis of recent work one can conclude that most of the apparently excited and overactive as well as depressed and underactive patients have normal amounts of sugar in their blood. , This was considered contrary to Cannon's findings of emotional hyperglycemia, in animals and of frequent glycosuria in students following examinations. Some of the psychiatrists, particularly Whitehorn and Bowman, were led to conclude that emotional hyperglycemia must be uncommon in man. Further studies, however, by Gildea, Mailhouse, and Morris have demonstrated that severe emotional reactions in normal people such as those produced by experiencing automobile accidents, by seeing a suicide, or by husbands' waiting for their wives to be delivered of babies, were commonly followed by hyperglycemia. It is probable therefore, that the abnormality in emotionally disturbed patients lies in their failure to mobilize sufficient sugar to raise the blood levels no matter how extreme their apparent emotional excitement may be.

Investigations of sugar tolerance curves have yielded conflicting results. It now appears however, that if allowance is made for the nutritional status, the nature of the diet, the disturbances in intestinal motility, (the latter can be ruled out by using the intravenous method) the sugar tolerance curves in excited and depressed patients are rarely remarkable. The defect therefore probably lies in the nervous regulation of carbohydrate mobilization. In one group patients have low levels of blood sugar and spells of weakness unless frequently fed. In another group patients eat large amounts of food to relieve emotional tension, although they are frequently obese and become increasingly so with the passage of years. Diabetes mellitus not infrequently develops in this latter group.

Although the biochemists have established the importance of protein in determining the fundamental activities of biological organisms, little work has been done in the metabolism of proteins in patients with mental diseases. Gjessing's meticulous studies of nitrogen balances in catatonic patients constitutes one of the few exceptions. His results suggest the presence of a disturbance in protein chemistry in these patients but give no clue to its nature.

The recent developments in still another branch of biochemistry, i.e., the vitamin field, have given us new insight into a certain small group of patients suffering from lethargy and hypochondriasis, for these symptoms may be due to partial deficiencies in B_1, nicotinic acid, and other factors of the B complex.[35] It must, however, be stressed that most of the patients with depression of mood and energy do not respond to vitamin therapy.

On the other hand, the fact that B complex deficiencies can produce symptoms of depression indistinguishable from those of unknown etiology should lead us to look for other as yet unknown enzyme deficiencies in our patients.

Ever since Thudichum's brilliant and meticulous work, revealed that a considerable portion of nervous tissue was lipoid in nature there has been much speculation as to the role that lipoid substances must play in brain activity. More recently Bloor and co-workers , and Brown, have contributed evidence that tissue activity and even the activity of the whole animal are in some way dependent on the amount and character of the lipoids which they may contain. These observations have led to numerous investigations of lipoid metabolism in normal and mentally ill patients and particularly in those who suffer marked fluctuation in activity and also commonly have remarkable variations in body weight. , , , These investigations may be summarized as follows Normal people were found to have amounts of serum lipoids, cholesterol, phospholipids, and fatty acids, which appear to be characteristic for each individual. Each individual undergoes rather wide fluctuations in lipoids in the course of a year but in spite of this retains his own level. Men with slender or leptosomic physique characteristically have low lipoids while those with stocky or pyknic physiques have high lipoids. In women the same relation appears to hold but is more difficult to demonstrate because of greater variation in fat deposits due to secondary effects of a different and more variable endocrine status. There is also some correspondence between high energy output and high lipoids, as well as the converse.

These findings proved of interest in connection with two groups of patients. The patients in group I although endowed with a high energy output, suffer from marked fluctuations in this output. These patients tend to be stocky in physique and to have high lipoids; and they contrast strikingly with the second group who appear never to have had much energy and who are leptosomic in build. These two groups of patients have different sorts of mental symptoms. Group 1, the manic depressive patients, have high lipoids, while the schizophrenic patients, Group II, have low lipoids. It is noteworthy that mixtures of these mental symptoms may occur in patients, in which cases the lipoids may or may not be high or low. These patients with manic depressive and schizophrenic disorders have been followed for five to ten years after the beginning of their illnesses in order to determine whether there was any relation between the level of serum lipoids and the outcome. The majority of the patients with high lipoids have experienced apparently complete remissions, while those with low lipoids have had incomplete or no remissions. These findings are necessarily empirical and the underlying biochemistry of these relations awaits clarification.

Some order may be introduced into this problem by utilizing a hypothesis introduced some years ago by Kahn. According to this conception people are endowed with certain constitutional potentialities which determine their development as regards physique, temperament, capacity for growth, output of energy, resistance to disease, ability to recover from disease, etc. Some of these may be recognized at present, but others must await advances in the biochemistry of constitution, for further definition. For the present there appear to be pyknophilic potentialities which are represented anatomically by pyknic body build, in terms of behavior by warmth of temperament, by high energy output, including sex drive, by strong resistance to disease, and by remarkable capacity for recovery if overwhelmed by disease. That high lipoids may well represent a biochemical measure of pyknophilic potentialities is suggested by the studies just described where they were found in the pyknic people with high energy output and in the patients who proved to have a high capacity for recovery. Deficiency in these qualities may be considered the result of the predominance of leptophilic over pyknophilic potentialities. The presence of leptophilic potentialities is indicated by leptosomic physique, cool temperament, low output of energy, weak sex drive, low resistance to disease, and poor capacity for recovery and restitution. In line with this evidence low serum lipoids would constitute another evidence of leptophilic potentialities.

From what we know of body chemistry, the structure of cells and tissues depends primarily on the structural arrangements of protein molecules. There is evidence, however, that the lipoids, are in some way fitted into the intricacies of these patterns. Consequently, the amount and nature of lipoids in a tissue may well depend in part on these protein arrangements. Thus, the serum lipoids may reflect fundamental constitutional qualities because of their intimate ties with the structural protein.

From these considerations it would appear obvious that the next step in attempting to estimate pyknophilic potentialities in biochemical terms would lie in the direction of protein chemistry and particularly in stereo-chemistry and in stoichio-chemistry.

While the field of structural protein chemistry is advancing with all its implications, the biochemist in psychiatry is not allowed to forget the rapid development of endocrinology with the discovery of new hormones, many of which are apparently derived from cholesterol. Deficiencies in some part of the adrenal cortical hormones may account for some of the chronic states of weakness and depression in patients. At present, however, the available adrenal cortical substances appear to have little effect in these patients. The effects of new fractions, like Kendall's E substance, should be investigated when such substances become available.

In discussing some of the ways of approaching the problem of energy exchange it has been necessary to leave out reference to much of the work on the effects of environment on body chemistry, such as prolonged emotional stimuli and changes in weather, particularly heat, and cold.

The problems relating to the nervous control of body chemistry have also scarcely been touched upon. The studies of Brobeck, Tepperman and Long the effects of hypothalamic lesions on appetite and, in turn, on the body weight of animals, have important implications for psychiatry where gross disturbances in appetite are common phenomena.

The examples chosen for discussion in this paper represent a few of the more concrete and simple contributions of biochemistry to psychiatry. Present-day biochemistry is actually in its infancy. Some of the cruder elements that make up the living animal have been discovered and roughly measured. But only a beginning has been made in the exploration of the delicate but extremely active and complex chemical systems that determine the functions of the living organism. This is particularly true of brain chemistry. An examination of Page's exhaustive review of Brain Chemistry in 1937 supports this point. There is also very little known of the chemistry of the degenerative diseases of the nervous system. Folling's discovery of phenylpyruvic acid in the urine of some mentally deficient patients has opened a new field of investigation. Jervis has confirmed and extended this work. The developments of recent years have been so extraordinary and numerous that we can only guess at their implications for psychiatry.

The brilliant experiments of Schoenheimer and his co-workers with isotopes of various kinds, including fatty acids and amino acids, have revealed continuous chemical changes to be taking place in the lipoid and protein structures of the body that were formerly looked upon as relatively inert tissues. Yet in spite of this continual change, the structure as well as the steady state of the living organism persists. In the light of these observations chemical analysis of the blood at suitable intervals will yield more information with regard to body structure, i.e., constitution, than was formerly possible. While these advances in chemistry of structure and integration in the living organism are occuring, many chemicals are being discovered or synthesized which are peculiarly effective in disturbing the organized forces of nature. At present the production of powerful enzymes and hormones appears to be out-stripping our knowledge of structure and function.

Puberty The stage in a humans life when acne occurs, sexual reproductive organs become active.

Pukateine An alkaloid, $C_{18}H_{17}O_3N$, having no pronounced physiological action; m.p., 200°C.

Pulegone A menthene, volatile oil from pennyroyal oil, b.p. 221°C.

Pulsar Unidentified celestial object or phenomenon that emits periodic pulses of energy; hypothesized to be rotating neutron stars.

Pulse Impulse felt throughout the circulatory system of vertebrates caused by the contraction of the heart.

Pulsed field gel electrophoresis (PFGE) A gel technique which allows size-separation of very large fragments of DNA, in the range of hundreds of kb to thousands of kb. As in other gel electrophoresis techniques, populations of molecules migrate through the gel at a speed related to their size, producing discrete bands. In normal electrophoresis, DNA fragments greater than a certain size limit all migrate at the same rate through the gel. In PFGE, the electrophoretic voltage is applied alternately along two perpendicular axes, which forces even the larger DNA fragments to separate by size.

Pupa In metamorphozing insects, a stage between the **larva** and adult during which the organism undergoes major developmental changes.

Purdie's reagent A mixture of silver oxide and methyl iodide, used in methylating sugars.

Pure line A genetically uniform (i.e., homozygous) population.

Purine A heterocyclic organic compound which is not found free in nature but whose derivatives occur in very important groups of substances, as nucleic acids, caffeine, uric acid.

Purine base A base derived from purine, eg. caffeine.

Purjesz Test For Bilirubin In Uriner 2-3 cc. of urine are shaken with 2 cc. 20% sulfosalicylic acid solution and 2-3 drops 30% hydrogen peroxide. On standing, a green color develops within a few minutes reaching a maximum in about ½ hour. A red color indicates a negative test.

Purpurin Isopurpurine, Anthrapurpurine; 1,2,4-Trihydroxy-anthraquinone; m.p., 256-259°C (anhydr.); occurs as glucoside in madder root; used as cotton dye, as stain in microscopy, and as calcium reagent.

Purshianin A glucoside from root Rhamnus purshiana melting at 237°C.

Putrescine Tetramethylene diamine, found in putrefying tissues, probably formed from arginine by bacterial action; also found in ergot and in urine in cystinuria.

Pyocyanin A blue, chloroform soluble pigment, produced by a group of bacteria now classified as Ps. aeruginosa, functioning as an oxidation-reduction system and having bactericidal properties; this pigment has now been synthesized; m.p. 133°C, nitrogen-linked cyclic compound, $C_{26}H_{20}O_2N_4$.

Pyorrhea alveolaris Rigg's disease; a suppurative inflammation of the membrane lining the tooth sockets caused probably by bacterial infection and acid fermentation. An alkalizing diet may be indicated. Autogenous vaccines have been used in difficult cases. Potassium chlorate is the favored oxidizing agent for disinfection.

Pyranose Referring to or having the pyran ring structure.

Pyrethrosin An amaroid, $C_{34}H_{44}O_{10}$, from flowers of Chrysanthemum cinerariaefolium melting at 188-189°C.

Pyrethrum Dried root of Anacyclus Pyrethrum,

used as an insecticide, delousing agent, rubefacient.

Pyridinoproteins Consists of diphosphopyridinoproteins as enzymes requiring coenzyme I, and triphosphopyridinoproteins which require coenzyme II.

Pyridium Phenylazo-a, a-diaminopyridine monohydrochloride, brick-red; used internally in urogenital infections.

Pyridoxin(E) Vitamin B_6; the term replaces also the expressions adermin, rat antidermatitis vitamin, factor Y, factor 1, antiacrodynia factor, and vitamin H. It has been isolated and synthesized and has the structure of 2-methyl-2-hydroxy4, 5 di-(hydroxymethyl) pyridine. Wheat germs, rice, fish liver and muscle are good sources and it is extremely wide spread in small quantities. Its deficiency is shown by a dermatitis, convulsions, microcytic hypochromic anemia and acceleration of hair graying. It is probably essential for hemoglobin formation, is "spared" by fat and will aggravate cortical necrosis of kidneys in rats on low choline. Its requirement is apparently proportional to metabolism and is higher in infectious fevers.

Pyrimidine A nitrogen- containing, double- ring, basic compound that occurs in nucleic acids. The pyrimidines in DNA are cytosine and thymine; in RNA, cytosine and uracil.

Pyrocalciferol An isomer of calciferol, m.p. 95°C; obtained from it by thermal treatment. It has 4 rings and 3 conjugated double bonds.

Pyrophosphatase A phosphatase which degrades organic pyrophosphates to phosphates, eg. in malt and in yeast.

Pyrrolidine Tetrahydropyrrole; a saturated, 4 carbon 1 nitrogen heterocyclic compound; b.p., 87-89°C; sp.gr., 0.852.

Pyruvic acid CH_3 COCOOH; a keto-fatty acid found in fermentation and as an intermediate of metabolism, b.p. 90°C C.

Pyruvic aldehyde Methyl glyoxal; b.p. 72°C; present in fermentations and is involved in breakdown of carbohydrates.

Pyruvic oxidase A thiaminoprotein enzyme of the avitaminous brain which catalyzes the aerobic oxidation of pyruvic acid to acetic acid and CO, or the anaerobic dismutation of pyruvic acid to CO_2, acetic and lactic acids; requires Mg or Mn.

Pyruvic oxidase (bacteria) A thiaminoprotein enzyme of Bact. Delbruckii which promotes an oxidative decarboxylation in which acetic acid is formed; similar products are derived from Gonococcus and Staphylococcus.

Q

Quassia Bitterwood; bitterash; wood of Picrena excelsa, a tree of Jamaica, contains "quassin," a bitter principle used in tonics and as a stomachic, also against pinworms.

Quebrachamine $C_{19}H_{26}$, N_2; m.p. 145-147°C; from bark of apidoserma quebrachoblanco, Schlecht., Apocynaceae; paralyzes respiration.

Quebrachitol Methyl ether of 1-inositol, found in quebracho bark and rubber latex.

Quercetagetin An amaroid, $C_{15}H_{10}O_8$, from flowers of African marigold; m.p., 318-320°C.

Quercetin $C_{15}H_{10}O_7H_2O$, present in the rind of many fruits; m.p. 312-314°C.

Quercimeritrin A chnomo-glucoside, $C_{21}H_{20}O_{12}$; a compound of glucose and quercetin occuring in certain flowers; m.p. 247-249°C.

Quercitol Acorn sugar; m.p. 234-235°C (dform); isomers of inositols (10); d-quercitol occurs in acorns and palm leaves.

Quercitrin $C_{21}H_{22}O_{122}H_2O$; quercimelin; quercitrinic acid; a glucoside of the bark of oak consisting of rhamnose and quercitin. Bright yellow needles or plates, m.p. anhydrous form 168°C.

Quick's test A test of liver detoxicating function by the administration of sodium benzoate and determination of hippuric acid in the urine.

Quinhydrone electrode An electrode containing an equimolecular mixture of quinone and hydroquinone. It is an oxidationreduction system, the developed potential of which varies with the hydrogen ion concentration of the solution. It is therefore used to measure hydrogen ion concentration. It gives accurate results only when there is no other oxidationreduction potential in the system.

Quinic acid Chinic acid, kinic acid; m.p., 162163°C; carboxylic acid of tetrahydroxycyclohexane, found in cinchona, coffee, etc.

Quinidine Conquinine; ß-quinine; pitayine; $C_{20}H_{24}$ O2N_2 ; dextrarotary isomer of quinine, found in cinchona to the extent of 0.25-1.5%, resembles quinine in action; m.p. 170-171°C.

Quinine $C_{20}H_{24}O_2N_2$, $3H_2O$, m.p. anhydrous form, 175°C; alkaloid of cinchona bark, a specific for malaria, also a tonic and cold cure; retards protoplasmic activity including fermentation and action of some enzymes like pepsin and trypsin.

Quinoidine A mixture of alkaloids of cinchona, other than quinine, quinidine, cinchonine and cinchonidine, obtained from the mother liquors of quinine manufacture; has a milder effect than quinine.

R

Racemic mixture or compound A mixture of equal amounts of dextro (d-) and levo (l-) rotatory isomers, with a resultant optical activity of zero.

Radial artery *Artery* that runs down a human's forearm, supplying the hand and fingers with blood and serving at the wrist as a useful place to detect the *pulse*.

Radial nerve Nerve located in a human's upper arm.

Radial velocity Speed of a celestial body moving toward or away from an observer, as measured along the line of sight from the body to the observer.

Radiant heat Invisible *infrared radiation* emitted by all warm or hot objects.

Radiation Any energy that is emitted in the form of particles or waves.

Radiation, biological effects of The term radiation includes energy in the form of the ã rays of radium, X-rays, ultraviolet radiation, visible light, infra-red radiation or heat, and electric or radiowaves. The shorter the wavelength the more energy is contained in a quantum or unit of radiation. Quanta of radio-waves and infra-red radiation do not contain enough energy to affect ordinary chemical reactions. X-rays and ã rays contain enough energy to completely ionize molecules, but they are so penetrating that only a small portion of the incident energy is absorbed in an ordinary reaction system. Ultraviolet radiation is readily absorbed and is consequently very effective in bringing about chemical reactions. The absorption of radiation in the ultraviolet and to a lesser extent in the visible region results in the displacement of electrons in the absorbing molecules and these activated molecules are then capable of entering into photochemical reactions.

Primary and Secondary Light Reactions. In a primary light reaction a substance A is changed to form B on absorption of radiation. If B is stable the amount formed is proportional to the energy absorbed. Usually B is unstable and may be represented only by the condition of activation in the absorbing molecule. B may revert to A with production of fluorescence. Or B may react with other substances present (i.e. B + C ? D). In that case the amount of the end product D formed is not proportional to the amount of energy absorbed. The fact that most of the photochemical reactions which are concerned in the biological effects of radiation are secondary reactions accounts for their failure to obey the Einstein Photochemical Equivalence Law which properly pertains only to the primary process.

Absorbed radiation, particularly in the short wave region, may initiate photochemical reactions of many types in gases, solids and liquids. These have been extensively studied. Those which concern us in considering the biological effects of radiation are the reactions produced by the absorption of radiation in the organic constituents of living cells. We may, therefore, consider first what is known in regard to the effect of radiation on proteins and sterols in vitro and secondly the effect produced on the living organism when these substances absorb radiation in the living cell.

Effect of radiation on organic substances Proteins, amino-acids and Related Substances. Proteins begin to absorb at approximately 310 mì. The absorption increases to a maximum at 280 mì, falls to a minimum at 250 mì and then rises again for shorter wavelengths. Wavelengths longer than approximately 310 mì produce no effect on proteins, but those shorter than 310 mì, which are present in variable amounts in sunlight, and to a much greater degree in carbon and metallic arcs and mercury vapor arcs, produce marked changes.

Recent work on the molecular weight of proteins has shown that they have a very large molecular weight and are very subject to dissociation or association with change of pH or on the addition of salts or other proteins. Proteins may exist in long-chain or globular form and are subject to change of configuration. On absorption of ultraviolet radiation active labile protein molecules are denatured and are no longer capable of forming part of a living cell. The exact nature of this denaturation change is still unknown.

A denatured protein is more easily precipitated than a native protein and denaturation may be thought of as a change in the configuration of the native protein resulting in a loss of solubility. Proteins are irreversibly denatured on exposure to ultraviolet radiation at any pH and at any temperature, but this denaturation is not necessarily followed by coagulation. The primary process, or light denaturation, is a unimolecular reaction with a temperature coefficient of 1. This primary process may be followed by secondary changes leading to coagulation. The second step is a reaction between the light-denatured molecule and water similar to heat-denaturation but occurring at a lower temperature, with a high temperature coefficient. The final step is the flocculation of the light and heat denatured molecules which takes place only near the isoelectric point in salt-free proteins, but over a wider pH range in the presence of certain salts.

Proteins are similarly denatured on exposure to X-rays and ã rays but as these radiations are so little absorbed by protein solutions it requires days of radiation to produce a measurable result.

Proteins after light-denaturation show a marked increase in absorption in the ultraviolet region.

In addition to the radiation change we call denaturation, proteins may be photo-oxidized with the formation of protein derivatives (proteoses). Denaturation and molecular splitting are due to different processes as the latter change requires oxygen and the former is independent of it.

Sulfhydryl groups that are not present in the native state appear when a protein is denatured by radiation. , The presence of these groups would promote oxidations and reductions in radiated tissues.

Amino-acids Many studies have been made of the effect of radiation on individual amino-acids. Melanin is believed to be the end product of the oxidation of tyrosin by ultraviolet radiation in the presence of an oxidase present in the skin cells.

It can be demonstrated that histamine can be produced by the irradiation of histidine at pH 12 in the absence of oxygen but the occurrence of this change in irradiated skin tissue is uncertain.

Enzymes and Hormones Since the activity of an enzyme depends upon the integrity of its colloidal protein component, the destruction of the activity of enzymes by ultraviolet radiation is intimately bound up with the effects of radiation on proteins. This subject has been studied extensively and the specific conditions of wavelength of radiation, pH of solution, and other factors determining the effect on individual enzymes are given in Ellis and Wells.

In the same book there is a concise account of the effect of radiation on hormones and on the substances concerned in immunological reactions. In general one may state that they are all inactivated by ultraviolet radiation on exposure in vitro.

Sterols Ergosterol, on exposure to ultraviolet radiation, goes through a series of isomeric changes and one of these radiation products is calciferol or vitamin D. Purified cholesterol does not produce vitamin D on radiation, but ordinary cholesterol, as found in the tissues, can be so activated. Despite early evidence that ergosterol is the provitamin in cholesterol later experiments indicated that such is not the case. In all probability 7dehydrocholesterol is the significant provitamin in the skin. Absorption spectra of 7-dehydrocholesterol and ergosterol are similar and show specific changes on radiation.

The structural similarity between carcinogenic hydrocarbons and sterols has suggested to some investigators that carcinogens may be formed from sterols in the skin by radiation. Experimental work has failed to substantiate this hypothesis.

Effect of Radiation on the Living Cell

Bacteria and Protozoa

Radiation shorter than 310 mì, which is absorbed by proteins and sterols, is highly destructive to all living cells. Consequently ultraviolet radiation has a marked lethal effect on bacteria and protozoa which has been extensively studied. The sensitivity curve of bacteria to radiation is so similar to the absorption curve of proteins that there seems little doubt that the lethal action on small organisms is linked with the denaturation of their protein constituents. Viruses have been found to be far more resistant to radiation than bacteria.

The lethal effect of X-rays and ã rays on cells has been summarized in recent reviews., Different types of cells vary greatly in their sensitivity.

Recent advances in nuclear physics have focussed the interest of many investigators on the biological effect produced by neutrons and the comparative effect of neutrons and X-rays. Neutrons, like X-rays and the a, ß and ã rays of radium produce biological effects because of the ionization they cause. Neutrons on collision give up their energy to hydrogen nuclei and produce intense ionization over a short path. They are, therefore, more effective biologically than X-rays and ã rays which cause a small degree of ionization over a long path.

Earge Organisms Ultraviolet radiation is not lethal to larger organisms because it is absorbed in such a very thin layer of the body surface. The direct effect therefore is on the eyes and skin only. Certain important indirect effects are produced by diffusion of the products of light injury in the skin.

Eye The absorption of the cornea begins at 300 mì so that if radiation shorter than 300 mì falls on the eye it is absorbed by the cornea and conjuctiva and produces a severe inflammation. The aqueous humor absorbs no radiation longer than 250 mì and the absorption of the vitreous humor is similar to that of the cornea. Therefore, the cornea screens the aqueous and vitreous from any effect. The absorption of the lens varies somewhat in different animals and increases with age. The limit of transmission varies from 313 mì in infants to 400 mì in elderly people, the average being 336 mì. Therefore, certain ultraviolet wavelengths between 300 and 336 mì penetrate the cornea, are absorbed by the lens, and denature the lens proteins. At the pH of the lens the denatured protein does not precipitate unless the salt content of the lens increases above normal but a combination of radiation and increased salt content may account for senile cataract.

Wavelengths from 400 to 760 mì, which we call visible light, penetrate to the retina, are absorbed there, and produce the photochemical changes which give rise to visual sensations.

Wavelengths longer than 760 mì are absorbed throughout the eye and are transformed into heat. Exposure of the eye to intense sources of infrared radiation, as in exposure to molten glass or metal, may produce cataract, probably by disturbing the nutrition of the lens.

Eclipse blindness is a scotoma due to heat necrosis of the retina. This is caused largely by infrared radiation but intense visible radiation may be a contributing factor. Snowblindness is caused by reflection of intense sunlight into the eye by snowfields or water and sometimes by sand. The reflected ultraviolet radiation produces a conjunctivitis and the reflected visible and infra-red radiation may produce temporary scotomata.

Skin, Direct Effects It may be stated in general that living skin 0.1 mm thick absorbs all ultraviolet radiation shorter than 290 mì and skin 1.0 mm thick absorbs rays shorter than 365 mì. It should be remembered however that the intensity reaching a certain depth d beneath the surface (Id) depends upon the intensity at the surface I0 as Id = I0e-md, where m is the absorption coefficient for the wavelength in question. So, although the stratum corneum may absorb all the incident ultraviolet radiation shorter than 290 mì, and reduce radiation of 300 mì to onehalf, still, if the incident radiation is very intense, an appreciable amount may reach the dermis.

Visible and infra-red radiation penetrate further having maximum transmission at 1150 mì. Unless the skin is sensitized the chief result is a rise in temperature.

Ultraviolet radiation shorter than 315 mì produces the familiar inflammatory reaction known as sunburn. After a latent period, which varies in length with the intensity of radiation and the temperature of the skin, there is an erythema due to capillary dilatation, followed by pigmentation and peeling. The effect begins at 315 mì, and the sensitivity increases to a sharp maximum at 297 mì, falls to a minimum at 280 mì and then increases towards shorter wavelengths. , , The erythema at first has a sharp outline but later becomes diffuse with radiating streaks. Lewis has, therefore, suggested that it is due to the release of an H-colloid from the damaged cells with a histaminelike action. Ellinger has suggested that this may in fact be histamine[10] but the length of the latent period suggests a macromolecule with a slow diffusion rate. Mitchell gives evidence for believing that the Hcolloid is a proteose formed by photolysis of proteins. By assuming that the stratum corneum acts as a screen or filter, and that the photolysis occurs in the stratum mucosum, he calculates the wavelength effectiveness at this depth for protein constituents and finds that it agrees well with the experimental sensitivity curve of the skin.

It has long been known that though X-rays and ã rays destroy cancer cells they may stimulate the formation of cancer in normal tissue. More recently it has been shown that excessive exposure to ultraviolet radiation results in the production of skin cancer. The wavelength sensitivity curve of the skin for this effect is the same as for skin erythema. The close structural similarity between carcinogenic hydrocarbons and sterols suggests, however, that skin cancer is associated with changes produced by radiation in skin sterols rather than in proteins. Roffo links the effect with an increase in the cholesterol content of the skin after radiation.[14] It has been suggested that radiated cholesterol may become carcinogenic but direct proof of this is lacking.[15]

X-rays and the ã rays of radium penetrate deeply into the body, ionize cellular constituents and produce injury where absorbed. This injury may be lethal or sublethal leading to degenerative changes such as faulty mitosis. A comprehensive summary of the effect of these radiations on organs and body systems is given by Warren.

Skin, Indirect Effects Though the direct effects of radiation are destructive some of the by-products of injury may be beneficial to the body. The activation of provitamin D in the skin results in the cure and prevention of rickets, through the effect of vitamin D on calcium and phosphorus metabolism.

The absorption of the products of tissue damage into the circulation apparently produces other less well defined results as after exposure to ultraviolet radiation the blood sugar is lowered, there is stimulation of the thyroid and an increased formation and excretion of androsterones. Temporary effects are produced on blood volume, blood pressure and the number and distribution of blood cells. Cells injured by radiation produce a substance or substances which stimulate growth.

Sensitization to Visible Light Ordinarily visible light has no effect other than to raise the skin temperature and increase peripheral circulation. It has, however, been shown that the third generation of rats raised in visible light show increased growth compared with those raised in the dark.

However, if body tissues are sensitized to visible radiation extensive effects are produced. A sensitized reaction is one in which there is a photocatalyst present which may itself be unaffected by the radiation. The way in which the catalyst functions may vary. In some cases it forms a loose compound with the reactants and this compound is excited by light in the absorption band of the catalyst or sensitizer. In other cases the catalyst is excited by absorbed radiation and transmits its energy to a reacting molecule by collision. Many photochemical reactions, which are ordinarily initiated by ultraviolet radiation only, proceed in visible light in the presence of a sensitizer. It has been found possible to sensitize bacteria, protozoa, red blood cells and enzymes to the action of visible light. The changes produced in biological materials by visible light plus a sensitizer differ from those produced by ultraviolet radiation in that the former require oxygen whereas this is not necessarily true of the latter.

Many substances act as sensitizers in vitro but only eosin, chlorophyll and certain derivatives of hemoglobin are effective in vivo. Hematoporyhyrin is the most potent of these sensitizers. Animals injected with small amounts of this substance are healthy in the dark but die in the light and remain sensitized a long time as the hematoporphyrin forms a stable photosensitive compound with some element of the skin tissue. Death in these sensitized animals is due to fall in blood pressure. As visible light penetrates further than ultraviolet radiation the depth of tissue damage and the extent of capillary dilatation is much greater. Exposure of a large sensitized area with extensive capillary dilatation brings about a fall in blood pressure and death.

Light dermatoses occur in animals following the ingestion of certain plants. The most familiar are buckwheat poisoning, hypericism (St. Johnswort poisoning) and geeldikkop.[36] Sheard and Chick have been able to produce buckwheat poisoning in experimental animals by means of fresh green buckwheat, flowers, or husks of seeds.

Apparently the body often contains potential sensitizers from derivatives of hemoglobin and chlorophyll but has a mechanism that ordinarily prevents them from reaching the skin in sensitizing form. Digestive processes in the case of ingested material and metabolic processes in the case of hemoglobin formation results in a destruction of sensitizing characteristics. When dietary disturbances are severe enough to interfere with these processes spontaneous light sensitization may occur.

Certain carcinogenic agents have been found to sensitize the skin and protozoa to visible light. One would expect this to favor cancer formation on exposure to visible light but two groups of observers have found that visible light, coincident with painting by benzpyrene, inhibits the production of tumors in the skin of Mice. Botanical Effects No mention has been made of the extensive researches on the effect of visible and ultraviolet radiation on plant growth. Reviews on radiation and plant physiology may be found in Duggar's, "Biological Effects of Radiation."Genetics Radiation has been a most valuable tool in the study of mutations. X-rays and ã rays of sublethal intensity increase the rate at which mutations occur in both animals and plants.

Radical Part of a molecule that cannot exist alone but that participates as a unit in chemical reactions without breaking down.

Radicle Structure within a plant *embryo* that develops into the plant's main root.

Radio wave Electromagnetic energy with long wavelength and low frequency; examples include microwaves, FM and AM radio transmissions, and some radiation emitted by celestial bodies.

Radioactive decay Continuous disintegration of the *nucleus* of radioactive atoms whereby one element is transformed into another and *radiation* is emitted.

Radioactivity Property possessed by elements whose atoms undergo spontaneous *radioactive decay.*

Radiocarbon dating Method of dating biological material, based on the occurrence rate and decay rate of a radioactive *isotope* of the element carbon.

Radioisotope *Isotope* of an element that exhibits *radioactive decay.*

Radiopaque Opaque to the Roentgen ray; not permitting the passage of radiant energy.

Raffinose A non-reducing tri-saccharide found in beet sugar. Not a useful supply of energy for man. It is 2-ß-fructofuranose-1-á-glucose6-á-galactoside. Also found in cotton seed meal (8%) and molasses, m.p. 118.

Random primed synthesis If you have a DNA clone and you want to produce radioactive copies of it, one way is to denature it (separate the strands), then hybridize to that template a mixture of all possible 6-mer oligonucleotides. Those oligos will act as primers for the synthesis of labeled strands by DNA polymerase (in the presence of radiolabeled precursors).

Ranwez reaction for phenobarbital 0.1 gm. phenobarbital is heated with 0.5 gm. potassium nitrate and 2 cc. sulfuric acid for 10 minutes; the nitrophenobarbital is reduced to aminophenobarbital with zinc. The addition of 0.1 gm. sodium nitrite and a sodium hydroxide solution of ß-naphthol gives a blood-red color.

Rapic acid An isomer of oleic acid, found in rape oil, the position of the double bond being unknown.

Raspail reaction for lecithin A solution of lecithin in concentrated sulfuric acid gives a yellow color changing to purple-red on the addition of sugar.

Rational number Any number capable of being expressed as a fraction.

Raybin reaction for sucrose A solution of 40-50 mg. of sugar in 5 cc. of 7% sodium hydroxide solution is shaken with 7-10 mg. diazouracil until dissolved, the solution being kept cold; a bluegreen color forms. Addition of a magnesium salt produces a stable precipitate.

Rbcl A gene which is located in the chloroplast of photosynthetic organisms. It codes for the large subunit of the protein rubisco, and its sequence has been useful in plant phylogenies.

Reading frame When mRNA is translated by the cell, the nucleotides are read three at a time. By starting at different positions, the groupings of three that are produced can be entirely different. The following example shows a DNA sequence and the three reading frames in which it could be read. Not only is an entirely different amino acid sequence specified by the different reading frames, but two of the three frames have stop codons, and thus are not open reading frames (asterisks indicate a stop codon). * A DNA open reading frame ...ATG ACA TGT AAA GAT AGA CTA ACC TTT TGG... * ...Met Thr Cys Lys Asp Arg Leu Thr Phe Trp... * Same bases, different grouping ...A TGA CAT GTA AAG ATA GAC TAA CCT TTT GG... * ... *** His Val Lys Ile Asp *** Pro Phe Gly.. * Same bases, another grouping ...AT GAC ATG TAA AGA TAG ACT AAC CTT TTG G.. * ... Asp Met *** Arg *** Thr Asn Leu Leu ... * If we shift the grouping again, we will just get the first reading frame again. The reading frame that is actually used is determined by the first methionine codon (the initiation codon). Once that first AUG is recognized, the pattern of triplet groupings follows unambiguously.

Real number Any *rational number* or irrational number (a number that cannot be expressed as a fraction), excluding *imaginary numbers.*

Recapitulation The appearance, during the development of a higher organism, of structures or functions characteristic of developmental stages of lower organisms.

Receptor Region on a cell surface or within a cell that is recognized (and typically temporarily joined) by a particular chemical messenger or other substance.

Recessive gene Form of a *gene* that leads to the

expression of a physical trait only when it is present in that form on both of a pair of *chromosomes*.

Recessive An allele whose expression is masked by the activity of another allele.

Reciprocal Value arrived at when 1 is divided by a given number; also called the inverse.

Recombinant clones Clones containing recombinant DNA molecules.

Recombinant DNA technologies Procedures used to join together DNA segments in a cell- free system (an environment outside a cell or organism). Under appropriate conditions, a recombinant DNA molecule can enter a cell and replicate there, either autonomously or after it has become integrated into a cellular chromosome.

Recombinant DNA The combination of foreign DNA inserts with vector DNA (e.g., plasmid, phage, cosmid, etc.) to produce a clone within a host.

Recombinase Enzymatic activity responsible for facilitating intra- or intermolecular recombination of DNA molecules.

Recombination The process by which DNA is exchanged between pairs of equivalent chromosomes (crossing over) during egg and sperm formation. Recombination has the effect of making the chromosomes of the offspring distinct from those of the parents.

Recombination-Repair A mode of filling a gap in one strand of duplex DNA by retrieving a homologous single strand from another duplex. Usually the underlying mechanism behind homologous recombination and gene conversion.

Rectangle *Quadrilateral* with parallel opposite sides of equal length and with four right angles.

Recuperation The reappearance of competence at late stages of development, as in e.g. the regenerative power of the adult ascidian after its mosaic embryonic development, or the lens competence believed to exist in the regenerating amphibian limb blastema.

Redox The phenomenon of mutual reduction and oxidation; oxidation-reduction (abbr.).

Reducase An obsolescent term for a dehydrogenase.

Reductase An obsolescent term for a dehydrogenase.

Reduction division One of the two meiotic divisions, usually the first one, in which the number of chromosomes is halved.

Reduction The gain, by the substance being reduced, of one or more electrons. Invariably accompanied by oxidation of some other member of the system under consideration. The term covers also the addition of hydrogen atoms and the loss of oxygen atoms.

Reductive division Meiosis 1, when the diploid chromosome number is halved.

Refraction Bending of a light ray or energy wave when passing from one medium into another.

Refractive index Measure of the amount of *refraction* undergone by a light ray passing betweentwo given transparent media.

Regeneration The formation of part or whole of a new organ in place of one that has been removed.

Regional differentiation The appearance of regional differences within the individuation field, e.g. the differentiation of an embryonic axis into cephalic and caudal ends, with all that that implies; or within organ districts such as that of the limb.

Regulation The power of continuing normal or approximately normal development or regeneration in spite of experimental interference by ablation, addition (implantation).

Reichard reactions for nicotine and coniine On addition of a drop of nicotine and a drop of hydrochloric acid to dry, finely powdered copper oxychloride, a violet-blue mass is obtained; coniine gives a bright green solution gradually becoming colorless when treated in the same way. Nicotine turns yellow when its solution with á-naphthol is treated with hydrochloric acid; coniine remains colorless. The addition of hydrochloric acid to a mixture of nicotine and bismuth subnitrate gives an intense yellow color; coniine remains white.

Reichert-meissl number A test usually applied to esterified fatty acids to determine the relative amounts of high and low molecular weight acids. It is defined as the number of cc. of 0.1N KOH required to neutralize the steam distillate of 5 grams of saponified fat (low mol. wt. fatty acids.).

Reinecke's salt $[Cr(CNS)_4(NH_3)_2]NH_4 \cdot H_2O$; forms very insoluble compounds with secondary and tertiary amines. Used to precipitate proline and oxyproline from protein hydrolysates. (The K salt is also known by this name).

Reintegration The restoration to the organism of control by its individuation field, through the action of hormonal and neural factors during the period of functional differentiation, after the period of selfdifferentiation.

Relative atomic mass Mass of an atom calculated as a fraction of the mass of one-twelfth an atom of carbon-12.

Relative density Density (mass per unit volume) of an object calculated as a fraction of the density of water.

Relative humidity Measure of the amount of water vapor in the air compared to the maximum amount the air can hold at that temperature and pressure.

Relativity Theory stating that time, motion, space, and mass are phenomena whose properties are interdependent rather than absolute; includes the special theory of relativity and the general theory of relativity.

Renal threshold The level of blood sugar at which it begins to appear in the urine; normally low in some people and in pregnancy, high in diabetics of long standing.

Renin An enzymatic protein prominent in hypertension which forms a pressor substance with a globulin of blood serum.

Repetitive DNA A surprising portion of any genome consists not of genes or structural elements, but of frequently repeated simple sequences. These may be short repeats just a few nt long, like CACACA etc. They can also range up to a few hundred nt long. Examples of the latter include Alu repeats, LINEs, SINEs. The function of these elements is often unknown. In shorter repeats like di- and tri-nucleotide repeats, the number of repeating units can occasionally change during evolution and descent. They are thus useful markers for familial relationships and have been used in paternity testing, forensic science and in the identification of human remains.

Replication The process of duplicating or reproducing, as replication of an exact copy of a polynucleotide strand of DNA or RNA.

Reporter gene The use of a functional enzyme, such as beta-galactosidase, luciferase, or chloramphenicol acetyltransderase, downstream of a gene, promoter, or translational control element of interest, to more easily identify successful introduction of the gene into a host and to measure transcription and/or translation.

Repression A form of gene regulation wherein the promoter is prevented from assembling an RNA polymerase complex, so that transcription does not occur.

Reproduction process by which living organisms give rise to offspring; includes either *sexual* or *asexual reproduction.*

Residue As applied to proteins, what remains of an amino acid after its incorporation into a peptide chain, with subsequent loss of a water molecule.

Resolution Degree of molecular detail on a physical map of DNA, ranging from low to high.

Respiration The dynamic equilibrium in which all living organisms maintain themselves, demands a steady supply of substances to the body cells. In these, the food (or partially split food) molecules are changed to release energy and to form other molecules useful in building the body tissues. The chemical changes (metabolism) which result in useful products mostly involve little energy gain or loss; the products may be larger molecules than the reactants; and any oxidation or reduction which does occur is between carbon containing molecules. The changes which release large amounts of energy for cellular use, on the contrary, depend on the complete or nearly complete disruption of the food molecules due to their oxidation by oxygen.

This burning of foodstuffs by oxygen in the cells is sometimes called internal respiration (as the partial oxidation and reduction of sugar molecules by each other or their products, which also yields some tree energy, is called fermentation). It is however, a phase of cell metabolism and need not be considered further here. What is important now is that large amounts of oxygen, as of food, must continuously be carried to the cells by the blood; and that the carbon dioxide formed as the foods are oxidized must be carried from them. Special mechanisms are present in the blood—hemoglobin to combine with oxygen and a reserve of alkali to combine with carbon dioxide-which enables this fluid to carry far more of both gases than could an equal amount of water. This duty is referred to as the respiratory function of the blood. But respiration proper, or breathing, or external respiration when paired with internal respiration, has to do only with the exchange of oxygen and carbon dioxide between the outside medium and the body.

This exchange ultimately depends, in all cases, on simple diffusion of each gas from a region in which its concentration (or, better, its partial pressure) is higheroxygen outside the body, carbon dioxide inside—to a region in which its concentration is lower. In the simplest organisms, the unicellulars and the loosely built multicellulars like the sponges, no more is needed. Larger or more compact bodies of massed cells require a circulating fluid, some sort of blood or sap, to distribute the gases between the surface of exchange with the outside and the deeper cells and tissues. (The insects, however, use a completely separate set of air tubes to supply the cells.) Further, since volume increases as the cube of linear dimensions and surface only as the square, the surface of exchange must itself eventually be augmented by some sort of folding or pouching. Finally, as this surface becomes more complex, means must be found for insuring an adequate renewal there of the external medium, as well as of the internal fluid.

In water-living animals, the membrane for gas exchange is turned out and projects into the outer medium. Finger-like or leaf-like folds of an especially thin membrane are extended from the main body wall as gills. They are largely filled with fine capillary blood vessels, to renew the medium inside them; and they either wave about in the water (marine worms, salamanders, etc.) or regularly move in-and-out along with mouth movements (fish), to renew the outside medium.

In land-living air-breathing animals, the reverse solution has been reached. The exchange membrane has been turned in, not out, and more or less subdivided sacks of air, the lungs, enter the body mass from the main body wall. This necessitates even more elaborate devices for renewing the outside medium, the air, in these lung sacks; and many special muscles, bones and other structures are devoted to the required bellows-like movements of the body wall. In each case, air or water, the alternate solution would not work. Gills dry out in air but not in water; lungs allow air but not the far denser water to pass freely through their fine-tubed passages.

The lungs of mammals constitute myriad microscopic air sacks, the alveoli, set like grapes on their stems at the ends of fine air tubes, the bronchioles. Still like a bunch of grapes, these passages join into ever larger twigs until, finally, a right and a left bronchus fuse to form the trachea, or wind pipe. This opens above through the voice box (larynx), a later addition, into the nose and back of the mouth. Each lung is sheathed in a smooth slippery coat, the pleura, which turns back on itself to line the chest wall and the diaphragm muscle. The latter forms the floor of the chest cavity, as the ribs and the muscles between them (intercostal muscles) form its sides and dome over its top. The lungs proper possess no muscle—they are elastic and collapse when the chest wall is opened—so all their movements are a passive following of the thorax (chest) in which they lie.

Inspiration—of a quart of air normally, of four times this on exertion—depends on the chest cavity enlarging as the ribs are pulled out and up by the intercostal muscles; and the diaphragm, which normally bows up in the center and is attached at its circumference, is pulled down by its own contraction. Expiration normally results from relaxation of these muscles; but other muscles (mainly internal intercostals, which oppose the external intercostals) can cause an active expiration of an extra three quarts of air normally held in the lungs. Another two quarts remain even then. Forced expiration occurs normally in such acts as coughing, sneezing, blowing, or straining, and abnormally in asthma, pneumonia, or other diseases with difficult breathing.

The respiratory movements depend on muscle action, which in controlled by nerves — the intercostal nerves, a dozen on each side to the rib muscles; the phrenic nerves, one on each side, to the diaphragm. Since the intercostal nerves come from the whole length of the spinal cord in the back of the cheat and the phrenic nerves come from the spinal cord in the neck, and since all these must be properly coordinated if breathing is to be effective in exchanging the lung air, a special region of the nervous system acts as a control center for the lot. This respiratory center (in the medulla of the brain) sends out messages for each normal inspiration and each forced expiration. It is under voluntary control —it is possible, within limits, to hold the breath or breathe rapidly-but ordinarily sends out its regular signals without attention. It is controlled mainly by the amount of carbon dioxide reaching the brainmore of this in the blood leads to increased respiration which, of course, tends to get rid of the excess carbon dioxide. It is also controlled reflexly by sensory nervesthe vagus from the pleura is stimulated as the lungs stretch on inspiration and its impulses stop the inspiratory discharge of the respiratory center before this goes too far. Similar reflexes are responsible for coughing, and the like.

In respiration, then, fresh outside air (or water) is kept in contact with a thin much-folded membrane by the respiratory movements. The blood within this membrane is kept circulating by the heart. Since oxygen is continuously removed from the blood (and carbon dioxide added to it) by cell metabolism, the oxygen in the blood is kept at a lower pressure (and carbon dioxide at a higher pressure) than that in the air in the alveoli. Simple diffusion across the extensive permeable membrane completes the gas exchange.

Respiratory pigments Oxygen carriers of various bloods, which are colored (so-far as investigated) which react reversibly with oxygen and are specifically adapted to carry as large an amount of oxygen as possible at the oxygen tension of the environment; e.g. hemoglobin, chlorocruorin, hemocyanin, hemerythrin, actinichematin, helicorubin.

Respiratory quotient The ratio between C02 output and O2 intake. To a large extent it is an indication of the type of food used, as R.Q. for fat is 0.70-0-71; carbohydrates 1.0; protein 0.800.82. It is used in indirect calorimetric measurements.

Response element By definition, a "response element" is a portion of a gene which must be present in order for that gene to respond to some hormone or other stimulus. Response elements are binding sites for transcription factors. Certain transcription factors are activated by stimuli such as hormones or heat shock. A gene may respond to the presence of that hormone because the gene has in its promoter region a binding site for hormone-activated transcription factor. Example the glucocorticoid response element (GRE).

Restriction enzyme cutting site A specific nucleotide sequence of DNA at which a particular restriction enzyme cuts the DNA. Some sites occur frequently in DNA (e.g., every several hundred base pairs), others much less frequently (rare-cutter; e.g., every 10,000 base pairs).

Restriction enzymes An endonuclease enzyme, isolated from bacteria, that recognizes specific base-pair sequences within DNA and causes endonucleolytic cleavage of the DNA at a site determined by the recognized DNA sequences. The sites vary from frequent to rare cutting, depending upon the length of the restriction site. For example, the restriction enzyme BamHI locates and cuts any occurrence of 5'-GGATCC-3', 3'-CCTAGG-5' Note that both strands contain the sequence GGATCC, but in antiparallel orientation. The recognition site is thus said to be palindromic, which is typical of restriction sites. Every copy of a plasmid is identical in sequence, so if BamHI cuts a particular circular plasmid at three sites producing three "restriction fragments", then a million copies of that plasmid will produce those same restriction fragments a million times over. Bacteria produce restriction enzymes for protection against invasion by foreign DNA such as phages. The bacteria's own DNA is modified in such a way as to prevent it from being clipped. Bacteria contain over 600 such enzymes that recognize and cut over 100 different DNA sequences.

Restriction fragment length polymorphism (RFLP) Variation between individuals in DNA fragment sizes cut by specific restriction enzymes; polymorphic sequences that result in RFLPs are used as markers on both physical maps and genetic linkage maps. RFLPs are usually caused by mutation at a cutting site.

Restriction fragment length polymorphism (RFLP) Variation in the distance between restriction enzyme cleavage sites that exist within a population producing unique DNA fingerprint patterns.

Restriction fragment The piece of DNA released after restriction digestion of plasmids or genomic DNA. "Restriction enzyme". One can digest a plasmid and isolate one particular restriction fragment (actually a set of identical fragments). The term also describes the fragments detected on a genomic blot which carry the gene of interest.

Restriction map A depiction of the locations within a stretch of known DNA where restriction enzymes will cut. The map usually indicates the approximate length of the entire piece (scale on the bottom), as well as the position within the piece at which designated enzymes will cut. This map happens to be of a plasmid, and the two ends are joined together with about 25 nt between the EcoRI and HindIII sites.

Retene An aromatic hydrocarbon obtained from abietic acid by debydrogenation with S; 1-methyl-7 isopropyl phenanthrene; b.p. 390-394°C.

Reticulin A scleroprotein found as fibres in reticular tissue. It is very resistant to the action of acids and alkalies. Pancreatic trypsin attacks it less readily than does pepsin.

Reticulocyte lysate A lysate of rabbit reticulocytes, which has been extensively digested with micrococcal nuclease to destroy the reticulocyte mRNAs. With the addition of an exogenous, usually synthetic, mRNA, amino acids and a source of energy (ATP), the translational machinery of the reticulocyte (ribosomes, eukaryotic translation factors, etc.) will permit in vitro translation of the added mRNA with production of a new polypeptide. This is only one of several available in vitro translation systems.

Retina (structure of the eye that is sensitive to light and that transmits nerve impulses to the brain for processing vision

Retinene Chromogen of the 'Visual Yellow' of the retina, insufficiency of which causes 'night blindness.' Closely related to Vitamin A.

Retrovirus Singel stranded RNA virus which is replicated and expressed via a double stranded DNA intermediate. Retroviruses contain enzyme reverse transcriptase (RNA dependent DNA polymerase); this enzyme converts viral RNA into DNA which can combine with DNA of host cell and produce more viral particles.

Reunition The reconstitution of an organism after it has been separated into its component cells, not all types of which may be available, as in sponges.

Reverse transcriptase An enzyme found in the retroviruses which catalyzes the RNA-dependent polymerization of DNA (DNA copy from RNA template). RT is used to make cDNA; one begins by isolating polyadenylated mRNA, providing oligo-dT as a primer, and adding nucleotide triphosphates and RT to copy the RNA into cDNA.

Reversible reaction Chemical reaction in which reactants are converted into products and products continually convert back into reactants.

Revertose An old name for a disaccharide formed by enzymatic synthesis (1889).

Revillon reactions for differentiating resorcinol nd hexylresorcinol Hexylresorcinol and resorcinol give different colors when treated with Denigès reagent, sucrose and sulfuric acid, ferric alum, ferric perchlorate, sodium carbonate, sulfuric and formic acids, and acetaldehyde.

Rflp Restriction fragment length polymorphism; the acronym is pronounced "riflip". Although two individuals of the same species have almost identical genomes, they will always differ at a few nucleotides. Some of these differences will produce new restriction sites (or remove them), and thus the banding pattern seen on a genomic Southern will thus be affected. For any given probe (or gene), it is often possible to test different restriction enzymes until you find one which gives a pattern difference between two individuals - a RFLP. The less related the individuals, the more divergent their DNA sequences are and the more likely you are to find a RFLP.

Rh A term representing the logarithm of the reciprocal of the pressure of the hydrogen gas in equilibrium with an oxidation-reduction system. Although Clark who introduced the term has advised against its continued use, it has found widespread entrance into the biological literature as a measure of the level of the oxidative or reductive intensity of biological redox systems. The latter should preferably be characterized by giving the Normal Potential (E_0') of the system at the pH obtaining during the measurements.

Rhamnetin $C_{16}H_{12}O_7$; m.p. above 300°C; quercetin-7-methyl ether; the aglycone of xanthorhamnin; occurs in berries of Rhamnus cathartica.

Rhamninose $C_{18}C_{34}O_{14}$; m.p. 135^0; a reducing trisaccharide consisting of one molecule of galactose and two of rhamnose; source Persian berries in which it is present as xanthorhamnin.

Rhamnoascorbic acid A synthetic substance which has about 1/5th the potency of ascorbic acid.

Rhamnose $CH_3(CHOH)_4CHO$; isodulcite; a methyl aldopentose, a constituent of many glycosides; non-fermentable, m.p. hydrate 94°C.

Rhein Rheic or parietic acid, rhubarb yellow; dihydroxyanthraquinonecarboxylic acid, $C_{15}H_8O_6$; m.p. 321-322°C; from Chinese rhubarb and senna leaves; used as purgative.

Rheinboldt test for mercaptans The test solution is poured over some sodium nitrite and some dilute hydrochloric acid or glacial acetic added. An intense red color is given by primary and secondary aliphatic mercaptans; in dilute solution, tertiary aliphatic or aromatic mercaptans give a green color, in concentrated solutions a dichroic green color.

Rheochrysin A glucoside, $C_{12}H_{22}O_{10}$, from root of Chinese rhubarb; m.p. 211°C.

Rheopexy Rapid solidification shown by a few thixotropic systems, induced by giving a slow circular to-andfro motion to the sol.

Rheotropism Reaction of a living form to the electric current.

Rhodanilic acid A modification of Reinicke's salt. Forms an insoluble complex with proline but not oxyproline. It is $[Cr(CNS)_4(C_6H_5NH_2)_2]H$.

Rhodopsin Visual purple; chromoprotein of the retina, insufficiency of which causes 'night blindness.' Consists of a protien and a Vitamin A derivative. On heating or subjecting it to light it is converted to 'visual yellow'.

Rhoeadine $C_{21}H_{21}O_6N$; m.p. 256-257°C; also given as 232^0; alkaloid of the red poppy or corn poppy.

Rhombus *parallelogram* with four sides of equal length.

Rhubarb Rhizome and roots of various species of rheum; used as cathartic, bitter tonic, stomachic.

Riboflavin Vitamin B_2; lactoflavin; Vitamin G; ovoflavin; 6,7-dimethyl isoalloxazine-9-d-riboside; m.p. about 275°C with decomposition. The vitamin concerned with cell respiration as a constituent of Warburg's yellow respiration enzyme, and with other bodily functions.

Deficiency in humans causes serious ill health, accompanied by a dermatitis and a gain in the water content of the tissues. It is found in meats, yeast, milk, greens, etc. The average dose is 3-5 mgm.

RIBOSE An aldopentose, with all the hydroxyl groups on the same side of the carbon chain, having a furanose configuration; constituent of nucleosides of plants and animals and of vitamin B2; m.p. 86°C.

Ribonucleic acid (RNA) A chemical found in the nucleus and cytoplasm of cells; it plays an important role in protein synthesis and other chemical activities of the cell. The structure of RNA is similar to that of DNA. There are several classes of RNA molecules, including messenger RNA, transfer RNA, ribosomal RNA, and other small RNAs, each serving a different purpose.

Riboprobe A strand of RNA synthesized in-vitro (usually radiolabeled) and used as a probe for hybridization reactions. An RNA probe can be synthesized at very high specific activity, is single stranded (and therefore will not self anneal), and can be used for very sensitive detection of DNA or RNA.

Ribosomal binding sequence (Shine-dalgarno sequence) In prokaryotic organisms, part or all of the polypurine sequence AGGAGG located on mRNA just upstream of an AUG initiation codon; it is complementary to the sequence at the 3' end of 16S rRNA; and involved in binding of the ribosome to mRNA. The internal ribosomal entry site found in some viruses may be an analogous eukaryotic genetic element.

Ribosomal rna (Rna) A class of RNA found in the ribosomes of cells.

Ribosome structure of living cells in which *proteins* are manufactured.

Ribosome binding site Sequences contained in an mRNA that organize the assembly of a ribosome to initiate translation of the mRNA into polypeptide.

Ribosome The large, multi-subunit ribonucleoprotein cellular complex responsible for translation of mRNA into polypeptide sequences. Ribosomes are a complex consisting of ribosomal RNAs (rRNA) and several proteins.

Ribozyme A catalytically active RNA. A good example is the hepatitis delta virus RNA which is capable of self-cleavage and self-ligation in the absence of protein enzymes.

Ricinoleic acid $CH_3 (CH_2)_5 CH(OH)CH_2 CH. CH (CH_2)_7 COOH$; an unsaturated monohydroxy fatty acid found in castor oil, m.p. 17^{U+00}B0C. (also given as 4-5°C); possibly the purgative principle of castor oil.

Ricinus lipase A fat hydrolyzing enzyme of the castor bean. Optimum PH for activity is 4.7-5.0.

Rickets Rachitis; osteomalacia; a deficiency disease due to lack of vitamin D or indirectly of sunshine, showing most markedly in children. Bone changes show weakening and deformation. Osteomalacia is the adult form of the disease. Inorganic phosphorus in the blood is low. The treatment requires administration of foods rich in vitamin D, irradiated ergosterol, sunshine or artificial irradiation to activate skin sterols.

Riegel-william reaction for procaine, butyn, tutocaine, butesin, benzocaine and orthoform The addition of a few drops of hydrochloric acid sodium nitrite solution and concentrated ammonia to a solution of 1 mg. of the substance in 1 cc. water gives an intense yellow color; this may be used for the colorimetric estimation of procaine.

Rigor mortis Rigidity of muscles in death due to myosin gelation.

Rivanol 2-Ethoxy-6,9-diaminoacridine-lacrate; used as bactericide, antiseptic; externally in acute abcesses, furuncles and other localized purulent processes.

Rivat reaction for oxyhemoglobin The test solution is treated with 2 drops of an alkaline phenolphthalein solution and a drop of 0.5% manganous albuminate solution. A rose-red color is obtained, changing to violet-red in about a minute. An excess of the albuminate must be avoided.

Rna polymerase Enzymatic activity responsible for DNA-dependent synthesis of RNA. In prokaryotes there is only one RNA polymerase. In eukaryotes, there are three, each of which transcribes a different group of genes.

Rna splicing A complex and incompletly understood series of reactions occuring in the nucleus of eukaryotic cells in which pre-mRNA transcribed from chromosomal DNA is processed such that noncoding regions of the pre-mRNA (introns) are excised, and coding regions (exons) are covalently linked to produce an mRNA molecule ready for transport to the cytoplasm. Because of splicing, eukaryotic DNA representing a gene encoding any given protein is usually much larger than the mRNA from which the protein is actually translated.

RNA Ribonucleic acid a category of nucleotides used in translating the genetic message of DNA into protein.

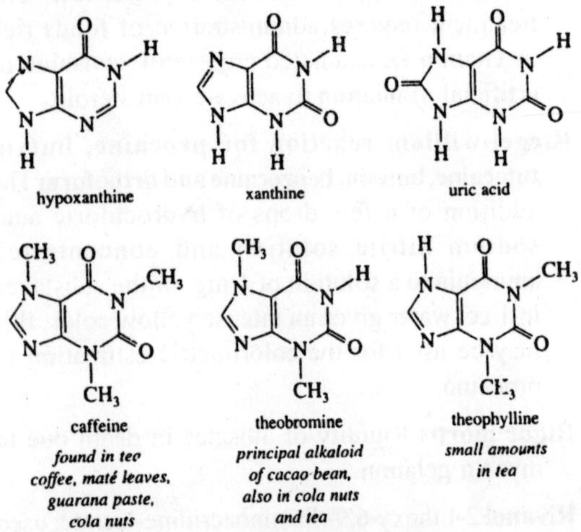

Fig. RNA

Rnase protection assay This is a sensitive method to determine.

(1) the amount of a specific mRNA present in a complex mixture of mRNA and/or.

(2) the sizes of exons which comprise the mRNA of interest. A radioactive DNA or RNA probe (in excess) is allowed to hybridize with a sample of mRNA (for example, total mRNA isolated from tissue), after which the mixture is digested with single-strand specific nuclease. Only the probe which is hybridized to the specific mRNA will escape the nuclease treatment, and can be detected on a gel. The amount of radioactivity which was protected from nuclease is proportional to the amount of mRNA to which it hybridized. If the probe included both intron and exons, only the exons will be protected from nuclease and their sizes can be ascertained on the gel.

Rnase Ribonuclease; an enzyme which degrades RNA. It is ubiquitous in living organisms and is exceptionally stable. The prevention of RNase activity is the primary problem in handling RNA. There are many different RNases, some of the more important include * RNase A Cleaves ssRNA 3' of pyrimidines * RNase T1 Cleaves ssRNA at guanasine nucleotides * RNase V1 Cleaves dsRNA (helical regions) * RNase H Degrades the RNA part of RNADNA hybrids.

Robinin A chromo-glucoside from flowers Robinia pseudacacia, melting at 195-197°C.

Robinose A reducing trisaccharide found in the glycoside robinine. Consists of two molecules of rhamnose and one of galactose.

Robison ester Glycopyranose-6-monophosphate, present in muscle and in fermentations.

Robison-embden ester An equilibrium mixture of 70% glucosemonophosphoric acid and 30% fructosemonophosphoric acid, found in muscles as an intermediate of hexose metabolism. It is formed reversibly from hexosediphosphoric acid (Harden-Young ester) by enzymatic hydrolysis in the presence of Mg ions.

Rods and cones Elements of nerve cells of retina, the former containing visual purple for sensitivity and the latter serving for color discrimination.

Romieu reaction for lecithin Freshly prepared egg or brain lecithin appears colorless in a thin film on glass. The addition of dilute hydrochloric acid and a concentrated iodine-potassium iodide solution produces a mahoganybrown or garnet color reminescent of glycogen in the same test.

By warming the film gently with 10% hydrochloric acid, glycogen is partially hydrolyzed and dissolved, differentiating it from lecithin.

Root value that can replace a *variable* in an equation and satisfy the equality expressed by the equation.

Ropy milk The result of a mucus fermentation of milk by a coccus.

Rosenberg test for bile pigments in urine A green color is obtained when 10cc. of urine are treated with 10cc. 20% potassium hydroxide solution and 2-3 drops of 10% copper sulfate solution.

Rosenheim reaction A test for tryptophane. A very dilute solution of formaldehyde is added to the solution to be tested, and sulfuric acid stratified under the mixture. A purple junction ring is positive. The test is made more sensitive by the addition of 5 rag. of potassium nitrite to 100 cc. of sulfuric acid.

Rosenheim reaction for choline A solution of choline-platinic chloride in 15% alcohol is evaporated to dryness and then treated with iodine-potassium iodide solution; large, dark brown dichroic and direfractive crystals of choline periodide are formed.

Rosenheim reaction for ergosterol or its esters When 1 mg. of ergosterol is brought into contact with 0.5 gm. of chloral hydrate (previously liquified) a carmine red color is produced. Anhydrous chloral does not react. Concentrated aqueous trichloracetic acid solution gives a red color changing to blue, with ergosterol in chloroform.

Rosenheim reaction for oxycholesterol An immediate purple color is obtained when oxycholesterol is moistened with dimethyl sulfate; cholesterol remains unchanged.

Rosenheim reaction for proteins Reagents — Formaldehyde solution, 12500. Pure sulfuric acid containing a few rag. of ferric chloride per 100 cc. A purple contact ring is obtained when the solution of sulfuric acid is stratified beneath the proteincontaining formaldehyde solution. The reaction is due to the presence of the tryptophane group. This is a modification of the Adamkiewicz reaction.

Rosenheim-callow reactions for sterols Reagent— Solution of 25 gm. of mercuric acetate in 100 cc. nitric acid, p. gr. 1.42. Equal volumes of reagent and a chloroform solution of the sterol are mixed and shaken immediately. Allo-cholesterol, allo-sitosterol, cholesterilene and ß-cholesterol give red colors while a blue following a transient pink is given by ergosterol and its eaters.

Rosenheim-drummond test for vitamin A One drop of liver oil, dissolved in petroleum ether, is treated with 1 cc. of arsenic trichloride and shaken; a blue color is obtained, changing to purple in a few seconds and becoming paler. This may be used for the colorimetric estimation of the vitamin.

Rosenthal reagent for hepatic function test 1. Sodium salt of tetrabromphenolphthalein sulfonic acid. 2. Disodium salt of phenoltetrachlorophthalein.

Rosenthal-erdelyl test for vitamin A Reagents — (a) 0.5% absolute chloroform solution of pyrocatecbol. (b) cold saturated absolute chloroform solution of antimony trichloride. 1-2 cc. of an absolute chloroform solution of the oil is mixed with 1 cc. of.

(a) and 2-3 cc. of.

(b), and the tube heated in a water bath at 60°C for 1-2 minutes. A potassium permanganate color is produced which is compared with a standard 0.01% permanganate solution.

Rosin The commercially important resin obtained as a residue after distilling off the volatile oils of the oleoresin of American pines; colophony; yellow resin; abietic anhydride; used medically as rubefacient in ointments and plasters.

Rotenone $C_{23} H_{22} O_6$, the principal active constituent of derris root, cubé, etc., m.p. 163°C; highly toxic to insects but not to man or animals (1 part in 20 million will kill a goldfish in 3 hours).

Rottlerin An amaroid from Rottlera tinctoria having a slight anthelmintic effect.

Rrna "ribosomal RNA"; any of several RNAs which become part of the ribosome, and thus are involved in translating mRNA and synthesizing proteins. They are the most abundant RNA in the cell (on a mass basis).

Ruberythric acid $C_{26}H_{26}O_{14}$ or $C_{25}H_{26}O_{13}$; rubianic acid; a glycoside of the roots of Rubia Tinctorum (madder). Consists of glucose and alizarin. Crystallizes as citron-yellow needles or prisms from water, m.p. 25860°C.

Rubiadin $C_{21}H_{20}O_9$; m.p. about 270°C; a glycoside of 24-dihydroxy-3-methyl anthroquinone, found in root of Rubia Tinctorum.

Rubisco — protein which fixes carbon in photosynthetic organisms. It binds molecules of carbon dioxide to a five-carbon molcule. Rubisco is the most common protein on earth.

Rubixanthin A carotenoid found in Rose leaves. Its structure is 3-hydroxy-ß-carotene-ß'-lycopene.

Ruff's method A method for removing a carbon atom from a sugar. The aldose is oxidized to the acid, the calcium salt of which, on treatment with hydrogen peroxide loses carbon dioxide.

Russmann reaction for adrenaline Four cc. of slightly acid adrenaline solution are treated with 1 drop of cold saturated mercuric chloride solution, 3 drops cold saturated sulfanilic acid solution and 1 cc. potassium biiodate solution. A brown color is obtained when the solution is boiled for 1 minute.

Rutaecarpine An alkaloid, $C_{18}H_{13}ON_3$, from fruit of evodia rutaecarpa; m.p. 260-262°C.

Rutin $C_{27}H_{32}O_{16}$-$2H_2O$; melin; phytomelin; osyritin; violaquercitrin; myrticolorin; eldrin; a glycoside of the leaves of Rista graveolens, rue, tobacco, tomato stems, and many flowers; consists of glucose, rhamnose, and quercetin, which.

S

S1 end mapping A technique to determine where the end of an RNA transcript lies with respect to its template DNA (the gene). Can't be described in a short paragraph.

S1 nuclease An enzyme which digests only single-stranded nucleic acids.

Sabadilline An alkaloid, $C_{34}H_{53}O_8N$, from sabadilla seed; melting at 200°C.

Sabadine An alkaloid, $C_{29}H_{51}O_8N$, from Sabadilla seed; m.p. 238-24^0°C.

Sabinene A dicyclic monoterpene; d-form found in oil of savin, etc.

Sabinic acid $C_{12}H_{24}O_3$; a saturated monohydroxy fatty acid found in conifer wax.

Sabinol A dicyclic monoterpene alcohol, found in oil of savin (d-form).

Saccharic Acied An oxidation product of glucose. It forms a slightly soluble salt with one molecule of KOH, which reaction is the basis of a test for glucose; COOH-$(CHOH)_4$ COOH; m.p. 125-126°C.

Sacuranin A glucoside from bark of Prunus yedoensis with no pronounced physiological action.

Safrol 3,4-Methylenedioxy-allylbenzene; constituent of various oils, notably that of sassafras; m.p. 11°C; b.p. 232-234°C; used instead of oil sassafras in subacute rheumatism; externally in pediculosis and ringworm; an insecticide.

Sah-ma reagent for organic halides When potassium -3- nitrophthalimide is refluxed with organic halides, well-crystallized sharp melting derivatives are obtained.

Sakaguchi reaction for guanidine After conversion to the carbonate, the guanidine is evaporated with a little water and 1 2 glycocoll. The glycocyamine formed can be detected with a-naphthol and sodium hypochlorite solution.

Sakaguchi test A test for a free guanidine group. A protein containing argimne or arginine itself will give the test. Treat the material with NaOH solution a-naphthol, and then sodium hypochlorite. An intense red color is positive. If specific conditions are followed, the test may be used for quantitative colorimetric determinations.

Salicin $C_{13}H_{18}O_7$; salicyl alcohol glucoside; a glycoside from the bark of the willow and poplar; consists of glucose and saligenin. Colorless leaflets, m.p. 201.5^{0}C.; used medicinally as antirheumatic and for neuralgia, chorea, etc.

Saligenin $C_7H_8O_2$; salicyl alcohol; saligenol; o-hydroxybenzyl alcohol; a hydrolytic product of the glycoside, salicin; local anaesthetic and remedy for rheumatism; m.p. 86-87^{0}C.

Salinigrin A glucoside, constituent of various willow barks.

Saliva fluid secreted by glands in the mouth of

vertebrates that assists the swallowing and digestion of food.

Salivary gland one of two *glands* in the mouths of mammals that produces and secretes *saliva.*

Salkowski reaction for indoleacetic acid A solution of indoleacetic acid, acidified with hydrochloric acid, is boiled with ferric chloride solution. A cherry-red color is obtained, A violet color is produced with very dilute solutions of the acid.

Salkowski reaction for inositol A trace of the substance, dissolved in nitric acid, sp. gr. 1.2, is treated with I drop of 10% calcium chloride solution and I drop of 1% platinic chloride solution. The solution is cautiously evaporated by blowing air over it; a rose-red to brick-red color is formed.

Salkowski reactions for cholesterol I. A few cgm. cholesterol in 2 cc. chloroform is treated with 2 cc. concentrated sulfuric acid. The sulfuric acid shows a green fluorescence, the chloroform becomes blood-red. When a few drops of the chloroform solution are placed in an evaporating dish, the solution rapidly becomes blue, green and finally yellow.

II. A solution of a trace of cholesterol in chloroform is allowed to evaporate from filter paper, giving a lemon-yellow color when the spot is touched with a drop of sulfuric acid.

Salkowski test for bile pigments The icteric urine, alkaline with sodium carbonate, is treated with calcium chloride solution; after filtration the precipitate is dissolved in alcohol containing hydrochloric acid and heated. Bile pigments give a green to blue color.

Salkowski test for proteins A few cc. of formaldehyde solution (1 55000 to 1 50000) is treated with an equal volume of concentarted hydrochloric acid, 3 drops of 3% ferric chloride and 0.1 gm. of protein are added and the solution heated to boiling. A violet color changing to deep violet-blue is obtained. This reaction is given by all proteins containing tryptophane.

Salkowski test for silicic acid in urine After partial evaporation, the urine, in a cylinder is precipitated with alcohol. The sticky precipitate is washed several times with alcohol and ether. It is digested with cold, 13 hydrochloric acid, filtered on ashless filter paper and ignited. The residue is almost pure silicic acid.

Salmine A protamine found in the sperm and testicles of the salmon, which hydrolyzes almost entirely to arginine.

Salting out The addition of a solid salt, or a concentrated solution of a salt to a solution or a colloid, causing the solute (liquid or solid) or the disperse phase to separate out. In colloids this is due to the neutralization of the micellar charge by an ion of the added salt. In liquid-liquid systems, the phenomenon take place only if the salt is soluble in just one of the liquids. If salt is added to an aqueous solution of a gas the solubility of the gas is decreased, the smaller the salt molecule the greater the effect.

Samaderin A glucosidal-amaroid, $C_{29}H_{34}O_{11}$, from the bark of Samadera indica; paralyzes the heart and brings it to a systolic standstill.

Sambunigrin A glycoside of Sambucus niger; m.p., 151-152°C; consists of glucose and 1-mandelonitrile; diuretic, sudorific and alterative.

Sanchez reactions for alanine I. 0.01 gm. alanine are heated with 20 drops of 1% potassium permanganate solution for one minute, cooled, treated with 0.05 gm. oxalic acid until decolorized. 2 cc. alcohol, 0.02-0.03 gm. o-nitrobenzaldehyde and 10 drops 10% sodium hydroxide are added and the whole shaken with chloroform. The chloroform becomes blue.

II. The alanine and sodium hypochlorite are heated for several seconds, 2 drops of 30% sodium hydroxide are added and iodinepotassium iodide solution added dropwise; a precipitate of iodoform is obtained.

III. Dry alanine is vaporize in a closed tube, cooled and dissolved in dilute hydrochloric acid; the solution given a reddish brown color with Bouchardt's reagent, a red precipitate with Dragendorff's reagent, and a white precipitate wtih Sanchez' molybdate reagent.

Sanchez reactions for cystine I. A mixture of 0.1-0.2 mg., 3 drops 10% sodium nitrite solution and 2 drops hydrochloric acid heated. After dilution with a few cc. water, extraction with ether and evaporation of the ether, an orange color is obtained with alkali hydroxides.

II. Cystine give a precipitate of iodoform when heated with sodium hydroxide and iodine-potassium iodide solution.

Sanchez test for choline and lecithin The nuclear choline molecule is converted into iodoform, by iodinepotassium iodide and sodium hydroxide. 0.1 gm. lecithin is boiled with 10 cc. 5% sodium hydroxide for 5 minutes, cooled, and 0.5% sulfuric acid added dropwise until acid. The fatty acids are filtered through a wet paper. 1 cc. of the aqueous filtrate can be used for carrying the above reaction.

Sanchez test for creatinine 0.02-0.03 gm. of substance are boiled with 2 cc. water and 0.5. gm. mercuric oxide. The decanted liquid is treated with 0.5 gm. resorcinol and 2 cc. sulfuric acid are added without mixing. A blue color is formed at the contact zone.

Sanguinarine An alkaloid, ø-Chelerythrine, $C_{20}H_{15}O_5N$; m.p. 240-265°C (depending on speed of heating); from roots of Papaveraceae.

Santalene Sesquiterpenes; alpha-form and beta-form found in Indian sandalwood.

Santalols Alpha and beta forms of a sesquiterpene alcohol found in Indian sandalwood oil.

Santene A terpene-like hydrocarbon, C_9H_{14}, found in sandalwood oil and in some pine oils, b.p. 140°C.

Santenone A saturated dicyclic ketone; l form found in sandalwood oil.

Santonin A lactone, $C_{15}H_{18}O_3$, m.p. 169-171°C, having a structure of three cycles, active principle of wormseed (Santonica); a vermifuge for roundworms and threadworms but not tapeworms; average dose 0.06 gm.

Sapogenin The nucleus of the non-sugar residue of the saponins.

Saponarin A glucoside from leaves of Saponaria, $C_{21}H_{24}0_{12}2H_2O$.

Saponification (1) The hydrolysis of fats by alkalies, forming glycerol and the alkali salts of the fatty acids.

(2) Sometimes the definition is extended to include any ester hydrolysis, whether by alkali, acid, steam, or enzyme.

Saponification number The number of milligrams of KOH needed to saponify I gram of a fat or oil.

Saponins Glycosides which have sapogenin as the non-sugar component; make strongly foaming aqueous colloids; include various types, e.g. sarsasapogenin, smilagenin, henderagenin. Some are strong fish poisons and hydrolyze red blood cells easily.

Saprophyte Organism living on organic decay.

Saprotroph any organism that obtains its nutrition and energy by feeding on the solid waste material or dead bodies of other organisms.

Sarcode An old term for protoplasm.

Sarcolobid A glucoside from inner bark of Sarcolobus narcoticus; a curarelike spastic poison.

Sarcosin(e) Methyl glycine, m.p. 212°C, product of alkaline hydrolysis of creatine.

Sarmentose A desoxymethyl pentose, obtained from sarmentocyrmarin, a glycoside.

Sarsasapogenin A component of the glycoside sarsasaponin, which is a sterol; $C_{27}H_{44}O_3$ (Sobotka); m.p. 199200°C; occurs in sarsaparilla, smilax, etc.

Sarsasaponin A saponin of sarsaparilla root and foxglove; consists of 3 molecules of glucose and one of sarsasapogenin.

Saxitoxin Neurotoxin found in a variety of dinoflagellates. If ingested, it may cause respiratory failure and cardiac arrest.

Scabiosin A glucoside from root of Scabiosa succisa melting at 108-110°C.

Scaevolin An amaroid from bark and leaves of Scaevola Königii; said to be slightly toxic.

Schall reagent for ultra-violet rays p-Phenylene diamine paper is colored blue by ultra-violet rays, while sun rays have no effect.

Scheringa reaction for neoarsphenamine A 1 1000 aqueous neoarsphenamine solution turns violet upon the addition of concentrated ammonium persulfate solution.

Schick test for diphtheria immunity A 1 500 toxin solution in physiological saline, preserved with 0.5% phenol is used for an intracutaneous reaction, using a dose 1/50 of the lethal dose for a guinea pig of 250 gm. weight. An area of redness and edema attaining a maximum in 48-72 hours, is produced at the injection site by patients whose blood contains insufficient antitoxin to neutralize the amount of toxin used.

Schiff reagent for aldehydes Aldehydes give a violet-red color when treated with a solution prepared by decolorizing a solution of 0.25 gm. of fuchsin in a liter of water with sulfur dioxide.

Schmalfuss-wernerkraul test for oxalic acid in urinary calculi A solution of the powdered material in 10% hydrochloric acid is filtered and extracted with ether; the ether extract is evaporated and 0.1-1 mg. of residue is gently fused with 0.4-2 mg. of resorcinol on white porcelain. After cooling the fusion is treated with a drop of concentrated sulfuric acid and heated several times for 3 seconds at 20 second intervals—a deep blue then a green color is formed.

Schmidt test for sulfanilamide in blood An acid solution of sodium ßnaphthoquinone-4-sulfonate is added to a tungstic acid blood filtrate, producing a straw-yellow to orange-red color which can be estimated colorimetrically.

Schoorl micro-reactions for choline Double salts are formed by the reaction of choline hydrochloride with the following mercuric chloride and iodide, sodium-gold chloride, potassium bismuth iodide, picric acid and picrolonic acid.

Schotten-baumann reaction The reaction of benzoyl chloride with amines in the presence of NaOH. With primary and secondary amines the product has the formula; $R_1 R_2 CNHCOC_6 H_5$; where the R's may be H or an organic radical. With tertiary amines one gets no reaction. In all cases one gets separable products, so that this is a means of identifying and separating amines.

Schulte test for pentamethylenetetrazole and blood A mixture of 50 cc. urine and 5 cc. 20% lead acetate solution is filtered and the filtrate saturated with ammonium sulfate and extracted with 20, 10 and 10 cc. of chloroform. After evaporation, the residue is dissolved in a small quantity of water and tested with cuprous chloride.

Schulz test for cacodylate in urine A few cc. of urine, treated with a small amount of crystalline phosphorous acid, gives rise to an odor of cacodyl oxide.

Schuntermann test for urochromogen in urine Reagent—Mixture of concentrated sulfuric acid and a little sodium perborate. 2 cc. of urine are mixed with 1 cc. of reagent, cooled and shaken with 3 cc. ether. Ether layer becomes intense lemon-yellow color.

Schweitzer's reagent Ammoniacal copper hydroxide, a solvent for cellulose.

Scientific notation (general) method of representing quantities as a number multiplied by a positive or negative power of 10; the number is written with one digit to the left of the decimal point (e.g., the quantity 8,500,000,000 is written as 8.5×10^8)

Scillain A glucoside from the bulb of squill, $(C_6 H_{10} O_3)x$; causes systolic heart standstill.

Scillaren A mixture of the A and B forms in the proportion in which they naturally occur in squill, 2 of A to 1 of B.

Scillaren a The crystalline constituent of the natural glucosides of squill; use as of digitalis.

Scillaren b The amorphous constituent of the glucosides of squill; use as of digitalis.

Scillipicrin An amaroid, active bitter principle from bulb of squill; cardiac action like that of digitalis.

Scillitin A glucoside, $C_{17} H_{25} O_6$, from squill; has been presumed to be the toxic constituent of squill.

Scombrine A protamine found in the testicles of the mackerel.

Scoparin A physiologically indifferent compound, $C_{22} H_{22} O_{11}$, from Cytisus scoparius, melting at 202-219°C.

Scopoletin The aglucone of scopolin occurring in the root of Scoplia japonica; m.p. 204°C.

Scopoline An alkaloid, a decomposition product of scopolamine, melting at 108-109°C, boiling at 248°C.

Scree (Earth science) chunks and sediment of rock that pile up at the base of a mountain or cliff

Screening To screen a library is to select and isolate individual clones out of the mixture of clones. For example, if you needed a cDNA clone of the pituitary glycoprotein hormone alpha subunit, you would need to make (or buy) a pituitary cDNA library, then screen that library in order to detect and isolate those few bacteria carrying alpha subunit cDNA. There are two methods of screening which are particularly worth describing screening by hybridization, and screening by antibody. Screening by hybridization involves spreading the mixture of bacteria out on a dozen or so agar plates to grow several ten thousand isolated colonies. Membranes are laid onto each plate, and some of the bacteria from each colony stick, producing replicas of each colony in their original growth position). The membranes are lifted and the adherent bacteria are lysed, then hybridized to a radioactive piece of alpha DNA (the source of which is a story in itself - When X-ray film is laid on the filter, only colonies carrying alpha sequences will "light up". Their position on the membranes show where they grew on the original plates, so you now can go back to the original plate (where the remnants of the colonies are still alive), pick the colony off the plate and grow it up. You now have an unlimited source of alpha cDNA. Screening by antibody is an option if the bacteria and plasmid are designed to express proteins from the cDNA inserts . The principle is similar to hybridization, in that you lift replica filters from bacterial plates, but then you use the antibody (perhaps generated after olde tyme protein purification rituals) to show which colony expresses the desired protein.

Scrotum The pouch that contains the testes and their accessory organs. It is composed of skin, the dartos, fascia, and the tunica vaginalis.

Scudi micro-tests for sulfanilamide I. A mixture, on a microscopic slide, of 1 drop sulfanilamide solution, 1 drop 37% formaldehyde solution and 1 drop 10% sodium carbonate, is evaporated to dryness on a water bath. Residue is freed from soluble salts by extraction with 2 drop portions of water. Product melts at 235-240°C C. with decomposition. Sensitivity—0.0012 mg.

II. One drop sulfanilamide solution and 1 drop iodine monochloride solution are evaporated to dryness; a drop of water is added to the residue, a second drop of hot water added, stirred and drawn off with absorbent filter paper. Alkali-soluble, water and acidsoluble needles are obtained, melting with decomposition at 265°C C.

III. A solution of 1 drop concentrated hydrochloric acid and 1 drop sulfanilamide solution is evaporated to dryness and treated with a small drop pf saturated picric acid solution. Long yellow needles, m.p. 179-180°C C. are obtained.

IV. One drop mercuric nitrate solution and 1 drop 10% sodium carbonate solution are added to 1 drop sulfanilamide solution, yielding a highly flocculated transparent white precipitate.

Scudi-ratish method for ascorbic acid Five cc. of a 5 mg. percent solution of sulfanilamide are treated with 1 cc. of 0.05% sodium nitrite and 1 cc. of 20% sulfoscalicylic acid. After standing for 1-3 minutes, the solution is treated with 1 cc. of 1% urea. 10 cc. of fresh 10% acetic acid solution of the vitamin are added after 5 minutes. After another 5 minutes. 7.4 cc. of 1-dimethylnaphylamine solution (1 cc. diluted to 500 cc. with alcohol) are added and the solution is mixed. Between 10 and 50 minutes, the colors developed are compared with standards prepared by lowering the sulfanilamide concentration and replacing the vitamin with vitamin-free 10% acetic acid.

Scurvy Avitaminosis C; a deficiency disease due to lack of vitamin C, showing latency in vague symptoms turning to hemorrhages, anemia, bone weakening, etc. Tests have been developed for capillary fragility and the saturation of the blood and urine with vitamin C. Sources rich in vitamin C or pure ascorbic acid are administered. Normal plasma ascorbic acid should run 0.70-1.0 mg. per 100 cc. If it is below 0.15 full scurvy sets in.

Scymol $C_{27}H_{43}O_5$, m.p. 187°C, a compound related to sterols found in shark bile as a sulphuric ester.

SDS Sodium dodecyl sulphate. It is an ionic detergent.

Seafloor spreading construction of ocean crust outward in opposite directions from ocean ridges (submerged mountains where two plates are moving away from each other.

Seborrhea Overactive secretion of the sebaceous glands.

Sebum Oily lubricating secretion from *glands* in the skin.

Sebum The secretion of the sebaceous glands of the skin, mostly cholesterol esters and waxes.

Secondary structure Local structure within a protein which is conferred by the nature of the side chains of adjacent amino acids (e.g., alpha helix, beta sheet, random coil); local structure within an RNA molecule which is conferred by base pairing of nucleotides which are relatively closely positioned within the sequence (e.g., hairpins, stem-loop structures).

Secondary This adjective has two senses; sometimes it is used as the equivalent of "second-grade", of organisers, but also for the embryo produced by an implanted organiser, as opposed to the host, or primary, embryo.

Secretin A hormone of the intestinal mucosa, which stimulates the pancreas to secrete digestive juices. A crystalline, basic protein, secretin (solution pH=7.5) with a molecular weight of 5000 has been isolated.

Secretion substance produced and released by a cell to perform a function outside the cell.

Secretogogue A substance which promotes secretion, e.g. gastric secretion by gastrin or histamine.

Sedentary—Living in a fixed location, as with most plants, tunicates, sponges, etc. Contrast with motile.

Sediment material that settles to the bottom of a *solution* or is deposited by wind, water, or ice.

Sedimentary rock Rock formed by the deposition of *sediment* and its gradual compaction and hardening.

Sedimentation potential The potential difference induced between the top and the bottom of a liquid when particles are allowed to fall through the liquid.

Seed A structure produced by seed plants which encapsulates the embryo. The seed often provides nourishment during germination, but may lie dormant for many years first.

Seed plant (life science) any plant that reproduces via seeds; also called a spermatophyte.

Selachyl alcohol Alpha-oleyl-glyceryl ether, b.p. 242°C at 5 mm.; a constituent of elasmobranch liver oils.

Selection The use of particular conditions, such as the presence of ampicillin, to allow survival only of cells with a particular phenotype, such as production of beta-lactamase.

Self-differentiation (Différentiation spontanée) The differentiation of a tissue solely as the result of factors existing within itself, i.e. differentiation after determination has been accomplished. Any differentiation in explanted tissue is self-differentiation.

Selfwise Behaving after transplantation in accordance with the old situation in which the graft had previously been, i.e. in accordance with its prospective significance, i.e. determined, self-differentiating, and no longer pluripotent. Refusing to "fall into step" with the new neighbour cells and go along with them to their morphogenetic destination.

Seliwanoff reaction for levulose and sucrose When levulose or sucrose is heated with a hydrochloric acid solution of resorcinol, a red color is produced which precipitates a darker amorphous precipitate on cooling. This is soluble in alcohol with a pale red color. Carbohydrates yielding levulose on hydrolysis give this reaction.

Sellard's test An approximate test of alkali reserve by the administration of sodium bicarbonate and testing the pH of the urine at intervals.

Semelparity participation in a single act of reproduction during an organism's lifetime.

Semen fluid that contains *sperm* (male sex cells of animals) and several other *secretions* from the male reproductive tract and that exits the penis during *copulation.*

Semiconductor material that conducts electricity better than an insulator but not as well as a metal; semiconductors are widely used in electronic devices.

Semidominance Intermediacy of the phenotype of the heterozygote between the phenotypes of the two homozygotes.

Semi-permeable membrane A membrane that allows a selective passage of molecules through it. Semi-permeable membranes can be prepared that have graded permeability, each one allowing only molecules of a certain maximum size to pass through it.

Senecifolidine An alkaloid, $C_{18} H_{25} O_7 N$, from Senecio latifolius, melting at 212°C; is toxic.

Senecifoline An alkaloid from Senecio latifolius, $C_{18} H_{27} O_8 N$, melting at 149195°C; is toxic.

Senecine An alkaloid from Senecio vulgaris and other allied compositae; in large doses attacks central nervous system, in small causes hemorrhages.

Senegin Saponin-like glucoside from root of Polygala Senega; m.p. about 240°C; used as expectorant, emetic.

Senna Dried leaves and pods of Cassia species; strong purgative and cathartic; constituents are crysophanic acid, sennaemodin, sennarhamnetin, sennanigrin, cathartomannite; dose 2-10 gm.

Sense organ Organ in animals that receives a stimulus and transmits a nerve impulse to the brain containing information about the animal's surroundings; examples include the ears, the eyes, the nose, the tongue, and various receptors on the skin.

Sepia Cuttle-fish bone; calcareous substance found under skin of back of Sepia officinalis, containing calcium carbonate and phosphate, and gluten; used as polishing agent and in tooth powders.

Septentrionaline An alkaloid of doubtful composition; m.p. 131°C; stated to have anaesthetic action; paralyses respiration.

Sequence homology Similarities in nucleic acid sequence and organization, and in their encoded products, that are sufficiently great as to imply common ancestral origins.

Sequence similarity Similarity in nucleic acid or polypeptide sequences, particularly in shorter segments, that may not be sufficient to imply common ancestral origins.

Sequence tagged site (STS) Short (200 to 500 base pairs) DNA sequence that has a single occurrence in the human genome and whose location and base sequence are known. Detectable by polymerase chain reaction, STSs are useful for localizing and orienting the mapping and sequence data reported from many different laboratories and serve as landmarks on the developing physical map of the human genome. Expressed sequence tags (ESTs) are STSs derived from cDNAs.

Sequence As a noun, the sequence of a DNA is a buzz word for the structure of a DNA molecule, in terms of the sequence of bases it contains. As a verb, "to sequence" is to determine the structure of a piece of DNA; i.e. the sequence of nucleotides it contains.

Sequencing Determination of the order of nucleotides (base sequences) in a DNA or RNA molecule or the order of amino acids in a protein.

Sequoyitol Methyl ether of i-inositol from redwood, Sequoyia sempervirens, m.p. 234°C.

1-SERINE $C_3 H_7 O_3 N$, ß-hydoxy-U+22ID-amino

propionic acid; $CH_2OH \cdot CHNH_2 \cdot COOH$; levorotatory; large prisms, m.p. 228°C., with decomposition. An amino acid of the silk scleroprotein sericin. Can be synthesized by the organism, so is not essential in the diet. Possibly interchangeable with alanine.

Serine metabolism Serine is oxidized to U+22ID-keto-ßhydroxy propionic acid in the body. Forms glucose in the diabetic.

Sertoli cells Lining epithelium cells, enlarged in connection with developing spermatozoa of testes.

Serum The straw-colored liquid, sp. gr. 1.027 in man, obtained by removal of fibrinogen and cells from the blood, slightly more than 90% water. It is the solvent for the food, waste and metabolic products carried by the blood.

Sessile 1) a leaf, flower, or fruit having no stalk and growing directly off of a plant's stem; 2) an animal that typically remains in one place for most of its life

Sex chromosomes The X and Y chromosomes in human beings that determine the sex of an individual. Females have two X chromosomes in diploid cells; males have an X and a Y chromosome. The sex chromosomes comprise the 23rd chromosome pair in a karyotype. Compare autosome.

Sex-linked A trait transmitted by one of the chromosomes determining sex of an individual.

Sexual reproduction — A type of reproduction in which two parents give rise to offspring that have unique combinations of genes inherited through the gametes of the two parents. Sexual reproduction involves **meiosis** and **syngamy**.

Sexual reproduction Type of *reproduction* in which male sex cells fertilize female sex cells; compare *asexual reproduction.*

Shafer test for ß-hydroxybutyric acid The acid is oxidized to acetone by potassium dichromate and sulfuric acid; the acetone is detected by any of the well-known tests for it.

Shaw test for bleached flour One kilo of flour is extracted with 95% alcohol for 4 hours. After evaporation of the extract, the residue is extracted with a mixture of equal volumes of ether and alcohol, filtered and evaporated to dryness. A diphenylamine solution in concentrated sulfuric acid is poured on the residue giving a blue color with flour bleached with nitrogen peroxide.

Shellac The resinous secretion of the insect, Tachardia lacca, contains aleuritic acid (q.v.).

Sherman-munsell vitamin a unit That amount of vitamin A, which when fed six times a week, just suffices to support a gain of weight of 3 grams a week in a standard test rat for a period of 4-8 weeks.

Sheterin A glucoside, $C_{26}H_{30}O_{13} \cdot H_2O$, from the berries of Rhamnus cathartica.

Shikimic acid A trihydroxy-cyclohexane carboxylic, acid, derived from Ilicium religiosum.

Shikimin A glucosidal amaroid, $C_7H_{10}O_5$, from the fruit of Illicium religiosum, melting at 175°C; is toxic, belonging to the picrotoxin group.

Short arm (p) One of the two prominent segments of a chromosome; the long or "q" arm is the other. The arms of a given chromosome join at its centromere.

Shotgun cloning The practice of randomly clipping a larger DNA fragment into various smaller pieces, cloning everything, and then studying the resulting individual clones to figure out what happened. For example, if one was studying a 50 kb gene, it "may" be a bit difficult to figure out the restriction map. By randomly breaking it into smaller fragments and mapping those, a master restriction map could be deduced.

Shotgun sequencing A way of determining the sequence of a large DNA fragment which requires little brain power but lots of late nights. The large fragment is shotgun cloned and then each of the resulting smaller clones ("subclones") is sequenced. By finding out where the subclones overlap, the sequence of the larger piece becomes apparent. Note that some of the regions will get sequenced several times just by chance.

Shuttle vector A type of cloning vector that contains

sequences enabling it to be propagated and maintained in more than one type of host (e.g. *E. coli* and mammalian cells). For this purpose shuttle vectors carry different origin of DNA replication which are characteristic of the desired host systems.

Sibling species Reproductively isolated but morphologically identical or nearly identical species.

Sicaloin An amaroid, $C_{14}H_{17}O_6(OCH_3)$, from Sicilian aloe.

Sidereal period time taken by a celestial body to complete one orbit as calculated with respect to a fixed star.

Sidereal time time as measured by Earth's rotation with respect to the stars.

Siemens unit of electrical conductance (the reciprocal of resistance) defined by 1 ampere per volt.

Sigma factor Subunit of bacterial RNA polymerase which controls the correct initiation of transcription. These proteins increase the binding affinity of RNA polymerase to a promoter. Different sigma factors recognize different promoter sequences.

Signal peptidase An enzyme present within the lumen of the endoplasmic reticulum which proteolytically cleaves a secreted protein at the site of a signal sequence.

Signal recognition particle (SRP) A chaperonin complex responsible for arresting polypeptide synthesis and facilitating the docking of a ribosome to the endoplasmic reticulum membrane. Normally, on ribosomes translating polypeptides destined for insertion into or across the endoplasmic reticulum membrane become associated with SRP.

Signal sequence A hydrophobic amino acid sequence which directs a growing peptide chain to be secreted into the endoplasmic reticulum.

Silent mutation A nucleotide substitution (never a single deletion or insertion) which does not alter the amino acid sequence of an encoded protein due to the degeneracy of the genetic code. Such mutations usually involve the third base (wobble position) of codons.

Silica — amorphous silicon dioxide (glass). It is a structural component in many organisms, such as diatoms and horsetails.

Simon reaction for pyruvic acid The acid solution is treated with ammonia, then sodium nitroprusside and heated gently. A blueviolet color forms which changes to orange on long standing.

Simon test for glycogen in urine A mixture of 90 cc. urine and 10 cc. 40% potassium hydroxide is filtered and the filtrate is treated with 10 gm. potassium iodide and 50 cc. 90% alcohol; a flocculent precipitate is formed.

Sinapine An alkaloid, the choline ester of sinapic acid; from black mustard seed.

Sine (*function* of an angle of a right triangle defined by dividing the length of the side of the triangle that is opposite the angle by the length of the triangle's hypotenuse (the side opposite the right angle).

Single- gene disorder Hereditary disorder caused by a mutant allele of a single gene (e.g., Duchenne muscular dystrophy, retinoblastoma, sickle cell disease). Compare polygenic disorders.

Single nucleotide polymorphism (SNP) A polymorphism caused by the change of a single nucleotide. Most genetic variation between individual humans is believed to be due to SNPs.

Single-strand conformational polymorphism (SSCP) Relies on secondary and tertiary structural differences between denatured and rapidly cooled amplified DNA fragments that differ slightly in their DNA sequence; different SSCP alleles are resolved on non-denaturing acrylamide gels, usually at low temperature; ability to resolve alleles depends on conditions of electrophoresis.

Singularity point in *space-time* where the known principles of physics break down; thought to exist at the center of black holes.

Sinigrin $C_{10}H_{16}O_9NS_2K{\cdot}H_2O$; m.p. 126129°C; sinigroside; potassium myronate; a mustard seed glycoside hydrolysed by the enzyme myrosin to glucose, allyl isothiocyanate and K hydrogen sulfate.

Sinomenine $C_{19}H_{23}O_4N$; needles, two m.p.'s, 160°C and 182°C; an alkaloid of Sinomenium Acutum, closely related to thebaine, used in rheumatism, large doses causing respiratory paralysis.

Site-Directed mutagenesis The introduction of a mutation, usually a point mutation or an insertion, into a particular location in a cloned DNA fragment. This mutated fragment may be used to "knock out" a gene in the organism of interest by homologous recombination.

Site-Specific recombination Occurs between two specific but not necessarily homologous sequences. Usually catalyzed by enzymes not involved in general or homologous recombination.

Sitosterol Plant mixture, $C_{29}H_{49}OH$, m.p. 137°C; found in wheat, legumes, cocoa, corn, figs, beet oil, cotton seed oil, laurel oil. Consists of a mixture of at least 3 components in varying proportions depending on the source, and difficult to separate from each other.

Skatole 3-methyl indole; m.p. 95°C; b.p. 265-266°C; an amine formed by bacterial decomposition of tryptophane. Partly responsible for the odor of feces and decaying protein. Occurs also in beet root, nectandra wood and coal tar.

Skeleton rigid framework, either external or internal, that gives structure and support to an animal's body and protects its internal parts.

Skimmianine An alkaloid, $C_{14}H_{13}O_4N$, from leaves of Skimmia Japonica, melting at 175-176°C.

Skull collection of bones or *cartilage* in vertebrates that surrounds and protects the brain.

Slot blot Similar to a dot blot, but the analyte is put onto the membrane using a slot-shaped template. The template produces a consistently shaped spot, thus decreasing errors and improving the accuracy of the analysis.

Smallpox Variola; infectious, contagious disease caused by a filtrable virus, with characteristic symptoms of skin eruption, papules, vesicles, pustules crusts and permanent scars. Types variola vera, hemorrhagic smallpox and mild smallpox. Vaccination immunity lasts only a limited time, about 10 years.

Snrnp "snerps", Small Nuclear RiboNucleoProtein particles, which are complexes between small nuclear RNAs and 7-10 proteins, and which are involved in RNA splicing and polyadenylation reactions.

Soft inheritance Inheritance during which the genetic material is not constant from generation to generation but may be modified by the effects of the environment, by use or disuse, or other factors.

Soil creep slow movement of soil down a slope as a result of *gravity.*

Soil depletion gradual decrease in the chemical and nutritional health of soil as a result of overfarming, erosion, and pollution.

Solangustine A gluco-alkaloid, $C_{33}H_{53}O_7N_2B_7H_2O$, from Solanum angustifolium, melting at 235°C.

Solanine A gluco-alkaloid of undetermined structure, obtained from potato sprouts and other Solanaceae.

Solar radiation *radiation* emitted by the Sun; composed mostly of visible light, *ultraviolet radiation,* and *infrared radiation*

Solm reagent for pepsin A solution of 0.5 gin. ricin in 50 cc. 5% sodium chloride solution is treated with 0.5 cc. of 0.1N hydrochloric acid after filtering. Pepsin, at 40°C, clarifies the milky turbidity of the reagent.

Solstice either of the two days per year when the Sun is farthest north or south from the celestial equator (the imaginary plane that cuts through Earth's equator); occurs around June 21 (summer solstice) and December 22 (winter solstice).

Solubility measure of the quantity of a substance that is able to dissolve in a given volume of a particular liquid.

Solute dissolved substance.

Solution mixture of two or more substances, at least one substance being dissolved in another.

Solution hybridization A method closely related to RNase protection. Solution hybridization is designed to measure the levels of a specific mRNA species in a complex population of RNA. An excess of radioactive probe is allowed to hybridize to the RNA, then single-strand specific nuclease is used to destroy the remaining unhybridized probe and RNA. The "protected" probe is separated from the degraded fragments, and the amount of radioactivity in it is proportional to the amount of mRNA in the sample which was capable of hybridization. This can be a very sensitive detection method.

Solvent substance in which another substance can be dissolved

somatic cells Body cells. All the cells except the germ-line cells.

Somatic cells Any cell in the body except gametes and their precursors.

Somatic hypermutation A very high frequency of mutational events that occur in specific loci, such as the variable segments of expressed immunoglobulin genes. Somatic hypermutation in immunoglobulin genes occurs after all rearrangements have occurred and provide additional potential for variation in immunoglobulin structure.

Somatic Refers to non-germline cells. Somatic cells may become terminally differentiated with alterations in their overall genetic compliment, because they are not responsible for passing along the organism's genetic material to the offspring.

Sorbic acid $CH_3CHCHCHCHCOOH$, mp. 133-134°C, found in the berries of Sorbus aucuparia.

d-SORBITE d-sorbite; a hexahydroxy hexane, m.p. 87-92°C, also at 55°C; obtainable by reduction of d-fructose and d-glucose. It is frequently found with d-mannitol and occurs in ripe mountain ash berries, cherries, plums, pears, apples, etc. On oxidation yields 1-sorbose, which can be oxidized to vitamin C.

Sorbitose d-SORBOSE Sorbitose; a ketose sugar formed by the bacterial oxidation of sorbitol by Bacterium xylinum; sweet non-fermentable, m.p. 154°C, also given as 165°C.

Sörensen titration A method for determining the carboxyl content of an amino acid mixture, by reacting the amino group with formaldehyde, thus allowing direct titration of the carboxyl group.

Sörensen's reversible dissociable component systems A theory of protein structure. Long polypeptide chains are combined in varying amounts with other groups, also mainly polypeptides, the forces holding these large groups together being residual valences.

Sorghum The name applied to a group of grasses (Andropogon Sorghum) from which sugar and sugary syrups may be prepared.

Southern blot The transfer (by absorption) of size-separated (by electrophoresis) DNA fragments to a synthetic membrane whereby the presence and rough size of one particular fragment of DNA can be ascertained by detection of specific base sequences ascertained by radiolabeled complementary probes. Initially described by E.M. Southern.

Southwestern Blot The binding of protein to a nucleic acid on a matrix similar to what is done for western, northern, and southern blots. This technique is used to identify DNA binding proteins and the recognition sites for these proteins.

Soya oil Soy Oil, Soya Bean or Chinese Bean Oil; made up of glycerides of certain fatty acids; used in mfg. of soaps, paints, varnishes; used as illuminant and as food.

Sp6 rna Polymerase A bacteriophage RNA polymerase which is commonly used to transcribe plasmid DNA into RNA. The plasmid must contain an SP6 promoter upstream of the relevant sequence.

Space-time System consisting of one dimension of time and three spatial dimensions, which can be used to locate any object or event; used in the theory of *relativity.*

Spaltung A twin embryo in which the two neural axes are united posteriorly; it is due to a splitting of the morphogenetic movements into two streams, and may simulate an induction.

Sparteine Lupinidine; $C_{15}H_{26}N_2$; b.p. 326°C; alkaloid of broom and yellow lupein seeds; resembles confine and digitalis in action and is less toxic; used as diuretic in certain arrhythmias.

Spasm Involuntary muscular paroxysm; tonic, if uniform; clonic, if in waves.

Spathulatine An alkaloid, $C_{33}H_{64}O_5N_4$, from seed hulls of lupinus spathulatis, melting at 227°C.

Spearmint Dried leaves and tops of Mentha spicata containing carvone as active principle; used as carminative, like peppermint.

Speciation Process by which a previously nonexistent *species* emerges during the course of *evolution;* a common method of speciation is the geographic isolation of a *population* that consequently undergoes a unique sequence of evolutionary events.

Species Group of organisms that can interbreed to produce fertile offspring and that possess a similar *genome* and similar physical characteristics.

Specific dynamic action of proteins The name applied to the increased metabolism observed when protein is fed. If a definite number of calories is fed to a dog which has been fasting, the heat production is about 5% more than the calorific value of the protein. Originally this was thought to be due to cell stimulation by the protein, causing them to burn more fat and carbohydrate. However, a large portion of the heat can be attributed to the extra work done in disposing of the N of the protein.

Specific heat capacity amount of heat required to raise the temperature of 1 kilogram of a substance by 1 degree (on either the Celsius or Kelvin scale).

Specific latent heat heat required to change the physical state of a given substance without changing its temperature.

Specific rotation A measure of the degree of optical rotation shown by a compound. Defined as the number of degrees of rotation of the plane of polarized light caused by a solution of 1 gram of a substance in 1 cc. of solution in a 1 decimeter tube.

Sperm (plural *spermatozoa*) male sex cell in animals; also called spermatozoon.

Sperm oil From the sperm whale containing fatty acid eaters, small quantities of spermaceti; saponification no., 123-147; Iodine no., 80-84; used as lubricant, in lamps, hardening steel, mfg. of soaps.

Spermaceti Cetaceum; m.p. 42-50°C; a waxy substance from the head of the sperm whale, mostly cetyl palmitate, esters of lauric, stearic and myristic acids and free cetyl alcohol; used as ointment base.

Spermatids The immature male gamete cells arising from the division of the spermatocytes. They give rise to spermatozoa.

Spermatocytes The immature male gamete cells arising from the division of the spermatogonia.

Spermatogonia In the development of male gametes, the first cells arising from the primordial sex cells.

Spermatophore capsule in some *invertebrates* that carries the sperm.

Spermidine $NH_2(CH_2)_3NH(CH_2)_4NH_2$; occurs in association with spermine in sperm and in animal organs.

Spermin An obsolete name for the male sex hormone.

Spermine $NH_2(CH_2)_3NH(CH_2)_4NH(CH_2)_3NH_2$; gerontine; musculamine; neuridine; strong base, a constituent of sperm (as phosphate), ox pancreas, yeast, etc.

Sphingomyelins A group of phospholipids, m.p. ranging from 196°C td 198°C, occurring in the brain, kidney and liver along with cerebrosides. Sol. in water and most organic solvents except ether, cold acetone and cold alcohol. On hydrolysis they yield phosphoric acid, choline and sphingosine, and a fatty acid -either lignoceric, stearic, nervonic, or cerebronic acid.

Sphingosine CH_3 $(CH_2)_{12}$ CHCHCH(NH_2) CH (OH)CH_2 OH; a basic component of the sphingomyelins and of cerebrosides.

Sphygmograph A recording instrument for pulse or pressure waves.

Spin characteristic of a subatomic particle similar to angular momentum but existing both when the particle is in motion and when it is at rest.

Spindle fibers Microtubules that form during mitosis or meiosis and that move the chromosomes.

Spine column of bones along the back in vertebrates; surrounds and protects the spinal cord.

Spirit of ammonia A soln. of 10% NH_3 in alcohol; formerly used as a diffusible indirect respiratory and circulatory stimulant; externally as counterirritant.

Spirit of ammonia, aromatic A soln. containing 34 gm. ammon. carbonate, 90 cc. 10% ammonia water, 10 cc. lemon oil, 1 cc. oil of lavender, 1 cc. oil of myristica, 700 cc. alcohol, dilute by water to one liter; used as stimulant and locally in insect bites.

Spirit of ether A mixture of alcohol and ether made by diluting 325 cc. ether U.S.P. to 1000 cc. with alcohol; used as anodyne, carminative, indirect circulation stimulant, antispasmodic.

Spleen vertebrate organ that destroys old cells, stores a portion of red blood cells, and produces *lymphocytes* (cells involved in the body's protection against disease).

Splice sequence A sequence within an mRNA molecule that is the site at which splicing occurs.

Splicing The process that removes introns (non-protein-coding portions) from transcribed RNAs. Exons (protein-coding portions) can also be removed. Depending on which exons are removed, different proteins can be made from the same initial RNA or gene. Different proteins created in this way are 'splice variants' or 'alternatively spliced'.

Splicosome A ribonucleoprotein structure responsible for splicing of primary transcripts.

Spondylitis deformans A form of arthritis characterized by progressive vertebral deformity including thickening and calcification of the anterior spinal ligament.

Spongin An albuminoid of the skeletons of the sponges and corals.

Spontaneous combustion Ignition of a material without the application of external heat; typically achieved through internal chemical processes.

Spontaneous generation The spontaneous origin of life from inanimate matter.

Spore Of fungi, a walled, resistant cell or multicellular structure, produced by mitosis or meiosis, that can germinate and give rise to a new mycelium. Of land plants, a reproductive cell formed by meiosis that can develop into gametophyte.

Sporophyte — The diploid stage in the life cycle of an organism undergoing an alternation of generations. The sporophyte is multicellular and develops from a zygote. The mature sporophyte meiotically produces haploid spores that later generate the gametophyte generation.

Sprintillamine An alkaloid, $C_{28}H_{45}O_4N$, from rhizome of Helleborus viridis, melting at 228-229°C.

Sprintilline An alkaloid $C_{25}H_{41}O_3N$, from rhizome of Helleborus viridis, melting at 141-142°C.

Spur ridge of rock that juts out to the side of a mountain or mountain range

Squalene Spinacene; $C_{30}H_{50}$, an open chain triterpene found in elasmobranch livers; a mixture.

Squill Sea onion; the fleshy inner bulb scales of the white variety of Urginea maritima, used as expectorant, nauseant, diuretic; action also resembles that of digitalis.

Stable transfection A form of transfection experiment designed to produce permanent lines of cultured cells with a new gene inserted into their genome. Usually this is done by linking the desired gene with a "selectable" gene, i.e. a gene which confers resistance to a toxin (like G418, Geneticin). Upon putting the toxin into the culture medium, only those cells which incorporate the resistance gene will survive, and essentially all of those will also have incorporated the experimenter's gene.

Stachydrine N-methylproline-methyl betaine, obtained from tubers of Stachys tubifera and leaves of orange tree.

Stachyose A tetrasaccharide, m.p. 167°C present in a number of plants, such as peas, ash manna, white jasmine twigs etc., yielding fructose and mannotriose on hydrolysis.

Stachytarphin An amaroid from fresh leaves of stachytarpheta dichotoma.

Stack vertical column of rock isolated from coastal headlands by oceanic water erosion.

Stamen collectively, the stalk, filament, and anther, which serve as the reproductive organ of flowering plants.

Stamm test for peroxide in ether One drop of Meyer phenolphthalein reagent is mixed with 2 drops of 12000 copper sulfate solution followed by I drop of the ether. A pink to intense red ring is formed at the interface.

Standard gravity acceleration of a body because of the force of gravity; on Earth, standard gravity is accepted as 9.81274 meters per second per second.

Standard metabolism Heat production of a resting animal at standard conditions Kept post absorptive between the two critical environmental temperature.

Standing wave wave along which the amplitude varies from place to place, is zero at particular fixed points, and reaches a maximum at other fixed points; water and electromagnetic radiation may travel in standing waves.

Staphisagroine An alkaloid of undetermined composition from the seeds of Delphinium staphisagria, melting at 276$_{°C}$.

Star anise Fruit of Illicium verum containing tannin, anisic acid, resin, pectin and other constituents, and used as a carminative like anise.

Starch The most important hexosan. It is the source of energy for plants. The starch molecule consists largely of glucosidal linked aglucopyranose, units, which are coiled back to form a closer unit, and are held in this position by secondary valences. Starch granules probably are composed of a readily water-dispersible a-amylose (granulose) and a difficultly dispersible phosphorus containing a-amylose (amylopectin), which are separable.

Starch value An index of digestibility of raw fodders; the percentage of utilization as compared to the total starch.

Starch,soluble A somewhat hydrolyzed starch showing an increase in reducing power and smaller molecular weight; used for determination of diastatic power of malt and as indicator for iodine.

Start codon That codon at which translation of an mRNA molecule begins. This is always an AUG (encoding methionine) in eukaryotes, and nearly always in prokaryotes. In prokaryotes N-formyl-methionine is used to initiate polypeptide synthesis.

Static electricity stationary electric charge acquired through *friction* or *induction.*

Steady state Dynamic equilibrium; the condition of balance between varying, shifting, and opposing forces, which is especially characteristic of living processes.

Steapsin Old name for lipase of pancreatic juice.

Stearic acid A saturated fatty acid, $CH_3(CH_2)_{16}COOH$, found in animal and vegetable fats, m.p. 69-70°C, b.p. 383°C; used in suppositories, ointments, enteric coatings for bitter drugs.

Stearidonic acid An 18 carbon unsaturated fatty acid with four double bonds.

Steatorrhea The appearance of fat in the feces.

Steigmann reactions for vitamin D With fuchsin-sulfurous acid, solid irradiated ergosterol acquires a violet color. Heating solid irradiated ergosterol with ammoniacal silver oxide yields a stable silver sol.

Steinle-kahlenberg reaction for cholesterol A chloroform solution of antimony trichloride gives a muddy brown precipitate with a dilute cholesterol solution. With excess chloroform, the precipitate dissolves to form a clear purple liquid quickly changing to cobalt-blue on exposure to light.

Stenohaline A term applied to animals that cannot withstand a material change in external salinity, e.g. spider crab Maia; opposite of euryhaline.

Steppihn-utkinljubowzow method for determining bayer 205 The determination in urine, serum and tissues depends upon decomposition of the substance in boiling alkali. The liberated components are coupled with diazo compounds; the resultant colored substances are determined colorimetrically.

Stepwise inhibition The successive inhibition of different organic processes by successively stronger applications of an externally acting agent.

Sternum flat bone in the chest of most vertebrates that lies between and is joined to the ribs; also called the breastbone.

Steroids Steroids comprise a group of isocyclic products of plant and animal origin with a tetracyclic hydromatic ring system, perhydrocyclopentenophenanthrene (CPP), as common nucleus. This carbon skeleton was correctly formulated, 1932, by Rosenheim and King as consisting of 1 five-membered and 3 six-membered rings, in angular annelation. It carries as sidechains 2 angular methyl groups and in many instances a side chain of 2 to 10 carbon atoms. Hydroxyl and carbonyl groups are the most important functional groups occurring in the cyclic and acyclic portions of the molecule, while a carboxyl group in the acyclic sidechain characterizes the bile acids, an important subdivision. Additional variety is caused by the presence in the cyclic and acyclic portions of double bonds which in the estrogenic hormones are accumulated in one or two rings, giving rise to a benzene or naphthalene system. An immense number of further isomeric possibilities is offered by the steric configuration on 4-10 asymmetric carbon atoms, but, remarkably enough, only a small number of them is realized in natural products. This selectivity of nature plays part in the great stability of the steroid skeleton which survives postmortally all other organic constituents of the animal body, carbohydrates, proteins, and fats, is found as ''adipocere" under certain conditions in skeletonized cadavers and has also been linked to the origin of ''mineral" oil.

As steroids are mostly of hydrophobic character, nature endeavors to render them watersoluble for various physiological functions by conjugation with hydrophilic groups. This is accomplished in the plant (neutral saponins, cardiac glycosides, and the steroid solanum alkaloids) by glycosidic linkage to hexoses and certain desoxyhexoses, in the animal by conjugation with the amino acids glycine and taurine in the case of the bile acids, with H_2SO_4 in scymnol in shark's bile, and with suberyl arginine in toad poison.

The biogenesis of the steroids is still obscure although the combination of several isoprene units offers a suggestive hypothesis. Experiments with heavy isotope tracers indicate the rapid synthesis from even smaller fragments, but it is undecided whether the general course of synthesis issues from amino acids such as leucine, containing a preformed isoprene skeleton, or from carbohydrate fragments.

Their catabolism in the animal body is obviously of an oxidative nature, first chopping off the sidechain and finally leading to dehydrogenation of the remaining cyclic rump. Whether complete dehydrogenation, leading to polycyclic aromatic hydrocarbons, closely related to the most carcinogenic compounds known, has any bearing on the pathogenesis of spontaneous tumors or is merely a formal coincidence, is doubtful. In any event, the powerful stimulating action of the steroid hormones on the various primary and secondary reproductive tissues seems a generic property whose specificity is delicately governed by minute structural and steric differences.

Many physiological functions of steroids may be traced to their surface activity. This, in turn, is due to the spatial configuration of the molecule and the polar distribution of hydrophilic and oleophilic groups, which offer themselves to study by the elegant methods of "twodimensional" chemistry in monomolecular layers. This surface activity endows bile acids with emulsifying power for fats in the digestive tract; it gives to cholesterol and its esters the rôle of a mediator between protoplasm and lipoid membranes, and contributes to the pharmacologically valuable properties of the cardiac glycosides and the cytolytic saponins.

Sterols Sterids; sometimes refers specifically to those sterids with a hydroxyl group in the structure. Compounds containing the perhydrocyclopentano phenanthrene nucleus.

Found in cholesterol, the saponins, certain hormones, especially the sex hormones, and in various plant and animal tissues.

Steudel reaction for nucleic acid Treatment of the acid, on a slide, with concentrated nitric acid leads to the formation of doubly refractive crystals of the nitrates of the purine bases. Concentrated hydrochloric acid may also be used.

Sthophanthus Dried ripe seeds of Strophanthus, kombé containing the cardiac glycoside strophanthin, kombic acid, choline and trigonelline, a heart tonic.

Sticky ends After digestion of a DNA with certain restriction enzymes, the ends left have one strand overhanging the other to form a short (typically 4 nt) single-stranded segment. This overhang will easily re-attach to other ends like it, and are thus known as "sticky ends". For example, the enzyme BamHI recognizes the sequence GGATCC, and clips after the first G in each strand * The overhangs thus produced can still hybridize ("anneal") with each other, even if they came from different parent DNA molecules, and the enzyme ligase will then covalently link the strands. Sticky ends therefore facilitate the ligation of diverse segments of DNA, and allow the formation of novel DNA constructs.

Stigma structure at the top of a flower's *carpel* (female reproductive organ) that receives *pollen* grains destined for *fertilization.*

Stigmasterol $C_{29}H_{48}O$, m.p. 170°C, forms a hydrate, a sterol of many plants, e.g. soya bean, calabar bean and other legumes.

Stilbestrol Stilboestrol; 44' -dihydroxy-alphabeta-diethyl-stilbene, a synthetic estrogenic substance which can be taken orally and is more potent than estrone and can be used for the same menstrual disorders, e.g. amonrrhea, dysmenorrhea.

Stimulus Any agent, act, or influence that produces functional or trophic reaction in an irritible tissue.

Stoeltzner test for vitamin d A 1% solution of the vitamin in olive oil is treated with phosphorus pentoxide yielding a reddishbrown color which develops outward from the pentoxide and gradually darkens until almost black. The reaction is also given by cod liver oil.

Stoma Opening in the outer cell layer of a plant through which oxygen and carbon dioxide pass.

Stop codon That codon at which translation of an mRNA molecule into a polypeptide is terminated. In the Universal Code this may be UGA, UAG, or UAA.

Stramonium Thorn apple, Jimson weed; stinkweed; Devil's apple; dried leaves of Datura. Stramonium containing the alkaloid hyoscyamine; used for relief of asthma and acts through the simultaneous presence of atropine and scopolamine; narcotic, antispasmodic, sedative; average dose 75 mgm.

Straub flavoprotein (Heart) Coenzyme factor; diaphorase; a heart flavoprotein enzyme, m.w. about 70,000, absorption bands at 274,359 and 451 mμ yellow-green; prosthetic group it, flavinadenine dinucleotide; participates in reoxidizing reduced coenzyme I (dihydrocoenzyme I) itself forming a leuco-flavoprotein which reduces a carrier, the reduced form of which finally reacts with molecular oxygen; universally distributed in animal tissues.

Streaming potential The potential difference induced between the ends of a capillary when a liquid is forced through it.

Streptavidin A bacterial analog of egg white avidin.

Stringency A term used to describe the conditions of hybridization. By varying the conditions (especially salt concentration and temperature) a given probe sequence may be allowed to hybridize only with its exact complement (high stringency), or with any somewhat related sequences (relaxed or low stringency). Increasing the temperature or decreasing the salt concentration will tend to increase the selectivity of a hybridization reaction, and thus will raise the stringency.

Strong nuclear force *fundamental force* that holds particles together within the nucleus of an atom.

Strophanthidin $C_{23}H_{32}O_6$, m.p. 169-170°C with$_2$ H_2 O, a cardiac aglucone of strophanthin glycosides, such as strophanthin, scillaren and digitoxin.

G-strophanthin The cardiac glycoside or mixture of glycosides of strophanthus; used like digitalis; dose 0.25-0.50 mgm.

Stroup test for differentiating caffeine and theobromine One drop of a 120 solution of potassium dichromate in sulfuric acid is added to the triturated sample in a porcelain dish. Caffeine produces a light blue-green color; theobromine gives a dark purple color, changing gradually to purple-green, olive-green and blue-green.

Strychnine $C_{21}H_{22}O_2N_2$, m.p. 286-288°C, alkaloid of nux vomica, etc., raises blood pressure, heart stimulant, promote appetite; dose 1.5 mgm. ee Toxicology.

Sturine A protamine of the testicles of the sturgeon.

Stylopene An alkaloid, $C_{19}H_{17}O_4N$, from the root of stylophoron diphyllum, and melting at 202°C.

Styracitol 15-anhydrosorbitol, m.p. 157°C found in the peel of Styrax obassia.

Sub-cloning If you have a cloned piece of DNA (say, inserted into a plasmid) and you need unlimited copies of only a part of it, you might "sub-clone" it. This involves starting with several million copies of the original plasmid, cutting with restriction enzymes, and purifying the desired fragment out of the mixture. That fragment can then be inserted into a new plasmid for replication. It has now been subcloned.

Subduction zone area where two of Earth's *plates* are pushing against each other, causing one plate to descend below the other.

Suberin Cork material of cell walls of higher plants.

Submicrons In the Siedentopf and Zaigmondy classification of particle size, those particles visible in the ultramicro. scope. Ultramicrons.

Substrate The substance acted on by an enzyme.

Succinic acid Amber acid, ethylenesuccinic acid; $HOOC^{U+00}B_7CH_2CH_2{}^{U+00}B_7COOH$; sp.gr. 1.56; m.p. 185-187°C; b.p. 235°C with partial conversion to anhydride; used in mfg. of dyes. perfumes, lacquers; salts are used medicinally.

Succinic cycle The theory of Szent- Györgyi and coworkers to explain the sustaining of oxygen uptake of minced muscle by succinate, fumarate, malate and oxaloacetate which are interconvertible and act as intermediators between metabolites and cytochrome-oxygen; triosephosphate is the chief metabolite concerned and is oxidized by oxaloacetate.

Succinic dehydrogenase A cytochrome-reducing dehydrogenase of heart muscle, etc., which catalyzes the oxidation of succinic acid to fumaric acid, especially in presence of cytochrome c.

Succus entericus Alkaline digestive juices produced by Brunner's and Lieberkuhn's glands; most abundant in the duodenum and in lesser amounts in the rest of the small intestine. It contains proteolytic enzymes, phosphatases, carbohydrases, and enterokinase.

Sucrase A carbohydrase of the intestine and of yeast, that hydrolyzes sucrose to fructose and gluocse. Optimum pH about 4.5. Enzymes, NonProteolytic.

Sucrose Cane sugar, beet sugar. It is the most important non-reducing diaaccharide. Obtained from the sugar cane or the sugar beet, as the sugar of commerce. It is a-glucopyranose-1-ß-fructofuranoside.

Sugahara reaction for levulose A mixture of 1 cc. of sugar solution, 2 cc. of 0.7N sulfuric acid and 10% sodium tungstate is treated with calcium oxide until faintly alkaline. A green precipitate forms, with large amounts of levulose the solution also turns green. The test gives a sharp differentiation between glucose and levulose.

Sugar — any of several small carbohydrates, such as glucose, which are "sweet" to the taste.

Suida reaction for aspartic acid Basic dyes, like crystal violet, Nile blue and safranine precipitate aqueous solutions of the hydrochloride of the acid.

Sulfapyridine Dagenan; 2-sulfanilyl aminopyridine; m.p. 190-193°C; used for pneumonia and gonnorhea.

Sulfatase A specific enzyme which splits organic sulfates (but not chondroitin sulfuric acid).

Sulfatides Compounds of sulfur with lipids, found in brain and nervous tissue. P may be absent.

Sulfhemoglobin Results from the ingestion of acetanilide and sulfanilic acid derivatives; may be made by the action of hydrogen peroxide and a sulfide on reduced hemoglobin.

Sulfhydryl The grouping of sulfur and hydrogen an SH, found in amino acids, proteins, oxidation-reduction systems.

Sulfoheme A hypothetical analogue of oxyheme considered as a component of sulfhemoglobin.

Sulfur bacteria (van niel) Facultative autotrophants which are strict anaerobes capable of oxidizing sulfur or sulfur compounds.

Sullivan reaction for guanidine One-half mg. of base in 1 cc. of aqueous solution is treated with 1 cc. freshly prepared 1% solution of sodium 1,2-naphthoquinone-4sulfonate and 0.3-0.5 cc. of N sodium hydroxide and placed in a boiling water bath for 1-2 minutes. After cooling, 0.3 cc. of 25% solution of urea are added, acidified with 1 cc. concentrated hydrochloric acid, mixed and 1 cc. concentrated nitric acid added. A distinctly red solution is obtained.

Sullivan tests for cystine and cysteine I. Cystine—Five cc. of test solution (0.1N with respect to hydrochloric acid) is treated with 1 cc. freshly prepared 1% sodium cyanide solution, shaken, and treated with 1 cc. freshly prepared aqueous 0.5% solution of sodium1,2-naphthoquinone-4-sulfonate. After shaking 5-10 seconds, 5 cc. of 10-20% solution of anhydrous sodium sulfite in 0.5N sodium hydroxide are added. After mixing and standing at 20°C for 10 minutes, a reddish-brown color appears. The addition of 1 cc. of 2% sodium hyposulfite solution in 0.5N sodium hydroxide changes the reddish color to a purer red.

II. Cystine—A mixture of 5 cc. test solution and 2 cc. freshly prepared aqueous 5% sodium cyanide solution is allowed to stand for 10 minutes. One cc. of 0.5% solution of sodium 1,2naphthoquinone-4-sulfonate is added and the mixture shaken 10 minutes. 5 cc. of 10% solution of anhydrous sodium sulfite in 0.5N sodium hydroxide are added, mixed and allowed to stand 30 minutes; 2 cc. 5N sodium hydroxide and 1 cc. of 2% sodium hyposulfite in 0.5N sodium hydroxide are added. Color reactions as given above are given if at least 100 p.p.m. of cystine are present. The test may be used for the colorimetric estimation of the compounds and is not given by glutathione and many other compounds.

Sumach Dried ripe fruit of Rhus glabra containing tannin, etc., and used as astringent. (The bark is utilized also for the same purpose.)

Sumner reagents and tests for glucose in urine I. In normal and diabetic urine sugar reduces 4.6-dinitroguaiacol to intensely colored 4-nitro-6aminoguaiacol when heated in the presence of sodium carbonate.

II. Ten grams crystalline phenol are mixed with 22 cc. 10% sodium hydroxide and diluted to 100 cc. with water. 6.9 gm. of sodium bisulfite are added. to 69 cc. of the phenol solution, followed by a solution containing 300 cc. 45% sodium hydroxide, 255 gm. Rochelle salts and 800 cc. of 1% 3,5-dinitrosalicylic acid. After mixing, the solution is kept lightly stoppered in a well-filled bottle.

III. Reagent—Solution of 2 gm. 3,5-dinitrosalicylic acid in 100 cc. 2% sodium carbonate solution. One cc. urine, 1 cc. reagent and 2 cc. 1.5% sodium hydroxide are heated in a boiling water bath for 5 minutes. Cool and compare the red color with a similarly treated mixture of 1 mg. glucose in 1 cc. water and 1 cc. reagent.

Sunspot Spot that appears dark on the surface of the Sun, where it is thought that strong magnetic fields prevent heat from flowing out.

Supercluster Group of several clusters of *galaxies* measuring 100 to 300 million light-years across.

Supercoil Double-stranded circular DNA which is twisted about itself. Commonly observed with plasmids and circular viral DNA genomes (such as that of hepatitis B virus). A nick in one strand of the plasmid may remove the twist, resulting in a relaxed, circular DNA molecule. A complete break in the DNA puts the plasmid in a linear form. Supercoils, relaxed circular DNA, and linear DNA all have different migration properties in agarose gels, even though they contain the same number of base pairs.

Supermetabolism A term suggested by Murlin to include the effects of all the known factors which cause a rate of heat production higher than basal in normal subjects; includes the specific dynamic action of protein (S.D.A.) and the effects of temperature, as in hibernation. M. K.

Superstring theory Theory that attempts to describe, through mathematics, the *elementary particles* and their interactions and that accommodates both the theory of *relativity* and *quantum theory*.

Supranephrine A commercial preparation of l epinephrine hydrochloride.

Suprarenalin A commercial preparation of l epinephrine hydrochloride.

Suprasterol I A sterol derivative m.p. 104°C C., obtained by the ultraviolet irradiation of calciferol, or the prolonged ultraviolet irradiation of ergosterol. It contains 4 rings, and three unconjugated double bonds.

Suprasterol II A product obtained by the ultraviolet irradiation of calciferol, m.p. 110°C.

Surface tension Cohesive force between surface molecules of a liquid that causes the molecules to assume the least amount of surface area and to behave as an elastic skin; surface tension enables some insects to stand without sinking on the surface of a body of water.

Swaminathan test for nicotinic acid in foodstuffs A hot, aqueous extract of the foodstuff is treated with lead acetate solution to precipitate the proteins, excess lead is removed with sulfuric acid, the pH adjusted to 10 and any color removed with charcoal. The clear, colorless filtrate is adjusted to pH 7.5 and made up to volume. An aliquot is diluted with water to 20-30 cc. in a 100 cc. separatory funnel, 4 cc. freshly prepared cyanogen bromide solution added, shaken and allowed to stand for ½ hour. The yellowish-green color is extracted three times with 10 cc., 5 cc., and 5 cc. portions of purified amyl alcohol and compared with a known nicotinic acid standard.

0.01 mg. of the acid is easily detected, and definite gradations of color are obtained for increments of 0.01 mg. up to 0.1 mg. nicotinic acid.

Symbiosis The living together or close association of two dissimilar organisms to the advantage of both; each organism known as a symbiont.

Sympathin A compound localized in the nerves and acting in the transmission process; may be adrenaline; acts with acetylcholine (q.v.). Possibly adrenaline combines with two cellular substances to form Sympathin E and Sympathin I.

Sympathomimetic Stimulating the sympathetic nerves, e.g. as by the action of epinephrine, tyramine and epinene.

Sympatric Coexisting at the same locality.

Synapse The anatomical relation of one nerve cell

to another; the region of contact between processes of two adjacent neurons, forming the place where a nervous impulse is transmitted from one neuron to another.

Synapses Moment by moment combining of excitatory and inhibitory signals arriving at a trigger zone of a neuron.

Synapsis The pairing of homologous chromosomes during the first division of meiosis.

Syncline downward fold in a *bed* of rock.

Syndrome A complex of symptoms; a set of symptoms which occur together; the sum of signs of any morbid state.

Syneresis The loss of dispersions medium by a colloidal sol.

Synergism (1) The joint action of drugs whose combined effect is greater than the sum of the individual effects;

(2) muscle coordination.

Syngamy The process of union of two gametes; sometimes called fertilization. It encompasses both **plasmogamy** and **karyogamy**.

Synovial fluid A clear, thick, alkaline liquid found bathing tendons, joints and bursae; acts as a cushioning and lubricating medium; similar, in most respects, to plasma composition.

Synthalin Decamethylene-diguanidine dihydrochloride; m.p. 193°C; has insulin-like effect, but operates merely by being toxic to the nervous system.

Synthesis Production of a substance through the combination of simpler compounds.

Syntonin Acid myosin; a metaprotein resembling various albuminates.

Syphilis A chronic, infectious disease due to Treponema pallidum or Spirochaeta pallida. It may be.

(1) acquired or.

(2) prenatal. Infection may be spread through any abrasion. Three stages are described primary, secondary and tertiary. In the first stage a chancre appears at the site of infection which, however, is systemic. It is an immunity reaction which becomes prevalent in the secondary stage. In the tertiary form the lesion is the gumma but every kind of symptom of other disease may be involved also. Many laboratory tests have been devised for diagnosis, of which the most famous is the Wassermann test which is a complement fixation test. Flocculation tests also have been used (Kahn, Kline, Eagle) as well as agglutination (Hinton).

The spectacular use of arsphenamine, Ehrlich's 606, and of neoarsphenamine and other arsenicals, in this disease gave a strong impetus to the modern science of chemotherapy. Bismuth and mercury are supplementary to such therapy. A popular compound for use in neurosyphilis is tryparsamide or sodium-N-phenylglycinamide-p-arsonate.

Syringin 5-methoxy coniferin, found in Syringa species, $C_{17}H_{24}O_9H_20$, m.p. 192°C. A glucoside yielding syringenin and glucose on hydrolysis.

Systolic pressure Pressure of the blood against the walls of the *arteries* as a result of the heart's contraction.

Szent-györgyi reaction for ascorbic acid A slightly alkaline 1% solution of the acid is mixed with ferrous sulfate solution and exposed to the air. A dark violet color, visible even after 100-fold dilution, forms rapidly. Reduction with sodium hyposulfite destroys the color, exposure to air restores it.

T

T cell One of three types of cell in the immune system that activate other immune-system cells, suppress immune reactions, or destroy harmful cells.

TN Turnover number, referring in enzyme study to the number of molecules of an enzyme or coenzyme reacting per minute at a given temperature.

a-TOCOPHEROL Vitamin E; the vitamin necessary to avoid reproductive failure and a type of male sterility. It also has a growth stimulating effect. It has been found in wheat germ oil, other seed oils and in various organs via the food. The leaves of alfalfa and lettuce are a good source.

T7 RNA Polymerase A bacteriophage RNA polymerase which is commonly used to transcribe plasmid DNA into RNA. The plasmid must contain a T7 promoter upstream of the relevant sequence.

Tabacin A glucoside from the leaves of Kentucky tobacco, converted by KOH to nicotine; toxic, paralyzes heart and respiration; lethal dose 80 mgm. per kilo.

Tachyphylaxis The appearance of decreased effects on repeated injection of a substance.

Tachysterol An unstable, non-anti-rachitic, toxic, irradiation product of ergosterol. On further irradiation yields calciferol. Contains three rings and four conjugated double bonds.

Taka-diastase An enzyme preparation obtained by growing the fungus Aspergillus oryzae on bran, leaching the mass of culture with water, and precipitating with alcohol; a mixture of enzymes, largely used for its diastatic properties; also known as koji; whitish-yellow, hygroscopic powder, converts over 300 times its weight of starch into maltose.

Takeuchi reagent for urea An aqueous soybean extract (containing the ferment urease) can convert urea into ammonia at ordinary temperatures. The amount of urea may be determined by titrating the resulting ammonium carbonate. The urea content of urine, blood serum and cerebrospinal fluid may be determined with this reagent.

Tamarind Partially dried ripe fruit of tamarindus indica preserved in sugar or syrup and used as laxative, refrigerant.

Tampicin A glucoside, $C_{34}H_{54}O_{14}$; m.p. 130°C; a constituent of tampico resin and similar pharmacol. to convolvulin.

Tanacetin An amaroid from seed, herb, and flowers of Tanacetum vulgare.

Tandem repeat sequences Multiple copies of the same base sequence on a chromosome; used as a marker in physical mapping.

Tangent *Function* of an angle of a right triangle defined by dividing the length of the side of the triangle that is opposite the angle by the length of the side that is adjacent to the angle

Tanghinin $C_{27}H_{40}O_8$; from the seed of Tanghinia madagascariensis Pet., Apocynaceae; a heart poison closely resembling strophanthin and ouabain but differing in its spastic action.

Tannase An esterase which hydrolyzes only esters the acid component of which has at least two phenolic hydroxy groups, neither of the hydroxyls being ortho to the carboxyl of the acid. Found in mould fungi.

Tannic acid Tannin; gallotannic acid; a glycoside of gallic acid; $C_{76}H_{52}O_{46}$; used as styptic and in the treatment of burns.

Tannins A group of naturally occurring complex glycosides found usually in tree barks. Vary in complexity from Turkish tannin, which is a glycoside containing one molecule of glucose and five of gallic acid to more complex ones as Chinese tannin, which contains 10 molecules of gallic acid to one glucose. Form colloids with hot water; inks with iron salts; insoluble precipitates with gelatin; precipitated by alkaloids and proteins; astringent.

Tanret reaction for ergotinine When a trace of ergotinine is treated with a few drops of ethyl acetate and concentrated sulfuric acid, a yellow-red color is obtained which changes rapidly to violet and blue; the color is not altered by water.

Taq polymerase A DNA polymerase isolated from the bacterium *Thermophilis aquaticus* and which is very stable to high temperatures. It is used in PCR procedures and high temperature sequencing.

Taraxanthin A carotenoid, $C_{40}H_{56}O_4$, m.p. 185°C, whose esters are present in dandelion petals; isomer of violaxanthin.

Target glands The glands controlled by the hormones of the pituitary.

Tartar A precipitate coating teeth, principally calcium phosphate, sulfomucin and cellular debris.

Tartaric acid Dextrotartaric acid, dihydroxysuccinic acid; $C_4H_4O_6$; a dihydroxy, dicarboxylic acid occurring in many fruits and in mfg. of wine; m.p. 168-170°C; sp. gr., 176; used as refrigerant, in preparing effervescent powders; industrially in baking powders, coloring and dyeing, as reagent in anal. chemistry.

TATA box A sequence found in the promoter (part of the 5' flanking region) of many genes. Deletion of this site (the binding site of transcription factor TFIID) causes a marked reduction in transcription, and gives rise to heterogeneous transcription initiation sites.

Tauber reaction for vitamin B1 Several mg. of vitamin and 5 mg. p-dimethylaminobenzaldehyde are mixed in a small crucible and evaporated with 0.1 cc. glacial acetic acid. After cooling, the addition of 1 cc. glacial acetic acid immediately produces an intense brick-red color. Proteins and amino-acids interfere.

Tauber test for ascorbic acid in plants, etc A few gm. of vegetable or organic tissue is macerated with 2 or 3 times as much hot 8% acetic acid for a few minutes. A drop of the liquid is placed on a double filter paper (the upper one for filtration, the lower for the test). One drop of a solution of 1 gm. ferric sulfate in 80 cc. boiling water and 18 cc. 85% phosphoric acid is placed on the lower filter paper and 1% potassium permanganate solution added until a faint pink color results. A blue color develops in ½ minute in the presence of 0.003 mg. vitamin in 0.5 cc. solution.

Tauber test for pentoses Reagent—1 gm. benzidine in 25 cc. glacial acetic acid. This is good for 4 days. One drop of test solution (containing 0.05 mg. or more of pentose) is boiled vigorously with 0.5 cc. of reagent and cooled; a very stable cherryred color is obtained. Glucose, fructose or galactose give yellow to brown colors.

Taurine Aminoethylsulphonic acid, combined with cholic acid forms the bile acid, taurocholic acid; originates from cystine in the liver.

Taurocholic acid $C_{26}H_{45}O_2NS$; m.p. about 125°C; cholaic acid; cholyltaurine; occurs in bile as Na salt; hydrolyzes to taurine and cholic acid.

Taxicatin A glucoside from leaves of taxus baccata melting at 164-165°C.

Taxine $C_{37} H_{51} O_{10} N$; alkaloid from needles and seed of Taxus baccata, m.p. 121-124°C, an abortive and paralyzant.

Taxis movement by a single cell (an organism or a sex cell) in response to an external stimulation.

Taxon A taxonomic group, at any categorical level.

Taxonomy Classification of living organisms into distinct groups.

Taxonomy The study of the laws and principles of classification.

Technology transfer The process of converting scientific findings from research laboratories into useful products by the commercial sector.

Tectonics Study of the movements of rocks and slabs of crust on Earth's surface.

Tectoridin A glucoside from the rhizome of Iris tectorum.

Teeth, biochemistry of In order to understand the biochemistry of teeth a brief discussion of their devolepment and histological structure is necessary.

The teeth are highly specialized cutaneous tissues. They consist of a calcified enamel layer covering the crown of a somewhat less calcified dentin, which in turn surrounds the pulp. The pulp is supplied with blood vessels and nerves. Supporting the teeth in the oral cavity and thereby giving them functional significance are the periodontal tissues (cementum, periodontal membrane, alveolar bone, and gingiva). The cementum covers the root dentin, and the tooth rests in its socket in the alveolar bone. The periodontal membrane lies between the cementum and the alveolar bone. Gingiva, or gum tissue, surrounds the alveolar bone and forms a collar through which the tooth projects into the oral cavity.

Formation of enamel and dentin by the ameloblasts (enamel forming cells) and odontoblasts (dentin forming cells) begins at the dentinoenamel junction. The ameloblasts lay down an organic enamel matrix as they recede outward from the dentino-enamel junction while the odontoblasts lay down an organic dentin matrix and recede inward from the junction thus constricting the pulp cavity. Following the formation of the initial portions of these matrices calcification takes place and proceeds in both directions from the dentino-enamel junction. The enamel and dentin are formed in definite incremental layers 16 micra in width. This 16 micra rhythm has been found to take place in man approximately every 4 days. The process of calcification of each increment of enamel is completed about 3 months after the initial depostion of the organic matrix by the ameloblasts. After eruption of the tooth into the oral cavity the ameloblasts are worn away and thus the enamel is incapable of further growth or regeneration. Since the odontoblasts are still present and functioning after the eruption of the tooth, dentin can continue to grow under pathological conditions. This secondary growth further constricts the pulp cavity.

During calcification of the enamel an inorganic calcium salt (primarily the phosphate) is deposited in, and largely replaces, the organic matrix so that in fully calcified enamel the organic portion amounts only to about 3 per cent of the total. In fully calcified dentin, on the other hand, the organic portion makes up some 25 per cent of the whole.

Methods of preparing teeth for analysis

Since the biochemistry of teeth is primarily concerned with the inorganic phases of enamel and of dentin the following methods are commonly employed in obtaining these fractions.

A large portion of the enamel layer may be mechanically removed from the surface of the crown with the aid of a dental drill, and similarly the dentin can be obtained from the inside of the tooth. These two fractions can then be treated by one of the methods listed below to obtain the inorganic phases when desired.

A second and much better method of separating enamel and dentin makes use of the different specific gravities of the two. The teeth are first cracked into several pieces, the pulps removed and the pieces extracted with a 50-50 alcoholether mixture in a Soxhlet extractor to remove any lipids. The lipid-free pieces are pulverized to pass through a 100-mesh screen and these powdered particles of enamel and dentin are suspended in a bromoform-acetone mixture of specific gravity 2.7. The enamel particles, most of which have a specific gravity between 2.89 and 3.00, will sink in this mixture, whereas the dentin and cementum particles with specific gravities of 2.14 and 2.03 respectively will float. The process of separation is facilitated by placing the powdered sample in a removable open cylinder contained in a centrifuge tube. The enamel particles are rapidly centrifuged through the cylinder and collect in the bottom of the centrifuge tube while the dentin and cementum particles float on the surface of the bromoform-acetone mixture in the cylinder and can be removed with the cylinder. The dentin and cementurn particles can subsequently be separated by employing a bromoform-acetone mixture of specific gravity 2.08.

These pure samples of enamel and dentin may now be analyzed as they are, or the organic fractions may be removed and analyses made on the inorganic residues. Ashing in a muffle furnace at 600°C-700°C C. will suffice to remove all organic matter, or the organic matter may be leached out by a boiling solution of KOH in ethylene glycol.

Chemical methods of analysis of enamel and dentin

The three elements most commonly analyzed for in enamel and dentin are calcium, phosphorus and fluorine.

A convenient and frequently employed method for calcium analysis is the micro titration method which has been applied to blood. In this procedure the calcium is precipitated as the oxalate at pH 5.0. The calcium oxalate is then washed, disolved in hot 1 normal H_2SO_4 and the resulting oxalic acid is titrated with potassium permanganate.

Phosphorus can be rapidly and accurately determined by the colorimetric method of Fiske and Subbarow. The color is developed by adding an aminonaphtholsulfonic acid reagent to an acid solution containing inorganic phosphate and ammonium molybdate and can be read with a Du Boscq or a photoelectric colorimeter. As little as 5 to 10 micrograms of phosphorus can be determined by this method.

The principle of the micro determination of fluorine involves the titration of hydrofluosilicic acid in a buffer at pH 2.8 with a standard thorium nitrate solution. The end point is determined by the lake which is formed when an excess of thorium ions react with the sodium alizarine sulfonate used as an indicator. As little as one microgram of fluorine can be determined.

The composition of enamel and dentin

The inorganic phase of enamel has been variously reported to make up between 95 and 97.5 per cent of whole enamel while for dentin it represents only 71 to 75 per cent.

Enamel and dentin have been found, on spectrographic examination, to contain traces of the following elements aluminum, barium, boron, chromium, copper, lead, manganese nickel, silicon, silver, strontium, tin, titanium, vanadium, and zinc.

Crystal structure

Several investigations ˉ have confirmed the fact, which had previously been recognized, that the powder X-ray diffraction patterns of the mineral of bone, enamel and dentin are very similar to those of the mineral apatites and that the differences in detail of the patterns are to be accounted for, in some degree, by variations in grain size of the crystallites. Actually the diffraction patterns and lattice constants of enamel and dentin more closely match those of hydroxyapatite than those of other members of the apatite group of minerals. It appears certain, therefore, that the fundamental crystal structure, of at least the predominant part of the mineral phase of calcified tissues, is that of an apatite. The apatites are a series of isomorphous minerals of wide distribution and variable composition. Hydroxyapatite has the formula $Ca_{10}(OH)_2(PO_4)_6$. A variety of substitutions are possible without disrupting the apatite structure. For example, chlorine or fluorine can replace the hydroxyl groups; and magnesium, sodium or potassium can replace some of the calcium atoms. Formerly it was assumed that the carbonate found in many apatites could occupy one of the hydroxyl positions in the crystal lattice but this possibility has now been excluded for structural reasons. One hypothesis[24] as to the position occupied by carbon in these crystals assumes that to a limited extent carbon can replace either calcium or phosphorus to give the following formula $(OH)_2\ Ca_6\ [[(P,C)O_4\ ]_6\ (Ca,C)_4\ ]$. On this hypothesis the degree of substitution of carbon in the crystal structure is higher in the case of dentin than in that of enamel. Another group of workers ˉ, prefer the view that carbonate apatites do not occur in calcified tissues, and that the sole apatite present in these substances is hydroxyapatite with "occluded, adsorbed, or interstitially crystallized carbonates."

The average size of enamel crystallites in 0.27 micra whereas dentin crystallites are about one-tenth this size.

Special methods of study

X-ray absorption method Roentgenograms of tooth slabs will depict the degree of calcification at various points in the enamel and dentin. Any points or areas of abnormal calcification can be detected when the plate is read on a densitometer. The densitometer employs a photoelectric cell and a galvanometer which accurately measure the intensity of light transmitted through various portions of the film. The more intense the light transmitted by a portion of the film, the more opaque, and therefore, the more highly calcified was that portion of the tooth slab.

Radioactive isotope method A much debated question in the field of the biochemistry of teeth has been concerned with the ability of erupted teeth to exhibit, or be influenced by metabolic processes. The advent of such radioactive isotopes as those of calcium, phosphorus, fluorine and sodium, has opened a new approach not only to this question but also to the question of the pre-eruption metabolism of teeth. The isotopes may be administered orally or by any parenteral route and the degree of deposit, exchange, or uptake of radioactive atoms may be measured with a Geiger-Müller counter. This approach has led to many interesting findings which will be discussed later.

Dietary methods The effect of nutritional factors on the development of teeth and on the preservation of their integrity can be studied by dietary methods. The rat is an extremely useful laboratory animal for such experiments because of the fact that the incisor teeth of the rat are continuously erupting teeth while the rat molars are mature erupted teeth. Such nutritional factors as the vitamins A, C, and D and the elements calcium, magnesium, phosphorus and fluorine have been extensively studied. It may be mentioned here that whereas the effect of dietary deficiencies on developing bone are qualitatively and quantitatively different from those produced in the completely formed skeleton, mature teeth, by contrast, appear to be almost, if not entirely, removed from the direct influence of nutrition. In several instances even developing teeth appear to be less liable to alteration by metabolic influences than forming bone.

Experimental Dental Caries in Rats The ability of teeth to resist dental caries, under a variety of experimental conditions, can be studied in the rat. The molar teeth of rats are sufficiently similar to human molars so that at least some conclusions of practical importance can be drawn from such animal experimentation. Caries may be produced in a rat molars by employing a diet containing coarse cereal particles. A convenient diet is as follows 60 parts of coarse, yellow cornmeal; 24 parts of skim milk powder; 12 parts of linseed meal; 3 parts of alfalfa meal; and 1 part of sodium chloride. The animals are fed this diet for 100 days at the end of which time the number of carious lesions are determined. Several methods are available for scoring dental caries. One of these which has been found useful is the method of Gomori.

Metabolism

Vitamin A Severe total dietary deficiencies of

vitamin A produce, in developing teeth, characteristic abnormalities of structure and growth. The primary dental disturbance in vitamin A deficiency is apparently on the developing enamel. This effect is, then, an example of the specific requirement for this vitamin by epithelial tissues and by those tissues derived from epithelium. Although an early effect of vitamin A deficiency is a result of failure of ameloblastic activity, which may result in enamel agenesis or hypoplasia, a prominent endresult is seen in a disorganized structure of the dentin. The differentiation of odontoblasts from pulpal cells fails in the absence of normally functioning odontogenic epithelium, which failure results in marked disturbances of the growth of dentin. That these effects of vitamin A deficiency are largely confined to developing teeth is nicely brought out by experiments in which rats are placed on the deficient diet at weaning when calcification of the molar teeth is nearly complete. In these experiments marked disturbances of the continuously erupting incisors are evident but few, if any, disturbances are noted in the molars.

Vitamin C

Another vitamin required for normal tooth formation is vitamin C. The most prominent dental effects of ascorbic acid deficiency are produced in the dentin with only minor, and possibly secondary, effects in the enamel. Normal dentin during its formation has a higher requirement for vitamin C than do many other tissues, and for this reason abnormalities in the dentin may be produced in the absence of obvious clinical signs of scurvy. As is well known, the gums and other investing tissues of the teeth of mature animals may be adversely affected in susceptible animals by a deficient supply of vitamin C. However, like vitamin A, there is no indication that ascorbic acid is required for the preservation and integrity of fully formed enamel and dentin.

Vitamin D, Calcium and Phosphorus On diets adequate in calcium and phosphorus but deficient in vitamin D, only slight local irregularities in enamel and dentin are observed in developing teeth. However when diets are deficient in either calcium or phosphorus as well as vitamin D more marked abnormalities occur.

Although some degree of enamel hypoplasia, that is, surface irregularities in the enamel, occurs in animals fed on vitamin D-free diets which are also deficient in calcium or phosphorus, the most pronounced effects are on developing dentin. On a rachitogenic diet which is low in phosphorus, free of vitamin D, but normal with respect to calcium, the band of predentin increases in width. Predentin is the dentin organic phase which has not yet been calcified. A single curative dose of vitamin D will, after a delay of 4 or 5 days, bring about normal calcification in the predentin laid down after the vitamin was administered. However, the predentin laid down during the period of vitamin deficiency remains essentially uncalcified.

Failure of predentin to calcify also occurs on diets which are low in calcium and vitamin D but adequate in phosphorus. Again, vitamin D therapy restores the calcification of subsequently formed predentin almost, if not completely, to normal. It should be emphasized again that, like vitamins A and C, deficiencies of vitamin D, of calcium, or of phosphorus exert practically no influence on preformed enamel and dentin. However, such nutritional deficiencies do produce alterations in the alveolar bone and other investing tissues and consequently do impair the function of fully formed teeth.

It must not be concluded from the above statements that fully formed enamel and dentin are incapable of interaction with substances present in the plasma and saliva. Experiments carried out with the aid of radioactive calcium , and radioactive phosphorus have amply demonstrated that both enamel and dentin of fully formed teeth can acquire significant amounts of these isotopes in vivo. The mechanism of acquisition is not clear and it cannot be assumed that the absolute amount of calcium and phosphorus in these tissues has been increased. It is, therefore, probable that an exchange takes place between the normal calcium and phosphorus atoms in the calcified tissues and the radioactive atoms in the plasma or saliva. Both in vitro and in vivo experiments have demonstrated that dentin acquires radioactive calcium or phosphorus about 6 to 10 times as rapidly as does enamel. This difference between enamel and dentin is probably partially due to the fact that the dentin crystallites are only about one-tenth the size of the enamel crystallites and therefore dentin has a much larger surface area. Another reason for this difference is that dentin is traversed by the so-called dentinal tubules which extend from the pulp cavity to the dentino-enamel junction. These tubules offer a more or less free path for movement of ions within the dentin.

It has been demonstrated by in vivo experiments with radioactive phosphorus that enamel can acquire this isotope both from the saliva and from the plasma via the dentin. Similarly, in vitro experiments have shown that phosphorus can traverse apparently normal enamel and thereby reach the dentin, so that presumably in vivo dentin acquires its radiophosphorus from the saliva as well as from the plasma though much the greater amount reaches it from the plasma.

Actually the amount of calcium and phosphorus which "exchanges" is such a small fraction of the total that it would require a great many years to renew completely all of the calcium and phosphorus atoms in teeth.

Magnesium This element exists only in small amounts in the mineral phase of teeth but it is not of adventitious occurrence. Diets deficient in magnesium will, within a few days, produce serious interferences in tooth formation. The function of the ameloblasts is disturbed and an atrophy of the odontoblasts occurs. However, the first and most characteristic effect of magnesium deficiency on the teeth is a sudden retardation of dentin calcification. Again, as in the case of some other deficiencies, the band of predentin increases in width. When animals on such a deficient diet are returned to an adequate diet, the degree and rhythm of dentin calcification return to normal in a few days but the abnormal dentin formed during the period of magnesium deficiency is not corrected. Long periods of inadequate magnesium intake cause marked disturbances in the character of the enamel.

Fluorine. As we have previously seen fluorine can replace the hydroxyl groups in hydroxyapatite. With one possible exception it is the fluorine which is carried by the drinking water, rather than that occurring in the food, which becomes incorporated in the mineral phases of enamel and dentin. The teeth of about 10 per cent of the individuals who have grown up in communities in which the drinking water contains one part per million of fluorine exhibit mottled enamel , which may be regarded as a clinical sign of the primary condition known as endemic dental fluorosis. In communities in which the fluorine concentration exceeds one p.p.m. the percentage of individuals showing mottled enamel and the severity of the mottling increases. Mottled enamel is characterized by patches of chalky white enamel which frequently become pigmented. Chemically the enamel is found to contain an increased amount of fluorine. In severe cases the mottled teeth are structurally inferior to normal teeth but at the same time they appear to be more resistant to dental caries. Evidence is at hand that the enamel of sound teeth which show no mottling contains significantly more fluorine than the sound enamel of carious teeth.[13], Also epidemiological studies have shown a lower incidence of dental caries in regions of endemic mottled enamel.

Fluorine is unique in that it is only in regard to this element that any evidence has been produced that the actual composition of fully developed enamel and dentin may be secondarily altered. Two investigations , have established the fact that in experimental animals the fluorine content of the enamel and dentin of mature erupted teeth can be increased by the use of water borne or dietary fluorine. It has also been demonstrated in vitro that the mineral phase of enamel is able to acquire radioactive fluorine from a solution. Thus it is probable that topical applications of fluorine to erupted teeth may increase the fluorine content of enamel. Two preliminary studies, on the topical application of fluorine have indicated a decreased incidence of dental caries in the treated patients but as yet no chemical analyses have been carried out on the enamel of the treated teeth.

Endrocrine factors parathyroidectomy in the rat results in poor calcification of the post-operative incisal dentin with the result that the incisors fracture in from 4 to 6 weeks after removal of the parathyroids. Similarly injections of parathormone, which produce a rise in blood calcium at the expense of bone calcium, produce a simultaneous hypocalcification of dentin followed by a hypercalcified band of dentin as the blood calcium returns to normal.

Administration of thyroxine induces precocious eruption of rat incisors whereas thyroidectomy results in retarded eruption of the teeth of rats, dogs, pigs and rabbits. , Defective tooth structure is common in cretins indicating a probable relationship between thyroid function and tooth development in man.

Adrenalectomv in the rat produces characteristic disturbances in the calcification of dentin. The incisal predentin contains disseminated globules of varying size, suggestive of premature calcification, and the post-operative dentin presents a picture of hypercalcification.

Castration apparently causes no significant chemical or histological changes in rat incisors but the incisors of similarly treated ground squirrels show a definite disturbance of calcification.

Hypophysectomy results, within a week, in a retardation in the eruption of rat incisors. , The eruption is progressively delayed so that a complete cessation can be observed in animals that survive the operation one year or longer. Appositional growth is also retarded but to a lesser extent than eruption so that some growth can be observed after eruption has ceased, with the result that the newly formed and uncalcified enamel and dentin at the basal end of the incisors become folded and misshapen. In hypophysectomy the predentin is globular in nature and the post-operative dentin is hypercalcified. Since this same picture is seen in adrenalectomy it is possible that this effect of hypophysectomy is mediated via the adrenals.

The problem of dental caries

Dental caries is the most prevalent chronic disease occurring in man and its control would be of great practical importance. The dental literature contains an abundance of conflicting opinion due largely to inadequately controlled experimentation.

The first detailed theory of the mechanism of dental caries was put forward by Miller. This theory does not attempt to answer the problem of the individual susceptibility to caries but aside from this it remains today as an adequate working basis for further research. The following quotations express the substance of Miller's theory.

Dental decay is a chemicoparasitical process consisting of two distinctly marked stages decalcification or softening of the tissue, and dissolution of the softened residue. In the case of enamel, however, the second stage is practically wanting, the decalcification of the enamel practically signifying its total destruction.

The acids which effect decalcification . . . are derived chiefly from particles of amylaceous and saccharine substances which lodge in the retaining-centers and there undergo fermentation . . .

All micro-organisms of the human mouth which possess the power of exciting an acid fermentation of foods may and do take part in producing the first stage of caries; . . . all possessing a peptonizing or digestive action upon albuminous substances may take part in the second stage; . . . whether there is any one bacterium which may always be found in decayed dentine, and which might therefore be entitled to the name of the bacterium of tooth-decay, or whether there are various kinds which occur with considerable constancy, we are not able to say.

At least two varieties of dental caries may be distinguished by their clinical and pathologic differences. The first and distinctly more common variety attacks the teeth of children and young persons predominately, occurs either in the pits or fissures of the crown, or in the neighborhood of points of contact of approximating teeth, penetrates the surface with only sharply localized destruction of enamel, spreads widely in dentin, and approaches the pulp relatively rapidly. The second and less common variety attacks the teeth principally of adults, appears in the gingival area of the crown or in the exposed roots, produces a large aperture at the surface, and penetrates at a varying rate, often very slowly.

Evidence from routine roentgenograms and from Grenz-ray photographs of incipient lesions indicates that radiolucent areas of enamel may be present before surface continuity is lost—a finding interpretable only as a demineralization. Microscopic examination of enamel partially disorganized by caries likewise reveals an altered appearance of enamel rods which can be duplicated closely by in vitro decalcification of enamel fragments in weak acids.

Several micro-organisms which generally inhabit the oral cavity, notably Lactobacillus acidophilus and Streptococcus salivarius, have the ability to ferment carbohydrates with the formation of acids. One or more of these organisms may well provide the acid which brings about enamel decalcification. In the case of L. acidophilus there is a distinct correlation between the occurrence of the organism in the saliva and the presence of active dental caries. Persons whose teeth have never been attacked by caries rarely, and apparently only accidentally, have these organisms in their saliva. Furthermore, it is impossible to infect these individuals with the organism and they therefore appear to possess some type of immunity. On the other hand, persons with active caries may have 50,000 or more organisms per cubic centimeter of saliva. Another interesting finding is that a high L. acidophilus count occurs several months before caries can be clinically diagnosed. However, as yet it is not known how long a period of time elapses between the onset of a carious process and the stage at which it may be clinically diagnosed.

Dietary factors can markedly alter the bacterial count. For instance, a number of cases have been reported in which the elimination of sugar from the diet has reduced the L. acidophlius count from 50,000 to 500 within 48 hours and restriction of carbohydrate in the diet has still further reduced the count. In connection with this dietary control of the bacterial count it will be recalled that caries can be experimentally induced in rats by the use of a diet containing coarse cereal particles. The particles apparently become packed in the occlusal fissures and furnish a substrate for acid formation by bacteria. Although this has not been adequately tested in man it remains a distinct possibility that hard or compact carbohydrate foods may lead to dental caries.

An interesting comparison has been made between endemic dental fluorosis and the salivary count of L. acidophilus. It was found that only 15 per cent of the children in a community in which the water supply contained 1.9 p.p.m. of fluorine had bacterial counts over 30,000 per cubic centimeter of saliva whereas 52 per cent of the children in a neighboring community whose water supply had only 0.2 p.p.m. of fluorine were found to have counts over 30,000. The incidence of caries in the latter community was three times as great as in the former.

Although as yet there has been no correlation found between caries activity and the number of streptococci found per cubic centimeter of saliva it has been pointed out that these streptococci are often 1000 times as prevalent as the lactobacilli. Furthermore, it has been shown that cultures of streptococci will dissolve as much tooth tissue in 8 hours as cultures of lactobacilli will dissolve in 24 hours. Consequently, we are not, in the present state of our knowledge, justified in indicating any single microorganism as being the sole one responsible for the decalcification of enamel.

For some time it has been known that lactic acid was formed during the fermentation of glucose by oral bacteria. Recently however, several of the intermediate products of carbohydrate degradation that are known to be formed in the cycles of glucose utilization by yeast and muscle have also been identified during the fermentation of glucose by salivary organisms. , Thus it is possible that such acids as pyruvic and phosphoglyceric, as well as lactic, may be important agents in the decalcification of dental tissues.Micromethods of analysis have shown caries of dentin to result in simultaneous decalcification and loss of organic matter. In the carious zone closest to normal dentin a proportionally greater loss of organic material than of inorganic substance has been observed. This raises some doubt that beginning caries is solely the result of decalcification. In this connection, another worker has claimed to have observed the production of enamel lesions by the action of proteolytic organisms in cultures in which the reaction never became acid.The complexity of the problem of dental caries may now be readily appreciated. Many avenues of study have been opened and partially explored. The necessity of further intense research will be realized when it is recalled that dental caries is our most prevalent chronic disease.

Telolecithal Having yolk accumulated at one point of the egg.

Telomerase A ribonucleoprotein complex that maintains the repeat sequence structures at the telomeric ends of chromosomes.

Telomere The natural distal end of a chromosome. Contain some form of simple repeating sequence, usually with a single stranded distal end that may form a hairpin. These specialized structures are involved in the replication and stability of linear DNA molecules.

Telophase Of mitosis, final stage when chromosomes decondense into threadlike structures and two daughter nuclei form.

Tendon (life science) length of strong tissue connecting muscle and bone

Teras A monster caused by developmental deviation.

Teratology The study of terata and their origin.

Teratoma A more or less malignant assembly of tissues, well differentiated histologically, but showing varying degrees of morphological differentiation, and embedded in the body of an otherwise normal fully developed organism. Some pathologists distinguish between the histologically differentiated type (*Embryoma*) and that showing both histological and morphological differentiation (*Teratoma* proper).

Terebratulas An extinct group of brachiopods (invertebrates).

Terminator A sequence downstream from the 3' end of an open reading frame that serves to halt transcription by the RNA polymerase. In bacteria these are commonly sequences that are palindromic and thus capable of forming hairpins. Sometimes termination requires the action of a protein, such as Rho factor in E. coli.

Terrestrial planet Refers to Mercury, Venus, Earth, and Mars, which are the four planets closest to the Sun and which have in common a rocky exterior.

Tertiary structure Refers to higher ordered structures conferred on proteins or nucleic acids by interactions between amino acid residues or nucleotides which are not closely positioned within the sequence (primary structure) of the molecule.

Tertiary The most recent of the major geological eras, extending from about 65 million years ago to the Recent.

Testicles (testes) Male gonad, primary reproductive organ in which male gametes and sex hormones are produced.

Testosterone *Hormone* in vertebrates that is responsible for the development of male sexual characteristics and regulates various other body processes.

Testosterone In male mammals, a major sex hormone that helps control male reproductive functions.

Tetanus A disease caused by Bacillus tetani (Clostriduim tetani) which may enter by a most trivial wound and whose toxins affect the central nervous system to produce spasm. Antitoxin or toxoid therapy is employed.

Tetanus antitoxin Purified antitetanic serum; concentrated or refined Tetanus antitoxin; antitetanic globulins; a sterile aq. soln. of antitoxic substances from blood serum or plasma from immunized horse.

Tetramethylammonium hydroxide May be the "sting" of jellyfishes, acts like curare.

Tetrandrine $C_{38} H_{42} O_6 N_2$; m.p. 217-218°C; an alkaloid present in the Chinese drug Han-Fang-Chi; affects respiratory and skeletal muscle.

Tetrose Any carbohydrate with four carbon atoms; there are four aldotetroses and two keto-tetroses.

Thalline An alkaloid, 6-methoxy-1,2,3,4tetrahydroquinoline, $C_{10} H_{13} ON$; white light-sensitive cryst.; m.p. 42-43°C; b.p. 283-285°C; its salts have antiseptic and antipyretic action.

Thapsiac acid $C_{18} H_{30} O_4$; a saturated dibasic fatty acid found in Conifer Wax.

Thebaine $C_{19} H_{21} O_3 N$; leaflets or prisms, m.p. 193°C; dimethylmorphine; paramorphine; an alkaloid occurring in opium to an extent of less than 1%; produces a bloodred color with $H_2 SO_4$; is a convulsant and not narcotic. Use in medicine is as a starting point for non-addicting morphine substitutes and as a diuretic.

Thelephoric acid A violet crystal pigment of the fungus thelephora. It is a dicarboxy, polyoxy phenanthrene derivative.

Theobromine Xantheose; $C_7 H_8 O_2 N_4$; 37-dimethylxanthine, sublimes 290°C; alkaloid of cacao, milder than cafieine, used as diuretic and myocardial stimulant.

Theophylline 13-dimethylxanthine, m.p. 269-272°C; found in tea; resembles caffeine and theobromine, diuretic.

Theorem Mathematical proposition that is shown to be true by logical deduction from a set of basic assumptions.

Therapic acid An 18 carbon fatty acid with four double bonds.

Thermal conductivity Movement of heat through a substance.

Thermal dissociation Heat-driven reversible breakdown of a chemical compound.

Thermal expansion Increase in length, area, or volume caused by an increase in temperature.

Thermodynamics, first law of conservation of energy, states that energy cannot be created or destroyed. Thus, the total amount of energy in the universe is constant.

Thermodynamics, second law of every time energy changes form, there is an increase in entropy (S), a measure of disorder, or randomness. Some energy is dissipated as heat, random molecular motion. Whenever matter becomes more ordered it does so at the expense of contributing to the disorder of the surroundings.

Thermogenetic hormones Hormones which promote liberation of heat so as to raise basal metabolic rate, e.g. epinephrine and sympathin.

Thermogenic effect Increase of animal's body temperature resulting from food drugs or other substances or from other influences such as diathermy or high environmental temperature.

Thermophene A curve relating quantitatively the effects of a gene to the temperature at which development has taken place.

Thermotropism Reaction of a living form to heat.

Thevetin $C_{24} H_{66} O_{18}$; a glucoside from the seed of Thevetia nereifolia of the digitalis group, b.p. 210^0.

Thiamin chloride $C_{12} H_{17} N_4 OSC1$; vitamin B_1; the anti-neuritic vitamin. The vitamin associated with carbohydrate metabolism. Found in yeast, meat, fish, milk, eggs, etc. Insufficiency causes syndrome and dermatitis. The hydrochloride, m.p. 232-234°C. is water soluble and is used therapeutically.

Thiaminoprotein enzymes Enzymes concerned with the breakdown of pyruvic acid which contain diphosphothiamine as the prosthetic group; include carboxylase (yeast), pyruvic oxidase (bacteria), pyruvic oxidase (animal tissue).

Thigmotropism Reaction of a living form to contact.

Thiochrome Coloring matter of yeast, $C_{12} H_{14} ON_4 S$, an oxidation product of vitamin B_1, fluoresces blue in alkaline or neutral solution, m.p. 228°C; physiologically almost inactive.

Thiocyanogen number The number of milligrams of thiocyanogen $(SCN)_2$, which will add on to 1 gram of a fat or a mixture of fatty acids. Only one double bond of linoleic acid reacts with thiocyanogen. This together with the iodine number enables one to calculate the amount of linoleic acid present.

Thioneine Ergothioneine. The betaine of thiolhistidine. Found in the red cells in human blood, where its concentration is 10-25 mg.%, in yeast and in ergot.

Thio-sugars Sugars in which a sulfur atom replaces an oxygen atom, e.g. methyl thiopentose of yeast.

Thixotropy A term introduced in 1927 to describe those colloidal systems in which a gel is isothermally converted into a sol by shaking or stirring the gel, the sol setting to a gel again when left undisturbed. Two examples are starch in water, and quicksand.

Threonine A-amino-ß-hydroxy-n-butyric acid. Needles or hexagonal plates. An indispensable amino acid found in fibrin, and many other proteins in small amounts.

d.threonine metabolism ß-oxidation occurs, yielding a-amino-ß-ketobutyric acid, which further oxidizes to acetic acid and glycine.

Threose One of the tetrose sugars, has land d- forms (aldehyde).

Thrombin Thrombase; the proteose enzyme which converts fibrinogen to fibrin in the clotting process; is formed from a precursor, prothrombogen) by the action of an activator, thrombokinase, and Ca ions.

Thrombokinase An activator in the formation of thrombin in the clotting process from prothrombin normally present in the blood.

Thrombolysis destruction of blood clots that is fueled by enzymes.

Thromboplastin An early name for cephalin, or a cephalin-protein combination.

Thromboplastin solution Soln. of brain extract in NaCl; used as hemostatic.

Thrombosis Clotting within a blood vessel. The clot is called a thrombus.

Thuja Dried leafy young twigs of Thuja occidentalis containing thujone and other components and used as antipyretic, expectorant, antithelmintic, irritant.

Thujene A dicyclic terpene present in alpha and beta-form in various oleoresins (Boswellia serrata).

Thujyl alcohol Secondary dicyclic terpene alcohol of wormwood oil.

Thyme Dried leaves and flowering tops of thymus vulgaris containing volatile oil, tannin, and gum; used as diaphoretic, stimulant, carminative.

Thymidine 3 - thymine - d - desoxyriboside, a nucleoside of desoxyribonucleic acids.

Thymine 5-methyl-26-dioxytetrahydropyrimidine, $C_5H_6O_2N_2$, m.p. 321-325°C; a constituent of nucleic acids of animal origin.

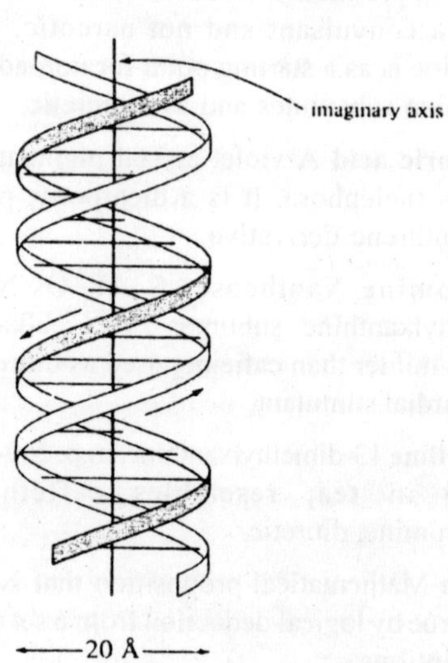

Fig. Thymine

Thyminose d-2-desoxyribose; a constituent of thymonucleic acid.

Thymol blue Thymolsulfonephthalein, $C_{27}H_{30}O_5$

S; a brownish-green, cryst. powder; used as acid-base indicator red at pH 1.2 to yellow at 2.8; also yellow at 8.0 to blue at 9.6.

Thymol iodide ($C_6 H_2 . CH_3 . O_I . C_3 H_7)_2$; used as antiseptic in place of iodoform.

Thymolphthalein A pH indicator; colorless 9.3 to blue 10.5; also used as reagent for blood.

Thymus A ductless glandlike body situated in the anterior mediastinal cavity which reaches its maximum development during the early years of childhood; greyish-red; usually has two longitudinal lobes joined across a median plane.

Thymus principle A hormone of the thymus gland which speeds up cell proliferation and growth up to puberty. Its injection in successive generations has been claimed to produce precociousness. A purified extract, called thymocrescin, is claimed to be a sulfur-containing polypeptide.

Thyr(e)oglobulin A conjugated protein of the thyroid gland, which to a very large extent is responsible for the activity of the gland. It consists of a protein in conjugation with the iodine containing amino acid thyroxin.

Thyroid Thyroid gland of domesticated animals that are used as food by man, freed from connective tissue and fat, dried and powdered; a yellowish powder, saline taste; used in thyroid deficiency; alterative; in myxedema, obesity, mental affections, etc.

Thyrotropic principle A protein-like hormone of the anterior lobe of the pituitary, having direct control of the formation of the thyroid hormone.

Thyroxin $C_{15}H_{11}O_4NI_4$; m.p. about 231-233°C; the active component of the hormone of the thyroid glands, appearing in l-form; promotes general metabolism and growth; dose 0.2-2 mgm.

Tide periodic rise and fall of sea level caused by the gravitational forces of the Moon and Sun.

Tiglic acid $C_5 H_8 O_2$; an unsaturated fatty acid found in croton oil. Has one double bond, m.p. 64.5°C.

Tigogenin $C_{27} H_{44} O_3$; 3-hydroxy-sapogenin; the non-sugar residue of the digitalis saponin, tigonin.

Tiliacorine An alkaloid, $C_{30} H_{27} NO_2 (CH_3O)_2$; m.p. 260-261°C; from the bark of tiliacora acuminate Miers, Menispermaceae; a cardiac and respiratory poison.

Tiliadin An amaroid, $C_{21} H_{32} O_2$, from fresh bark of Tilia ulmifolia; m.p. 228229°C; b.p. 360°C.

Timboin An amaroid from root of Serjania curassavica; is a fish poison.

Tincture iodine Contains 6.5-7.5 gm. iodine, 4.5-5.5 gm. potass. iodide per 100 cc. soln.; alcohol 80-85% by volume; used as antiseptic, disinfectant for skin surfaces, abrasions, ulcers, etc.; in glandular and lymphatic enlargements; also in arthritis, and other ailments.

Tincture iodine, mild 1.8-2.2 gm. Iodine and 2.1-2.5 gm. sod. iodide per 100 cc. of soln.; used as skin disinfectant.

Tissue Any of many distinct substances within living organisms each of whose constituents are cells having the same characteristics

Tissue-specific expression Gene function which is restricted to a particular tissue or cell type. For example, the glycoprotein hormone alpha subunit is produced only in certain cell types of the anterior pituitary and placenta, not in lungs or skin; thus expression of the glycoprotein hormone alpha-chain gene is said to be tissue-specific. Tissue specific expression is usually the result of an enhancer which is activated only in the proper cell type.

Titre In general the amount of one substance required to react with a given volume of another substance, as the amount of acid required to neutralize a definite amount of base. In the analysis of fats and oils, the titre refers to the temperature at which the fatty acids obtained by alkaline hydrolysis of the fat, resolidify.

Tm The midpoint of the temperature range over which DNA is melted or denatured by heat; the temperature at which a duplex nucleic acid molecule is 50% melted into single strands, it is dependent upon the number and proportion of

G-C base pairs as well as the ionic conditions. Often referred to as a measure of the thermal stability of a nucleic acid probetarget sequence hybrid.

Töpfer's reagent Dimethyl-amino-azo-benzene, an indicator that changes from red through pink to yellow, as pH changes from 3 to 4.

Topoisomerase An enzymatic activity responsible for relieving excessive supercoiling in DNA.

Topsoil Topmost layer of soil that contains decomposed vegetation

Toringin A glucoside, $C_{21} H_{20} O_9 \cdot 2H_2 O$, from bark of Pirus toringa; m.p. 135-137°C.

Torque Force that tends to produce rotation or twist.

Torsion Twisting of a bodily organ about the axis along which it is attached to other structures in the body, (*physics*) strain possessed by a twisted material.

Toxisterol A toxic sterol obtained by continued irradiation of calciferol.

Toxoflavin $C_6 H_6 N_4 O_2$; a yellow pigment of cocovenenans which is a violent heart poison, probably isomeric with methylxanthine.

Toxoids Anatoxins; toxins which have been treated with formaldehyde or altered in other ways, with the destruction of only the toxic properties but not of antigenic capacity, e.g. diphtheria toxoid.

TPN Triphosphopyridinenucleotide, a hydrogen carrier like cozymase.

Tracer Small amount of a radioactive element used to trace the path of one or more parts of a biological, chemical, or physical process.

Trachea Tube that serves as a passage for air into and out of the body of an air-breathing animal

Trade wind Wind that blows nearly continually toward Earth's equator from the northeast and southeast.

Tragacanth Gum Tragacanth, Goat's Thorn; dried gummy exudation from Asiatic species of Astragalus, Leguminosae; consists of bassorin, pectin, starch; used as demulcent, as base for skin medication; industrially as emulsifier, adhesive, excipient for pills.

Trait A characteristic found in your genes.

Trans- Located on two physically dis-contiguous DNA molecules.

Transcription Synthesis of a strand of *RNA* within a cell, using the cell's *DNA* as a template.

Transcription factor A protein which controls the transcription of genes. These usually bind to DNA as part of their function (but not necessarily). A transcription factor may be general (i.e. acting on many or all genes in all tissues), or tissue-specific (i.e. present only in a particular cell type, and activating the genes restricted to that cell type). Its activity may be constitutive, or may depend on the presence of some stimulus; for example, the glucocorticoid receptor is a transcription factor which is active only when glucocorticoids are present.

Transcription start site The point in a DNA sequence at which transcription of a gene into RNA begins.

Transcription The process of copying a gene into RNA transcript. This is the first step in the expression of any gene. The resulting RNA, if it codes for a protein, will be spliced, polyadenylated, transported to the cytoplasm, and by the process of translation will produce the desired protein molecule, although not all transcripts lead to proteins. Compare translation.

Transcriptome The complete set of RNAs transcribed from a genome.

Transducer Any device that converts energy of one form into energy of another form.

Transduction The incorporation of a cellular gene into a viral genome, that can then be introduced into other cells.

Transfection A method by which foreign DNA may be put into a cultured mammalian cell. Such experiments are usually performed using cloned DNA containing coding sequences and control regions (promoters, etc) in order to test whether the DNA will be expressed. Since the cloned DNA may have been extensively modified (for example, protein binding sites on the promoter may have been altered or removed), this procedure is often used to test whether a particular modification affects the function of a gene.

Transfer RNA A class of RNA having structures with triplet nucleotide sequences that are complementary to the triplet nucleotide coding sequences of mRNA. The role of tRNAs in protein synthesis is to bond with amino acids and transfer them to the ribosomes, where proteins are assembled according to the genetic code carried by mRNA.

Transformation (with respect to cultured cells) A change in cell morphology and behavior which is generally related to carcinogenesis. Transformed cells tend to exhibit characteristics known collectively as the "transformed phenotype" (rounded cell bodies, reduced attachment dependence, increased growth rate, loss of contact inhibition, etc). There are different "degrees" of transformation, and cells may exhibit only a subset of these characteristics. Not well understood, the process of transformation is the subject of intense research.

Transformation A process by which the genetic material carried by an individual cell is altered by incorporation of exogenous DNA into its genome. The cancerous alteration of mammalian cells

Transformed phenotype Acquisition of a phenotype characteristic of oncogenic transformation.

Transgenic mouse A mouse which carries experimentally introduced DNA. The procedure by which one makes a transgenic mouse involves the injection of DNA into a fertilized embryo at the pro-nuclear stage. The DNA is generally cloned, and may be experimentally altered. It will become incorporated into the genome of the embryo. That embryo is implanted into a foster mother, who gives birth to an animal carrying the new gene. Various experiments are then carried out to test the functionality of the inserted DNA.

Transgenic Creation of an individual in which genetic modification has occured in the germline tissues and may be transmitted to subsequent generations.

Transient transfection When DNA is transfected into cultured cells, it is able to stay in those cells for about 2-3 days, but then will be lost (unless steps are taken to ensure that it is retained. During those 2-3 days, the DNA is functional, and any functional genes it contains will be expressed. Investigators take advantage of this transient expression period to test gene function.

Transit Passage of a celestial object in front of a larger celestial object, as seen from an observer's point of view.

Transition A nucleotide substitution in which one pyrimidine is replaced by the other pyrimidine, or one purine replaced by the other purine (e.g., A is changed to G, or C is changed to T) (contrast with Transversion).

Translation The process of converting the genetic code into polypeptides, catalyzed by the ribosome and a host of soluble factors. mRNA codons are recognized by tRNA anti-codons. Each tRNA codes for a single amino acid, resulting in synthesis of polypeptide wherein the amino acid sequence is dictated by and matches the order of the codons in the mRNA. Sometimes, however, people speak of "translating" the DNA or RNA when they are merely reading the nucleotide sequence and predicting from it the sequence of the encoded protein. This might be more accurately termed "conceptual translation". Compare transcription.

Translational frame shifting A mechanism used by certain viruses and even higher organisms to change the reading frame used during translation of an mRNA molecule, in a controlled manner, so that the polypeptide product is the result of translation in more than one reading frame.

Translocation The process by which a newly synthesized protein is directed toward a specific cellular compartment (i.e, the nucleus, the endoplasmic reticulum).

Translocations (DNA) Of cells, the transfer of part of one chromosome to a nonhomologous chromosome. Of vascular plants, the conduction of organic compounds through the plant body by the way of the phloem.

Transplantation The transfer of a piece of tissue from one embryo to another.

Transposible elements Genetic elements characterized by their abilities to insert into and withdraw from a given location within the genome, resulting in movement from site to site within the genome over a period of time. Transposable elements may cause epigenetic changes in phenotype.

Transposition The movement of DNA from one location to another location on the same molecule, or a different molecule within a cell.

Transudate The name applied to the filtrate of blood through capillaries, when the normal filtration is modified in a non-inflammatory manner, in the direction of increased accumulation of fluid in the tissues.

Transversion A nucleotide substitution in which a purine replaces a pyrimidine, or vice versa (e.g., A is changed to T, or T is changed to G).

Trasentin Diphenylacetyl - diethylaminoethanol hydrochloride, melting at 112-114°C; used as antispasmodic.

Traumatic acid The "wound hormone" of plants, a dibasic acid, HOOCCH=CH $(CH_2)_8$ COOH, which induces cell multiplication and healing.

Travaille test for bile pigments Reagent—2 gm. p-dimethylaminobenzaldehyde, 50 gm. hydrochloric acid and 50 cc. water. The addition of 4 cc. of reagent to 10 cc. of urine gives rise to the formation of a blue color.

Trephones Proteoses which stimulate the growth of cells in tissue cultures.

Trichloracetic acid CCl_3 COOH; colorless deliquescent crystals which are very caustic and whose aq. soln. is very acid; m.p. 57-58°C; b.p. 196-197°C; sp.gr., $1.629^{61}/_4$; produced by oxidation of chloral with nitric acid; used as escharotic, astringent, hemostatic; also as decalcifier and fixative in microscopy and as protein precipitant.

Trifolianol Sitosterol-d-glucoside belonging to the phytosterol glucoside group from the flowers of red clover; m.p. 295°C with decomposition.

Trifoliin A chromo-glucoside from flowers of red clover, melting at 260°C; on hydrolysis yields the coloring matter trifolitin.

Trigonelline $C_7H_7O_2N$; coffearin; gynesin; trigenolline; Betaine of nicotinic acid. Inert alkaloid of peas, potatoes and many other plants. Hygroscopic prisms, mp. with one molecule of water 130°C., anhyd. 218°C.

Trilobine $C_{36}H_{36}O_5,N_2$; m.p. 235°C; an alkaloid from the root Cocculus trilobus; paralyzes respiration and the heart, lowers blood pressure.

Trimethylamine $(CH_3)_3$ N; a colorless gas with a fishy odor; b.p. 3-4°C, sp.gr., 0.67090/4; occurs in beet sugar residues and herring brine; used in synthesis of organic amino compounds.

Trimethylamine oxide C_3H_9ON; m.p. 96°C; a nitrogenous base found in marine types, in the muscles of Crustacea and in the urine of cephalopods, also in fish and animal tissues.

Triose enzyme The pyridinoprotein enzyme which catalyzes the oxidation of glyceraldehyde, 1(+), or glycollic aldehyde by coenzyme I; optimum pH about 8, very sensitive to iodoacetic acid.

Triosephosphoric enzyme A pyridinoprotein enzyme which catalyzes the initial oxidation of 3-phosphoglyceraldehyde, m.w. about 100,000, found in brain, skeletal and cardiac muscle, yeast and bacteria.

Trioses Carbohydrates with 3 carbon atoms; dihydroxyacetone and two stereoisomers of glyceric aldehyde.

Triphal Sodium auro-thiobenzimidazolcarboxylate; used as antitubercular.

Triphosphopyridinium enzymes Pyridinoprotein enzymes containing coenzyme II, which catalyze the oxidation of substrates such as the hexosemonophosphoric, isocitric and glutamic acids.

Trisaccharides $C_{18}H_{32}O_{16}$; sugars yielding three monosaccharides on hydrolysis; classified as reducing and nonreducing. Their structures have been worked out as follows.

I. Reducing Trisaccharides Mannotriose is glucose-6—1-galactose-6—1-galactoside. Rhamninose is galactose-rhamnose-rhamnoside. Robinose is galactose-rhamnoserhamnose.

II. II. Non-reducing Trisaccharides Melezitose is l-*a*-glucose-2-ßfructofuranose-6-*a*-glucoside. Gentianose is 2-ß-fructofuranosel-a-glucose-6-ß-glucoside. Raffinose is 2-ß-fructofuranosel-a-glucose-6-a-galactoside.

Triticin Fructosan from the honey of Triticum repens.

Trityl An abbreviation for the triphenylmethyl group, used particularly in studies of sugar structure.

Trna (Transfer RNA) Special RNAs that are charged with amino acids and which carry anticodons for recognition of the codons in mRNA. tRNAs are responsible for translation from the language of nucleic acids into the language of polypeptides. Each tRNA can be charged with only one type of amino acid, although multiple tRNAs exist for many of the amino acids.

Troensegaard's pyrrole hypothesis Theory of protein structure. Postulates the existence in proteins of a large number of pyrrole rings which hydrolyze at the nitrogen atom, yielding amino acids. This theory cannot account for the optical isomers obtained on hydrolysis, nor is the postulated ring structure sufficiently liable to account for the comparative ease of acid hydrolysis.

Tropacocaine $C_{15}H_{19}O_2N$; benzoyl-pseudotropine; m.p. 49°C; an alkaloid of coca leaves, acts like cocaine, less toxic, more transient.

Trophallaxis The exchange of secretions etc. between individual members of a community of social insects, whereby social cohesion and caste differentiation are thought to be effected.

Trophic level Position of a given species within a *food chain.*

Trophic An action of the nervous system in the absence of which muscle tonus fails and regeneration is impossible.

Trophoblast The non-embryonic part of the mammalian blastocyst, which effects its implantation into the uterine wall, and part of which later develops into the foetal portion of the placenta.

Tropic acid HO-CH_2-CH(C_6H_5)-COOH; m.p. 117-118°C; alpha-phenyl-beta-hydroxypropionic acid; occurs as an ester in atropine and is a mixture of optical isomers.

Tropine 36-endo-aminoethyl-cycloheptanone-1; $C_8H_{15}ON$; m.p. 63°C, b.p. 233°C, a component of atropine and a product of the hydrolysis of atropine.

Tropism A turning of an animal resulting from a change in its invironment; taxis.

Truxine An alkaloid, $C_{21}H_{30}O_4N_2$, from the seed of Strychnos Ignati and melting at about 250°C with decomposition.

Trypan blue Sodium ditolyldiazobis-8-amino1-naphthol-3,6-disulfonate; used in trypanosomiasis in cattle, horses, dogs, foot and mouth disease.

Trypan red A reddish-brown powder used subcutaneously or perorally in trypanosomiasis.

Tryparsamide Na phenylglcinamide-p-arsonate, used as a trypanocide and a mild spirillicide in paresis, tabes, neurosyphilis, etc.

Trypsin A proteolytic enzyme of the pancreatic juice. Hydrolyzes proteins to proteoses and peptones. The crystalline enzyme has an optimum pH of 8-9 with casein. Activated by enterokinase of the small intestine and the pancreas.

Trypsinogen The crystallizable zymogen occurring in the pancreas from which trypsin is formed by the pancreatic juice.

Tryptamine ß-indole-ethylamine; a poisonous amine formed by bacterial decomposition of tryptophane.

l-TRYPTOPHANE $C_{11}H_{12}O_2N_2$; ß-Indol-a-amino propionic acid; an indispensable amino acid found in most cell proteins. Shiny hexagonal or rhombic plates, readily sol. hot, slightly sol. cold water, m.p. 289°C (corr.).

Tryptophane metabolism There are two theories, neither of which has been proven beyond question.

(1) Tryptophane goes to indolepyruvic acid, which breaks at the double bond, then recombines, finally yielding kynurenic acid.

(2) Tryptophane, after a several step oxidation, finally yields kynurenin.

The further metabolism is indefinite.

Tschirch reaction for ergot A red aqueous solution is obtained when 20 cc. ether is poured on 1 gm. ergot, followed by the addition of 10 drops ammonia and 20 cc. water. After swirling and standing for 2 hours, the ether is separated and evaporated, the residue taken up with acetic acid and, after filtering, floated on concentrated sulfuric acid containing a trace of ferric chloride. A blue color forms at the contact zone. The acetic acid layer should show a green fluorescence.

Tschopp test A test for determining male sex hormone activity by the capon test.

Tschugaev test for hydroxyl compounds The formation of easily identifiable methane by the reaction of methyl magnesium iodide with hydroxyl compounds such as acids, alcohols, oximes and phenols is proof of the presence of the hydroxy group.

Tuberculosis A disease caused by Bacillus tuberculosus (Mycobacterium tuberculosis) which may be acute or chronic, congenital or due to infection, affecting the lungs, liver, nervous system, genital organs, etc. Human and bovine types predominate. Principal forms.

(1) acute miliary tuberculosis.

(2) pulmonary tuberculosis (phthisis, consumption). Sputum diagnoses of several kinds may be made and tuberculin skin tests. The disease can be arrested with great care. Collapse therapy (of the lung) requires the use of artificial pneumothorax.

Tuberculo-stearic acid 10-methyl stearic acid, isolated from tubercle bacilli.

Tuberin A globulin of the potato.

Tubocurarine An alkaloid used as an antitetanic.

Tumor Overproduction of cells of a particular tissue leading to an abnormal lump.

Tumor suppressor A gene that prevents tumor formation until deleted or mutated. The best-known examples of tumor suppressors are the proteins p53 and Rb.

Tunica Vaginalis The thin and watery membrane covering the front and sides of testis and epididymus.

Turacin Red feather pigment of turaco bird; a copper salt of uroporphyrin.

Turanose Fructopyranose-6-alpha-glucoside, reducing hydrolytic product of melezitose, m.p. 157°C.

Turicine Hydroxy-stachydrine; from herb of Betonica officinalis; m.p., 260°C with decomp. when anhydrous.

Turnover The balance between synthesis and degradation of a product.

Turpethin A glucoside, composition doubtful; a constituent of turpenthum resin; is stated to be a cathartic like jalapin.

Tutin A glucoside, either $C_{17}H_{20}O_7$ or $C_{15}H_{18}O_6$ from the fruit of Coriaria ruscifolia; m.p. 209-212°C; pharmacol. action places it in picrotoxin group.

Tutocaine Butamin, p-aminobenzoyl-dimethylamino-1,2-dimethyl propanol hydrochloride; m.p. 212-215°C; a local anesthetic.

Twitchell reagent for fat hydrolysis Preparation—A mixture of naphthalene and oleic acid is treated with excess concentrated sulfuric acid, excessive temperature rise being avoided by cooling and slow addition of the acid. The crude product formed is dissolved in ether and washed several times with dilute hydrochloric acid and then extracted with water. The aqueous extract is made to contain about 10% naphthalenestearosulfonic acid, used as a catalyst for ester hydrolysis.

Tyndall beam (Cone) The band of reflected light that is visible when a beam of light is passed through a medium containing dispersed particles capable of reflecting the light. Frequently an apparently homogeneous medium is shown to have dispersed particles by the use of this test.

Tyndall meter An instrument for measuring the intensity of a Tyndall Cone.

Typhoid prophylactic Enteric or typhoid vaccine; bacterial vaccine made from typhoid bacillus; a sterile suspension of killed typhoid bacilli (Eberthella typhi) in physiol. sod. chloride soln. or other suitable diluent; used in prophylaxis of typhoid.

Tyrian purple A bromine containing dye of the shell fish Murex brandaris.

Tyrocidin A polypeptide of the soil bacillus which is hemolytic and bactericidal in vitro.

Tyrosinase An oxidase of potato, bran, fungi, etc., responsible for darkening of cut surface of plants by conversion of tyrosine to melanin. $C_9 H_{11} O_3 N$; p-hydroxyphenylalanine; an essential amino acid, and may be a precursor of adrenaline, thyroxine and melanin; m.p. 314318°C also given as 295°C l-rotatory, i.p. 5.41.

Tyrosinosis An abnormality in which tyrosine metabolism stops at the formation of p-hydroxyphenylpyruvic acid.

Tyrosinuria The presence of tyrosine crystals in the urine in cases of destruction of liver tissue, as in acute yellow atrophy.

Tyrosol ß-(p-hydroxyphenyl)-ethyl-alcohol.

U

Ukambin A glucoside from the East African arrow poison of the Wakambas; m.p., 179°C; a heart poison.

Ulcer Inflamed break in the skin, *mucous membrane,* or organ lining caused by infection, irritation, *tumor*, or possibly viruses.

Ulexoside A glucoside, from fresh flowers of Ulex europaeus; m.p., 247°C.

Ultracentrifuge A centrifuge, originally developed by The Svedberg, having very high speed, and developing centrifugal forces as high as 7,000,000 times gravity.

Ultrafiltration Process of filtering substances out of a solution by the size of their constituent molecules.

Ultrafiltration The filtration of a colloidal dispersion through a filter with pores sufficiently fine to retain the dispersed phase, and yet allow the dispersions medium to pass through.

Ultramicroscope A microscope, used largely in the study of colloids, where the source of light is a sharp beam at right angles to the line of vision. What is seen is not the particle itself, but the reflection of light from the particle.

Ultrasonic waves Sound waves of a frequency too high to be audible.

Ultraviolet radiation Electromagnetic radiation that is just outside the blue (short-wavelength) end of the visible range; extended exposure causes skin burns and may damage *DNA* in skin cells, leading to cancer.

Umbellic acid 24-dihydroxycinnamic acid, found in asafoetida.

Umbelliferone $C_9 H_6 O_3$; m.p. 233-234°C; 7-hydroxycoumarin; anhydride of umbellic acid, found widely in plants.

Umbellulone Unsaturated dicyclic ketone of minty odor from Umbellularia. californica, $C_{10}H_{14}O$.

Umbilical cord Cord that connects the *embryo* or fetus to the *placenta* during pregnancy in female mammals.

Umhüllung The wrapping of an inductor in sheets of competent ectoderm, in order to test inductive power.

Unified field theory Incomplete theory that physicists are formulating in an attempt to accommodate the four *fundamental forces* as individual parts of a single force.

Universal Genetic Code The "standard" codon usage that is common to most organisms.

Unna-golodetz sulfhydryl reaction An aqueous suspension of sulfur is converted to hydrogen sulfide by the reducing action of proteins. The product is identified by its odor and reaction with lead acetate paper.

Upstream activator sequence A binding site for

transcription factors, generally part of a promoter region. A UAS may be found upstream of the TATA sequence (if there is one), and its function is (like an enhancer) to increase transcription. Unlike an enhancer, it can not be positioned just anywhere or in any orientation.

Upstream/Downstream In an RNA, anything towards the 5' end of a reference point is "upstream" of that point. This orientation reflects the direction of both the synthesis of mRNA, and its translation - from the 5' end to the 3' end. In DNA, the situation is a bit more complicated. In the vicinity of a gene (or in a cDNA), the DNA has two strands, but one strand is virtually a duplicate of the RNA, so it's 5' and 3' ends determine upstream and downstream, respectively. NOTE that in genomic DNA, two adjacent genes may be on different strands and thus oriented in opposite directions. Upstream or downstream is only used on conjunction with a given gene.

Uracil 26-dioxytetrahydropyrimidine, m. p. 338°C with decomposition; a pyrimidine constituent of plant nucleic acids.

Urea Carbamide, Carbonyldiamide; occurs in urine, tissue fluids and blood of all mammals; m.p., 13233°C sp. gr., 1.335; used in ascites, as test of renal function; externally to promote healing; employed industrially to stabilize explosives and celluloid compositions, in calico printing, mfg. of barbital and allied products.

Urease The enzyme which splits urea into ammonia and carbon dioxide, probably by way of ammonium carbamate.

Urechitin A glucoside, $C_{28}H_{42}O_8$, from leaves of Urechites suberecta; a heart poison causing salivation, vomiting, diarrhea, and cardiac paralysis.

Ureotelic Pertaining to nitrogen metabolism when the main end product is urea.

Urethane Ethyl carbamate, $NH_2CO{\cdot}OC_2H_5$; m.p. 48-50°C; b.p. 182-184°C; sp.gr. 1.1; hypnotic antispasmodic, sedative; used internally in convulsions, mild insomnia, also as antidote for certain poisons.

Urethra The small opening through which secretions pass in the penis

Urginin A mixture of the two active squill glucosides, Scillaren A and B; used as myocardial stimulant.

Uric acid $C_5H_4O_3N_4$; 2,6,8-trioxypurine, an excretion product of the oxidation of nitrogen compounds, e.g. the nucleic acids.

Uricase An enzyme of animal tissues which catalyzes the oxidation of uric acid by molecular oxygen with the formation of peroxide and allantoin, probably formed via 45glycol uric acid and oxyacetylenediurein carboxylic acid; optimum pH 9.3, attacks other purine compounds; found in liver and kidney.

Urinary system System in animals that removes waste materials and excess water from the body.

Uronic acids Acids obtained by the oxidation of a primary alcohol group of a sugar without a change of the aldehyde group, e.g. glucuronic acid.

Uroporphyrin $C_{40}H_{38}O_{16}N_4$; a porphyrin derivative found in the urine of people with congenital porphynuria.

Uropterin(E) A constituent of human urine probably identical with xanthopterin(e),

Urorosein The precursor of the red coloring matter of certain pathological urine, probably indole acetic acid, which turns red with conc. HCl and Na nitrite.

Urorubin A pathological red urine pigment.

Ursodesoxycholic acid A bile acid, m.p. 198°C, found in the bear. It is a dihydroxycholanic acid, $C_{24}H_{40}O_4$.

Uterus Organ in female mammals in which the *embryo* develops during pregnancy.

Uterus Chamber in which the developing embryo is contained and nurtured during pregnancy.

Uva ursi Bearberry; dried leaves of Arctostaphylos Uva-ursi; used as diuretic, astringent.

Uzarin A glucoside, $C_{75}H_{108}O_{30} \cdot 9H_2O$, from uzara root; m.p., 200-210°C.

V

Vaccenic acid An isomer of oleic acid with double bond at 7th and 8th carbon atom.

Vaccine brucella melitensis Suspension of killed B. melitensis (B. abortus) in physiol. soln. sod. chloride; used in undulant fever.

Vacciniin A glucoside, 6-benzoyl-d-glucose; from various species of Vaccinium berries; m.p. 104-6°C.

Vacuole membrane-bound structure in a cell that stores substances and transports other substances to the cell membrane to be excreted from the cell.

Vagina Part of a female reproductive system that receives sperm, forms part of the birth canal, and channels menstrual flow to the exterior.

Vagotonin A supposed hormone which controls the excitability of the vagal nerve center; affects heart and general blood pressure.

Valdivin A glucosidal amaroid, from fruit of Simaba valdivia, melting at 230°C when anhydrous; claimed to be identical with cedrin by some and more toxic than cedrin by others.

Valenta reagent for fats Depending upon their source, fats and fatty oils produce clear solutions with glacial acetic acid at various temperatures.

Valerian Dried rhizome and roots of Valeriana officinalis L. Valerienaceae; contains valerine, chatinine (alkaloids); used as sedative, antispasmodic.

Valerianic acid Four structural isomerides are possible and known.

(1) n-valeric acid, b.p. 185°C;

(2) isovaleric acid, $(CH_3)_2$ $CHCH_2$ COOH, b.p. 175°C;

(3) methyl ethylacetic acid, b.p. 177°C;

(4) trimethyl acetic acid (pivalic acid) b.p. 164°C. Isovaleric acid occurs in the free state in many plants, e.g. valerian root. The acidum valerianicum of pharmacy is isovaleric acid with some methyl ethyl acetic acid.]

VALINE-1 Alpha-amino-isovaleric acid, m.p. 93-96°C amino acid, d-form in nature, found in plant seedling protein and others.

Vallarin A glucoside from an undetermined Valaris species; in frog 2 mgm. causes miosis, paralysis, dyspnea, and death.

Valonia An alga much used in permeability studies because of its large cells.

Valton test for methylamine in presence of mmonia The test solution, mixed with 30 cc. 2N sodium hydroxide is placed in a flask and diluted with water to 80 cc. Using a Kjeldahl apparatus, the mixture is distilled into 10 cc. of a 0.5% alcoholic solution of 2,4-dinitrochlorobenzene. After distilling 10 cc., the mixture is allowed to stand for 20 hours; a crystalline precipitate of dinitromethylaniline is formed.

Valve Structure in a blood vessel that maintains the direction of blood flow.

Van deen reaction for blood Treatment of dilute blood solution with several drops of tincture of guaiac resin and red (ozonized) turpentine oil yields a blue color.

Van den bergh test for bilirubin in blood One part of blood serum is mixed with 2 parts 95% alcohol, the precipitated albumin is removed by filtering or centrifuging and, to the clear colution, ¼ its volume of Ehrlich's diazo reagent is added; a red color is obtained.

Van der haar reagent for d-galactose d-Galactose gives a crystalline hydrazone with o-tolylhydrazine, in contrast to other sugars.

Van itallie-steenhauer microchemical reactions for differentiating tyramine and histamine The formation of the following double salts serves to distinguish easily between the two bases chloroplatinate, iodobismuthate, iodoplatinate, phosphotungstate, picrate and silicotungstate.

Van slyke's method A method for determining amino nitrogen of proteins or amino acids by titration with nitrous acid and the liberation of nitrogen gas.

Van slyke's method of protein analysis A method which combines the Hausmann procedure with amino nitrogen determinations, etc., to determine individual amino acids.

Van zijp reaction for cholesterol A thin layer of cholesterol, obtained by evaporation of a chloroform solution, gives blue or bluish-black crystals upon 1-4 hours exposure to vapors of concentrated hydriodic acid containing iodine. Ergosterol and phytosterols do not react, nor does cholesterol in the presence of more than 5% phytosterols.

Vanadium chromogen A respiratory pigment of the blood of Tunicata such as Ascidia, containing about 10% of vanadium in the form of various oxides combined with a pyrrol containing protein.

Vanillin Methylprotocatechuic aldehyde; Vanillic aldehyde; 3-Methoxy-4hydroxybenzaldehyde; m.p., 8081°C; occurs in vanilla, potato parings, Siam benzoin etc.; synthesis from eugenol or guaiacol; used as flavoring agent in confectionery, foods, beverages; also as reagent in anal. chemistry.

Vapor density Density of a gas, defined by the ratio of the mass of a unit volume of the gas to the mass of an equal volume of either hydrogen or air at the same temperature and pressure.

Vapor pressure Pressure of the vapor that evaporates from the surface of a liquid or solid.

Variable Quantity within a mathematical expression that can take various values; compare *constant.*

Variation Differences among individuals belonging to a single *species.*

Vector A construct used to propagate DNA in a host (bacteria, yeast, or cultured cells). The vector provides all sequences essential for replicating the test DNA. Typical vectors include plasmids, cosmids, phages and YACs.

Vegetal pole That pole of the egg which contains most yolk and least cytoplasm.

Vegetalisation The shifting of the presumptive fate of tissue which should have been ectodermal, in echinoderm development.

Vein Vessel that transports blood from parts of the body to the heart.

Velicogna test for adrenaline A cardinal-red color is developed in 5-10 minutes when a mixture of 3 cc. 1% solution of phenol in 50% alcohol, 0.6 cc. saturated potassium iodate solution, and 0.1 cc. saturated sulfanilic acid solution is heated on a water bath. The addition of a trace of adrenaline results in a more rapid development of color; sensitivity—112 million.

Vellarin An amaroid from leaves and root of Hydrocotyle asiatica, and used in leprosy and syphilis.

Vellosine An alkaloid from bark of Geissospermum Vellosii Allem., Apocynaceae; m.p. 189°C; very toxic, physiol. action corresponding with that of brucine.

Velocity Measure of the rate of movement of an object in a given direction.

Venn diagram Diagram of overlapping circles representing sets of numbers and the relationships between the sets.

Ventricle Either of two chambers of the heart that receive blood from the atria (plural of *atrium*) and pump it out of the heart into circulation.

Ventriculin A commercial name for defatted, desiccated hog stomach used in the treatment of anemia.

Veratric acid Dimethyl-protocatechuic acid; 3, 4-Dimethoxybenzoic acid; obtained synthetically and from seed of Sabadilla officinalis; m.p., 180-181°C when anhydr., sublimes in rhombic crystals.

Veratridine An alkaloid, $C_{36}H_{51}O_{11}N$; from seed of Sabadilla officinalis and from rhizome of Veratrum album; m.p. 180°C; is hydrolyzed into. amorphous cevine (verine) and veratric acid.

Veratrine alkaloids A group of alkaloids of which the outstanding ones are veratrine, or cevadine, and protoveratrine, occurring in various hellebores; soften pulse and lower blood pressure; emetic.

Verbenalin A glucoside, $C_{17}H_{25}O_{10}$; from whole plant, particularly the flowers of Verbena Officinalis; m.p., 178°C; hastens blood coagulation and is a uterine contractor.

Verbenone An unsaturated ketone in Spanish verbena oil found in d- and 1-forms.

Verdohemochromogen Bile pigment hematin-protein.

Verdohemoglobin Choleglobin; pseudohemoglobin; a green chromoprotein obtainable from hemoglobin by hydrogen peroxide oxidation, contains verdohematin, a biliverdin-iron complex or prosthetic group.

Vernonin $C_{10}H_{24}O_7$; a glucosidal amaroid of the root of Veronia nigritiana Oliv., Compositae, which is a heart poison less toxic than digitalin.

Verschmelzung Fusion of two eggs together.

Vertebral column Same as *spine* and backbone

Vertebrata A subphylum of the phylum Chordata; characterized by having a sectional bony column running down the center of the dorsal side of the animal.

Vertebrate Any of the group of animals that have backbones.

Vertigo Dizziness; giddiness; a disorder of the sense of equilibrium.

Vestigial Structures A structure that seems not to have a function in an organism but resembles a functional organ in another type of organism.

Vestlin test for saponin A solution of Gypsophila saponin hemolyzes horse blood at 37-42°C; a 11 million solution produces only slight hemolysis.

Viable count A method of counting bacteria by diluting, plating and counting colonies.

Viburnum prunifolium Dried root bark of viburnum prunifolium containing viburnin, bitter resin, tannin and certain organic acids; is a supposed uterine sedative; internally for afterpains, habitual abortion, menorrhagia.

Vicianin $C_{19}H_{25}O_{10}N$; m.p. 147-148°C; a glycoside of the seeds of Vicia angustifolia yielding vicianose and HCN by the action of the enzyme vicianase.

Vicianose Glucosc-6-beta-1-arabinoside, a constituent of the glycosides, vicianin, violutin and gein; m.p. 210°C, unfermentable.

Vicilin A globulin of various legumes.

Vicine C10H16O7N4 · 2H2O; m.p. 239-240°C; 3-divicine-d-glucoside, a nucleoside found free in plant tissues, e.g. seed of Vicia sativa.

Viehover test for marihuana One-third or more of the wrapperless cigarette or at least 0.1 gm. of the weed is placed in a stoppered flask and treated with 3 cc. of isopropyl (or any common) alcohol containing 0.05% sodium hydroxide and 0.05 gm. activated charcoal. An immediate color change to pinkish, deepening after a short time, is positive.

After evaporation, the residue dissolves in concentrated ammonia with an orange-red color and in acetone with violet to blueviolet color.

Villikinin A substance stimulating motility of the villi and secretion of the intestinal mucosa, found also in the Bios fraction of yeast.

Villosin A glucoside from root bark of Rubus villosus, melting at 173-175°C.

Villus Any of millions of hairlike structures that project from the inner wall of the small intestine and that vastly increase the area capable of absorbing nutrients from food.

Vilter-spies-mathews reaction for nicotinic acid and nicotinamide A mixture of the acid with twice its weight of 2,4-dinitrochlorobenzene is melted over a flame and treated with alcoholic potassium hydroxide; a brilliant purple color is obtained even with 0.3 mg. of acid.

To test urine, 15 cc. are boiled with 0.1-0.3 gm. decolorizing charcoal, filtering and evaporating 3 cc. to dryness at 80-100°C. One cc. of 1% alcoholic 2,4dinitrochlorobenzene is added, kept at room temperature for 1-3 hours, again evaporated to dryness and heated at 105°C for 10 minutes. After cooling to 25°C, add 10 cc. of a clear cold 0.1% alcoholic potassium hydroxide solution. Nicotinamide gives a red color, nicotinic acid a clear purple color.

Vincent's angina Trench mouth; an ulcerative disease of the mouth due possibly to a spirochaete at least as a secondary infection. It may resemble diphtheria. Strong oxidizing agents or arsphenamine derivatives are used persistently.

Vincetoxin A glucoside; m.p., 132°C with decomp.; from root of Vincetoxicum officinale.

Violanin A chromo-glucoside, $C_{27}H_{28}O_{15}$, from flowers of viola tricolor.

Violaxanthin $C_{40}H_{56}O_4$, a carotenoid pigment, present in ester form in viola petals, m.p. 208°C.

Violutin $C_{19}H_{26}O_{12}$; m.p. 169-172°C; violutiside; a glucoside from Viola cornuta which on hydrolysis yields methyl salicylate, l-arabinose and d-glucose.

Viosterol A commercial name for irradiated ergosterol used in the prophylaxis and treatment of rickets and infantile tetany.

Virion A replication-competent virus particle. Virulence Ability to infect or cause disease.

Virus A noncellular biological entity that can reproduce only within a host cell. Viruses consist of nucleic acid covered by protein; some animal viruses are also surrounded by membrane. Inside the infected cell, the virus uses the synthetic capability of the host to produce progeny virus.

Viscosity The internal friction of a liquid; the resistance to shear or flow. The unit of viscosity is the poise.

Visual purple Rhodopsin; reddish purple pigment of retinal rods.

visual yellow The compound formed from visual purple (rhodopsin) by light; actually yellow retinene and a protein.

Vitaids A term used by Lepeshkin (1936) for the continuous phases of living protoplasm, which disintegrate on death.

Vitamin A A growth promoting organic substance, lack of which leads to keratinization of the epithelial tissues, with ensuing bacterial invasion. Synthesized in liver cells from U+22ID-and ß-carotene; found in fish liver oils, egg yolk, etc. 70 Kg. adult needs 1400-2000 U.S.P. units, and 2-4000 is desirable. No absorption spectra bands in visible, broad region in ultraviolet light, with maximum at 328 mm. in CHC13. Stored to varying

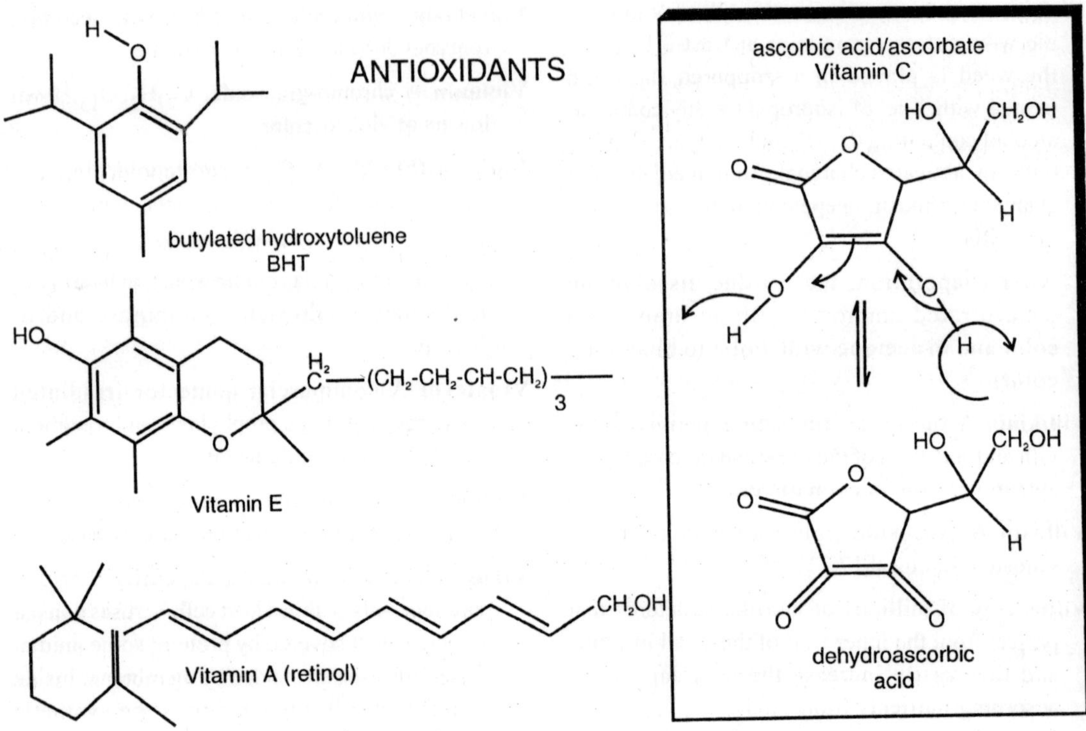

Fig. Anti Oxidant Vitamins

extent in animals, largely in liver. Absorbed as an ester from the thoracic duct. Forms a blue compound with SbC13, max. abs. band at 620 mm.

Vitamin a international unit The growth promoting activity in rats of 0.0006 mg. of a standard crystalline ß-carotene.

Vitamin A, crystalline $C_{20}H_{30}O$; yellow prisms, m.p. 63-64°C, from shark liver, cod liver, jewfish liver and halibut viscera; provisional potency 4,300,000 international units.

Vitamin A2 Another naturally occurring form of Vitamin A. The principal vitamin A of fresh water fish, and found to a lesser extent in the liver oils of other fish. Differs spectroscopically from vitamin A. Probably physiologically active in mammals.

Vitamin B Complex A name for the mixture of water soluble vitamin factors found together in yeast and elsewhere, which have been separated in at least 9 components. The nomenclature is now in a highly fluid state.

Vitamin B1 effects The administration of the vitamin has a sparing action on fat, it aids in the conversion of carbohydrates via pyruvic acid to fat (rat, pigeon) and accelerates fatty liver development in rats and chicks on low choline.

Vitamin B4 Obsolescent term for a factor of vitamin B complex supposedly necessary for the growth of rats.

Vitamin B5 An obsolescent name for the heat stable component of vitamin B complex which prevents the loss of weight in pigeons. Vitamin B3, however, is necessary for a gain of weight.

Vitamin D There are possibly 9 forms, including calciferol, 2,2-dihydroxycalciferol, irradiation products of 7-dehydrocholesterol, 7-hydroxycholesterol, 7-dehydrositosterol, their isomers, cholesterilene sulfonic acid, an ergosterol derivative with alkyl nitrites and alkyl amines, natural forms not quite identified.

Vitamin D1 A molecular compound of calciferol (vitamin D2) and lumisterol.

Vitamin D2 The name used for calciferol especially in German literature, C28H44O, m.p. 115-117°C; 1 milligram equals 40,000 international units.

Vitamin D3 The irradiation product of dehydrocholesterol, found in fish liver oils and in the skin; C27H44O, m.p. 82-4°C.

Vitamin D4 The irradiation product of 22-dihydroergosterol.

Vitamin k unit The equivalent of I gamma of 2methyl-1,4-naphthoquinone, or 3 and 1/3 gamma of vitamin K1.

Vitamin K1 2-methyl-3-phytyl-1,4-naphthoquinone; alphaphylloquinone; a yellow colored substace found in plant tissues and vegetable oils, promoting blood coagulation; also synthesized by bacteria.

Vitamin K2 A 2-methyl-1,4-naphthoquinone derivative with a longer unsaturated side chain than vitamin K1, which has similar action.

Vitamin P A factor found in lemon juice, the lack of which results in hemorrhages of the mucous membranes and severe constitutional disturbances. It may be a mixture of the flavones, hesperidin and eriodicytol.

Vitellin Ovovitellin; a phosphoprotein of egg yolk, yellow.

Vitelline membrane The non-cellular membrane surrounding the yolk before development in the hen's egg, eventually superseded by the yolk-sac.

Vividiffusion A technic used for applying diffusion and ultrafiltration to the blood stream of a living animal. It consists of passing the carotid arterial blood, after addition of an anticoagulant, through collodion tubes suspended in water or physiological salt solution, and then back to the femoral vein.

Viviparity The carrying of the young till birth in the maternal body.

Viviparity The maintenance of embryos during their development within the maternal body in relatively close contact with the maternal tissues or blood stream.

Voisenet's reaction A test for proteins, supposedly due to the presence of tryptophane. The protein solution is treated with a trace of formaldehyde. On adding hydrochloric acid with a bit of nitrous acid to this, a violet color is developed.

Volemitol Sedoheptitol; a 7 carbon sugar alcohol of Lactarius volemus.

Volume Three-dimensional space taken up by a solid object.

Vosges-proskauer diacetyl reaction for proteins Bacteria in a culture form acetylmethylcarbinol from the sugar. The test is based on the oxidation of the carbinol to diacetyl. In the presence of alkali, proteins and diacetyl form a violet-red product with a green fluorescence. The aerogenes group, which gives the reaction, may thus be distinguished from the colon group.

Votocek test for ketoses A drop of test solution is added to a mixture of 2 cc. glacial acid, a small amount of carbazole and I drop fuming hydrochloric acid and heated on a water bath for 10 minutes; a cherry-red color is obtained.

Vulpian reaction, The green color formed by reaction of adrenaline with ferric chloride in slightly acid solution.

Vuzin dihydrochloride Isooctylhydrocupreine dihydrochloride, $C_{19}H_{23}ON_2 \cdot OC_8H_{17} \cdot 2HCl \cdot 2H_2O$; used as powerful bactericide and externally for boils, abscesses, carbuncles, and infected wounds.

W

Wagenaar reaction for caffeine Deep brown crystals are obtained when a caffeine solution is treated with cesium chloride, a drop of potassium iodide solution and a drop of sulfuric acid.

Wagenaar test for acetone in urine A mixture of 10 cc. urine, 10 drops glacial acetic acid, 10 drops 20% tartaric acid solution, 20 drops fresh 20% sodium nitroprusside solution is layered with ammonia. A distinct purple ring is produced, even with as little as 0.5 mg. acetone. The test is specific.

Wagenaar test for distinguishing citric, oxalic and tartaric acids A mixture of a few drops of ferric chloride solution and 20 drops of water is treated with 0.3 gm. aminopyrine and 0.03 gm. resorcinol; 0.05 gm. of the acid and 3 cc. concentrated sulfuric acid are added and heated over a small flame. Citric acid gives a yellowish-brown color; oxalic acid a dark green color changing to pale green; tartaric acid gives deep red changing to raspberry.

Wagenaar test for hydrochloric acid in gastric contents A solution of 0.2 gm. of vanillin in 5 cc. of boiling water is crystallized by cooling. One drop of the crystalline suspension is evaporated on a watch glass, treated with 1 drop of gastric juice and evaporated at a low temperature; a violet zone forms if the hydrochloric acid contents exceeds 0.0015N. Of the organic acids, only oxalic interferes.

Wagenaar tests for fructose and sucrose in lactose A. suspension of a few mg. of sugar in a drop of a 2% glycerin solution of D-naphthol is treated with a drop of concentrated sulfuric acid. Fructose gives an immediate blue color. A suspension of lactose in a drop of the glycerin solution of D-naphthol is treated with a drop of concentrated sulfuric acid and stirred with a platinum wire. Sucrose forms a blue-violet color within 10 minutes.

Wait reaction for vitamins A, C and D Vitamins A and C give a blue color with a 1% acetic acid solution of phosphomolybdic acid, vitamin D gives a green color. Substances containing both vitamins and D give a green color first, followed by the characteristic blue of vitamin A.

Wall pressure Pressure exerted by the contents of a plant cell against the cell wall.

Warburg's respiratory ferment An intracellular, structure-linked catalyst present in all aerobic cells, catalyzing the reaction of cytochrome with oxygen. Consists of a pheohemin in combination with a high-molecular protein; the ferri form reacts with HCN, the ferro form binds CO in a reversible, photodissociable manner. . K.G.St.

Warburg-christian flavoprotein A flavoprotein enzyme from bottom yeast, m.w. about 70,000; promotes the reaction of hexosemonophosphate and coenzyme II which forms phosphohexonate and reduced coenzyme II by reoxidizing the reduced coenzyme and itself being reduced; the reduced flavoprotein is oxidized by molecular oxygen with the formation of peroxide.

Wart Verucca; a virus infection showing as a thickened callosity. Warts may be removed by trichloracetic acid, salicylic acid, freezing and cauterizing, electrodesiccation by sparking, X-ray therapy and even by internal administration of arsenic.

Waser test for amino acids A few cc. of solution (or a few mg. of solid) are boiled with 3-4 cc. of 10% sodium carbonate solution and treated with a pinch of p-nitrobenzoyl chloride; amino acids form a dark winered to violet-blue color. By boiling 0.5-1 cc. of solution with 1 cc. of pyridine and a crystal of the p-nitro compound and slowly adding sodium carbonate solution, a similar color may be obtained.

Wasserman reaction This reaction is an application of the complement fixation method (q.v.) which employs as antigen a non-specific lipoid, usually the alcoholic extract of beef heart.

Water intoxication Convulsions produced by forcing very large amounts of water due to effect on brain, edema, loss of chloride and alkalosis.

Water-soluble b A name employed in 1916 by McCollum for thiamin. Fell into disuse after 1920.

Wats-ghosh test for atebrin in urine A mixture of 100 cc. of urine and 10 gm. of potassium carbonate is shaken with 20 cc. amyl alcohol. The alcohol layer is decanted and washed with a saturated aqueous solution of potassium carbonate; atebrin gives a yellow color and the passage of a beam of light through the liquid against a black background gives a distinct green fluorescence. Sensitivity—110 million.

Wave Disturbance that transfers energy through a medium.

Wavelength Distance between two points on a *wave* that correspond to the same position of two adjacent phases of the wave.

Waxes Solid or liquid esters of the higher fatty acids with mono- and sometimes dihydroxy-alcohols or with sterols. They frequently appear as protective coatings in the plant and animal kingdoms.

Weak acid A*cid* in which the ratio of hydrogen ions (H^+) to hydroxide ions (OH^-) is smaller than in the average acid; when weak acids are added to a neutral solution (such as water), they cause less of a change in the ratio of hydrogen ions to hydroxide ions than the average acid does

Weak base B*ase* in which the ratio of hydroxide ions (OH^-) to hydrogen ions (H^+) is smaller than in the average base; when weak bases are added to a neutral solution (such as water), they cause less of a change in the ratio of hydroxide ions to hydrogen ions than the average base does.

Weak nuclear force F*undamental force* that mediates *radioactive decay* and other chemical reactions within the nucleus of an atom.

Weight Force that an object experiences because of *gravity;* weight is distinct from mass.

Weighting Assigning a value to a taxonomic character.

Weltmann-barrenscheen test for urea An aqueous solution of urea gives, with Ehrlich's aldehyde reagent, an intense yellowishgreen color. Serum, dealbuminated with trichloracetic acid, is filtered and 1 cc. of filtrate is treated with 4 drops of Ehrlich's reagent; when the residual nitrogen value exceeds 36-40 mg., a yellow-green color develops. Sensitivity of reaction corresponds to pathological threshold value.

Wender reactions for diastase Five cc. of 11000 diastase solution are mixed with 1 cc. dilute hydrogen peroxide and 10 drops of one of the following. (a) Alcoholic guaiac resin solution and hydrogen peroxide, intense blue color; (b) guaiac wood tincture and hydrogen peroxide, dark blue color; (c) pyrogallol solution and peroxide, orange-red color; (d) naphthol solution and peroxide, violet-blue; (e) dilute pphenylene diamine hydrochloride solution and peroxide, violetbrown.

Western blot A technique for analyzing mixtures of proteins to show the presence, size and abundance of one particular type of protein.

Similar to Southern or Northern blotting, except that.

(1) a protein mixture is electrophoresed in an acrylamide gel, and.

(2) the "probe" is an antibody which recognizes the protein of interest, followed by a radioactive secondary probe (such as 125I-protein A).

Weyl's reaction The reaction of creatinine with sodium nitroprusside to give a red color which changes to yellow.

Wharton's jelly A preparation of the umbillical. cord.

Wheeler-johnson reaction for uracil, cytosine and isocytosine Bromine water is added to uracil or cytosine until a yellow color is obtained; excess bromine is removed by bubbling in air followed by barium hydroxide solution. A purple-red to violet color or precipitate is formed. Isocytosine gives a blue color. When the reagent is mixed with sodium chloride, it turns blue due to agglomeration of the gold particles, but certain Colloids prevent this change. The amount required to prevent the color change may be used to classify colloids and is called the gold number.

Whiskey A liquid produced by distillation of the fermented mash of malted cereal grains and which has been stored in wood containers for not less than 4 years; contains 0.050.16% acid calculated as acetic acid and 0.038-0.15% esters as ethyl acetate; absolute alcohol is 47-53% by volume; is stated to have intoxicating effects when taken internally, attacking the central nervous system.

Whitehorn reagent for amines Organic bases are taken up from aqueous solutions and solutions in organic solvents by Permutit, a sodium aluminum silicate.

Widmark & larsson's titration Electrometric titration of the carbonyl group in amino acids or proteins.

Wijs solution for iodine number A solution of 13 gm. iodine in I liter of 95% acetic acid is treated with hydrogen chloridefree chlorine gas until the titer of the solution is doubled. The solution changes only very slightly in titer. Since the solution contains 7.8 gm. iodine trichloride and 8.5 gm. iodine per liter, the solution may be prepared by dissolving these substances.

Wild cherry Wild black cherry bark; from Prunus serotina, Ehrhart, Rosaceae; contains prunasin, glucose, hydrocyanic acid and other organic components; used in form of syrup as vehicle for cough medicine.

Wildtype The native or predominant genetic constitution before mutations, usually referring to the genetic consitution normally existing in nature.

Willstaedt reaction for vitamin b1 2,4-dichlorobenzenediazonium chloride reacts with the vitamin forming a yellow-red color, quantitatively extracted with ether. The dye is separated from other dyes by filtration through a column of calcium hydroxide, followed by elution of the adsorbent with alcohol.

Windaus reaction for cholesterol I. A solution of cholesterol in the least volume of ether is treated with bromine in glacial acetic acid until distinctly yellowbrown. Cholesterol dibromide, melting at 123-1240, precipitates immediately. II. Cholesterol, in alcoholic solution, is precipitated almost quantitatively by an alcoholic solution of digitonin in the form of an equimolecular complex.

Winkler reactions for differentiating theobromine, theophylline and caffeine I. 0.1 gm. theophylline is shaken with a mixture of 2 drops 1100 phenolphthalein, I drop of 0.1N sodium hydroxide and 5 cc. water. The solution is decolorized immediately while caffeine and theobromine do not react. II. One cc. of ammonia will dissolve 0.1 gm. theophylline, while the other two will not dissolve.

Winternitz test for pancreatic insufficiency Five cc. of ethyl iodobenzoate are ingested by the patient during a meal. On the non-appearance of iodine in the urine within 24 hours, insufficiency of the pancreas may be safely assumed.

Wistarin A chromo-glucoside from bark of Wistaria sinensis; a yellow crystalline powder; m.p. 204°C; toxic, causing diarrhea followed by collapse.

Witches milk Milk produced by infants or by men with endocrine disturbances.

Withanolides A group of C28 plant steroids based on the parent hydrocarbon, ergostane and containing a characteristic withanolide ring system with a lactone side chain.

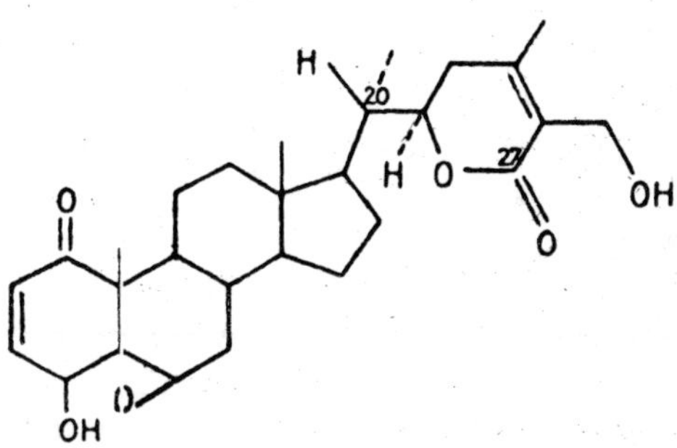

Fig. Withaferol A

Wobble Position The third base position within a codon, which can often (but not always) be altered to another nucleotide without changing the encoded amino acid.

Wolffian body Mesonephros; the primitive excretory organ of the embryo.

Wolffian regeneration The regeneration of a new lens in the eye after the removal of the former one, from the upper margin of the iris.

Wood gum The crude extract of wood and straw with caustic soda, of which xylan is the chief component.

Worenine An alkaloid, $C_{18}H_{13}NO(OCH_2O)_2$, from root of Coptis japonica; it is probably ß-methyl-coptisin.

Worms, intestinal Parasitic infections of the intestine. These may be.

(1) tapeworm.

(2) roundworm or.

(3) pinworm. There are many types of tapeworm originating from beef, pork, fish or adapted to other animals than man. They range from 1 inch to 20 feet in length. After dietary preparation, etc., anthelmintics are administered. These are oleoresin of aspidium, a chlorinated hydrocarbon like carbon tetrachloride, pelletierine tannate, pumpkin seeds. Roundworms or eelworms (ascaris lumbricoides) invade in large numbers and are treated with hexylresorcinol, oil of chenopodium, and the specific, santonin. Pinworms (seatworms, threadworms, oxyuris) caused strong anal itching. The remedies used are gentian violet, hexylresorcinol and other anthelmintics if necessary.

Wyss determination of insulin Insulin retards or prevents the hydrogen peroxide oxidation of glucose or of phenols.

X

Xan Abb. for Xanthine.

Xanthophylls A group of Carotenoids.

X-anthopterin 2-amino-4, 6-dioxotetrahydropteridine. It is the yellow pigment of bees, wasps and hornets. It is biosynthesized from guanine and two carbon atoms of a pentose.

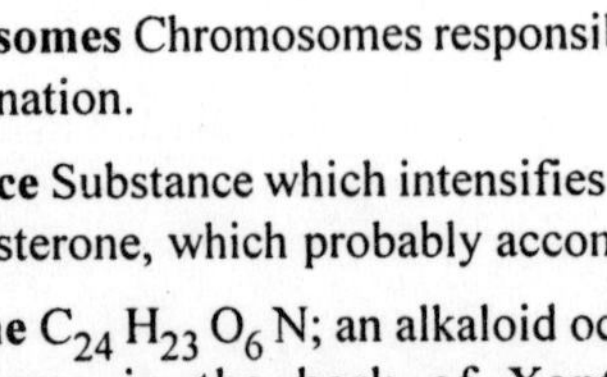

Fig. Xanthoptenin

X chromosomes Chromosomes responsible for sex determination.

X substance Substance which intensifies the action of testosterone, which probably accompanies it.

Xantherine $C_{24}H_{23}O_6N$; an alkaloid occurring in two forms in the bark of Xanthoxylum ochroxylum; paralyzes the nervous system.

Xanthine oxidase Milk flavoprotein; aldehyde oxidase; dihydrocoenzyme I oxidase; Schardinger enzyme; a flavoprotein enzyme catalyzing the oxidation of certain purines, aldehydes, hypoxanthine and xanthine to uric acid, the dismutation of xanthine to uric acid and hypoxanthine, etc., m.w. about 280,000 (milk); various differences depending on the source remain to be explained.

Xanthine Ureous acid; xanthine oxide; 26oxypurine, an intermediate in nucleoprotein metabolism, formed from guanine by the enzyme guanase or from hypoxanthine by oxidation, and itself oxidizing to uric acid.

Xanthones The name given to the class of plant pigments derived from xanthone or benzophenone oxide, anthoxanthones.

Xanthophyll (1) A name for the companion of carotene consisting of two isomers, lutein and zeaxanthin, oxidation products of alpha and beta carotene respectively; $C_{40}H_{56}O_2$;

(2) a general name for yellow pigments which are oxidized carotenoids, soluble in methanol.

Xanthoproteic test A test for proteins containing amino-acids with a benzene nucleus, depending on the yellow color produced with conc. nitric acid, turning to orange on being made alkaline.

Xanthopterin Yellow pigment of butterfly wings, $C_{19}H_{13}O_6N_{16}$.

Xanthorhamnin Rhamnin; U+22ID-rhamnegin; glycoside of Persian berries, hydrolyzable to rhamnetin and rhamninose, a triose.

Xanthosine Xanthose riboside; a nucleoside consisting of d-ribose and xanthine, formed by the deamination of guanosine.

Xanthotoxin $C_{11}H_5O_3 \cdot OCH_3$; m.p. 145-146°C; a cyclic "bitter principle" of fruit peel of Fragaria xanthoxyloides, a fish poison.

Xanthoxylum Dried bark and full grown berries of Xanthoxylum americanum; used as carminative and mild diaphoretic.

Xanthydrol A precipitant for urea, which also precipitates allantoin and allantoic acid.

Xenia Effect of pollen on the characters of the endosperm.

Xeno-immunization Stimulation of immune response to antigens from different species.

Xenoplastic transplantation Transplantation between individuals of different genera or widely distant species.

Xerophytes Plants adapted to growth and survival under a scarcity of water or draught. They are often thorny and supplied with thick rinds or cuticles.

X-ray Electromagnetic radiation of wavelength shorter than *ultraviolet radiation;* used in medicine to image internal body structures.

X-ray diffraction Use of x-rays to determine the structure of the atoms or molecules that make up a substance.

Xylan A hemicellulose of bran, straw, wood, which yields xylose on hydrolysis.

Xylanase An enzyme which hydrolyzes xylans to xylose.

Xylem Woody portion of plants between pith and cambium

Xylenol blue p-Xylenolsulfonphthalein; 1,4Dimethyl - 5 - hydroxybenzenesulfonphthalein; used as indicator changes being red at pH 1.2, yellow at 2.8, and blue at 9.6.

Xylinic acid An acid produced from xylose by Bact. xylinum.

Xyloketose A pentose of straw, bran, hemicelluloses, some glycosides (primeverosides), m.p. 153°C, not fermentable.

Y

Yajeine A toxic alkaloid, $C_{14}H_8N_3O$.

Yangonin An amaroid, $C_{15}H_{14}O_4$, from Kava root, crystallizing in yellowish prisms; m.p., 153-154°C.

Yaws Frambesia; an infectious disease caused by Treponema pertenue (Castellani) resembling the spirochete of syphilis. The lesions resemble raspberries (=frambesia). Contagiousness is very high. Arsphenamine or neoarsphenamine with bismuth and iodine preparations are used as specific therapy.

Yeast artificial chromosome (YAC) This is a method for cloning very large fragments of DNA. Genomic DNA in fragments of 200-500 kb are linked to sequences which allow them to propagate in yeast as a mini-chromosome (including telomeres, a centromere and an ARS - an autonomous replication sequence). This technique is used to clone large genes and intergenic regions, and for chromosome walking. Compare cloning vector, cosmid.

Yeast gum Probably chiefly a mannose polysaccharide, made by enzymatic degradation of yeast or by extraction with caustic and precipitation with Fehling's solution.

Yeast The living cells of the species of Saccharomyces; in commercial form, it is usually combined with a starchy or absorbent base appearing as white, soft, easily broken masses of a characteristic slightly sour odor; used as antineuritic in B avitaminosis; also effective in treatment of acne, indigestion, flatulence, furunculosis, constipation; when irradiated acquires antirachitic properties; stimulates leukocytosis.

Yeastone A purified, concentrated extract of yeast containing the active principles of yeast but without vitamin; used in same way as yeast except in B avitaminosis.

Yellow fever An infectious disease due to a virus transmitted by the mosquito, Aedes aegypti (Stegomyia). A condition of jaundice follows fever, albuminuria and hemorrhages. Quarantine against the mosquito is practiced as prophylaxis, but there is no specific treatment for the disease.

Yellow respiration enzyme Lactoflavinphosphate ester with a specific protein.

Yerbe maté Leaves of a Paraguay tree which produce beverage like tea.

Ylang ylang oil Cananga Oil; a volatile oil distilled from the flowers of Canaga Odorata containing geraniol and linalool esters of acetic and benzoic acids and other compounds; used in perfumes.

Yocca A caffeine containing drug of the bark and stems of Yocca.

Yoghurt Bulgarian curdled milk; said to displace harmful intestinal bacteria.

Yohimbine $C_{21}H_{26}O_3N_2$, an alkaloid of the African corynanthe yohimbe and Argentine quebracho, m.p. 235°C; used as aphrodisiac; other names quebrachine, corynine, aphrodine, hydraergotocin.

Yolk Nutrient material of egg; vitellus, ooplasm; deutoplasm.

Yolk Sac Of many vertebrates, an extraembryonic membrane that provides nourishment to the developing embryo; of humans, the sac does not include yolk but helps give rise to a digestive tube.

Yolk-sac The extension of the blastoderm. into a membrane surrounding the yolk in amniotes, specialised for absorption of nutritive materials.

Young test for cocaine in the presence of novocaine Four-five drops of 2% cobalt thiocyanate are poured over a small portion of the solid. Dark blue flocks form if only cocaine is present. If novocaine is also present, the whole solution turns blue. Four-five drops of freshly prepared stannous chloride solution (10 gm. stannous chloride and 5 gm. metallic tin in 100 cc. 11 hydrochloric acid) are added; in the presence of novocaine alone, the precipitate dissolves; if only cocaine is present the blue flocks remain unchanged.

Yucca fructosan A fructosan of Yucca mohavensis, Sarg., similar to graminin of rye flour.

Z

Zappacosta test for indole and skatole in blood An acetic acid solution of blood serum is treated with trichloracetic acid; indole and skatole are indicated by a reddish-blue color which may also be used for colorimetric estimation.

Z-DNA Alternative left-handed form of the double helix which retains Watson-Crick type base pairing.

Zea Corn Silk; consists of maizenic acid, fixed oil, resin, mucilage; found in fresh styles and stigmas of Zea Mays; used as demulcent, i.e. in urogenital diseases.

Zeaxanthin A carotenoid found in yellow Zea. Mays. Its structure is 3-3'-dihyroxy-ß-ß'-carotene, m.p. 207°C.

Zedoary Dried rhizome of Curcuma Zedoaria containing resin, volatile oil, starch, mucilage; used as a stomachic like ginger in dyspepsia and flatulence.

Zein A prolamine of maize seeds, lacks lysine and trytophane, m.w. about 35,000.

Zenith Point on the *celestial sphere* directly above an observer; compare *nadir*

Zeolites Minerals which are essentially hydrous silicates of aluminum which usually contain sodium or calcium and sometimes barium, strontium, potassium and magnesium. They are employed in ion exchange or for the adsorption of complex organic compounds, like vitamins, from solutions.

Zephiral An aqueous solution of a mixture of high-molecular alkylbenzylammonium chlorides; used as a bactericide in surgery, obstetrics and dermatology.

Zimmerman precipitant for physiologically important bases Arginine, creatinine, betaine, choline, histidine, etc., form difficultly soluble salts with rufianic acid (1,4-dihydroxyanthraquinone2-sulfonic acid).

Zimmerman reaction for histidine 0.2 mg. histidine hydrochloride, in a 1 cc. Thunberg tube, is treated with 3 drops of 1% cobalt nitrate solution. A small tube containing 10 drops of 2N sodium hydroxide is inserted, the tube evacuated and the solutions mixed. A violet color develops.

Zimmet reagent for reduced glutathione and other sulfhydryl compounds Sodium nitroprusside reacts with glutathione but not with acetone within a very narrow pH range (about 9). A reagent having the proper pH is made up of 5 gm. sodium nitroprusside, 50 gm. sodium phosphite and 100 cc. water and gives a red color with very dilute glutathione solutions.

Zinc finger A protein structural motif common in DNA binding proteins. Four Cys residues are found for each "finger" and one finger can bind a molecule of zinc. A typical configuration is CysXxxXxxCys—(intervening 12 or so aa's)—CysXxxXxxCys.

Zona pellucida A membrane surrounding the ovum; ovolemma.

Zooerythrin Red pigment of feathers.

Zoo-fulvin Yellow pigment of feathers.

Zoogamy Sexual reproduction in animals.

Zoomaric acid Palmitoleic acid or hexadecenoic, acid.

Zoomelanin Black pigment of feathers.

Zoophyte An animal with plant habits.

Zoosterols Sterols found in animal cells and tissues, e.g. cholesterol, dihydrocholesterol, 7-dehydrocholesterol, agnosterol.

Zootoxin Animal poison, as of snakes, wasps, etc.

Zooxanthin Yellow pigment of feathers.

Zsigmondy gold number The concentration of colloid required to stabilize a standard gold colloid against precipitation by a standard NaCl solution.

Zsigmondy gold reagent for colloids Reagent—When auric chloride is reduced by formaldehyde in the presence of a weak alkali, a red colloidal solution of metallic gold is obtained.

Zulkowsky starch reagent for iodine Sixty gm. of starch and 1 kilo of glycerin are stirred and gradually heated to 190°C, until the mass is clearly soluble in water; this aqueous solution gives a blue color with iodine.

Zwikker reaction for pentamethylenetetrazol (cardiazol, metrazol) A hydrochloric acid solution of cuprous chloride gives a crystalline precipitate with Metrazol even a dilutions of 140000.

Zwitterion An ion that is charged both positively and negatively; other terms are amphoteric ion and dipole ion, e.g. glycine which in solution has the form $-OOC \cdot CH_2 \cdot NH_3+$.

Zygadenine $C_{39} H_{63} O_{10} N$; m.p. 200-201°C; an alkaloid of the leaves of Zygadenus intermedius, Liliaceae; acts like veratrine.

Zygote The product of gamete fusion. In organisms with a haploid life cycle, the zygote immediately undergoes meiosis, but in organisms with a multicellular diploid stage, the zygote is merely the first stage in the diploid portion of the life cycle.

Zymase The old name for a mixture of enzymes and coenzymes obtained from yeast juice free of cells, responsible for the alcoholic fermentation of glucose.

Zymo-excitor A substance which activates a zymogen into the formation of an enzyme, e.g. HCl in the of pepsinogen to pepsin.

Zymogen A precursor of an enzyme, e.g. as pepsinogen is to pepsin.

Zymohexase A pyridinoprotein enzyme present in many plant, animal tissues and bacteria, which catalyzes the rereversible splitting of fructose-16diphosphate to 3- phosphoglyceraldehyde and dihydroxyacetonephosphate, and also the aldol condensation of dihydroxyacetone phosphate (dap).

Zymoid (1) Any poison derived from a decaying tissue;

(2) a ferment or enzyme which has lost its power of decomposing the substratum but not its power of uniting with it.

Zymosterol $C_{27} H_{44} O$ or Ä814,2425 cholestadienol, a secondary yeast sterol, dextrarotary, m.p. 108-109°C; found in ergosterol residues.

Zymotic (1) A germ disease;

(2) pertaining to enzymes.

Zytase A malt enzyme which splits hemicelluloses at pH 5 optimum; transforms xylan to xylose.1.

3' end/5' end A nucleic acid strand is inherently directional, and the "5 prime end" has a free hydroxyl (or phosphate) on a 5' carbon and the "3 prime end" has a free hydroxyl (or phosphate) on a 3' carbon (carbon atoms in the sugar ring are numbered from 1' to 5'. That's simple enough for an RNA strand or for single-stranded (ss) DNA. However, for double-stranded (ds) DNA it's not so obvious - each strand has a 5' end and a 3' end, and the 5' end of one strand is paired with the 3' end of the other strand (it is "antiparallel"). One would talk about the 5' end of ds DNA only if there was some reason to emphasize one strand over the other - for example if one strand is the sense strand of a gene. In that case, the orientation of the sense strand establishes the direction.

3' Flanking region A region of DNA which is not copied into the mature mRNA, but which is present adjacent to 3' end of the gene. It was originally thought that the 3' flanking DNA was not transcribed at all, but it was discovered to be transcribed into RNA, but quickly removed during processing of the primary transcript to form the mature mRNA. The 3' flanking region often contains sequences which affect the formation of the 3' end of the message. It may also contain enhancers or other sites to which proteins may bind.

3' untranslated region A region of the DNA which is transcribed into mRNA and becomes the 3' end or the message, but which does not contain protein coding sequence. Everything between the stop codon and the polyA tail is considered to be 3' untranslated . The 3' untranslated region may affect the translation efficiency of the mRNA or the stability of the mRNA. It also has sequences which are required for the addition of the poly(A) tail to the message (including one known as the "hexanucleotide", AAUAAA).

5' Flanking region A region of DNA which is not transcribed into RNA, but rather is adjacent to 5' end of the gene. The 5'-flanking region contains the promoter, and may also contain enhancers or other protein binding sites.

5' Untranslated region A region of a gene which IS transcribed into mRNA, becoming the 5' end of the message, but which does not contain protein coding sequence. The 5'-untranslated region is the portion of the DNA starting from the cap site and extending to the base just before the ATG translation initiation codon. While not itself translated, this region may have sequences which alter the translation efficiency of the mRNA, or which affect the stability of the mRNA.

7-Methyl-guanosine triphosphate The methylated form of guanosine triphosphate used to cap an mRNA molecule in most eukaryotes.